Eigenvalue Problems in Power Systems

Eigenvalue Problems in Power Systems

Federico Milano (federico.milano@ucd.ie)
Ioannis Dassios (ioannis.dassios@ucd.ie)
Muyang Liu (muyang.liu@ucdconnect.ie)
Georgios Tzounas (georgios.tzounas@ucdconnect.ie)

CRC Press
Taylor & Francis Group
Boca Raton London New York

CRC Press is an imprint of the
Taylor & Francis Group, an **informa** business

First edition published 2021
by CRC Press
6000 Broken Sound Parkway NW, Suite 300, Boca Raton, FL 33487-2742

and by CRC Press
4 Park Square, Milton Park, Abingdon, Oxon OX14 4RN

ISBN-13: 978-0-367-34367-5 (hbk)
ISBN-13: 978-0-429-32531-1 (ebk)

Typeset in Computer Modern font
by KnowledgeWorks Global Ltd.

To Yolanda.

F.M.

To my sister and parents.

I.D.

To Junru.

M.L.

To my family.

G.T.

Every existence above a certain rank has its singular points; the higher the rank the more of them. At these points, influences whose physical magnitude is too small to be taken account of by a finite being may produce results of the greatest importance.

J.C. Maxwell,
Cambridge, February 1873.

Latent roots of a matrix – latent in a somewhat similar sense as a vapour may be said to be latent in water or smoke in a tobacco leaf.

J.J. Sylvester,
On the equation to the secular inequalities in the planetary theory,
Philosophical Magazine, series 5, vol. 16, pp. 267–269, 1883.

The eigenvalue problem has a deceptively simple formulation and the background theory has been known for many years; yet the determination of accurate solutions presents a wide variety of challenging problems.

J.H. Wilkinson,
The algebraic eigenvalue problem,
Clarendon Press, Oxford, 1965.

Contents

List of Figures

List of Tables

List of Mathematical Statements

List of Examples

Authors

Federico Milano received, from the University of Genoa, Italy, the ME and PhD in Electrical Engineering in 1999 and 2003, respectively. From 2001 to 2002 he was with the University of Waterloo, Canada, as a Visiting Scholar. From 2003 to 2013, he was with the University of Castilla-La Mancha, Spain. In 2013, he joined the University College Dublin, Ireland, where he is currently Professor of Power Systems Control and Protections and Head of Electrical Engineering. He has authored or co-authored six monographs and over 200 journal and conference papers. He was elevated IEEE Fellow in 2016 for his contributions to power system modeling and simulation, and IET Fellow in 2017. He is or has been an editor of several international journals published by IEEE, IET, Elsevier, and Springer, including the IEEE Transactions of Power Systems and the IET Generation, Transmission & Distribution. Since 2020, he is an IEEE PES Distinguished Lecturer.

Ioannis Dassios is currently a UCD Research Fellow at AMPSAS, University College Dublin (UCD), Ireland. He studied Mathematics and completed a two-year MSc in Applied Mathematics & Numerical Analysis at the National and Kapodistrian University of Athens (NKUA), Greece with grade "Excellent" (highest mark in the Greek system). As a MSc student he received a travel grant to visit the University of North Texas, USA. In July 2013 he received his PhD in Mathematics from NKUA. Since then, he has been a Post-Doctoral Researcher at MACSI, University of Limerick, Ireland and a Senior researcher at ERC/ESIPP at UCD. Previously, he worked as a Post-Doctoral Research & Teaching Fellow at the University of Edinburgh, UK, and for six months as a Post-Doctoral Research Associate at the University of Manchester, UK. He has published 67 articles in internationally leading academic journals. He is an Editor in several journals including Mathematics and Computers in Simulation (MATCOM) by Elsevier, Applied Sciences, Mathematics, MDPI, Symmetry, Experimental Results by Cambridge University Press, Signals, etc. He has also served as a reviewer more than 700 times in 90 different journals.

Muyang Liu received, from University College Dublin, Ireland, the ME in Electrical Energy Engineering in 2016 and the PhD degree in Electrical Engineering in 2019. She is currently a post-doctoral researcher with University College Dublin. Her current research interests are the stability analysis and robust control of power system with inclusion of measurement delays and the modeling of relays for transient stability analysis.

Georgios Tzounas received, from the National Technical University of Athens, Greece, the Diploma in Electrical and Computer Engineering in 2017. He is currently a Ph.D. candidate at University College Dublin, Ireland. His scholarship is funded through the SFI Investigator Award "Advanced Modelling for Power System Analysis and Simulation" (AMPSAS). His research interests include stability analysis and robust control of power systems.

Acronyms

AESOPS	Analysis of Essentially Spontaneous Oscillations in Power Systems
AIITS	All-Island Irish Transmission System
AVR	Automatic Voltage Regulator
CDS	Constant Delay System
CI	Contour Integration
CI-RR	Contour Integration with Rayleigh-Ritz
CoI	Center of Inertia
CSR	Compressed Sparse Row
DAE	Differential Algebraic Equation
DDAE	Delay Differential Algebraic Equation
DDE	Delay Differential Equation
DDS	Distributed Delay System
DER	Distributed Energy Resource
EMF	Electromotive Force
ENTSO-E	European Network of Transmission System Operators for Electricity
ERD	Explicitly Restarted and Deflated
FACTS	Flexible AC Transmission System
FDE	Functional Differential Equation
FOC	Fractional-Order Controller
FO-PSS	Fractional-Order Power System Stabilizer
GD	Generalized Davidson (method)
GEP	General Eigenvalue Problem
GPU	Graphical Processor Unit
HVDC	High-Voltage Direct Current
IC	Initial Conditions
IEEE	Institute of Electrical and Electronics Engineers
IR	Implicitly Restarted

JD	Jacobi-Davidson (method)
KCL	Kichhoff current law
KVL	Kichhoff voltage law
LE	Lyapunov Exponent
LEP	Linear Eigenvalue Problem
LLE	Largest Lyapunov Exponent
LM	Largest Magnitude
LR	Largest Real (part)
MCDS	Multiple Constant Delay System
MKL	Math Kernel Library
MPI	Message Passing Interface
NDDE	Neutral Delay Differential Equation
NEP	Non-linear Eigenvalue Problem
NG	Next Generation
NNZ	Number of Nonzero (elements of a sparse matrix)
OMIB	One-Machine Infinite-Bus
PCC	Point of Common Coupling
PD	Proportional Derivative
PDF	Probability Density Function
PEALS	Program for Eigenvalue Analysis of Large Systems
PF	Participation Factor
PFC	Primary Frequency Control
PI	Proportional Integral
PMU	Phasor Measurement Unit
POD	Power Oscillator Damper
PSS	Power System Stabilizer
QEP	Quadratic Eigenvalue Problem
QP	Quenching Phenomenon
QPMR	Quasi-Polynomial Mapping based Root-finder
QSS	Quasi-Steady-State
RCI	Reverse Communication Interface
RDDE	Retarded Delay Differential Equation
R-L	Riemann-Liouville
RoCoF	Rate of Change of Frequency
RQ	Rayleigh Quotient
SDM	Sparse Discretization Method

SM	Smallest Magnitude
TG	Turbine Governor
TR	Target Real (part)
TVDS	Time-Varying Delay System
ULTC	Under-Load Tap Changer
WAMS	Wide-Area Measurement System
WF	Wash-out Filter
ZOH	Zero-Order Holder

Notation

This section states the high-level notation adopted in the book. Where a different notation is used, quantities are defined in the text.

Scalars, Vectors and Matrices

v, V, $\mathcal{V}$	scalar
$\boldsymbol{v}$, $\mathbf{v}$, $\boldsymbol{V}$	vector
$v(t)$	time domain quantity (v if the context is unambiguous)
$\dot{v}$	time derivative ($\dot{v} = \frac{dv}{dt}$)
$\ddot{v}$	second time derivative ($\ddot{v} = \frac{d^2v}{dt^2}$)
$\dddot{v}$	third time derivative ($\dddot{v} = \frac{d^3v}{dt^3}$)
$v^{[n]}$	higher order time derivative ($v^{[n]} = \frac{d^nv}{dt^n}$)
$v^{(i)}$	i-th iteration of a numerical method
$v(s)$	frequency domain quantity (Laplace transform)
$s\,v$	time derivative in frequency domain (Laplace transform)
$\bar{v}_{\rm dq}$	Park vector in dq-axis reference frame, i.e. $\bar{v}_{\rm dq} = v_{\rm d} + \jmath\, v_{\rm q}$
$\bar{v}$	phasor
v^*	complex conjugate
$\mathbf{V}$	matrix
$\mathbf{V}^{\rm T}$	matrix transpose
$\mathbf{V}^{\rm H}$	matrix conjugate transpose

Eigenvalues, Eigenvectors, Eigenspaces

f_d	damped frequency of an eigenvalue
f_n	natural (or undamped) frequency of an eigenvalue
$\mathbf{u}$	right eigenvector (column vector)
$\langle\mathbf{u}\rangle$	eigenspace
$\mathbf{w}$	left eigenvector (row vector)
ζ	damping ratio of an eigenvalue
λ	eigenvalue

Quantities

a	coefficient of a polynomial, transformer tap ratio

B	susceptance
C	capacitance
D	damping
e	electromotive force (EMF)
e	Euler's number
$\boldsymbol{f}$	vector of differential equations
$\boldsymbol{g}$	vector of algebraic equations
G	conductance
$\mathcal{H}$	Heaviside unit step function
$\imath$	current
$\mathbf{I}_m$	identity matrix of order m
$\jmath$	imaginary unit
$\mathbf{J}_m$	Jordan matrix of order m
K	controller gain
L	inductance
$\mathcal{L}$	Laplace transform
M	mechanical starting time
$\boldsymbol{\mathcal{M}}$	Monodromy matrix
p	multiplicity of finite eigenvalues
P	active power
$\mathcal{P}$	per-unit time derivative of Park vector
q	multiplicity of infinite eigenvalues
Q	reactive power
R	resistance
$\mathcal{R}$	droop of primary frequency control
$\bar{S}$	complex power
s	Laplace transform variable
t	time (r within integrals)
T	time constant
$\boldsymbol{u}$	vector of input signals
v	voltage
$\mathcal{V}$	Lyapunov function
$\boldsymbol{x}$	vector of state variables
X	reactance
$\boldsymbol{y}$	vector of algebraic variables
$\bar{Y}$	admittance
Y_c	characteristic admittance
z	inverse or transformed s
$\bar{Z}$	impedance
α	attenuation factor
δ	angular position
θ	phase angle of voltage phasors
ϑ	distributed delay
$\boldsymbol{\xi}$	generalized state variable vector
π	participation factor

ϖ	polynomial function
$\mathbf{\Pi}_{\rm PF}$	participation factor matrix
ϱ	continuation parameter
τ	time delay
$\mathrm{\tau}$	mechanical torque
ϕ	phase shift
$\boldsymbol{\varphi}$	vector of implicit differential equations
ψ	magnetic flux
ω	angular speed

Superscripts and Subscripts

b	base quantity
d	direct-axis quantity of the dqo transform
D	demand
dq	dq-frame (Park) vector
e	electrical
G	generator
L	transmission line
m	mechanical
max	maximum value
min	minimum value
o	reference, initial or base-case condition
q	quadrature-axis quantity of the dqo transform
r	rotor
ref	reference or set point
s	stator
T	transformer

Units

The units of absolute quantities follow the International System of Units. Unless explicitly indicated, however, the equations that describe ac circuits are in per unit values, as usual in power system analysis. The bases are the three-phase apparent power, $S_{\rm b}$, the line-to-line voltage $V_{\rm b}$ and the frequency $f_{\rm b}$. All other bases are derived from these three quantities. For example, the bases of the impedance and the line current are, respectively:

$$Z_{\rm b} = \frac{V_{\rm b}^2}{S_{\rm b}} , \qquad I_{\rm b} = \frac{S_{\rm b}}{\sqrt{3}\, V_{\rm b}} .$$

Preface

This book is about eigenvalue analysis and power systems. It originates from the collaboration between electrical engineers and a mathematician. The result is, we trust, a unique monograph that blends state-of-art theoretical insights on the eigenvalue problem with practical applications to circuit analysis and real-world electric power systems.

Small-Signal Stability Analysis

In power systems, the solution of the eigenvalue problem is often referred to as *small-signal stability analysis* and is mainly concerned with low frequency, poorly damped oscillations. This problem, in the past known as *hunting*, was already studied in the 1920s [174]. An example of hunting is the spontaneous oscillations of a synchronous machine with high armature resistance, which can be conveniently solved through amortisseur windings [175]. The multi-machine hunting problem was tackled in the 1950s by Gabriel Kron [86] and the matrix-based eigenvalue analysis as we know today was introduced in the 1960s, e.g. [85].

Small-signal stability is defined as the ability of a system to maintain all variables within their acceptable limits after a small disturbance [76]. Small-signal stability is assumed to be a subset of the rotor angle stability, as synchronous machines and their automatic voltage regulators (AVRs) are the principal cause of instabilities that arise after a small disturbance. The well-known blackout of the WSCC system in 1996 is an emblematic example of such a phenomenon.

In recent years, power systems all around the world have undergone deep changes, among others, the commissioning of a wide variety of devices and controllers that have substantially changed the dynamics and, hence, possible instability routes that the systems can undergo. Relevant examples are FACTS devices, converter-interfaced generation and energy storage systems, VSC-HVDC links, wide-area measurement and control systems, and microgrids.

On one hand, the *granularity* of the system has dramatically increased, thus significantly increasing the possible interactions among the dynamics of devices and controllers. An interesting example are the phase-locked loops, which are little, apparently insignificant electronic chips used in electronic converters for their interconnection to the ac grid. The microprocessor of a phase-locked loop has the size of a lentil, but if poorly tuned or if coupled to a weak distribution system can make an entire wind power plant unstable [95].

On the other hand, the increasing trend to substitute conventional synchronous machines with non-synchronous devices, often based on renewable energy resources, such as wind and solar energy, reduce the dominant effect of synchronous machines on the dynamics of the systems [121]. Back in 2004, when [76] was published, it made sense to classify the different kinds of (in)stability based on the synchronous machine. Such a classification has been recently revised and expanded to include more devices, in particular those based on power electronics.

In this ever-changing context, a general mathematical approach to the small-signal stability analysis seems to us very appropriate and timely.

Given an equilibrium point, the eigenvalue analysis indicates whether a system is stable or not. It is possible to extend such a definition to periodic orbits (eigenvalues are in this case called *Floquet multipliers*) and even to chaotic motions (eigenvalues are in this case called *Lyapunov exponents*). Thus the eigenvalue analysis provides information on the stability of any *stationary condition* of a dynamic system. And except for very peculiar cases, e.g. limit points, the stability conditions are necessary and sufficient. If one runs a parametric eigenvalue analysis, i.e. studies how eigenvalues change as a parameter of the system is varied, then it is possible to define the stability margin of the system, i.e. the difference between the actual value of the parameter and the value for which the system becomes unstable.

It is relevant to observe that the eigenvalue analysis is independent from the physical meaning of the parameter and the devices that compose the system. It does not matter whether the parameter is the loading level or whether the device that drives the system to instability is the automatic voltage regulator of a synchronous machine or a phase-locked loop of a voltage-source converter. Such a general approach is thus ready for any current and future technology that can be developed in power systems. This is, ultimately, the strength of mathematical opposed to device-based approaches.

Small-signal stability analysis is generally associated to first order sensitivities or *linearization*. While this is generally the case, the eigenvalue problem *per se* does not imply linearity. Several non-linear eigenvalue problems are of interest in physics and engineering. In this book, we discuss three relevant examples: the quadratic eigenvalue problem that is widely utilized to study mechanical vibrations and suits particularly well the study of the electromechanical modes of synchronous machines; fractional order systems whose eigenvalue problem involves polynomial matrix pencils with non-integer exponents; and the transcendental eigenvalue problem that originates from delayed differential equations and is appropriate to study measurement and communication delays of wide-area controllers in power systems.

The main limitation of the eigenvalue analysis is that it can assess exclusively stationary conditions and, often, only the behavior of the system *close* to such stationary conditions. Indeed, there are very few methods that can successfully assess the transient stability of power systems: time domain integration and the *extended area criterion*, which is, as the eigenvalue analysis, another the Lyapunov stability criterion. The numerical time domain integration is undoubtedly the workhorse of the dynamic assessment of power systems. However, it is intrinsically a *qualitative* approach: a time domain simulation allows determining whether a system is stable or unstable after a contingency but not *why* it is so.

When a power system collapses, the synchronous machines lose synchronism, bus voltage phase angles separate and bus voltage magnitudes go to zero. This happens regardless the fact that the *cause* of the collapse is an angle, voltage or frequency instability. There is thus no clue, by running exclusively a time domain simulation, on what is the most appropriate remedial action that would prevent the collapse. Moreover, even if the simulation shows that the system is stable, one has no information on the stability margin of the system just by looking at how the trajectories of the system variables evolve in time.

On the other hand, the Lyapunov stability criterion is *quantitative* as it provides the stability margin of the system for a given contingency. This method has been quite successful in power system analysis in the past century, also because it involves the construction of the so-called Lyapunov function which in power systems is generally build as an *energy function* and has thus a quite clear and profound physical meaning which is easy to understand for the engineers.

Interestingly several books have been written on the construction and applications of the energy function (see, for example, the excellent monograph [135]), well before than books on small-signal stability analysis appeared on the shelves of the libraries. The conventional energy function, however, has a fundamental and unresolvable flaw: it provides only sufficient conditions for stability, not necessary and sufficient. This means that, in many cases, it is inconclusive. Its applications to power systems have slowly but inexorably faded in the last two decades also for other three reasons, as follows:

- The definition of the energy function tends to be very involved and time consuming even for relatively simple power system models, cannot take into account losses and, more importantly, cannot keep up with the relentless commissioning of new devices and controllers in modern power systems.

- In the last two decades the computational power of workstations and personal computers has increased so much that time domain simulations have become inexpensive.

- The energy function is effectively utilized, in power systems, almost exclusively in *transient stability analysis*, i.e. the problem of determining whether synchronous machines lose synchronism following a large disturbance. This problem, however, is virtually solved thanks to the short response time of modern protections (as fast as three or even only two cycles). Then, again, in low-inertia systems, synchronous machines are not expected to play the most relevant role.

One can therefore argue that the energy function was a reasonable tool when computers were slow and expensive and when power system dynamics were dominated by the electromechanical modes of synchronous machines but are not so relevant or useful today.

Eigenvalue analysis appears thus as a relevant complementary method to time domain simulation. First, eigenvalue analysis, being a consequence of the Lyapunov stability criterion as the energy function approach, is also a quantitative method. As discussed above, it allows calculating parametric stability margin and, through the determination of the *participation factors* which are obtained based on left and right eigenvectors associated to the eigenvalues, allows defining which state variables and, thus, which device(s), participate the most to a given mode. This is particularly useful for the design of corrective actions and the tuning of the parameters of the controllers. Then, since the eigenvalue analysis studies the stability of equilibrium points, it is a required preliminary step to any time domain simulation. It does not make much sense, in fact, to solve a time domain analysis if the initial starting point is unstable, among other reasons, because the system can never reach an unstable operating condition and remain there.

Matrix Pencils

We have discussed so far well-known aspects of the eigenvalue problem in the context of power system stability analysis. This book, however, discusses eigenvalue problems from a novel perspective, i.e. the concept of *matrix pencils.* This is an emerging and advanced topic in mathematics, especially regarding *singular* matrix pencils. The co-authorship with a mathematician is key for the development of a consistent and systematic treatise of eigenvalue problems from the point of view of matrix pencils rather than specific engineering problems. The book, in fact, proceeds in the other way round: in each chapter, we discuss the properties of a particular case of eigenvalue problem and then illustrate its applications to power systems. We find that this approach is more general and not easily prone to obsolescence. Power system applications, in fact, can change, but mathematical techniques are *ære perennius.*

Matrix pencil theory is used in many branches of mathematics including differential equations, network theory, and in applied mathematics in general. Matrix theory is also used as an important tool in engineering. Some examples are in material science where linear and non-linear models are used and in electrical engineering were circuits and dynamics of electrical power systems are often modeled as systems of differential equations.

The first interest in matrix theory was raised in the eighteenth century when there was an attempt to provide formulas of solutions for a linear system of differential equations. Jean d'Alembert and Joseph Louis Lagrange proved and used properties of matrices to provide such solutions.

By using the fact that the Solar System is modeled as a system of differential equations, in 1780, Lagrange further explored and used these results to study the stability of this system, see [60]. He extended the initial results on properties of matrices, and used spectral theory of a matrix to obtain solutions of the system and then studied the stability of the system through these solutions. However, he did not develop a complete theory, and did not provide definitions to terms such as eigenvalues, eigenvectors, algebraic-geometric multiplicity. Pierre Simon Laplace supported and continued this initiative and the research started by Lagrange.

The next two important results came out in 1801 and 1802. This time there was a first attempt by Carl Friedrich Gauss, see [60] and *Disquisitiones arithmeticæ* (1801), to provide properties on rectangular matrices and also there was a first attempt by Joseph Fourier to model the heat diffusion as a set of differential equations. In his work he applied previous published results on matrix theory to study the solutions of the model.

Interested in the matrix theory established in previous years, especially by Lagrange and Laplace, in 1829 Sturm initially rewrote, taught the theory on systems of differential equations and emphasized on the importance that matrix properties play on the solutions of this type of systems. In fact he introduced the generalized eigenvalue problem, seeking the eigenvalues of the pencil $s\mathbf{E} - \mathbf{A}$, for $\mathbf{E}$, $\mathbf{A}$ square matrices. Again, however, although he worked on the characteristic equation $\det(s\mathbf{E} - \mathbf{A}) = 0$, he did not provide fundamental definitions and also did not refer to the importance of the pencil in matrix theory and linear algebra. At the same period Carl Jacobi continued on this framework but worked mostly on properties of determinants.

The next contributions on matrix theory were breakthroughs. These results were provided and developed by Leopold Kronecker and Karl Weierstrass between 1850 and 1860, see [60]. Kronecker and Weierstrass worked and proved new properties of determinants and introduced the eigenvalues and eigenvectors as the invariants of a matrix pencil. They then managed to prove how all these properties are connected in matrix algebra. They also provided several new important definitions at the time such as symmetry of a matrix, the singular matrix and its properties. They finally introduced the diagonal matrix and proved that its existence is related to the eigenvalues and eigenvectors of a matrix. Later, in 1868 and 1874, they provided the first canonical forms of matrices for specific cases and also outlined properties of a matrix in order to have a canonical form.

Regarding new definitions, Augustin-Louis Cauchy first mentioned the matrix as "tableau" in his work[1] see [60]. In the end of the same century that Kronecker and Weierstrass published their first results, Arthur Cayley published his work *Memoir on the theory of matrices* (1858) where he referred to basic but very important properties in matrix theory such as matrix multiplications, and other matrix computations. Also important work was his theory

[1]It was James Joseph Sylvester who officially defined the term matrix in 1850.

on relating functions to matrices and a first attempt to refer to the importance of bilinear transformations. Actually his interest on matrix functions was much later used by other researchers for the study of matrix polynomials, a generalization of the linear pencil which at the time was considered as matrix pencil. However it is important to outline that Cayley's work was snubbed by important mathematicians of that period and especially researchers working in geometry and analysis [60].

In 1878, Georg Frobenius published his first important work on the bilinear transformation. In this work it seems that he was not aware of several definitions and properties published by Cayley in previous years. However in 1896 while publishing results on matrix functions and especially results of algebraic linear matrix equations, he cited for the first time Cayley and the work he published back in 1958. Besides acknowledging his work he also used the term 'matrix' as defined by Cayley. Frobenius was also influenced by the work of Kronecker and Weierstrass of that period. He was especially interested on canonical forms and studied and extended the results on special cases published by Kronecker and Weierstrass. Based on these results he also provided new insight on equivalence between matrices which is fundamental in matrix algebra and is connected with canonical forms if seen as a special case of equivalence.

The age of the development of canonical forms had come. Kronecker and Weierstrass introduced the idea and proved computational results on the first canonical forms of special cases mostly using a diagonal matrix. But it was Camille Jordan who generalized these special cases by introducing the, well known today, Jordan canonical form. It has to be noted that, although the Jordan canonical form was established later than the canonical forms introduced by Kronecker and Weierstrass, Jordan did not cite these previously proved results. Actually it has to be acknowledged that even earlier (1952) James Joseph Sylvester proved the important theorem that if a matrix is symmetric it will always have an equivalent unique diagonal matrix.

Beside the results on canonical forms of matrices, John Stephen Smith introduced the concept of canonical forms of pencils and later Weierstrass (1868) wrote an important relevant paper in a more fundamental way. However his results included matrices with real elements while Jordan's work which came later (1870) included matrices defined in complex space. Again Jordan did not cite papers by Smith and Weierstrass who first mentioned canonical forms of pencils. In a review paper on canonical forms by Frobenius he cited mostly Jordan but acknowledged also the work by Smith and Weierstrass.

One of the major difficulties in matrix theory is the computation and study of the algebraic and geometric multiplicity of eigenvalues of a pencil. This was studied in the book *Introduction to Higher Algebra* by the Harvard mathematician Maxime Bacher (1907), who included theory on determinants, abstract theory on matrix pencils; eigenvalues and their algebraic multiplicities; and theory on the corresponding eigenvectors.

While most studies referred to square matrices, Cuthbert Edmund Cullis, a mathematician from Cambridge (1913–1925), focused on rectangular matrices.

There was a period from 1920 until 1950 where besides some textbooks in early 1930, there was not much interest and no research done in matrix theory. The methodologies used in these textbooks were repetitive and always in a specific order, namely the definition of a matrix, determinants and calculation of eigenvalues-eigenvectors.

The situation changed only after 1950 and especially with Felix Gantmacher and his two-volume book "Theory of Matrices." In this book, he introduced new concepts of regular, and singular pencils including some preliminary results on other invariants besides the eigenvalues such as the minimal row and column indices for singular pencils. Despite being very practical in introducing these new concepts for pencils and invariants the book lacked on having computational methodologies.

Meanwhile, research of course has been proceeding, in established and in new directions. In particular, the arrival of computers has led to a rise in numerical linear algebra, where the properties, especially spectral, have been investigated and studied.

Numerical Eigensolvers

The latter observation leads us to a crucial aspect that we have taken very seriously in the book, namely the implementation in computer-code of the techniques discussed in theoretical chapters. Our focus is on large scale non-Hermitian eigenvalue problems, which are the most difficult to solve and computationally demanding among all eigenvalue problems. Also for this reason, available software libraries that are free and open source and are able to solve this particular eigenvalue problem are a small subset of all existing libraries. The vast majority deals with symmetric, Hermitian or tridiagonal eigenvalue problems, which are way more common in engineering, physics and computer applications. Moreover, we further restrict our analysis to software tools that are able to solve "large" eigenvalue problems. Algorithms and techniques that do not scale well are not discussed in this book. We hope that the book will help the reader find a valuable reference of eigenvalue problems that can be actually utilized to solve large high-granularity power system models.

Book Organization

The book is organized in five parts with fifteen chapters and three appendices, as follows:

Part I – Introduction

This part consists of two chapters.

Chapter 1 provides an overview of differential-algebraic equations in both implicit and explicit forms and in both continuous and discrete representations; power system modeling; the Lyapunov stability criterion and its particularization for linear systems; and the conventional definition of small-signal stability. The chapter also discusses a novel approach to the classification of the stability of non-linear dynamic systems.

Chapter 2 provides key definitions related to *matrix pencils*, which is the fundamental mathematical object utilized throughout the rest of the book and allows classifying the eigenvalue problem with a generalized and consistent framework. The canonical forms of matrix pencils are also provided in this chapter along with plenty of examples that clarify the theory and make the treatise accessible for engineers.

Part II – Linear Eigenvalue Problems

The linear eigenvalue problem is the conventional eigenvalue problem utilized in small-signal stability analysis. This part, is based on the taxonomy given in Chapter 2.

The first four chapters of Part II explore different matrix pencils, all of them with relevant applications to power system models. Chapter 3 discusses systems of differential equations with regular pencils; Chapter 4 discusses explicit differential algebraic equations; Chapter 5 discusses implicit differential-algebraic equations; and Chapter 6 discusses systems of differential equations with singular pencils and their relevant application to discrete maps.

The last two chapters of Part II, namely Chapter 7 and Chapter 8, discuss the Möbius transform and participation factors, respectively. The Möbius transform is an interesting variable transformation with relevant applications to numerical methods for the determination of the eigenvalues. Participation factors allow evaluating the participation of system states to system modes and have relevant applications in power system control and stability analysis.

Part III – Non-Linear Eigenvalue Problems

Part III discusses non-linear eigenvalue problems. The application of such problem to power systems is unconventional and constitute another distinguished feature of the book. This part is thus a novel contribution of the book. Chapter 9 discusses the quadratic eigenvalue problem and puts it in the more general context of polynomial eigenvalue problems. The quadratic eigenvalue problem has a special application to mass-spring-friction systems, which include the classical synchronous machine model. Chapter 9 also discusses the non-linear eigenvalue problem that arises from the theory of fractional order systems.

Chapters 10 to 12 discuss systems with delays. These represent a special class of non-linear eigenvalue problems where the characteristic equation is a transcendental function. Both delay differential and differential-algebraic equations and both lumped and distributed delays are considered. Chapter 12 extends the treatise to systems with delays that vary with time either periodically or stochastically, which are shown to be tractable as a special case of distributed delays.

Part IV – Numerical Methods

Part IV is dedicated to the numerical solution of the characteristic equation. The most relevant numerical methods for the generalized eigenvalue problem are outlined in Chapter 13. We have carefully selected only methods for non-Hermitian matrices for which efficient open-source numerical libraries are available and that we were able to test.

The mathematical background and the algorithms underlying numerical libraries are briefly described in Chapter 14 and comprehensively tested through two large power system models in Chapter 15. These are a model of the all-island Irish system with 1,443 states and 7,197 algebraic variables) and a model of the ENTSO-E system with 49,396 states and 96,768 algebraic variables.

Part V – Appendices

Appendix A provides the complete set of static and dynamic data for the three-bus system that is utilized in several illustrative examples included in the book. Finally, Appendices B and C provide the state matrix and Jacobian matrices for the standard and generalized eigenvalue problem, respectively, of the three-bus system. Appendix B also provides the left and right eigenvector matrices for the standard eigenvalue problem.

Software Tools

For the reader interested in software technicalities, all simulations included in the book are obtained using the Python-based software tool Dome [114]. The Dome version utilized for this book is based on Fedora Linux 28, Python 3.6.8, CVXOPT 1.1.9 and KLU 1.3.9. Detailed information on the libraries for eigenvalue analysis is given in Chapter 14. The hardware consists of two quad-core Intel Xeon 3.50 GHz CPUs, 1 GB NVidia Quadro 2000 GPU, 12 GB of RAM.

Lessons Learned

The germ of the idea of this book was born about five years ago, when the first author started collaborating with the second author on the small-signal stability analysis of dual eigenvalue problems. Since then, the collaboration has grown and led to a variety of papers on mathematical, circuit theory and power system journals and conference proceedings.

Interdisciplinary is a nice adjective that is often used in applications to appeal the reviewers and satisfy the requirements of the funding agencies. However, whoever has ever attempted to publish a truly interdisciplinary manuscript has certainly struggled to receive fair reviews and, ultimately, to get their manuscript accepted.

The main difficulty is that a truly interdisciplinary work requires an interdisciplinary reviewer and this is often a chimera. There is however a workaround to this issue, namely, to publish more than a paper per each idea. For example, in our case, we often managed to publish in both mathematical and power systems journals. In the former the theoretical aspect of a given idea were formally developed and proved, whereas, in the latter, practical applications of that idea were discussed through simulations.

Nevertheless, we believe that a truly interdisciplinary collaboration should lead to combine in a single document the different fields of knowledge, different technical languages and, ultimately, different methodological approaches. This book is a tangible evidence that this fusion is possible and can lead to relevant advances in all involved research fields.

We hope that the reader will find the intertwined theory and practical applications discussed in the following pages as stimulating as it was for us to learn from each other.

Federico Milano, Ioannis Dassios, Muyang Liu & Georgios Tzounas
Dublin, July 2020

Part I

Introduction

1

Power System Outlines

1.1 Differential Equations

Dynamic systems are often modeled through a set of differential equations, whose general form is [22]:

$$\mathbf{0}_{m,1} = \boldsymbol{\chi}\left(\boldsymbol{\xi}(t), \dot{\boldsymbol{\xi}}(t), \boldsymbol{u}(t), t\right), \tag{1.1}$$

where t is the independent variable, which models the time; $\boldsymbol{\chi} : \mathbb{R}^{(2m+r+1)\times 1} \mapsto \mathbb{R}^{m\times 1}$, is a vector of non-linear non-autonomous equations, e.g. $\boldsymbol{\chi}$ depends explicitly on t; $\boldsymbol{u} \in \mathbb{R}^{r\times 1}$, are input quantities; $\boldsymbol{\xi} \in \mathbb{R}^{m\times 1}$, is the vector of state variables and $\dot{\boldsymbol{\xi}}$ is the vector of the first time derivatives of the state variables. For economy of notation, in the following, we drop the dependency on time of variables and assume that variables depend on time, while parameters do not. With this notation, (1.1) can be written more compactly as:

$$\mathbf{0}_{m,1} = \boldsymbol{\chi}(\boldsymbol{\xi}, \dot{\boldsymbol{\xi}}, \boldsymbol{u}, t)\,. \tag{1.2}$$

The formulation in (1.2), which is generally referred to as *implicit*, can be often particularized for classes of physical systems. These alternative formulations address special cases of the dependency of $\boldsymbol{\chi}$ on $\dot{\boldsymbol{\xi}}$.

A special case that is utilized in the formulation of electronic circuits, e.g. in SPICE-like software tools [164], is the *semi-implicit* form, as follows:

$$\mathbf{0}_{m,1} = \mathbf{E}(\boldsymbol{\xi}, \boldsymbol{u}, t)\,\dot{\boldsymbol{\xi}} - \boldsymbol{\varphi}(\boldsymbol{\xi}, \boldsymbol{u}, t)\,, \tag{1.3}$$

or, equivalently:

$$\mathbf{E}(\boldsymbol{\xi}, \boldsymbol{u}, t)\,\dot{\boldsymbol{\xi}} = \boldsymbol{\varphi}(\boldsymbol{\xi}, \boldsymbol{u}, t)\,, \tag{1.4}$$

where the elements of $\mathbf{E} \in \mathbb{R}^{m\times m}$ are non-linear functions of the state variables; and $\boldsymbol{\varphi} : \mathbb{R}^{(m+r+1)\times 1} \mapsto \mathbb{R}^{m\times 1}$ is a vector of non-linear non-autonomous equations. Note that, in dimensional physical systems, the equations $\boldsymbol{\varphi}$ can be always arranged in such a way that the elements of the matrix $\mathbf{E}$ have the dimension of the independent variable t, i.e. seconds. In the remainder of the book, for simplicity, we will refer to (1.4) as "implicit" differential-algebraic equations (DAEs) rather than "semi-implicit."

Apart from the assumption that the dependency on the first time-derivative vector is linear, the formulation in (1.4) is still quite general as it allows setting to make the equations independent from any given element of the vector $\dot{\boldsymbol{\xi}}$. If the h-th column of $\mathbf{E}$ is null, in fact, the differential equations (1.4) do not depend on $\dot{\xi}_h$.

A relevant question is how one can interpret such a situation, i.e. the fact that (1.4) or, more in general, (1.2), might not depend on some of $\dot{\xi}_h$, with $h \in \mathbb{H} \subset \{1, 2, \dots, m\}$.

In some mathematical papers and books, i.e. [22], no particular meaning is assigned. And, in fact, the implicit formulation (1.2) or its special case (1.4) do not require any specific dependency on any variable of the system, including their time derivatives. It is relevant to note, however, that, if at least one column of $\mathbf{E}$ is null, then $\mathbf{E}$ is a singular matrix. This property leads to interpret the independence of (1.4) from some $\dot{\xi}_h$ as a *singularity*, or, which is the same, a set of equations in the form of (1.4) with not full-rank $\mathbf{E}$ can be interpreted as a *singular system*. This interpretation is discussed in many works, e.g. [19] and [24].

In this book, we utilize the concept of *singular pencil* as it shows relevant properties from the point of view of the definition and calculation of eigenvalues and eigenvectors, which directly follow from the singularity of $\mathbf{E}$. Note that, in the study of matrix pencils, $\mathbf{E}$ is not necessarily square (see Chapter 2). A rectangular $\mathbf{E}$ can be obtained, for example, by removing null columns. To avoid confusion, in the remainder of this book, the dimension of matrix pencils are defined case by case, depending on the problem to be solved.

Further elaborating on (1.4), for some physical systems, such as power systems, it is convenient, while not strictly necessary from the formal point of view, to distinguish, using a different notation, the state variables whose time derivatives have a non-null sensitivity, i.e. have some non-zero elements in the columns of $\mathbf{E}$, and those who do not. The latter are called *algebraic variables.* Splitting the vector $\boldsymbol{\xi}$ into state and algebraic variables leads to the form:

$$\mathbf{E}(\boldsymbol{x}, \boldsymbol{y}, \boldsymbol{u}, t) \begin{bmatrix} \dot{\boldsymbol{x}} \\ \dot{\boldsymbol{y}} \end{bmatrix} = \boldsymbol{\varphi}(\boldsymbol{x}, \boldsymbol{y}, \boldsymbol{u}, t) \, , \tag{1.5}$$

or, equivalently:

$$\begin{bmatrix} \mathbf{T}(\boldsymbol{x}, \boldsymbol{y}, \boldsymbol{u}, t) & \mathbf{0}_{n,l} \\ \mathbf{R}(\boldsymbol{x}, \boldsymbol{y}, \boldsymbol{u}, t) & \mathbf{0}_{l,l} \end{bmatrix} \begin{bmatrix} \dot{\boldsymbol{x}} \\ \dot{\boldsymbol{y}} \end{bmatrix} = \begin{bmatrix} \boldsymbol{f}(\boldsymbol{x}, \boldsymbol{y}, \boldsymbol{u}, t) \\ \boldsymbol{g}(\boldsymbol{x}, \boldsymbol{y}, \boldsymbol{u}, t) \end{bmatrix} , \tag{1.6}$$

where $\mathbf{T}$ and $\mathbf{R}$ are $n \times n$ and $l \times n$ matrices, respectively, with $n + l = m$ and whose elements are non-linear functions of the state and algebraic variables; $\boldsymbol{x} \in \mathbb{R}^{n \times 1}$, are the state variables; $\boldsymbol{y} \in \mathbb{R}^{l \times 1}$ are the algebraic variables; $\boldsymbol{f} : \mathbb{R}^{(m+r+1) \times 1} \mapsto \mathbb{R}^{n \times 1}$ are the differential equations; and $\boldsymbol{g} : \mathbb{R}^{(m+r+1) \times 1} \mapsto \mathbb{R}^{l \times 1}$ are the algebraic equations.

Note that, while in general $\mathbf{E}$ and, hence, $\mathbf{T}$ and $\mathbf{R}$ depend on system variables and time, the vast majority of the models of power system devices and controllers can be described by constant matrices. For this reason, $\mathbf{E}$, $\mathbf{T}$ and $\mathbf{R}$ are hereinafter assumed to have constant elements.

The form (1.6) allows discussing a very common interpretation of algebraic variables as states with *infinitely fast dynamics*. With this regard, it is convenient to provide first the following definition:

Definition 1.1 (Stiff differential equations)**.** Dynamic systems whose dynamics span several orders of magnitude are called *stiff*.

Stiff dynamic systems are studied apart as they generally harder to numerically integrate than non-stiff ones [62]. Then, according to such a definition, a set of DAEs in the form of (1.6) is intrinsically extremely stiff.

Example 1.1 (Singular perturbation)**.** Consider the following system:

$$\begin{aligned} T_1\,\dot{x}_1 &= f_1(x_1, x_2)\,, \\ \epsilon\,\dot{x}_2 &= f_2(x_1, x_2)\,, \end{aligned} \tag{1.7}$$

where ϵ is a small positive quantity. If $\epsilon \to 0$, the dynamic behavior of x_2 becomes infinitely fast and for $\epsilon = 0$, f_2 is always in steady-state ($f_2 = 0$) or, which is the same, x_2 will change instantaneously to accommodate any change of x_1 and satisfy the condition $f_2 = 0$. This interpretation of algebraic variables as states with vanishing transients is called *singular perturbation* approach, which has had some applications in power system analysis, e.g. [180].

To further illustrate this approach, we consider a proportional integral (PI) controller:

$$\begin{aligned} \dot{x} &= K_i\,u\,, \\ 0 &= x + K_p\,u - y\,, \end{aligned} \tag{1.8}$$

In (1.9), the algebraic variable $y \equiv x + K_p u$, $\forall t$. Applying the singular perturbation approach, we can rewrite (1.9) as:

$$\begin{aligned} \dot{x} &= K_i\,u\,, \\ \epsilon\,\dot{y} &= x + K_p\,u - y\,, \end{aligned} \tag{1.9}$$

with $\epsilon \ll 0$. Note that, even if very small, ϵ introduces a delay in the output and thus modifies the behavior of the controller.

It is important to note that algebraic variables cannot always be easily interpreted as infinitely fast states. For example, the power flow equations (1.79) of an electric grid assume quasi-steady-state (QSS) phasors and are intrinsically algebraic (see Section 1.3.2.4). Moreover, in many software tools, algebraic variables are also utilized in auxiliary equations to link different devices. For example, the last two equations of (1.127) that set the values of $\tilde{v}_{\rm ef}$ and $v^{\rm ref}$ link the synchronous machine to the AVR and the AVR to the power system stabilizer (PSS), respectively. These equations are simple identities and do not approximate any "very fast" dynamic behavior. □

The formulation given in (1.6) has been introduced in power systems only very recently (see [115]) and it is still not widely utilized. Slightly more common is a formulation that preserves the ability to remove the dependency from some $\dot{x}_h$ but imposes $\mathbf{R} = \mathbf{0}_{l,n}$ and $\mathbf{T} = \mathbf{T}_\Delta$, where $\mathbf{T}_\Delta$ is diagonal, with diagonal elements either 1 or 0. Such an approach has been utilized, for example, in [8] and leads to the form:

$$\begin{bmatrix} \mathbf{T}_\Delta & \mathbf{0}_{n,l} \\ \mathbf{0}_{l,n} & \mathbf{0}_{l,l} \end{bmatrix} \begin{bmatrix} \dot{\boldsymbol{x}} \\ \dot{\boldsymbol{y}} \end{bmatrix} = \begin{bmatrix} \boldsymbol{f}(\boldsymbol{x}, \boldsymbol{y}, \boldsymbol{u}, t) \\ \boldsymbol{g}(\boldsymbol{x}, \boldsymbol{y}, \boldsymbol{u}, t) \end{bmatrix} . \tag{1.10}$$

The latter form (1.10) is basically just a modification of the most common formulation of power system models, where $\mathbf{T}_\Delta = \mathbf{I}_n$, i.e. the identity matrix of order n. This form is *explicit* and reads as follows:

$$\begin{bmatrix} \mathbf{I}_n & \mathbf{0}_{n,l} \\ \mathbf{0}_{l,n} & \mathbf{0}_{l,l} \end{bmatrix} \begin{bmatrix} \dot{\boldsymbol{x}} \\ \dot{\boldsymbol{y}} \end{bmatrix} = \begin{bmatrix} \boldsymbol{f}(\boldsymbol{x}, \boldsymbol{y}, \boldsymbol{u}, t) \\ \boldsymbol{g}(\boldsymbol{x}, \boldsymbol{y}, \boldsymbol{u}, t) \end{bmatrix} . \tag{1.11}$$

Example 1.2 (Explicit DAE form)**.** Let us consider again the system (1.7) given in Example 1.1. Its explicit form according to (1.11) is:

$$\begin{aligned} \dot{x}_1 &= \frac{1}{T_1} f_1(x_1, x_2) , \\ \dot{x}_2 &= \frac{1}{\epsilon} f_2(x_1, x_2) . \end{aligned} \tag{1.12}$$

The condition $\epsilon \to 0$ is less straightforward to interpret using the explicit form (1.12) than it is if one considers the implicit form (1.7). Let us discuss various cases. If f_2 is finite and not null, then $\dot{x}_2 \to \pm\infty$ as $\epsilon \to 0$. This, however, cannot happen in physical systems. An infinite rate of change of a physical variable would in fact imply an infinite energy (for example, think of x_2 as the variable that describes the position of an object with a non-null mass). Another case is that $f_2 = 0$, $\forall t$, which, from (1.12), leads to $\dot{x}_2 = 0$, independently from the value of ϵ. However, considering

$$0 = f_2(x_1, x_2) ,$$

and differentiating with respect to time, one obtains:

$$0 = \frac{\partial f_2}{\partial x_1} \dot{x}_1 + \frac{\partial f_2}{\partial x_2} \dot{x}_2 ,$$

or, equivalently

$$\dot{x}_2 = - \left(\frac{\partial f_2}{\partial x_2} \right)^{-1} \frac{\partial f_2}{\partial x_1} \dot{x}_1 , \tag{1.13}$$

which is defined for $\frac{\partial f_2}{\partial x_2} \neq 0$ and indicates that $\dot{x}_2 \neq 0$ in general. The only case left is:

$$\lim_{\epsilon \to 0} \frac{f_2}{\epsilon} = a ,$$

with a finite and non-null in general. This is consistent with (1.13) from where, in fact, one can deduce that $a = - \left(\frac{\partial f_2}{\partial x_2} \right)^{-1} \frac{\partial f_2}{\partial x_1} \dot{x}_1$. □

A relevant consequence of the fact that algebraic variables can vary in a infinitely small lapse of time without needing an infinite energy is that algebraic variables can *jump* following discrete events, e.g. a fault or a line outage. On the other hand, state variables are always *continuous* unless subject to impulses, e.g. Dirac delta function. The difference between state and algebraic variables is illustrated in Figure 1.1.

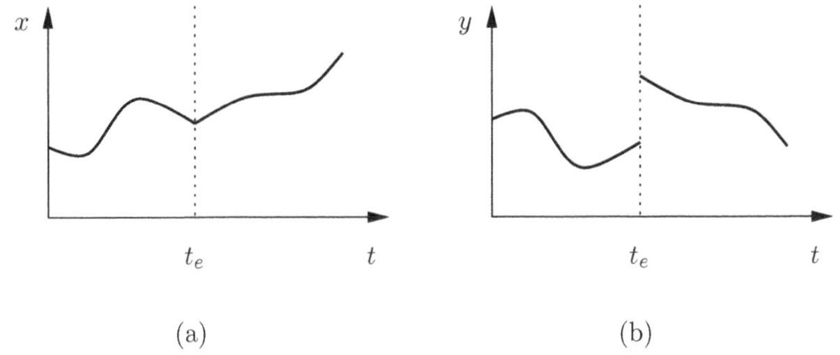

FIGURE 1.1: Transient behavior of state and algebraic variables. (a) State variables are continuous. A discontinuous event in the right-hand side occurring at t_e leads to a change of the rate of change of the variable. (b) Algebraic variables can jump following a discrete event.

Based on the discussion above, we give the following operative Definitions of state and algebraic variables as well as of state space:

Definition 1.2 (State variable). A state variable $x(t)$ is a variable that can show a first-order discontinuity, i.e. $x(t)$ is continuous and $\dot{x}(t)$ can jump.

Definition 1.3 (Algebraic variable). An algebraic variable $y(t)$ is a degenerate state variable that can show a zero-order discontinuity, i.e. $y(t)$ can jump.

Definition 1.4 (State space). The sets of the combined vectors of state and algebraic variables define the state space of a dynamic system.

In the remainder of the book, the set of DAEs is assumed to be autonomous. This assumption does not affect generality, as, in turn, the book discusses mainly equilibria of the set of equations (1.2) to (1.11) and small perturbations around such equilibria. The explicit dependency on time can be thus safely dropped. For example, the autonomous versions of (1.6) and (1.11) read:

$$\mathbf{E} \begin{bmatrix} \dot{\boldsymbol{x}} \\ \dot{\boldsymbol{y}} \end{bmatrix} = \begin{bmatrix} \boldsymbol{f}(\boldsymbol{x}, \boldsymbol{y}, \boldsymbol{u}) \\ \boldsymbol{g}(\boldsymbol{x}, \boldsymbol{y}, \boldsymbol{u}) \end{bmatrix}, \tag{1.14}$$

and

$$\mathbf{E}_{\mathrm{I}} \begin{bmatrix} \dot{\boldsymbol{x}} \\ \dot{\boldsymbol{y}} \end{bmatrix} = \begin{bmatrix} \boldsymbol{f}(\boldsymbol{x}, \boldsymbol{y}, \boldsymbol{u}) \\ \boldsymbol{g}(\boldsymbol{x}, \boldsymbol{y}, \boldsymbol{u}) \end{bmatrix}, \tag{1.15}$$

respectively, where

$$\mathbf{E} = \begin{bmatrix} \mathbf{T} & \mathbf{0}_{n,l} \\ \mathbf{R} & \mathbf{0}_{l,l} \end{bmatrix}, \qquad \mathbf{E}_{\mathrm{I}} = \begin{bmatrix} \mathbf{I}_n & \mathbf{0}_{n,l} \\ \mathbf{0}_{l,n} & \mathbf{0}_{l,l} \end{bmatrix}.$$

In (1.14), the elements of the matrices $\mathbf{T}$ and $\mathbf{R}$ are assumed to be constant. The implicit DAE form (1.14) and the explicit DAE form (1.15) are the ones utilized in the remainder of this book. It is important to note that, if (1.14) and (1.15) model the same system, then $\boldsymbol{f}$ and $\boldsymbol{g}$ in the two sets of DAEs are not the same equations (see Examples 1.1 and 1.2).

The explicit DAE form (1.15) suggests yet another interpretation of algebraic variables $\boldsymbol{y}$ and equations $\boldsymbol{g}$. These, in fact, can be viewed as *constraints* of the differential equations. If one were able to solve $\boldsymbol{g}$ for $\boldsymbol{y}$, i.e.

$$\boldsymbol{y} = \hat{\boldsymbol{g}}(\boldsymbol{x}, \boldsymbol{u}), \tag{1.16}$$

then, (1.15) could be rewritten as a set of ordinary differential equations (ODEs):

$$\dot{\boldsymbol{x}} = \boldsymbol{f}\big(\boldsymbol{x}, \hat{\boldsymbol{g}}(\boldsymbol{x}, \boldsymbol{u}), \boldsymbol{u}\big). \tag{1.17}$$

In practice, $\hat{\boldsymbol{g}}$ cannot be determined explicitly for any power system model except for the simplest one-machine infinite-bus (OMIB) with lossless line and classical machine model. The only special case where $\boldsymbol{g}$ can be inverted is the case where $\boldsymbol{g}$ consists of a set of linear functions or is linearized around an equilibrium point. This case is discussed in Section 1.5 and further elaborated in Chapter 4.

Example 1.3 (Summary of DAE forms)**.** It is useful to illustrate the various formulations presented above with an example. Figure 1.2 shows a simple control diagram scheme, which is composed of two lead-lag blocks.

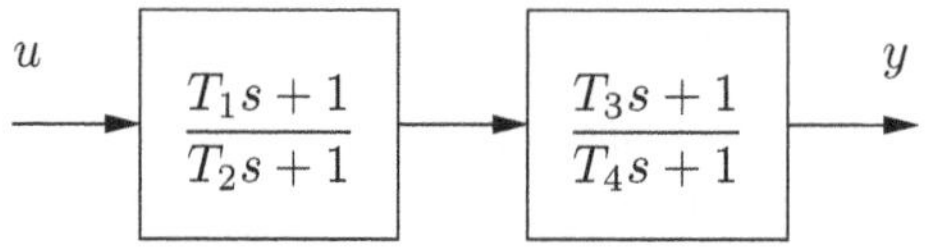

FIGURE 1.2: Series of two lead-lag blocks.

There are infinite sets of implicit and explicit DAEs that describe the scheme in Figure 1.2.[1] We pick two possible representations that illustrate the forms (1.14) and (1.15).

The implicit form can be written as:

$$\begin{aligned} T_2\,\dot{x}_1 &= u - x_1\,, \\ T_4\,\dot{x}_2 - T_1\,\dot{x}_1 &= x_1 - x_2\,, \\ -T_3\,\dot{x}_2 &= x_2 - y\,, \end{aligned} \tag{1.18}$$

[1]This can be readily shown by considering the ODE: $k\dot{x} = -kx$, with $k \in \mathbb{R}_{\neq 0}$. The properties of such an equation are clearly the same independently from k, which can take infinitely many values.

which leads to:

$$
\begin{aligned}
\mathbf{T} &= \begin{bmatrix} T_2 & 0 \\ -T_1 & T_4 \end{bmatrix}, & \mathbf{R} &= [0 \quad -T_3] \,, \\
\boldsymbol{f} &= \begin{bmatrix} u - x_1 \\ x_1 - x_2 \end{bmatrix}, & \boldsymbol{g} &= [x_2 - y] \,.
\end{aligned}
\tag{1.19}
$$

The explicit formulation can be written as:

$$
\begin{aligned}
\dot{x}_1 &= \frac{1}{T_2}(u - x_1) \,, \\
\dot{x}_2 &= \frac{1}{T_4}\left(\frac{T_1}{T_2}(u - x_1) + x_1 - x_2\right) \,, \\
0 &= \frac{T_3}{T_4}\left(\frac{T_1}{T_2}(u - x_1) + x_1 - x_2\right) + x_2 - y \,,
\end{aligned}
\tag{1.20}
$$

or, equivalently:

$$
\begin{aligned}
\dot{x}_1 &= \frac{1}{T_2}(u - x_1) \,, \\
\dot{x}_2 &= \frac{1}{T_4}(c_{12}u + (1 - c_{12})x_1 - x_2) \,, \\
0 &= c_{34}\big(c_{12}u + (1 - c_{12})x_1\big) + (1 - c_{34})x_2 - y \,,
\end{aligned}
\tag{1.21}
$$

where $c_{12} = T_1/T_2$ and $c_{34} = T_3/T_4$ and, according to the notation of (1.15):

$$
\boldsymbol{f} = \begin{bmatrix} (u - x_1)/T_2 \\ \big(c_{12}u + (1 - c_{12})\,x_1 - x_2\big)/T_4 \end{bmatrix}
\tag{1.22}
$$

and

$$
\boldsymbol{g} = \big[c_{34}\big(c_{12}u + (1 - c_{12})\,x_1\big) + (1 - c_{34})x_2 - y\big] \,.
\tag{1.23}
$$

From this simple example, one can deduce that the implicit form (1.14) is, in general, more compact than (1.15). It is also interesting to note that the input signal does not *propagate* through the equations in implicit form, i.e. the input signal u in (1.18) appears only in the first differential equation, while it appears everywhere in (1.21). These properties make the implicit form computationally more efficient, although this efficiency comes with the price of having to define and store matrices $\mathbf{T}$ and $\mathbf{R}$.

Finally, in (1.18), it is straightforward to set to zero any time constant T_1 to T_4. On the other hand, T_2 and T_4 cannot be set to zero in (1.21). One has to rewrite the equations or set T_2 and T_4 to very small quantities. The latter solution, however, is prone to numerical issues. So, assuming for example that $T_4 = 0$, the formulation (1.18) does not need to be updated, whereas (1.21) has to be rewritten as:

$$
\begin{aligned}
\dot{x}_1 &= \frac{1}{T_2}(u - x_1) \,, \\
0 &= c_{12}u + (1 - c_{12})\,x_1 - x_2 \,, \\
0 &= c_{34}\big(c_{12}u + (1 - c_{12})\,x_1\big) + (1 - c_{34})\,x_2 - y \,,
\end{aligned}
\tag{1.24}
$$

with

$$\boldsymbol{f} = \left[(u - x_1)/T_2\right] \tag{1.25}$$

and

$$\boldsymbol{g} = \begin{bmatrix} c_{12}u + (1 - c_{12})\,x_1 - x_2 \\ c_{34}\big(c_{12}u + (1 - c_{12})\,x_1\big) + (1 - c_{34})\,x_2 - y \end{bmatrix} . \tag{1.26}$$

□

A remark on the meaning of input variables $\boldsymbol{u}$ is relevant. These are *exogenous* quantities that are imposed from an external source, i.e. a source that is not modeled in the system, or *endogenous* quantities, i.e. outputs other devices within the same system. Depending on its origin, thus, $\boldsymbol{u}$ can be state or algebraic variables of the system obtained by combining two device models, or just independent quantities whose values are imposed to the system. The rotor speed measurement utilized by turbine governors to regulate the frequency is an example of endogenous variable shared between two devices of the same system. On the other hand, the reference synchronous frequency of the turbine governor is an example of exogenous input quantity, which is also typically constant. The wind speed that feeds the model of a wind turbine is another relevant example of time-dependent stochastic independent input variable. This concept of endogenous and exogenous $\boldsymbol{u}$ is further discussed in the following example.

Example 1.4 (Series of two transfer functions)**.** Consider the system of Figure 1.3, which is composed of two transfer functions similar to the ones shown in Figure 1.2 of Example 1.3.

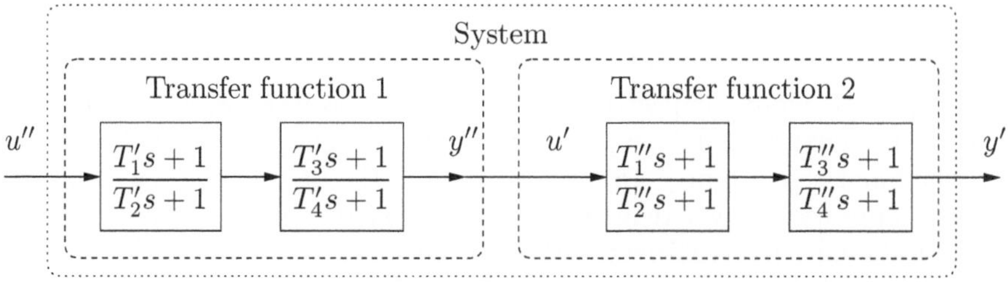

FIGURE 1.3: System composed of the series of two transfer functions.

In this system, the algebraic variable y' is the input variable u'' for the transfer function 2. Considering the system as a whole, the set of algebraic variables is thus $\boldsymbol{y} = [y' \equiv u'', y'']^{\mathrm{T}}$, and there is only one input, $\boldsymbol{u} = [u']$. On the other hand, each transfer function, considered alone, shows one input and one output, namely u', y' and u'', y'', respectively. □

Each device and controller model discussed in the next section, while described one by one, is an element of the power system model. Thus, all devices are connected to others by sharing state and/or algebraic variables. In the following, only variables that are not described by an equation are identified as $\boldsymbol{u}$. Variables shared among devices, on the other hand, are included in either $\boldsymbol{x}$ or $\boldsymbol{y}$ vectors.

1.2 Discrete Maps

The models described so far are continuous except for a finite number of points which corresponds to the discontinuities introduced by the inputs $\boldsymbol{u}$, e.g. line disconnections or the operations of under-load tap changers (ULTCs). In general, however, one never really utilized the models above. Because of their non-linearity, in fact, either one integrates the models using a discrete integration scheme or linearizes the equations and studies the local stability of the system. This book focuses mainly on the latter option. It is relevant however to provide quick outlines of the discretization of continuous DAEs for integration schemes as they are, ultimately, all we have to obtain the trajectories of non-linear DAEs.

We first define discrete maps. These are of the form:

$$\boldsymbol{\xi}^{(i+1)} = \boldsymbol{\eta}(\boldsymbol{\xi}^{(i)}, \boldsymbol{u}^{(i)}) , \tag{1.27}$$

where $\boldsymbol{\eta} : \mathbb{R}^{(2m+r)\times 1} \mapsto \mathbb{R}^{m\times 1}$, is a vector of non-linear equations; $\boldsymbol{u}^{(i)} \in \mathbb{R}^{r\times 1}$, are input quantities at the i-th step; $\boldsymbol{\xi}^{(i)} \in \mathbb{R}^{m\times 1}$, are the states of the system at the i-th step. The map (1.27) is written in an explicit form. There exist implicit maps, of course, but for the purposes of this book, it suffices to consider exclusively the explicit formulation.

Discrete maps are useful to describe some physical phenomena, e.g. the well-known logistic map that has been used to describe chaotic motions or the fascinating Mandelbrot set. However, in the scope of the book, it is relevant to describe a relevant class of maps, namely those that define numerical integration schemes of continuous DAEs.

The simplest integration scheme is the forward Euler method (FEM), which is an explicit first order method. Given a set of explicit ODEs in the form:

$$\dot{\boldsymbol{x}} = \boldsymbol{f}(\boldsymbol{x}, \boldsymbol{u}) , \tag{1.28}$$

the FEM is:

$$\boldsymbol{x}^{(i+1)} = \boldsymbol{x}^{(i)} + \Delta t \, \boldsymbol{f}(\boldsymbol{x}^{(i)}, \boldsymbol{u}^{(i)}) , \tag{1.29}$$

where Δt is the time step and $\boldsymbol{x}^{(i)}$ is known from the previous step. The initial condition is given $\boldsymbol{x}_o = \boldsymbol{x}^{(0)}$ and $\boldsymbol{u}_o = \boldsymbol{u}^{(0)}$. The FEM is known for being the least accurate and numerically stable of all integration methods. Its implicit counterpart, namely the backward Euler method (BEM), on the other hand, is the most numerically stable, and is actually, *hyperstable*. This point is further discussed in Section 6.2. The BEM for the system (1.28) has the form:

$$\boldsymbol{x}^{(i+1)} = \boldsymbol{x}^{(i)} + \Delta t\, \boldsymbol{f}(\boldsymbol{x}^{(i+1)}, \boldsymbol{u}^{(i+1)})\,. \tag{1.30}$$

Since $\boldsymbol{f}$ is evaluated at $\boldsymbol{x}^{(i+1)}$, it requires some numerical methods to be solved if $\boldsymbol{f}$ is non-linear, e.g. the Newton method.

With this aim, let us rewrite (1.30) as:

$$\mathbf{0}_{n,1} = \boldsymbol{x}^{(i)} + \Delta t\, \boldsymbol{f}(\boldsymbol{x}^{(i+1)}, \boldsymbol{u}^{(i+1)}) - \boldsymbol{x}^{(i+1)}\,, \tag{1.31}$$

or, equivalently:

$$\mathbf{0}_{n,1} = \boldsymbol{h}(\boldsymbol{x}^{(i)}, \boldsymbol{x}^{(i+1)}, \boldsymbol{u}^{(i+1)}, \Delta t) = \boldsymbol{h}^{(i+1)}\,, \tag{1.32}$$

where $\boldsymbol{x}^{(i)}$, $\boldsymbol{u}^{(i+1)}$ and Δt are parameters and $\boldsymbol{x}^{(i+1)}$ are the unknowns. Then, the k-th iteration of the Newton method for (1.32) is:

$$\boldsymbol{x}^{(i+1)}_{(k+1)} = \boldsymbol{x}^{(i+1)}_{(k)} - \mathbf{H}^{(i+1)}_{(k)}\, \boldsymbol{h}^{(i+1)}\,, \qquad \text{for } k = 1, 2, \dots\,, \tag{1.33}$$

with initial condition $\boldsymbol{x}^{(i+1)}_{(0)} = \boldsymbol{x}^{(i)}$, i.e. the value of the states at the previous integration step and where:

$$\mathbf{H}^{(i+1)}_{(k)} = \frac{\partial \boldsymbol{h}^{(i+1)}}{\partial \boldsymbol{x}^{(i+1)}_{(k)}} = \Delta t\, \frac{\partial \boldsymbol{f}^{(i+1)}}{\partial \boldsymbol{x}^{(i+1)}_{(k)}} - \mathbf{I}_n\,. \tag{1.34}$$

The iterations of the Newton method stop when either $|\boldsymbol{x}^{(i+1)}_{(k+1)} - \boldsymbol{x}^{(i+1)}_{(k)}| < \epsilon$ (convergence) or $k > k_{\max}$ (divergence).

Implicit methods such as the BEM are particularly suited to integrate implicit sets of DAEs in a simultaneous iteration. If the DAEs are in the form of (1.4), we can define $\boldsymbol{h}$ as:

$$\begin{aligned} \mathbf{0}_{m,1} &= \boldsymbol{h}(\boldsymbol{\xi}^{(i+1)}, \boldsymbol{\xi}^{(i)}, \boldsymbol{u}^{(i+1)}, \boldsymbol{u}^{(i)}, \Delta t) \\ &= \mathbf{E}^{(i)}\, \boldsymbol{\xi}^{(i)} + \Delta t\, \boldsymbol{\varphi}^{(i+1)} - \mathbf{E}^{(i+1)}\, \boldsymbol{\xi}^{(i+1)}\,, \end{aligned} \tag{1.35}$$

where, for simplicity, $\mathbf{E}^{(i)}$, $\mathbf{E}^{(i+1)}$ and $\boldsymbol{\varphi}^{(i+1)}$ indicate:

$$\begin{aligned} \mathbf{E}^{(i)} &= \mathbf{E}(\boldsymbol{\xi}^{(i)}, \boldsymbol{u}^{(i)}, t^{(i)})\,, \\ \mathbf{E}^{(i+1)} &= \mathbf{E}(\boldsymbol{\xi}^{(i+1)}, \boldsymbol{u}^{(i+1)}, t^{(i)} + \Delta t)\,, \\ \boldsymbol{\varphi}^{(i+1)} &= \boldsymbol{\varphi}(\boldsymbol{\xi}^{(i+1)}, \boldsymbol{u}^{(i+1)}, t^{(i)} + \Delta t)\,, \end{aligned}$$

and $\mathbf{H}^{(i+1)}$:

$$\begin{aligned} \mathbf{H}^{(i+1)} &= \frac{\partial \boldsymbol{h}^{(i+1)}}{\partial \boldsymbol{\xi}^{(i+1)}} \\ &= \Delta t\, \frac{\partial \boldsymbol{\varphi}^{(i+1)}}{\partial \boldsymbol{\xi}^{(i+1)}} - \mathbf{E}^{(i+1)} \\ &= \Delta t\, \mathbf{A}^{(i+1)} - \mathbf{E}^{(i+1)}\,. \end{aligned} \tag{1.36}$$

It is interesting to note that $\mathbf{H}^{(i+1)}$ has formally the same structure as a matrix pencil, as defined in Chapter 2. We leave to the interested reader the determination of the expressions of $\boldsymbol{h}$ and $\mathbf{H}$ for (1.14) and (1.15).

Since the FEM is inaccurate and potentially numerically unstable and the BEM is hyperstable, it makes sense to look for alternative integration schemes that have better performance and stability properties. A well-known approach is the Implicit Trapezoidal Method (ITM), which, in turn, is nothing else than the weighted sum of the FEM and the BEM, with equal weights for the two methods. The functions $\boldsymbol{h}^{(i+1)}$ and $\mathbf{H}^{(i+1)}$ for the ITM are as follows:

$$\boldsymbol{h}^{(i+1)} = \mathbf{E}^{(i)}\,\boldsymbol{\xi}^{(i)} + \frac{1}{2}\,\Delta t\,\boldsymbol{\varphi}^{(i)} + \frac{1}{2}\,\Delta t\,\boldsymbol{\varphi}^{(i+1)} - \mathbf{E}^{(i+1)}\,\boldsymbol{\xi}^{(i+1)}\,, \tag{1.37}$$

and

$$\mathbf{H}^{(i+1)} = \frac{1}{2}\,\Delta t\,\mathbf{A}^{(i+1)} - \mathbf{E}^{(i+1)}\,. \tag{1.38}$$

The combined effect of FEM and BEM, makes the ITM one of the most stable (A-stable, in fact) and robust methods available for numerical integration, especially of electrical circuits and systems. To properly illustrate the features and stability of numerical integration schemes, we need further definitions and theoretical results. We thus leave the examples to a later stage of the book, in particular Sections 3.3 and 6.2.

1.3 Power System Models

This section provides the models of power system devices that compose the three-bus system described in Appendix A and that is utilized throughout the book. All models are presented in two forms, namely the implicit DAE form (1.14) and the explicit DAE form (1.15). Unless otherwise indicated, equations are written assuming per-unit values and the angular frequency base, indicated with ω_o and expressed in rad/s is also the reference angular frequency of the system.

In the models discussed in this section, the ac system is assumed balanced and abc phases are transformed using a Park dq-axis frame according to notation shown in Figure 1.4.

The assumption of balanced conditions allows neglecting the homopolar axis and defining the *Park vector* for voltages, currents and fluxes [123]. For example, the Park vector of the voltage at bus h is defined as:

$$\bar{v}_{h,\mathrm{dq}} = v_{h,\mathrm{d}} + \jmath\, v_{h,\mathrm{q}}\,, \tag{1.39}$$

where $\jmath$ is the imaginary unit, the current injected at the same bus as:

$$\bar{\imath}_{h,\mathrm{dq}} = \imath_{h,\mathrm{d}} + \jmath\, \imath_{h,\mathrm{q}}\,, \tag{1.40}$$

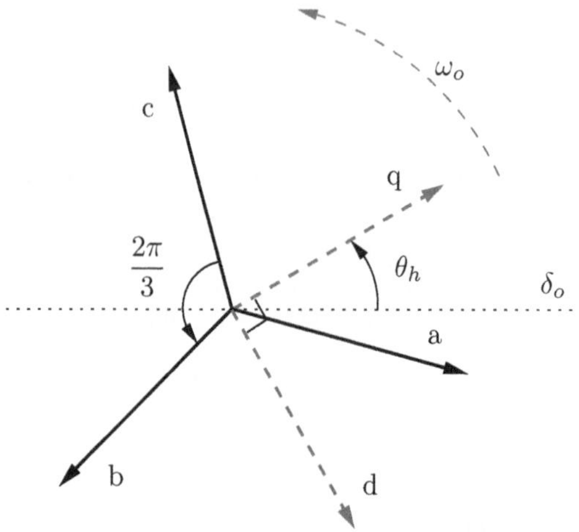

FIGURE 1.4: dq-axis of the Park transform of grid abc quantities. δ_o and ω_o are the reference phase and angular speed of the Park transform. Conventionally $\delta_o = 0$ is assumed.

and the complex power is thus defined as:

$$\bar{S}_h = \bar{\imath}_{h,\mathrm{dq}}\, \bar{\imath}^*_{h,\mathrm{dq}} = \big(\upsilon_{h,\mathrm{d}}\imath_{h,\mathrm{d}} + \upsilon_{h,\mathrm{q}}\imath_{h,\mathrm{q}}\big) + \jmath\big(\upsilon_{h,\mathrm{q}}\imath_{h,\mathrm{d}} - \upsilon_{h,\mathrm{d}}\imath_{h,\mathrm{q}}\big)\,, \tag{1.41}$$

where:

$$\begin{aligned} P_h &= \upsilon_{h,\mathrm{d}}\imath_{h,\mathrm{d}} + \upsilon_{h,\mathrm{q}}\imath_{h,\mathrm{q}}\,, \\ Q_h &= \upsilon_{h,\mathrm{q}}\imath_{h,\mathrm{d}} - \upsilon_{h,\mathrm{d}}\imath_{h,\mathrm{q}}\,. \end{aligned} \tag{1.42}$$

Equation (1.39) can be also written using the conventional polar form of the network bus voltage phasors, namely $\bar{\upsilon}_h$, at bus h. With the dq-axis reference indicated in Figure 1.4, one has:

$$\upsilon_h \angle \theta_h = (\upsilon_{h,\mathrm{d}} + \jmath\, \upsilon_{h,\mathrm{q}})\, \mathrm{e}^{-\jmath\pi/2}\,, \tag{1.43}$$

and, hence:

$$\begin{aligned} \upsilon_{h,\mathrm{d}} &= -\upsilon_h \sin(\theta_h)\,, \\ \upsilon_{h,\mathrm{q}} &= \upsilon_h \cos(\theta_h)\,. \end{aligned} \tag{1.44}$$

A relevant aspect of the Park transform is that it is rotating at a given angular speed, i.e. ω_o. This feature, while often allowing simplifying the equations, most notably those of the synchronous machines, complicates the calculation of the time derivatives of transformed quantities. With this aim, let define the following time derivative operator for per unit Park-vector quantities:

$$\mathcal{P} = \omega_o^{-1}\frac{d}{dt} + \jmath\,, \tag{1.45}$$

where ω_o is expressed in rad/s. Note that the imaginary unit multiplies the per unit value of the reference angular frequency, which is equal to 1 pu(rad/s). For example, the time derivative of the voltage in (1.39) is:

$$\begin{aligned} \mathcal{P}\,\bar{\upsilon}_{h,\mathrm{dq}} &= (\omega_o^{-1}\frac{d}{dt} + \jmath)\,(\upsilon_{h,\mathrm{d}} + \jmath\,\upsilon_{h,\mathrm{q}}) \\ &= (\omega_o^{-1}\dot{\upsilon}_{h,\mathrm{d}} - \upsilon_{h,\mathrm{q}}) + \jmath\,(\omega_o^{-1}\dot{\upsilon}_{h,\mathrm{q}} + \upsilon_{h,\mathrm{d}})\,. \end{aligned} \tag{1.46}$$

Example 1.5 (Park model of an *RL* circuit)**.** Before describing conventional power system models, it is convenient to illustrate the utilization of the Park vector with the *RL* circuit shown in Figure 1.5.

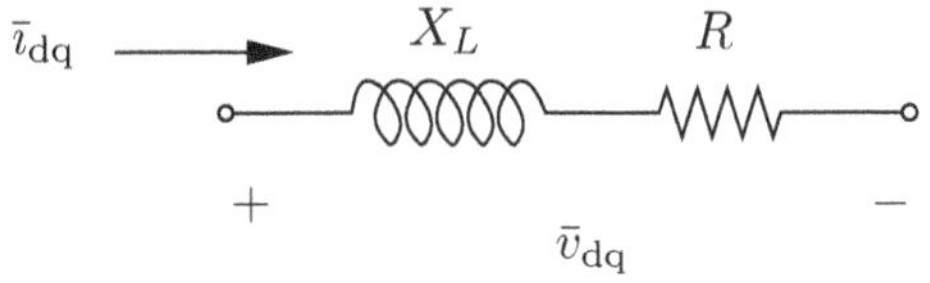

FIGURE 1.5: *RL* circuit to illustrate the Park vector and its time derivative.

The voltage at the terminal nodes of the circuit is given by:

$$\bar{\upsilon}_{\mathrm{dq}} = (R + \mathcal{P}\,X_L)\,\bar{\imath}_{\mathrm{dq}}\,, \tag{1.47}$$

where R and X_L are the per-unit resistance and reactance, respectively. Expanding (1.47) and using the notation of (1.18) lead to:

$$\begin{aligned} \omega_o^{-1}X_L\,\dot{\imath}_{\mathrm{d}} &= \upsilon_{\mathrm{d}} - R\,\imath_{\mathrm{d}} + X_L\,\imath_{\mathrm{q}}\,, \\ \omega_o^{-1}X_L\,\dot{\imath}_{\mathrm{q}} &= \upsilon_{\mathrm{q}} - R\,\imath_{\mathrm{q}} - X_L\,\imath_{\mathrm{d}}\,, \end{aligned} \tag{1.48}$$

with:

$$\begin{gathered} \boldsymbol{x}^{\mathrm{T}} = [\imath_{\mathrm{d}},\,\imath_{\mathrm{q}}]\,, \qquad \boldsymbol{y} = \emptyset\,, \qquad \boldsymbol{u}^{\mathrm{T}} = [\upsilon_{\mathrm{d}},\,\upsilon_{\mathrm{q}}]\,, \\ \boldsymbol{f} = \begin{bmatrix} \upsilon_{\mathrm{d}} - R\,\imath_{\mathrm{d}} + X_L\,\imath_{\mathrm{q}} \\ \upsilon_{\mathrm{q}} - R\,\imath_{\mathrm{q}} - X_L\,\imath_{\mathrm{d}} \end{bmatrix}, \qquad \boldsymbol{g} = \emptyset\,, \\ \mathbf{T} = \omega_o^{-1}\begin{bmatrix} X_{\mathrm{L}} & 0 \\ 0 & X_{\mathrm{L}} \end{bmatrix}, \qquad \mathbf{R} = \emptyset\,. \end{gathered} \tag{1.49}$$

In a 50 Hz system, $\omega_o = 314.16$ rad/s. Assuming a per-unit value of X_L of typical high-voltage transmission lines, i.e. $X_L \approx 0.1$ pu(Ω), the diagonal elements of matrix $\mathbf{T}$ are of the order of 10^{-4} s, hence much smaller than the conventional time scale of transient stability analysis. This justifies the common assumption of steady-state conditions for electromagnetic elements, thus leading to:

$$\begin{aligned} 0 &= \upsilon_{\mathrm{d}} - R\,\imath_{\mathrm{d}} + X_L\,\imath_{\mathrm{q}}\,, \\ 0 &= \upsilon_{\mathrm{q}} - R\,\imath_{\mathrm{q}} - X_L\,\imath_{\mathrm{d}}\,, \end{aligned} \tag{1.50}$$

which, in turn, is the well-known phasor representation:

$$\bar{\upsilon}_{\mathrm{dq}} = (R + \jmath\,X_L)\,\bar{\imath}_{\mathrm{dq}}\,. \tag{1.51}$$

From the developments above, it descends that, in steady-state, the Park vector coincides with the phasor representation of voltages and currents. It is important to note, however, that the Park vector representation is valid for any transient condition, not just in steady state. Equation (1.51) is thus a set of QSS differential equations where fast dynamics have been neglected. □

1.3.1 Nodes

The study of an electrical circuit requires the definition of its topology. Since the nodal approach is definitively more common in computer implementations than the meshed one, the definition of the grid generally starts with the definition of its nodes, or buses.

The Kirchhoff current law (KCL) at each bus of the grid imposes that (see also Figure 1.6):

$$\sum_{h \in \mathbb{D}_r} \bar{\imath}_{h,\mathrm{dq}} = 0\,, \qquad r = 1, 2, \dots, b\,, \tag{1.52}$$

where $\mathbb{D}_r$ is the set of devices and/or branches that are connected to the r-th bus and b is the number of nodes that form the grid. Since each (1.52) is a complex equation, one can write $2b$ real algebraic equations that express the energy conservation of the grid. These equations, in turn, determine the real and imaginary components (or the magnitude and phase angles) of the voltages $\bar{\boldsymbol{v}}_{\mathrm{dq}}$ at the network nodes.

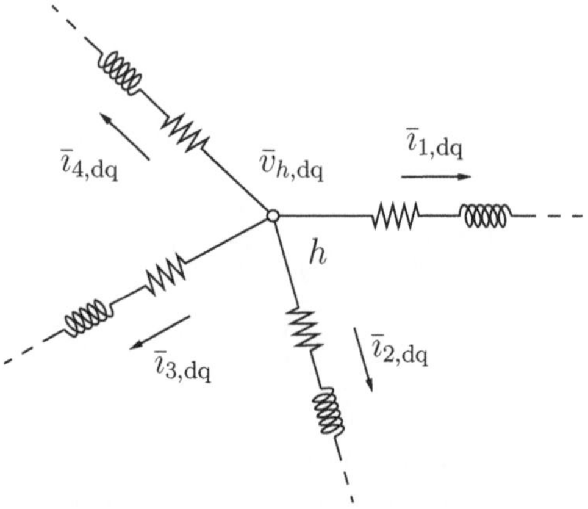

FIGURE 1.6: Illustration of the KCL.

In power flow analysis, it is also common to write the KCL in terms of powers rather than currents:

$$\sum_{h \in \mathbb{D}_r} \bar{S}_h = 0\,, \qquad r = 1, 2, \dots, b\,, \tag{1.53}$$

where $\bar{S}_h$ is the complex power injected into node r from branch h.

In the remainder of this chapter and then in all following chapters, we always assume that the set of $2b$ algebraic equations (1.52) or (1.53) are utilized to link the electrical devices that form the grid. Moreover, we assume that the currents injected by each device into the nodes, namely $\bar{\imath}_{h,\rm dq}$ and $\bar{\imath}_{k,\rm dq}$, belong to the vector $\boldsymbol{u}$. In fact, from a computer-programming point of view, it is not necessary to define explicitly $\bar{\imath}_{h,\rm dq}$ and $\bar{\imath}_{k,\rm dq}$ as algebraic variables, as their expressions can be summed directly into the equations (1.52) when forming the set of DAEs that model the whole system.

1.3.2 Transmission System

The transmission system is the "glue" that connects all devices, in particular, loads and generators, and is composed mainly of overhead ac transmission lines and transformers. In recent years, also dc connections have gained momentum, thanks to the improved reliability and reduced cost of voltage source converters. However dc connections are not discussed in this book. The interested reader can find an introduction to converters in [181] and on dc connections in [31].

Even just focusing on conventional ac connections, detailed dynamic models of these devices can take hundreds of pages. Hence, this section focuses exclusively on lumped, balanced, dynamic and steady-state models that are commonly used in the transient stability analysis of power systems. Long transmission lines that can be modeled using distributed parameters and partial differential equations are described in Chapter 10. The *admittance matrix* is also defined in this section.

1.3.2.1 Transmission Lines

A lumped-parameter Π-model of transmission line is adequate for sufficiently short distances. This is the model shown in Figure 1.7.

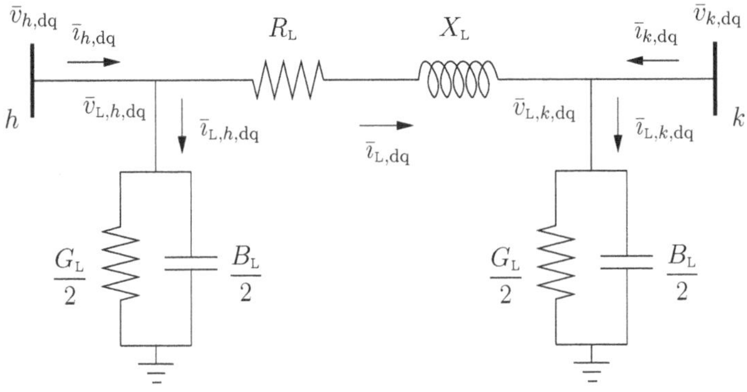

FIGURE 1.7: Single-phase equivalent Π-model of a transmission line.

The per-unit differential equations of a transmission line are:

$$\begin{aligned}
\bar{\upsilon}_{\mathrm{L},h,\mathrm{dq}} &= \bar{\upsilon}_{\mathrm{L},k,\mathrm{dq}} + (R_{\mathrm{L}} + \mathcal{P}\, X_{\mathrm{L}})\, \bar{\imath}_{\mathrm{L},\mathrm{dq}} \,, \\
\bar{\imath}_{\mathrm{L},h,\mathrm{dq}} &= \frac{1}{2}(G_{\mathrm{L}} + \mathcal{P}\, B_{\mathrm{L}})\, \bar{\upsilon}_{\mathrm{L},h,\mathrm{dq}} \,, \\
\bar{\imath}_{\mathrm{L},k,\mathrm{dq}} &= \frac{1}{2}(G_{\mathrm{L}} + \mathcal{P}\, B_{\mathrm{L}})\, \bar{\upsilon}_{\mathrm{L},k,\mathrm{dq}} \,,
\end{aligned} \tag{1.54}$$

and the current injections to be included in the node equations (1.52) are:

$$\begin{aligned}
\bar{\imath}_{h,\mathrm{dq}} &= \bar{\imath}_{\mathrm{L},\mathrm{dq}} + \bar{\imath}_{\mathrm{L},h,\mathrm{dq}} \,, \\
\bar{\imath}_{k,\mathrm{dq}} &= -\bar{\imath}_{\mathrm{L},\mathrm{dq}} + \bar{\imath}_{\mathrm{L},k,\mathrm{dq}} \,.
\end{aligned} \tag{1.55}$$

Note that, formally, since the bus voltages $\bar{\bar{\upsilon}}_{\mathrm{dq}}$ are algebraic variables, one has to impose:

$$\begin{aligned}
\bar{\upsilon}_{h,\mathrm{dq}} &= \bar{\upsilon}_{\mathrm{L},h,\mathrm{dq}} \,, \\
\bar{\upsilon}_{k,\mathrm{dq}} &= \bar{\upsilon}_{\mathrm{L},k,\mathrm{dq}} \,.
\end{aligned} \tag{1.56}$$

The latter equations make consistent and properly determined the number of equations and variables that describe the transmission line.

Using the notation of (1.14), the DAEs in (1.54), (1.55) and (1.56) become:

$$\boldsymbol{x}^{\mathrm{T}} = \left[\imath_{\mathrm{L,d}},\, \imath_{\mathrm{L,q}},\, \upsilon_{\mathrm{L},h,\mathrm{d}},\, \upsilon_{\mathrm{L},h,\mathrm{q}},\, \upsilon_{\mathrm{L},k,\mathrm{d}},\, \upsilon_{\mathrm{L},k,\mathrm{q}}\right], \tag{1.57}$$

$$\boldsymbol{y}^{\mathrm{T}} = \left[\imath_{\mathrm{L},h,\mathrm{d}},\, \imath_{\mathrm{L},h,\mathrm{q}},\, \imath_{\mathrm{L},k,\mathrm{d}},\, \imath_{\mathrm{L},k,\mathrm{q}},\, \upsilon_{h,\mathrm{d}},\, \upsilon_{h,\mathrm{q}},\, \upsilon_{k,\mathrm{d}},\, \upsilon_{k,\mathrm{q}}\right], \tag{1.58}$$

$$\boldsymbol{u}^{\mathrm{T}} = \left[\imath_{h,\mathrm{d}},\, \imath_{h,\mathrm{q}},\, \imath_{k,\mathrm{d}},\, \imath_{k,\mathrm{q}}\right], \tag{1.59}$$

$$\boldsymbol{f} = \begin{bmatrix}
\upsilon_{\mathrm{L},h,\mathrm{d}} - \upsilon_{\mathrm{L},k,\mathrm{d}} - R_{\mathrm{L}}\imath_{\mathrm{L,d}} + X_{\mathrm{L}}\imath_{\mathrm{L,q}} \\
\upsilon_{\mathrm{L},h,\mathrm{q}} - \upsilon_{\mathrm{L},k,\mathrm{q}} - R_{\mathrm{L}}\imath_{\mathrm{L,q}} - X_{\mathrm{L}}\imath_{\mathrm{L,d}} \\
\imath_{\mathrm{L},h,\mathrm{d}} - \dfrac{G_{\mathrm{L}}}{2}\upsilon_{\mathrm{L},h,\mathrm{d}} + \dfrac{B_{\mathrm{L}}}{2}\upsilon_{\mathrm{L},h,\mathrm{q}} \\
\imath_{\mathrm{L},h,\mathrm{q}} - \dfrac{G_{\mathrm{L}}}{2}\upsilon_{\mathrm{L},h,\mathrm{q}} - \dfrac{B_{\mathrm{L}}}{2}\upsilon_{\mathrm{L},h,\mathrm{d}} \\
\imath_{\mathrm{L},k,\mathrm{d}} - \dfrac{G_{\mathrm{L}}}{2}\upsilon_{\mathrm{L},k,\mathrm{d}} + \dfrac{B_{\mathrm{L}}}{2}\upsilon_{\mathrm{L},k,\mathrm{q}} \\
\imath_{\mathrm{L},k,\mathrm{q}} - \dfrac{G_{\mathrm{L}}}{2}\upsilon_{\mathrm{L},k,\mathrm{q}} - \dfrac{B_{\mathrm{L}}}{2}\upsilon_{\mathrm{L},k,\mathrm{d}}
\end{bmatrix}, \tag{1.60}$$

$$\boldsymbol{g} = \begin{bmatrix}
\imath_{h,\mathrm{d}} - \imath_{\mathrm{L,d}} - \imath_{\mathrm{L},h,\mathrm{d}} \\
\imath_{h,\mathrm{q}} - \imath_{\mathrm{L,q}} - \imath_{\mathrm{L},h,\mathrm{q}} \\
\imath_{k,\mathrm{d}} + \imath_{\mathrm{L,d}} - \imath_{\mathrm{L},k,\mathrm{d}} \\
\imath_{k,\mathrm{q}} + \imath_{\mathrm{L,q}} - \imath_{\mathrm{L},k,\mathrm{q}} \\
\upsilon_{h,\mathrm{d}} - \upsilon_{\mathrm{L},h,\mathrm{d}} \\
\upsilon_{h,\mathrm{q}} - \upsilon_{\mathrm{L},h,\mathrm{q}} \\
\upsilon_{k,\mathrm{d}} - \upsilon_{\mathrm{L},k,\mathrm{d}} \\
\upsilon_{k,\mathrm{q}} - \upsilon_{\mathrm{L},k,\mathrm{q}}
\end{bmatrix}, \tag{1.61}$$

$$\mathbf{T} = \omega_o^{-1} \begin{bmatrix} X_{\mathrm{L}} & 0 & 0 & 0 & 0 & 0 \\ 0 & X_{\mathrm{L}} & 0 & 0 & 0 & 0 \\ 0 & 0 & B_{\mathrm{L}}/2 & 0 & 0 & 0 \\ 0 & 0 & 0 & B_{\mathrm{L}}/2 & 0 & 0 \\ 0 & 0 & 0 & 0 & B_{\mathrm{L}}/2 & 0 \\ 0 & 0 & 0 & 0 & 0 & B_{\mathrm{L}}/2 \end{bmatrix}, \tag{1.62}$$

$$\mathbf{R} = \mathbf{0}_{8,6} \, . \tag{1.63}$$

Since $\mathbf{T}$ is diagonal and full rank, the explicit form of the differential equations as in (1.15) is easily obtained as $\mathbf{T}^{-1}\boldsymbol{f}$.

The QSS model of the transmission line is obtained by assuming $\mathbf{T} = \mathbf{0}_{6,6}$ in (1.62), which leads to a purely algebraic model. In QSS models, however, it is unusual to use the expressions given in (1.54). Since the currents entering in the line are inputs, a notation exclusively based on admittances is preferred, as follows:

$$\begin{bmatrix} \bar{\imath}_{h,\mathrm{dq}} \\ \bar{\imath}_{k,\mathrm{dq}} \end{bmatrix} = \begin{bmatrix} \bar{Y}_{\mathrm{L},hk} + \bar{Y}_{\mathrm{L},h} & -\bar{Y}_{\mathrm{L},hk} \\ -\bar{Y}_{\mathrm{L},hk} & \bar{Y}_{\mathrm{L},hk} + \bar{Y}_{\mathrm{L},k} \end{bmatrix} \begin{bmatrix} \bar{v}_{h,\mathrm{dq}} \\ \bar{v}_{k,\mathrm{dq}} \end{bmatrix}, \tag{1.64}$$

where

$$\begin{aligned} \bar{Y}_{\mathrm{L},hk} &= (R_{\mathrm{L}} + \jmath X_{\mathrm{L}})^{-1} \, , \\ \bar{Y}_{\mathrm{L},h} &= \bar{Y}_{\mathrm{L},k} = \frac{1}{2}(G_{\mathrm{L}} + \jmath B_{\mathrm{L}}) \, . \end{aligned} \tag{1.65}$$

1.3.2.2 Transformers

Transformer models include a series inductive impedance that models resistive losses and the flux leakages of the windings. The magnetization reactance (X_μ) and iron losses (R_{Fe}) are modeled through shunt elements which, however, are often neglected when studying power systems. Figure 1.8 shows the resulting single-phase equivalent per-unit circuit of the transformer obtained assuming that the magnetization current is small and that primary and secondary winding losses and leakage can be merged together in a single impedance, namely $R_{\mathrm{T}} + \jmath X_{\mathrm{T}}$. Figure 1.8 also shows an ideal (lossless) transformer that takes into account off-nominal tap ratio (a_{T}) and phase shift (ϕ_{T}) between primary and secondary voltages. The off-nominal tap ratio is assumed to be on the primary winding, hence the need to multiply by a_{T}^2 the series impedance of the transformer [113].

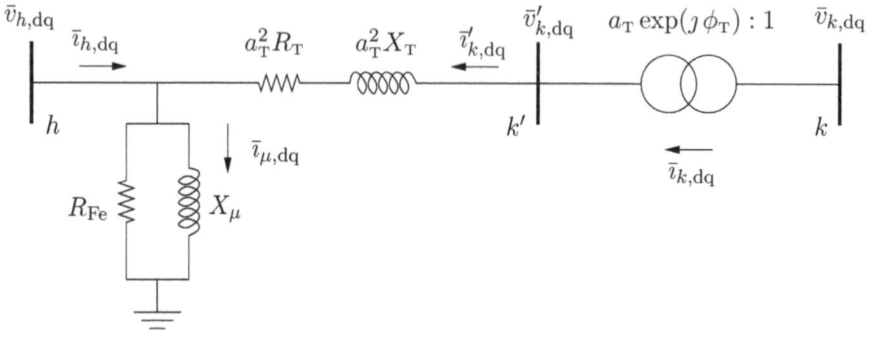

FIGURE 1.8: Single-phase equivalent circuit a transformer with off-nominal tap ratio and phase shift.

The dynamic per-unit equations of a transmission line are as follows:

$$
\begin{aligned}
\bar{v}_{h,\rm dq} &= \bar{v}'_{k,\rm dq} - a_{\rm T}^2 (R_{\rm T} + \mathcal{P}\, X_{\rm T})\, \bar{\imath}'_{k,\rm dq}\,, \\
\bar{v}_{h,\rm dq} &= \mathcal{P}\, X_\mu\, \bar{\imath}_{\mu,\rm dq}\,, \\
\bar{\imath}_{h,\rm dq} &= -\bar{\imath}'_{k,\rm dq} + \frac{1}{R_{\rm Fe}} \bar{v}_{h,\rm dq} + \bar{\imath}_{\mu,\rm dq}\,, \\
\bar{v}'_{k,\rm dq} &= \bar{a}_{\rm T}\, \bar{v}_{k,\rm dq}\,, \\
\bar{\imath}_{k,\rm dq} &= \bar{a}_{\rm T}^*\, \bar{\imath}'_{k,\rm dq}\,,
\end{aligned}
\tag{1.66}
$$

where $\bar{a}_{\rm T} = a_{\rm T}\, e^{\jmath\phi_{\rm T}}$ is the complex tap ratio of the transformer. Note that the ideal transformer is power invariant:

$$
\bar{v}_{k,\rm dq}\, \bar{\imath}^*_{k,\rm dq} = \bar{a}\, \bar{v}_{k,\rm dq}\, \frac{\bar{\imath}^*_{k,\rm dq}}{(\bar{a}^*)^*} = \bar{v}'_{k,\rm dq}\, \bar{\imath}'^*_{k,\rm dq}\,. \tag{1.67}
$$

If iron core magnetization and losses are neglected, as usual in power system studies, $X_\mu \to \infty$, which implies $\bar{\imath}_{\mu,\rm dq} = 0$, and $1/R_{\rm Fe} \to 0$, which leads to $\bar{\imath}_{h,\rm dq} = -\bar{\imath}'_{k,\rm dq}$.

Using the notation of (1.14), (1.66) becomes:

$$
\boldsymbol{x}^{\rm T} = [\imath'_{k,\rm d},\, \imath'_{k,\rm q},\, \imath_{\mu,\rm d},\, \imath_{\mu,\rm q}]\,, \tag{1.68}
$$

$$
\boldsymbol{y}^{\rm T} = [v_{h,\rm d},\, v_{h,\rm q},\, v_{k,\rm d},\, v_{k,\rm q}, v'_{k,\rm d},\, v'_{k,\rm q}], \tag{1.69}
$$

$$
\boldsymbol{u}^{\rm T} = [\imath_{h,\rm d},\, \imath_{h,\rm q},\, \imath_{k,\rm d},\, \imath_{k,\rm q}]\,, \tag{1.70}
$$

$$
\boldsymbol{f} = \begin{bmatrix}
v'_{k,\rm d} - v_{h,\rm d} - a_{\rm T}^2 R_{\rm T}\, \imath'_{k,\rm d} + a_{\rm T}^2 X_{\rm T}\, \imath'_{k,\rm q} \\
v'_{k,\rm q} - v_{h,\rm q} - a_{\rm T}^2 R_{\rm T}\, \imath'_{k,\rm q} - a_{\rm T}^2 X_{\rm T}\, \imath'_{k,\rm d} \\
v_{h,\rm d} + X_\mu\, \imath_{\mu,\rm q} \\
v_{h,\rm q} - X_\mu\, \imath_{\mu,\rm d}
\end{bmatrix}, \tag{1.71}
$$

$$\boldsymbol{g} = \begin{bmatrix} \imath_{h,\mathrm{d}} + \imath'_{k,\mathrm{d}} - \frac{1}{R_{\mathrm{Fe}}} v_{h,\mathrm{d}} - \imath_{\mu,\mathrm{d}} \\ \imath_{h,\mathrm{q}} + \imath'_{k,\mathrm{q}} - \frac{1}{R_{\mathrm{Fe}}} v_{h,\mathrm{q}} - \imath_{\mu,\mathrm{q}} \\ a_{\mathrm{T}} \cos(\phi) v_{k,\mathrm{d}} - a_{\mathrm{T}} \sin(\phi_{\mathrm{T}}) v_{k,\mathrm{q}} - v'_{k,\mathrm{d}} \\ a_{\mathrm{T}} \cos(\phi) v_{k,\mathrm{q}} + a_{\mathrm{T}} \sin(\phi_{\mathrm{T}}) v_{k,\mathrm{d}} - v'_{k,\mathrm{q}} \\ a_{\mathrm{T}} \cos(\phi) \imath'_{k,\mathrm{d}} + a_{\mathrm{T}} \sin(\phi_{\mathrm{T}}) \imath'_{k,\mathrm{q}} - \imath_{k,\mathrm{d}} \\ a_{\mathrm{T}} \cos(\phi) \imath'_{k,\mathrm{q}} - a_{\mathrm{T}} \sin(\phi_{\mathrm{T}}) \imath'_{k,\mathrm{d}} - \imath_{k,\mathrm{q}} \end{bmatrix} , \tag{1.72}$$

$$\mathbf{T} = \omega_o^{-1} \begin{bmatrix} a_{\mathrm{T}}^2 X_{\mathrm{T}} & 0 & 0 & 0 \\ 0 & a_{\mathrm{T}}^2 X_{\mathrm{T}} & 0 & 0 \\ 0 & 0 & X_\mu & 0 \\ 0 & 0 & 0 & X_\mu \end{bmatrix} , \tag{1.73}$$

$$\mathbf{R} = \mathbf{0}_{6,4} \ . \tag{1.74}$$

Similarly to the model of the transmission line, since $\mathbf{T}$ is diagonal and full rank, the explicit form of the differential equations as in (1.15) is easily obtained as $\mathbf{T}^{-1}\boldsymbol{f}$.

The QSS model of the transformer is obtained by assuming $\mathbf{T} = \mathbf{0}_{4,4}$ in (1.73), which leads to a purely algebraic model. And, again, in the same vein as the model of transmission lines, a notation based on admittances is preferred. Neglecting iron core magnetization and losses, the currents injected into the transformer are linked to the bus voltages as follows:

$$\begin{bmatrix} \bar{\imath}_{h,\mathrm{dq}} \\ \bar{\imath}_{k,\mathrm{dq}} \end{bmatrix} = \bar{Y}_{\mathrm{T}} \begin{bmatrix} 1/a_{\mathrm{T}}^2 & -1/\bar{a}_{\mathrm{T}}^* \\ -1/\bar{a}_{\mathrm{T}} & 1 \end{bmatrix} \begin{bmatrix} \bar{v}_{h,\mathrm{dq}} \\ \bar{v}_{k,\mathrm{dq}} \end{bmatrix} , \tag{1.75}$$

where

$$\bar{Y}_{\mathrm{T}} = (R_{\mathrm{T}} + \jmath X_{\mathrm{T}})^{-1} \ . \tag{1.76}$$

1.3.2.3 Admittance Matrix

The QSS models of transmission lines and transformers enable the merger of these two devices in compact formulation based on the matrix notation of (1.64) and (1.75). The result is the so-called *admittance matrix* $\bar{\mathbf{Y}}_{\mathrm{bus}}$, that links current injections and bus voltages at all network buses:

$$\bar{\boldsymbol{\imath}}_{\mathrm{dq}} = \bar{\mathbf{Y}}_{\mathrm{bus}} \, \bar{\boldsymbol{v}}_{\mathrm{dq}} , \tag{1.77}$$

where, according to the assumptions of the QSS model, the elements of $\bar{\mathbf{Y}}_{\mathrm{bus}}$ are effectively constant as they depend exclusively on transmission line and transformer parameters, on the system impedance and frequency bases.

For every pair of buses h and k connected by one or more branches, the elements of the admittance matrix are given by:

$$\begin{aligned}
\bar{Y}_{hh} &= \sum_{m \in \mathbb{L}_{hk}} (\bar{Y}^m_{\mathrm{L},hk} + \bar{Y}^m_{\mathrm{L},h}) + \sum_{n \in \mathbb{T}_{hk}} \frac{1}{a^2_{\mathrm{T},n}} \bar{Y}^n_{\mathrm{T}} \,, \\
\bar{Y}_{hk} &= - \sum_{m \in \mathbb{L}_{hk}} \bar{Y}^m_{\mathrm{L},hk} - \sum_{n \in \mathbb{T}_{hk}} \frac{1}{\bar{a}^*_{\mathrm{T},n}} \bar{Y}^n_{\mathrm{T}} \,, \\
\bar{Y}_{kh} &= - \sum_{m \in \mathbb{L}_{hk}} \bar{Y}^m_{\mathrm{L},hk} - \sum_{n \in \mathbb{T}_{hk}} \frac{1}{\bar{a}_{\mathrm{T},n}} \bar{Y}^n_{\mathrm{T}} \,, \\
\bar{Y}_{kk} &= \sum_{m \in \mathbb{L}_{hk}} (\bar{Y}^m_{\mathrm{L},hk} + \bar{Y}^m_{\mathrm{L},k}) + \sum_{n \in \mathbb{T}_{hk}} \bar{Y}^n_{\mathrm{T}} \,,
\end{aligned} \tag{1.78}$$

where $\mathbb{L}_{hk}$ and $\mathbb{T}_{hk}$ are the sets of transmission lines and transformers, respectively, connecting buses h and k, and line and transformer parameters are defined in (1.65) and (1.76), respectively.

Example 1.6 (Admittance matrix of the three-bus system). The admittance matrix of the three-bus system described in Appendix A is shown in Table 1.1. Since the three-bus system does not include transformers with off-nominal tap ratios or phase shifts, $\bar{\mathbf{Y}}_{\mathrm{bus}}$ is symmetric. □

TABLE 1.1: Admittance matrix of the three-bus system.

Bus	1	2	3
1	$1.27 - \jmath 13.1$	$-0.45 + \jmath 4.50$	$-0.82 + \jmath 9.02$
2	$-0.45 + \jmath 4.50$	$1.35 - \jmath 13.1$	$-0.90 + \jmath 9.00$
3	$-0.82 + \jmath 9.02$	$-0.90 + \jmath 9.00$	$1.72 - \jmath 17.6$

1.3.2.4 Power Flow Equations

The power flow equations are a relevant set of equations that can be derived from the admittance matrix, as follows. The net power injections at network buses are defined as:

$$\bar{\boldsymbol{S}} = \bar{\boldsymbol{v}}_{\mathrm{dq}} \circ \bar{\boldsymbol{\imath}}^*_{\mathrm{dq}} \,, \tag{1.79}$$

where "∘" denotes the Hadamard product, i.e. the element-wise vector multiplication. Recalling the expression of the current given in (1.77), one obtains:

$$\bar{\boldsymbol{S}} = \bar{\boldsymbol{v}}_{\mathrm{dq}} \circ \left(\bar{\mathbf{Y}}^*_{\mathrm{bus}} \, \bar{\boldsymbol{v}}^*_{\mathrm{dq}}\right) \,, \tag{1.80}$$

which are the sought power flow equations. The expression (1.80) is valid in time as it is deduced from the definition of complex power and KCL. Very often, however, the power flow equations are used in steady-state to solve the well-known power flow problem.

Before defining such a problem, it is convenient to expand (1.80) into its real and imaginary components, as follows:

$$P_h = \sum_{k \in \mathbb{B}} \big[v_{h,\mathrm{d}}(G_{hk}\, v_{k,\mathrm{d}} - B_{hk}\, v_{k,\mathrm{q}}) + v_{h,\mathrm{q}}(G_{hk}\, v_{k,\mathrm{q}} + B_{hk} v_{k,\mathrm{d}})\big],\ h \in \mathbb{B}\ ,$$
$$Q_h = \sum_{k \in \mathbb{B}} \big[v_{h,\mathrm{q}}(G_{hk}\, v_{k,\mathrm{d}} - B_{hk}\, v_{k,\mathrm{q}}) - v_{h,\mathrm{d}}(G_{hk}\, v_{k,\mathrm{q}} + B_{hk} v_{k,\mathrm{d}})\big],\ h \in \mathbb{B}\ ,$$
$$(1.81)$$

where $\mathbb{B}$ is the set of network buses. These equations can be conveniently rewritten in polar form:

$$\begin{aligned} P_h &= v_h \sum_{k \in \mathbb{B}} v_k \big(G_{hk} \cos(\theta_{hk}) + B_{hk} \sin(\theta_{hk})\big), \quad h \in \mathbb{B}\ , \\ Q_h &= v_h \sum_{k \in \mathbb{B}} v_k \big(G_{hk} \sin(\theta_{hk}) - B_{hk} \cos(\theta_{hk})\big), \quad h \in \mathbb{B}\ , \end{aligned} \tag{1.82}$$

where

$$v_h = \sqrt{v_{h,\mathrm{d}}^2 + v_{h,\mathrm{q}}^2}\ , \tag{1.83}$$

$$\theta_h = \angle \bar{v}_{h,\mathrm{dq}} = \mathrm{atan}\left(-\frac{v_{h,\mathrm{d}}}{v_{h,\mathrm{q}}}\right)\ , \tag{1.84}$$

and $\theta_{hk} = \theta_h - \theta_k$.

The conventional power flow problem is set up by defining three kinds of buses:

- *Slack bus*, where the voltage magnitude v and the phase angle θ are known and the active and reactive power injections, P and Q, are unknown.
- *PV bus*, where the voltage magnitude v and the active power injection P are known and the voltage phase angle θ and reactive power injection Q are unknown.
- *PQ bus*, where the active and reactive power injections, P and Q, are known and the voltage magnitude v and phase angle θ are unknown.

1.3.3 Loads

Loads play a substantial role in the stability and transient behavior of power systems. On the other hand, detailed load models are difficult to obtain in practice. At the high- and medium-voltage levels, in fact, loads rarely consist of a single device. The power consumption at a node of the network is thus the combined effect of several devices, each of them with a potentially different dynamic behavior. Another difficulty is given by the stochastic nature of loads, which makes their power consumption and dynamic behavior volatile. To further complicate the problem, the trend of recent years to install distributed energy resources (DERs) at the medium and low-voltage levels introduces new dynamics and increases the randomness of the power consumption as seen at the load buses at transmission system level.

For the reason above, there are in the literature a large variety of load models, that take into account all kinds of dynamics and stochastic effects. However, most of these models are very specific of a certain network and cannot be adopted for general-purpose analysis. It is thus more common and practical to consider a simple voltage-dependent steady-state model, as follows:

$$\bar{\imath}_{h,\mathrm{dq}} = -\frac{\bar{v}_{h,\mathrm{dq}}}{v_h^2}\left(P_{\mathrm{D}} - \jmath Q_{\mathrm{D}}\right), \tag{1.85}$$

where:

$$\begin{aligned} P_{\mathrm{D}} &= P_{\mathrm{D},o}\left(\frac{v_h}{v_{h,o}}\right)^{\gamma_p}, \\ Q_{\mathrm{D}} &= Q_{\mathrm{D},o}\left(\frac{v_h}{v_{h,o}}\right)^{\gamma_q}; \end{aligned} \tag{1.86}$$

v_h is the voltage magnitude at bus h, where the load is connected:

$$v_h^2 = v_{h,\mathrm{d}}^2 + v_{h,\mathrm{q}}^2 \, ; \tag{1.87}$$

$P_{\mathrm{D},o}$ and $Q_{\mathrm{D},o}$ are the load active and reactive powers at the reference voltage magnitude $v_{h,o}$; and γ_p, and γ_q are parameters that depend on the load behavior. The minus sign in (1.85) indicates that $\bar{\imath}_{h,\mathrm{dq}}$ is injected into the grid.

With the notation of (1.14), the set of equations describing the load is:

$$\begin{aligned} &\boldsymbol{x} = \emptyset\,, \quad \boldsymbol{y}^{\mathrm{T}} = [v_{h,\mathrm{d}},\, v_{h,\mathrm{q}},\, P_{\mathrm{D}},\, Q_{\mathrm{D}},\, v_h]\,, \quad \boldsymbol{u}^{\mathrm{T}} = [\imath_{h,\mathrm{d}},\, \imath_{h,\mathrm{q}}]\,, \\ &\boldsymbol{f} = \emptyset\,, \quad \boldsymbol{g} = \begin{bmatrix} -P_{\mathrm{D}} v_{h,\mathrm{d}} - Q_{\mathrm{D}} v_{h,\mathrm{q}} - \imath_{h,\mathrm{d}} \\ -P_{\mathrm{D}} v_{h,\mathrm{q}} + Q_{\mathrm{D}} v_{h,\mathrm{d}} - \imath_{h,\mathrm{q}} \\ Q_{\mathrm{D},o}\,(v_h/v_{h,o})^{\gamma_p} - P_{\mathrm{D}} \\ Q_{\mathrm{D},o}\,(v_h/v_{h,o})^{\gamma_q} - Q_{\mathrm{D}} \\ v_{h,\mathrm{d}}^2 + v_{h,\mathrm{q}}^2 - v_h^2 \end{bmatrix}, \\ &\mathbf{T} = \emptyset \quad \mathbf{R} = \emptyset\,. \end{aligned} \tag{1.88}$$

Since $\mathbf{T} = \mathbf{R} = \emptyset$, the explicit formulation – as in (1.15) – of the load coincides with the above.

For example, $\gamma_p = \gamma_q = 0$ gives constant power consumption, whereas $\gamma_p = \gamma_q = 2$ leads to a constant admittance. Empirical values of γ_p, γ_q, β_p, and β_q for some industrial loads can be found, for example, in [15]. The coefficients of aggregated load models are determined through measurements and often vary depending on the hour of the day or the season. Hence, it is often convenient to consider *worse* scenarios. In this vein, constant admittance and constant power models are the most utilized, for the following reasons.

The constant admittance model is a good representation of loads, in most cases, during short-term transient conditions, e.g., in the few seconds after a short circuit. The constant power model, on the other hand, is mostly used in long-term analysis, in particular, voltage stability studies.

These modeling assumptions are consistent, in transmission and distribution systems, with the common practice to connected subnetworks to the grid through an ULTC that regulates the voltage at the low-voltage winding. Since the voltage regulators of the ULTCs is *slow* (time constants of the order of minutes), in the short term, the transformer tap ratio does not move, and since the vast majority of loads can be well represented with a constant admittance, it is reasonable to assume $\gamma_p \approx \gamma_q \approx 2$. On the other hand, in the long term, the controllers of the ULTC control keeps the voltage constant or almost constant and, hence, even voltage-dependent loads can be approximated with a constant-power model. Further discussion on the dynamic response of ULTCs and their effect on load modeling and behavior can be found in [113].

For the reason above, and given the main objective of the book is the study of equilibria and stationary conditions, $\gamma_p = \gamma_q = 0$, i.e. a constant power model, is assumed in all examples and simulations, unless indicated otherwise. In eigenvalue analysis, this is also the model that generally leads to the lower stability margin or most critical conditions and, thus, it can be assumed as the worst scenario.

Finally, whenever it is required to define the equivalent impedance of a load power consumption, the following expression is utilized:

$$\bar{Y}_{\mathrm{D},o} = \frac{\bar{S}^*_{\mathrm{D},o}}{v^2_{h,o}} , \tag{1.89}$$

where $\bar{S}_{\mathrm{D},o} = P_{\mathrm{D},o} + \jmath Q_{\mathrm{D},o}$.

1.3.4 Synchronous Machine

The synchronous generator is the most important device of high-voltage transmission grids as its dynamic behavior and controls dominate the dynamics of the whole system. This situation may change in the future due to the penetration of non-synchronous power-electronic based devices but, arguably, the synchronous machine will keep its predominant role still for a (long) while. It is thus important to dedicate a significant part of this chapter to the modeling of this machine and its controllers.

In the following, the description of the synchronous machine model is based on well-known Park-Concordia model of the synchronous machine which is based on the Park transform [123, 151].

1.3.4.1 Network Interface

As discussed at the beginning of this section, the common choice for the angular frequency of the dq-axis is the constant reference angular frequency

ω_o. For synchronous machines, however, it is convenient to use the machine rotor angular speed ω_r, as this choice allows obtaining a set of electrical and magnetic equations of the machine with constant parameters. This choice has the drawback that each machine has its own angular reference and hence, a dq-axis rotation is needed to allow each machine stator voltage and currents to be referred to the common dq-axis reference frame of the system, which is rotating at ω_o. The relationship between system and machine dq-axis frames is shown in Figure 1.9.

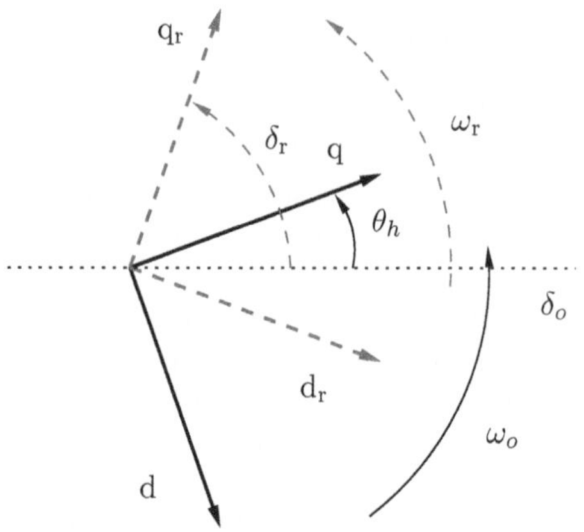

FIGURE 1.9: Machine-grid interface. Axis d_r-q_r are on the machine Park reference frame and rotate with angular speed ω_r, whereas axis d-q are on the grid reference and rotate with angular speed ω_o.

According to the notation of Figures 1.4 and 1.9, the following link between the Park components of stator voltage, namely $v_{s,d}$ and $v_{s,q}$, of the machine on the machine dq-axis frame rotating at ω_r and the same voltage on the dq-axis frame of the system rotating at ω_o is given by:

$$\bar{v}_{s,dq} = \bar{v}_{h,dq}\, e^{-\jmath\,\delta_r}\,, \tag{1.90}$$

or, equivalently:

$$\begin{aligned} v_{s,d} &= v_{h,d} \cos\delta_r + v_{h,q} \sin\delta_r\,, \\ v_{s,q} &= v_{h,q} \cos\delta_r - v_{h,d} \sin\delta_r\,, \end{aligned} \tag{1.91}$$

or, equivalently:

$$\begin{aligned} v_{s,d} &= v_h\, \sin(\delta_r - \theta_h)\,, \\ v_{s,q} &= v_h\, \cos(\delta_r - \theta_h)\,, \end{aligned} \tag{1.92}$$

where the subscript h indicates the index of the bus at which the machine is connected and δ_r is the angle position of the rotor of the synchronous machine whereas θ_h is the phase angle of the system dq-axis at the terminal bus of the machine. Both angles, δ_r and θ_h, are referred to the reference angle δ_o of the system dq-axis reference frame, which is conventionally assumed $\delta_o = 0$.

Similarly, the stator current of the machine that is rotating at ω_r and the same current, assumed to be injected into bus h, referred to the dq-axis frame of the system that is rotating at ω_o is given by:

$$\begin{aligned} \imath_{h,d} &= \imath_{s,d} \cos \delta_r - \imath_{s,q} \sin \delta_r \,, \\ \imath_{h,q} &= \imath_{s,q} \cos \delta_r + \imath_{s,d} \sin \delta_r \,. \end{aligned} \tag{1.93}$$

Regardless from the choice of the dq-axis reference frame, the active and reactive power of the machine is given by (see also (1.42)):

$$\begin{aligned} P_h &= v_{h,d}\, \imath_{h,d} + v_{h,q}\, \imath_{h,q} \\ &= v_{s,d}\, \imath_{s,d} + v_{s,q}\, \imath_{s,q} \,, \\ Q_h &= v_{h,q}\, \imath_{h,d} - v_{h,d}\, \imath_{h,q} \\ &= v_{s,q}\, \imath_{s,d} - v_{s,d}\, \imath_{s,q} \,. \end{aligned} \tag{1.94}$$

Equation (1.91) can be also written using the polar form of the network bus voltage $\bar{v}_h$ at the terminal bus of the machine.

1.3.4.2 Mechanical Equations and Electromagnetic Torque

The mechanical "swing" equations of the machine are:

$$\omega_o^{-1} \dot{\delta}_r = \omega_r - 1 \,, \tag{1.95}$$

where 1 is again the per-unit value of the reference angular frequency, and

$$M \dot{\omega}_r = \tau_m - \tau_e - D\,(\omega_r - 1) \,, \tag{1.96}$$

where M is the mechanical starting time; D is the damping coefficient; 1 is the reference angular frequency in per unit; τ_m is the mechanical torque as imposed by the turbine connected to the machine shaft; and τ_e is the electromagnetic torque, which is a function of the dq-axis components of stator flux and current, as follows:

$$\tau_e = \psi_{s,d}\, \imath_{s,q} - \psi_{s,q}\, \imath_{s,d} \,. \tag{1.97}$$

Note that, in (1.95), ω_o is expressed in rad/s whereas ω_r is in per unit.

1.3.4.3 Stator Electrical Equations

The per-unit electrical equations that link stator fluxes, voltages and currents are:

$$\bar{v}_{s,dq} = -R_a\, \bar{\imath}_{s,dq} - \left(\omega_o^{-1} \frac{d}{dt} + \jmath\, \omega_r \right) \bar{\psi}_{s,dq} \,, \tag{1.98}$$

or, equivalently:

$$\begin{aligned} \omega_o^{-1} \dot{\psi}_{s,d} &= R_a\, \imath_{s,d} + \omega_r\, \psi_{s,q} + v_{s,d} \,, \\ \omega_o^{-1} \dot{\psi}_{s,q} &= R_a\, \imath_{s,q} - \omega_r\, \psi_{s,d} + v_{s,q} \,. \end{aligned} \tag{1.99}$$

Since stator flux dynamics are much faster than the transient and sub-transient dynamics of the machine, a common simplification is to assume $\omega_o^{-1} \approx 0$ in (1.99), which leads to:

$$\begin{aligned} 0 &= R_a \imath_{s,d} + \omega_r \psi_{s,q} + v_{s,d} , \\ 0 &= R_a \imath_{s,q} - \omega_r \psi_{s,d} + v_{s,q} . \end{aligned} \tag{1.100}$$

A further simplification in conventional transient stability power system models, is to assume that the variations of the rotor speed are "small" with respect to the reference, i.e. $\omega_r \approx 1$ pu(rad/s). With this assumption, (1.100) becomes:

$$\begin{aligned} 0 &= R_a \imath_{s,d} + \psi_{s,q} + v_{s,d} , \\ 0 &= R_a \imath_{s,q} - \psi_{s,d} + v_{s,q} . \end{aligned} \tag{1.101}$$

Note that the approximation $\omega_r \approx 1$ pu(rad/s) in (1.101) *cannot* be extended to the mechanical equations (1.95) and (1.96), otherwise the main dynamic behavior of the synchronous machine would be lost.

In the remainder of this book, (1.101) is assumed in the machine model, unless stated otherwise.

1.3.4.4 Rotor Electrical Equations

Conventional models include four rotor dynamics: the dc field winding, excited by the excitation field voltage v_{ef} and three fictitious damper windings that represent rotor-core induced currents, one on the direct and two on the quadrature axis. The model presented in this section combines two models: M1 that is described in [151]; and M2 that is described in [106]. This is achieved through the introduction of proper coefficients in the rotor electrical equations.

Using the implicit form of DAEs discussed in Section 1.1, the equations that describe the transient (′) and sub-transient (″) flux dynamics are, respectively:

$$\begin{aligned} T'_{do}\dot{e}'_{r,q} + \widehat{T}''_{do}\dot{e}''_{r,q} &= -e'_{r,q} - (X_d - X'_d - \alpha_d)\, \imath_{s,d} + \tilde{v}_{ef} , \\ T'_{qo}\dot{e}'_{r,d} - \widehat{T}''_{qo}\dot{e}''_{r,d} &= -e'_{r,d} + (X_q - X'_q - \alpha_q)\, \imath_{s,q} , \end{aligned} \tag{1.102}$$

and:

$$\begin{aligned} T''_{do}\dot{e}''_{r,q} &= -e''_{r,q} + e'_{r,q} - (X'_{d\ell} + \beta_d)\, \imath_{s,d} , \\ T''_{qo}\dot{e}''_{r,d} &= -e''_{r,d} - e'_{r,d} - (X'_{q\ell} + \beta_q)\, \imath_{s,q} , \end{aligned} \tag{1.103}$$

where

$$\begin{aligned} X'_{d\ell} &= X'_d - X_\ell , \\ X'_{q\ell} &= X'_q - X_\ell . \end{aligned}$$

For M1, $\alpha_d = \alpha_q = \beta_d = \beta_q = 0$ and:

$$\begin{aligned} \widehat{T}''_{do} &= \frac{(X_d - X'_d)(X'_d - X''_d)}{(X'_{d\ell})^2}\, T''_{do} , \\ \widehat{T}''_{qo} &= \frac{(X_q - X'_q)(X'_q - X''_q)}{(X'_{q\ell})^2}\, T''_{qo} . \end{aligned}$$

For M2, $\widehat{T}''_{\mathrm{d}o} = \widehat{T}''_{\mathrm{q}o} = 0$ and:

$$\alpha_{\mathrm{d}} = \frac{T''_{\mathrm{d}o}}{T'_{\mathrm{d}o}} \frac{X''_{\mathrm{d}}}{X'_{\mathrm{d}}} (X_{\mathrm{d}} - X'_{\mathrm{d}}), \qquad \alpha_{\mathrm{q}} = \frac{T''_{\mathrm{q}o}}{T'_{\mathrm{q}o}} \frac{X''_{\mathrm{q}}}{X'_{\mathrm{q}}} (X_{\mathrm{q}} - X'_{\mathrm{q}}),$$

$$\beta_{\mathrm{d}} = X_{\ell} - X''_{\mathrm{d}} + \alpha_{\mathrm{d}} , \qquad \beta_{\mathrm{q}} = X_{\ell} - X''_{\mathrm{q}} + \alpha_{\mathrm{q}} .$$

Machine variables and parameters are defined in Tables 1.2 and 1.3, respectively.

TABLE 1.2: Variables of the synchronous machine model.

Variable	Description	Unit
$\bar{e}'_{\mathrm{r,dq}}$	Rotor dq-axis transient voltage	pu(kV)
$\bar{e}''_{\mathrm{r,dq}}$	Rotor dq-axis sub-transient voltage	pu(kV)
$\bar{\imath}_{\mathrm{s,dq}}$	Stator terminal-bus dq-axis current	pu(kA)
$\tilde{v}_{\mathrm{ef}}$	Excitation field voltage	pu(kV)
$\bar{v}_{\mathrm{s,dq}}$	Stator terminal-bus dq-axis voltage	pu(kV)
δ_{r}	Rotor angular position	rad
τ_{e}	Electromagnetic torque	pu(MNm)
τ_{m}	Mechanical torque	pu(MNm)
$\bar{\psi}_{\mathrm{s,dq}}$	Stator dq-axis flux	pu(kWb)
ω_{r}	Rotor angular speed	pu(rad/s)

TABLE 1.3: Parameters of the synchronous machine model.

Parameter	Description	Unit
D	Damping coefficient	pu(MW)
M	Mechanical starting time	s pu(MW)
R_{a}	Armature resistance	pu(Ω)
$T'_{\mathrm{d}o}$	d-axis transient time constant	s
$T''_{\mathrm{d}o}$	d-axis sub-transient time constant	s
$T'_{\mathrm{q}o}$	q-axis transient time constant	s
$T''_{\mathrm{q}o}$	q-axis sub-transient time constant	s
X_{d}	d-axis synchronous reactance	pu(Ω)
X'_{d}	d-axis transient reactance	pu(Ω)
X''_{d}	d-axis sub-transient reactance	pu(Ω)
X_{ℓ}	Leakage reactance	pu(Ω)
X_{q}	q-axis synchronous reactance	pu(Ω)
X'_{q}	q-axis transient reactance	pu(Ω)
X''_{q}	q-axis sub-transient reactance	pu(Ω)

Equations (1.102) are written using the implicit notation. Substituting the expressions of $\dot{e}''_{\mathrm{r,q}}$ and $\dot{e}''_{\mathrm{r,d}}$ from (1.103) and the expression of $\widehat{T}''_{\mathrm{d}o}$ and $\widehat{T}''_{o}$, one obtains the explicit form for transient and sub-transient rotor dynamics, as follows:

$$\begin{aligned}
\dot{e}'_{\mathrm{r,q}} &= \Big(- e'_{\mathrm{r,q}} - (X_{\mathrm{d}} - X'_{\mathrm{d}}) \left(\kappa_{\mathrm{d}}\, \imath_{\mathrm{s,d}} - \gamma_{\mathrm{d}}(e''_{\mathrm{r,q}} - e'_{\mathrm{r,q}}) \right) + \tilde{v}_{\mathrm{ef}} \Big) / T'_{\mathrm{d}o} \,, \\
\dot{e}'_{\mathrm{r,d}} &= \Big(- e'_{\mathrm{r,d}} + (X_{\mathrm{q}} - X'_{\mathrm{q}}) \left(\kappa_{\mathrm{q}}\, \imath_{\mathrm{s,q}} - \gamma_{\mathrm{q}}(e''_{\mathrm{r,d}} + e'_{d}) \right) \Big) / T'_{\mathrm{q}o} \,, \\
\dot{e}''_{\mathrm{r,q}} &= (-e''_{\mathrm{r,q}} + e'_{\mathrm{r,q}} - X'_{\mathrm{d}\ell}\, \imath_{\mathrm{s,d}}) / T''_{\mathrm{d}o} \,, \\
\dot{e}''_{\mathrm{r,d}} &= (-e''_{\mathrm{r,q}} - e'_{\mathrm{r,d}} - X'_{\mathrm{q}\ell}\, \imath_{\mathrm{s,q}}) / T''_{\mathrm{q}o} \,,
\end{aligned} \tag{1.104}$$

where

$$\kappa_{\mathrm{d}} = 1 - X'_{\mathrm{d}\ell}\, \gamma_{\mathrm{d}} \,, \qquad \kappa_{\mathrm{q}} = 1 - X'_{\mathrm{q}\ell}\, \gamma_{\mathrm{q}} \,,$$

and, for M1:

$$\gamma_{\mathrm{d}} = \frac{X'_{\mathrm{d}} - X''_{\mathrm{d}}}{(X'_{\mathrm{d}\ell})^2} \,, \qquad \gamma_{\mathrm{q}} = \frac{X'_{\mathrm{q}} - X''_{\mathrm{q}}}{(X'_{\mathrm{q}\ell})^2} \,,$$

whereas, for M2, $\gamma_{\mathrm{d}} = \gamma_{\mathrm{q}} = 0$.

1.3.4.5 Magnetic Equations

The following algebraic equations link stator and rotor magnetic fluxes and stator currents:

$$\begin{aligned}
0 &= \psi_{\mathrm{s,d}} + X''_{\mathrm{d}}\, \imath_{\mathrm{s,d}} - \eta_{\mathrm{d}}\, e'_{\mathrm{r,q}} - (1 - \eta_{\mathrm{d}})\, e''_{\mathrm{r,q}} \,, \\
0 &= \psi_{\mathrm{s,q}} + X''_{\mathrm{q}}\, \imath_{\mathrm{s,q}} + \eta_{\mathrm{q}}\, e'_{\mathrm{r,d}} - (1 - \eta_{\mathrm{q}})\, e''_{\mathrm{r,d}} \,,
\end{aligned} \tag{1.105}$$

where, for M1, $\eta_{\mathrm{d}} = \kappa_{\mathrm{d}}$ and $\eta_{\mathrm{q}} = \kappa_{\mathrm{q}}$; whereas, for M2, $\eta_{\mathrm{d}} = \eta_{\mathrm{q}} = 0$.

1.3.4.6 Complete Model

The resulting model of the synchronous machine is a two direct- and two quadrature-axis machine model that consists of the following equations: (1.96), (1.97), (1.91), (1.93), (1.94), (1.95), (1.99), (1.102), (1.103) and (1.105).

With the notation of (1.14), one has a DAE system where the vectors of variables are:

$$\begin{aligned}
\boldsymbol{x}^{\mathrm{T}} &= [\delta_{\mathrm{r}},\, \omega_{\mathrm{r}},\, e'_{\mathrm{r,q}},\, e'_{\mathrm{r,d}},\, e''_{\mathrm{r,q}},\, e''_{\mathrm{r,d}},\, \psi_{\mathrm{s,d}},\, \psi_{\mathrm{s,q}},\, \tilde{v}_{\mathrm{ef}}] \,, \\
\boldsymbol{y}^{\mathrm{T}} &= [\tau_{\mathrm{e}},\, v_{\mathrm{s,d}},\, v_{\mathrm{s,q}},\, \imath_{\mathrm{s,d}},\, \imath_{\mathrm{s,q}},\, P_h,\, Q_h,\, v_{h,\mathrm{d}},\, v_{h,\mathrm{q}},\, \tau_{\mathrm{m}}] \,, \\
\boldsymbol{u}^{\mathrm{T}} &= [\imath_{h,\mathrm{d}},\, \imath_{h,\mathrm{q}}] \,,
\end{aligned} \tag{1.106}$$

with equations:

$$
\boldsymbol{f} = \begin{bmatrix}
\omega_{\rm r} - 1 \\
\tau_{\rm m} - \tau_{\rm e} - D\,(\omega_{\rm r} - 1) \\
-e'_{\rm r,q} - (X_{\rm d} - X'_{\rm d} - \alpha_{\rm d})\,\imath_{\rm s,d} + \tilde{v}_{\rm ef} \\
-e'_{\rm r,d} + (X_{\rm q} - X'_{\rm q} - \alpha_{\rm q})\,\imath_{\rm s,q} \\
-e''_{\rm r,q} + e'_{\rm r,q} - (X'_{{\rm d}\ell} + \beta_{\rm d})\,\imath_{\rm s,d} \\
-e''_{\rm r,d} - e'_{\rm r,d} - (X'_{{\rm q}\ell} + \beta_{\rm q})\,\imath_{\rm s,q} \\
R_{\rm a}\,\imath_{\rm s,d} + \hat{\omega}\,\psi_{\rm s,q} + v_{\rm s,d} \\
R_{\rm a}\,\imath_{\rm s,q} - \hat{\omega}\,\psi_{\rm s,d} + v_{\rm s,q}
\end{bmatrix} , \tag{1.107}
$$

and

$$
\boldsymbol{g} = \begin{bmatrix}
\psi_{\rm s,d}\,\imath_{\rm s,q} - \psi_{\rm s,q}\,\imath_{\rm s,d} \\
\psi_{\rm s,d} + X''_{\rm d}\,\imath_{\rm s,d} - \eta_{\rm d}\,e'_{\rm r,q} - (1 - \eta_{\rm d})\,e''_{\rm r,q} \\
\psi_{\rm s,q} + X''_{\rm q}\,\imath_{\rm s,q} + \eta_{\rm q}\,e'_{\rm r,d} - (1 - \eta_{\rm q})\,e''_{\rm r,d} \\
v_{h,{\rm d}}\cos(\delta_{\rm r}) + v_{h,{\rm q}}\sin(\delta_{\rm r}) - v_{\rm s,d} \\
v_{h,{\rm q}}\cos(\delta_{\rm r}) - v_{h,{\rm d}}\sin(\delta_{\rm r}) - v_{\rm s,q} \\
\imath_{\rm s,d}\cos(\delta_{\rm r}) - \imath_{\rm s,q}\sin(\delta_{\rm r}) - \imath_{h,{\rm d}} \\
\imath_{\rm s,q}\cos(\delta_{\rm r}) + \imath_{\rm s,d}\sin(\delta_{\rm r}) - \imath_{h,{\rm q}} \\
v_{\rm s,d}\,\imath_{\rm s,d} + v_{\rm s,q}\,\imath_{\rm s,q} - P_h \\
v_{\rm s,q}\,\imath_{\rm s,d} - v_{\rm s,d}\,\imath_{\rm s,q} - Q_h
\end{bmatrix} , \tag{1.108}
$$

where $\hat{\omega} = \omega_{\rm r}$, and left-hand-side matrices:

$$
\mathbf{T} = \begin{bmatrix}
\omega_o^{-1} & 0 & 0 & 0 & 0 & 0 & 0 & 0 & 0 \\
0 & M & 0 & 0 & 0 & 0 & 0 & 0 & 0 \\
0 & 0 & T'_{\rm do} & 0 & \widehat{T}''_{\rm do} & 0 & 0 & 0 & 0 \\
0 & 0 & 0 & T'_{\rm qo} & 0 & -\widehat{T}''_{\rm qo} & 0 & 0 & 0 \\
0 & 0 & 0 & 0 & T''_{\rm do} & 0 & 0 & 0 & 0 \\
0 & 0 & 0 & 0 & 0 & T''_{\rm qo} & 0 & 0 & 0 \\
0 & 0 & 0 & 0 & 0 & 0 & T_\psi & 0 & 0 \\
0 & 0 & 0 & 0 & 0 & 0 & 0 & T_\psi & 0
\end{bmatrix} , \tag{1.109}
$$

$$
\mathbf{R} = \mathbf{0}_{9,9} \ ,
$$

where $T_\psi = \omega_o^{-1}$. Note that the algebraic variable $\tau_{\rm m}$ is shared with the primary frequency control (PFC) (see Section 1.3.5), This is why, in (1.108), vectors $\boldsymbol{f}$ and $\boldsymbol{g}$ have orders $n = 8$ and $l = 9$, respectively, and consequently the number of columns and rows of $\mathbf{T}$ and $\mathbf{R}$ in (1.109).

Similarly, the stator current of the machine stator flux dynamics and angular speed variations in the electrical equations, thus $\hat{\omega} \approx 1$ pu(rad/s) and $T_\psi \approx 0$.

The set of explicit DAEs depends on which time constants are null or non-null in $\mathbf{T}$, there is a specific. Several models are indeed utilized in the literature (see for example Chapter 15 in [113]). A common simplification in transient stability analysis is to assume negligible the stator flux dynamics ($T_\psi \approx 0$) and the sub-transient rotor dynamics ($T''_{\rm do} = T''_{\rm qo} \approx 0$), which leads to the well-known one d- and one q-axis model. It is worth mentioning that other common assumptions consist in setting $\alpha_{\rm d}$, $\alpha_{\rm q}$, $\beta_{\rm d}$, $\beta_{\rm q}$, $\eta_{\rm d}$, $\eta_{\rm q}$, $\widehat{T}''_{\rm do}$, and $\widehat{T}''_{\rm qo} = 0$. These approximations lead to the machine model described, for example, in [101] and [83].

For the sake of example, the explicit model of the machine in the form of (1.15) for the case where $T_\psi \approx 0$ and $\hat{\omega} \approx 1$ is presented below. This model is characterized by the following vectors of variables:

$$
\begin{aligned}
\boldsymbol{x}^{\rm T} &= [\delta_{\rm r},\, \omega_{\rm r},\, e'_{\rm r,q},\, e'_{\rm r,d},\, e''_{\rm r,q},\, e''_{\rm r,d},\, \tilde{\upsilon}_{\rm ef}]\,, \\
\boldsymbol{y}^{\rm T} &= [\psi_{\rm s,d},\, \psi_{\rm s,q},\, \tau_{\rm e},\, \upsilon_{\rm s,d},\, \upsilon_{\rm s,q},\, \imath_{\rm s,d},\, \imath_{\rm s,q},\, P_h,\, Q_h,\, \upsilon_{h,\rm d},\, \upsilon_{h,\rm q},\, \tau_{\rm m}]\,, \\
\boldsymbol{u}^{\rm T} &= [\imath_{h,\rm d},\, \imath_{h,\rm q}]\,,
\end{aligned}
\tag{1.110}
$$

with equations:

$$
\boldsymbol{f} = \begin{bmatrix}
\omega_o(\omega_{\rm r} - 1) \\
(\tau_{\rm m} - \tau_{\rm e} - D\,(\omega_{\rm r} - 1))/M \\
(-e'_{\rm r,q} - (X_{\rm d} - X'_{\rm d})\,(\kappa_{\rm d}\,\imath_{\rm s,d} - \gamma_{\rm d}(e''_{\rm r,q} - e'_{\rm r,q})) + \tilde{\upsilon}_{\rm ef})/T'_{\rm do} \\
(-e'_{\rm r,d} + (X_{\rm q} - X'_{\rm q})\,(\kappa_{\rm q}\,\imath_{\rm s,q} - \gamma_{\rm q}(e''_{\rm r,d} + e'_{d})))/T'_{\rm qo}\,, \\
(-e''_{\rm r,q} + e'_{\rm r,q} - X'_{{\rm q}\ell}\,\imath_{\rm s,d})/T''_{\rm do} \\
(-e''_{\rm r,q} - e'_{\rm r,d} - X'_{{\rm q}\ell}\,\imath_{\rm s,q})/T''_{\rm qo}
\end{bmatrix}, \tag{1.111}
$$

and

$$
\boldsymbol{g} = \begin{bmatrix}
R_{\rm a}\,\imath_{\rm s,d} + \psi_{\rm s,q} + \upsilon_{\rm s,d} \\
R_{\rm a}\,\imath_{\rm s,q} - \psi_{\rm s,d} + \upsilon_{\rm s,q} \\
\psi_{\rm s,d}\,\imath_{\rm s,q} - \psi_{\rm s,q}\,\imath_{\rm s,d} \\
\psi_{\rm s,d} + X''_{\rm d}\,\imath_{\rm s,d} - \eta_{\rm d}\,e'_{\rm r,q} - (1 - \eta_{\rm d})\,e''_{\rm r,q} \\
\psi_{\rm s,q} + X''_{\rm q}\,\imath_{\rm s,q} + \eta_{\rm q}\,e'_{\rm r,d} - (1 - \eta_{\rm q})\,e''_{\rm r,d} \\
\upsilon_{h,\rm d}\cos(\delta_{\rm r}) + \upsilon_{h,\rm q}\sin(\delta_{\rm r}) - \upsilon_{\rm s,d} \\
\upsilon_{h,\rm q}\cos(\delta_{\rm r}) - \upsilon_{h,\rm d}\sin(\delta_{\rm r}) - \upsilon_{\rm s,q} \\
\imath_{\rm s,d}\cos(\delta_{\rm r}) - \imath_{\rm s,q}\sin(\delta_{\rm r}) - \imath_{h,\rm d} \\
\imath_{\rm s,q}\cos(\delta_{\rm r}) + \imath_{\rm s,d}\sin(\delta_{\rm r}) - \imath_{h,\rm q} \\
\upsilon_{\rm s,d}\,\imath_{\rm s,d} + \upsilon_{\rm s,q}\,\imath_{\rm s,q} - P_h \\
\upsilon_{\rm s,q}\,\imath_{\rm s,d} - \upsilon_{\rm s,d}\,\imath_{\rm s,q} - Q_h
\end{bmatrix}. \tag{1.112}
$$

1.3.4.7 Classical Model

For historical reasons but also for its unquestionable didactic value, a very commonly used model in transient stability analysis is the so-called *classical model.* This model removes all dynamics except those of the mechanical (*swing*) equations. This model is clearly very approximated but it can be useful if one is only interested in studying whether any synchronous machine goes out of step after the clearance of a fault.

The classical machine model assumes that: (i) all stator and rotor dynamics as well as stator losses are null; and (ii) the mechanical and electrical torque can be approximated with the mechanical and electrical power, respectively, as the variations of the rotor angular speed are small.

These assumptions lead to rewrite the swing equations as:

$$\begin{aligned} \omega_o^{-1}\dot{\delta}_{\rm r} &= \omega_{\rm r} - 1\,, \\ M\dot{\omega}_{\rm r} &= P_{\rm m} - P_{\rm e}(\delta_{\rm r}) - D(\omega_{\rm r} - 1)\,, \end{aligned} \tag{1.113}$$

as well as the expressions of the stator currents:

$$\begin{aligned} 0 &= \upsilon_{\rm s,q} - e'_{\rm r,q} + X'_{\rm d}\,\imath_{\rm s,d}\,, \\ 0 &= \upsilon_{\rm s,d} - X'_{\rm d}\,\imath_{\rm s,q}\,, \end{aligned} \tag{1.114}$$

where $e'_{\rm r,q}$ is the EMF "behind the reactance" and is assumed to be constant.

Finally, from (1.44) and (1.114), the expressions of the active and reactive powers injected by the machine into the grid are:

$$\begin{aligned} P_h &= \frac{e'_{\rm r,q}\upsilon_h}{X'_{\rm d}}\sin\left(\delta_{\rm r} - \theta_h\right)\,, \\ Q_h &= \frac{e'_{\rm r,q}\upsilon_h}{X'_{\rm d}}\cos\left(\delta_{\rm r} - \theta_h\right) - \frac{\upsilon_h^2}{X'_{\rm d}}\,, \end{aligned} \tag{1.115}$$

and $P_{\rm e}(\delta_{\rm r}) = P_h$.

1.3.4.8 Center of Inertia

In the model discussed so far, the machine rotor angles and angular speeds are referred to the system synchronous reference, which is defined by δ_o and ω_o. While this is certainly a reasonable choice, using a constant reference angle leads often to undesirable *drifts* in time domain simulations of the rotor angles and bus voltage phase angles of the whole system [46]. While this book is fundamentally interested in steady-state conditions of power systems, it is still convenient to define another common option for the reference frame of synchronous machines.

A conventional solution is the utilization of the center of inertia (CoI) which is a weighted mean of the rotor speeds of synchronous machines, as follows [138, 161, 177]:

$$\delta_{\text{CoI}} = \frac{\sum_{k \in \mathbb{G}} M_k \, \delta_{\text{r},k}}{\sum_{k \in \mathbb{G}} M_k} \, , \qquad \omega_{\text{CoI}} = \frac{\sum_{k \in \mathbb{G}} M_k \, \omega_{\text{r},k}}{\sum_{k \in \mathbb{G}} M_k} = \dot{\delta}_{\text{CoI}} \, , \tag{1.116}$$

where $\mathbb{G}$ is the set of synchronous generators, $\omega_{\text{r},k}$ are the rotor speeds and M_k are the normalized mechanical starting times. Using the CoI, the swing equations (1.95) and (1.96) become:

$$\omega_o^{-1} \, \dot{\omega}_{\text{r}} = \omega_{\text{r}} - \omega_{\text{CoI}} \, , \tag{1.117}$$

and

$$M \, \dot{\omega}_{\text{r}} = \tau_{\text{m}} - \tau_{\text{e}} - D \, (\omega_{\text{r}} - \omega_{\text{CoI}}) \, , \tag{1.118}$$

respectively.

1.3.5 Primary Frequency Control

All synchronous generators connected to the high-voltage transmission grid are requested to provide PFC. This is implemented through a Turbine Governor (TG) that regulates the mass flow of the thermodynamic fluid (water, steam or gas) in the turbine that is connected to the rotor shaft of the machine.

The ultimate goal of the TG is to recover the synchronous speed after a contingency or a power imbalance occurs in the grid. However, typically, TGs include a dead band on the frequency measurement and are not perfect tracking controllers, so they can effectively reduce the frequency deviation using the rotating power reserve of the turbine but never really recover the synchronous speed.

Figure 1.10 shows a scheme of the PFC and turbine of a steam power plant that is commonly utilized in benchmark power systems. The model includes a governor, a servo-motor and a re-heater block. While specifically a steam turbine, this model is general enough to be able to approximate hydro and gas turbines, at least in their most basic behavior. The interested reader is referred to the IEEE Report [74] for a recent review of current TG modeling practices in industry.

The set of per-unit DAEs of the scheme shown in Figure 1.10 is:

$$\begin{aligned} T_{\text{g}} \, \dot{x}_{\text{g}} &= P_{\text{in}} - x_{\text{g}} \, , \\ T_{\text{sm}} \, \dot{x}_{\text{c}} - T_{\text{t}} \, \dot{x}_{\text{g}} &= x_{\text{g}} - x_{\text{c}} \, , \\ T_{\text{rh}} \, \dot{x}_{\text{rh}} &= x_{\text{c}} - x_{\text{rh}} \, , \\ -\kappa_{\text{rh}} \, T_{\text{rh}} \, \dot{x}_{\text{rh}} &= x_{\text{rh}} - \tau_{\text{m}} \, , \end{aligned} \tag{1.119}$$

where $P_{\rm in}$ is the output of windup limiter, as follows:

$$P_{\rm in} = \kappa_{\rm in}(\hat{P}_{\rm in}) = \begin{cases} \hat{P}_{\rm in} & \text{if } P^{\rm min} \leq \hat{P}_{\rm in} \leq P^{\rm max}\,, \\ P^{\rm max} & \text{if } \hat{P}_{\rm in} > P^{\rm max}\,, \\ P^{\rm min} & \text{if } \hat{P}_{\rm in} < P^{\rm min}\,, \end{cases} \tag{1.120}$$

where

$$\hat{P}_{\rm in} = P^{\rm ref}_{\rm G} + \frac{1}{\mathcal{R}}(\omega^{\rm ref} - \omega_{\rm r})\,. \tag{1.121}$$

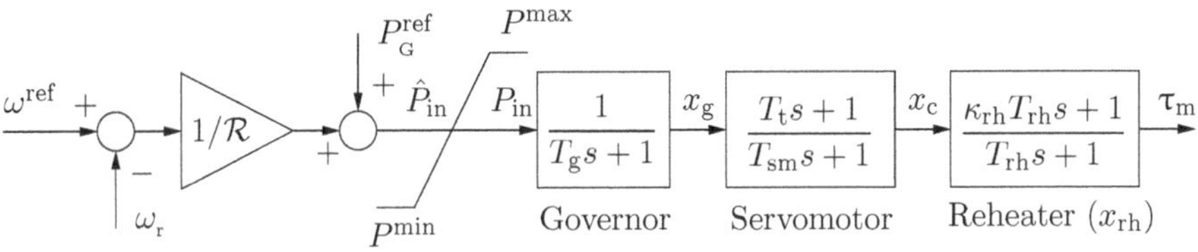

FIGURE 1.10: Scheme of a simple steam turbine governor model.

Using the notation of (1.14), the set (1.119)–(1.121) becomes:

$$\boldsymbol{x}^{\rm T} = [x_{\rm g},\, x_{\rm c},\, x_{\rm rh},\, \omega_{\rm r}]\,, \quad \boldsymbol{y}^{\rm T} = [\tau_{\rm m},\, \hat{P}_{\rm in},\, P_{\rm in}]\,, \quad \boldsymbol{u}^{\rm T} = [P^{\rm ref}_{\rm G},\, \omega^{\rm ref}]\,, \tag{1.122}$$

$$\boldsymbol{f} = \begin{bmatrix} P_{\rm in} - x_{\rm g} \\ x_{\rm g} - x_{\rm c} \\ x_{\rm c} - x_{\rm rh} \end{bmatrix}, \quad \boldsymbol{g} = \begin{bmatrix} x_{\rm rh} - \tau_{\rm m} \\ P^{\rm ref}_{\rm G} + (\omega^{\rm ref} - \omega_{\rm r})/\mathcal{R} - \hat{P}_{\rm in} \\ \kappa_{\rm in}(\hat{P}_{\rm in}) - P_{\rm in} \end{bmatrix}, \tag{1.123}$$

$$\mathbf{T} = \begin{bmatrix} T_{\rm g} & 0 & 0 & 0 \\ -T_{\rm t} & T_{\rm sm} & 0 & 0 \\ 0 & 0 & T_{\rm rh} & 0 \end{bmatrix}, \quad \mathbf{R} = \begin{bmatrix} 0 & 0 & -\kappa_{\rm rh}T_{\rm rh} \\ 0 & 0 & 0 \\ 0 & 0 & 0 \end{bmatrix}, \tag{1.124}$$

or, equivalently, using the notation of (1.15):

$$\boldsymbol{f} = \begin{bmatrix} (P_{\rm in} - x_{\rm g})/T_{\rm g} \\ \big(P_{\rm in}/T_{\rm g} + (1 - 1/T_{\rm g})\,x_{\rm g} - x_{\rm c}\big)/T_{\rm sm} \\ (x_{\rm c} - x_{\rm rh})/T_{\rm sh} \end{bmatrix}, \tag{1.125}$$

$$\boldsymbol{g} = \begin{bmatrix} \kappa_{\rm rh}x_{\rm c} + (1 - \kappa_{\rm rh})\,x_{\rm rh} - \tau_{\rm m} \\ P^{\rm ref}_{\rm G} + (\omega^{\rm ref} - \omega_{\rm r})/\mathcal{R} - \hat{P}_{\rm in} \\ \kappa_{\rm in}(\hat{P}_{\rm in}) - P_{\rm in} \end{bmatrix}, \tag{1.126}$$

where $\omega_{\rm r}$ is the rotor speed of the synchronous machine to which the TG is connected; $\omega^{\rm ref}$ is the reference speed of the regulator, typically equal to the synchronous reference speed and thus $\omega^{\rm ref} = 1$ pu(Hz); $P^{\rm ref}_{\rm G}$ is the power set point of the TG as defined by the system operator, for example, through the electricity market, and adjusted by the automatic generation control; and other variables are defined in Figure 1.10. Table 1.4 defines the parameters of the turbine and TG.

TABLE 1.4: Parameters of the steam turbine governor model.

Parameter	Description	Unit
$P^{\max}$	Maximum turbine output	pu(MW)
$P^{\min}$	Minimum turbine output	pu(MW)
$\mathcal{R}$	Droop of the turbine governor	pu(MW)
$T_{\rm g}$	Governor time constant	s
$T_{\rm rh}$	Re-heater time constant	s
$T_{\rm sm}$	Servo-motor time constant	s
$T_{\rm t}$	Transient gain time constant	s
$\kappa_{\rm rh}$	Re-heater fraction	–

The main characteristic of the scheme of Figure 1.10 is that it is not perfect tracking and thus, after a disturbance, it introduces a steady-state frequency error. This behavior is commonly known as *droop control*. With the proper choice of the parameter $\mathcal{R}$, i.e. using similar values for all machines in per unit on machine bases, the droop control allows the power plants to compensate power imbalance proportionally to their capacity and optimize the utilization of the available spinning reserve.

Even if a perfect tracking TG was used, in practice, the frequency error $\omega^{\rm ref} - \omega_{\rm r}$ undergoes a dead band, whose typical value is 0.0006 pu(Hz). The dead band is generally included in real-world PFC to reduce the stress of the mechanical parts that compose the TG. Since this book focuses on small-signal stability analysis, the dead band is not considered in the model as dead bands effectively decouple the controller from the system.

For the same reason, in the numerical examples discussed in the book, windup and anti-windup limiters are assumed not to be binding to avoid decoupling the control from the system. For example, in the case of (1.120), the condition $P^{\min} < \hat{P}_{\rm in} < P^{\max}$ always holds.

1.3.6 Automatic Voltage Regulator

The automatic voltage regulator (AVR) is the primary voltage control of synchronous machines. It measures the bus voltage at the terminal bus of the machine or at some relevant bus of the grid, compares such a measurement to a reference voltage and varies the field voltage of the machine to regulate the reactive power produced by the machine itself.

Three main types of AVRs are identified based on the technology used for the excitation system: dc generator, ac generator with rectifier and transformer with rectifier [75]. Figure 1.11 shows a control scheme commonly utilized in benchmark systems, namely the IEEE Type DC1 exciter.

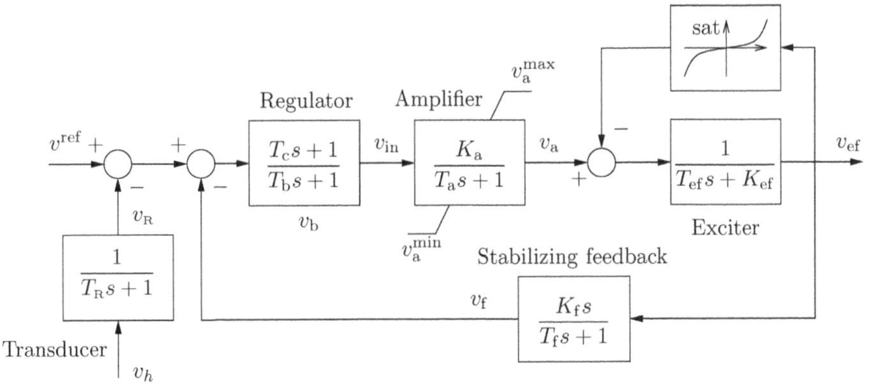

FIGURE 1.11: Control diagram of the IEEE Type DC1 exciter [75].

The AVR depicted in Figure 1.11 is described by the following equations:

$$
\begin{aligned}
T_R\,\dot{v}_R &= v_h - v_R\ , \\
T_b\,\dot{v}_b &= v^{ref} - v_R(t) - v_f - v_b\ , \\
y_1\,y_2\,T_a\,\dot{v}_a &= y_1\,y_2\,K_a v_{in} + \hat{y}_1\,v_a^{max} + \hat{y}_2\,v_a^{min} - v_a\ , \\
T_f\,\dot{v}_f - K_f\,\dot{v}_{ef} &= -v_f\ , \\
T_{ef}\,\dot{v}_{ef} &= v_a - \big(K_{ef} + \mathrm{sat}(v_{ef})\big)\,v_{ef}\ , \\
0 &= \tilde{v}_{ef} - v_{ef}, \\
0 &= v_o^{ref} - v_{ref},
\end{aligned}
\tag{1.127}
$$

where v_h is the regulated voltage at the generator terminal bus h, and v_{in} is the amplifier input signal, which, for the IEEE Type DC1 is:

$$
v_{in} = T_c\,\dot{v}_b + v_b\ , \tag{1.128}
$$

and the anti-windup limiter of the amplifier is implemented as:

$$
\begin{aligned}
y_1 &= \kappa_1(v_a, \dot{v}_a)\ , \\
y_2 &= \kappa_2(v_a, \dot{v}_a)\ , \\
\hat{y}_1 &= 1 - y_1\ , \\
\hat{y}_2 &= 1 - y_2\ ,
\end{aligned}
\tag{1.129}
$$

where κ_1 and κ_2 are anti-windup limiter functions:

$$
\kappa_1(v_a, \dot{v}_a) = \begin{cases} 0, & \text{if } v_a \geq v_a^{max} \text{ and } \dot{v}_a \geq 0\ , \\ 1, & \text{otherwise}\ , \end{cases} \tag{1.130}
$$

TABLE 1.5: Parameters of the IEEE Type DC1 exciter model.

Parameter	Description	Unit
A_{ef}	1st ceiling coefficient	–
B_{ef}	2nd ceiling coefficient	$\mathrm{pu(kV)}^{-1}$
K_{a}	Amplifier gain	–
K_{ef}	Field circuit integral deviation	–
K_{f}	Stabilizer gain	–
T_{a}	Amplifier time constant	s
T_{b}	Pole of the regulator inherent dynamic	s
T_{c}	Zero of the regulator inherent dynamic	s
T_{ef}	Field circuit time constant	s
T_{f}	Stabilizer time constant	s
T_{R}	Measurement time constant	s
$v_{\mathrm{a}}^{\max}$	Maximum regulator voltage	pu(kV)
$v_{\mathrm{a}}^{\min}$	Minimum regulator voltage	pu(kV)

$$\kappa_2(v_{\mathrm{a}}, \dot{v}_{\mathrm{a}}) = \begin{cases} 0, & \text{if } v_{\mathrm{a}} \leq v_{\mathrm{a}}^{\min} \text{ and } \dot{v}_{\mathrm{a}} \leq 0\,, \\ 1, & \text{otherwise}\,. \end{cases} \tag{1.131}$$

The ceiling function approximates the saturation of the excitation winding:

$$\mathrm{sat}(v_{\mathrm{ef}}) = A_{\mathrm{ef}}\ \mathrm{e}^{B_{\mathrm{ef}}|v_{\mathrm{ef}}|}\,. \tag{1.132}$$

Further details on the definition of the ceiling function can be found in [151] and [113]. Table 1.5 defines the parameters that appear in (1.127)–(1.132).

Using the notation of (1.14), the set (1.127)–(1.132) becomes:

$$\begin{aligned} \boldsymbol{x}^{\mathrm{T}} &= [v_{\mathrm{R}}, v_{\mathrm{b}}, v_{\mathrm{a}}, v_{\mathrm{f}}, v_{\mathrm{ef}}]\,, \\ \boldsymbol{y}^{\mathrm{T}} &= [v_h, v_{\mathrm{in}}, y_1, y_2, \hat{y}_1, \hat{y}_2, \tilde{v}_{\mathrm{ef}}, v^{\mathrm{ref}}], \\ \boldsymbol{u} &= [v_o^{\mathrm{ref}}]\,, \end{aligned} \tag{1.133}$$

$$\boldsymbol{f} = \begin{bmatrix} v_h - v_{\mathrm{R}} \\ v^{\mathrm{ref}} - v_{\mathrm{R}} - v_{\mathrm{f}} - v_{\mathrm{b}} \\ y_1 y_2 K_{\mathrm{a}} v_{\mathrm{in}} + \hat{y}_1 v_{\mathrm{a}}^{\max} + \hat{y}_2 v_{\mathrm{a}}^{\min} - v_{\mathrm{a}} \\ -v_{\mathrm{f}} \\ v_{\mathrm{a}} - \left(K_{\mathrm{ef}} + \mathrm{sat}\right) v_{\mathrm{ef}} \end{bmatrix}, \quad \boldsymbol{g} = \begin{bmatrix} v_{\mathrm{b}} - v_{\mathrm{in}} \\ \kappa_1(v_{\mathrm{a}}, \dot{v}_{\mathrm{a}}) - y_1 \\ \kappa_2(v_{\mathrm{a}}, \dot{v}_{\mathrm{a}}) - y_2 \\ 1 - y_1 - \hat{y}_1 \\ 1 - y_2 - \hat{y}_2 \\ \tilde{v}_{\mathrm{ef}} - v_{\mathrm{ef}} \\ v_o^{\mathrm{ref}} - v^{\mathrm{ref}} \end{bmatrix}, \tag{1.134}$$

$$\mathbf{T} = \begin{bmatrix} T_{\mathrm{R}} & 0 & 0 & 0 & 0 \\ 0 & y_1 y_2 T_{\mathrm{a}} & 0 & 0 & 0 \\ 0 & 0 & T_{\mathrm{b}} & 0 & 0 \\ 0 & 0 & 0 & T_{\mathrm{f}} & -K_{\mathrm{f}} \\ 0 & 0 & 0 & 0 & T_{\mathrm{ef}} \end{bmatrix}, \quad \mathbf{R} = \begin{bmatrix} 0 & -T_{\mathrm{c}} & 0 & 0 & 0 \\ 0 & 0 & 0 & 0 & 0 \\ 0 & 0 & 0 & 0 & 0 \\ 0 & 0 & 0 & 0 & 0 \\ 0 & 0 & 0 & 0 & 0 \\ 0 & 0 & 0 & 0 & 0 \\ 0 & 0 & 0 & 0 & 0 \end{bmatrix}. \tag{1.135}$$

The explicit DAE formulation is, in this case, slightly more complicated to obtain as matrix $\mathbf{T}$ is not diagonal and matrix $\mathbf{R}$ is not null. Another issue is that one cannot simply divide the right-hand side of $\boldsymbol{f}$ by the variable y_1 and y_2. The effect of the anti-windup limiter must thus be taken into account imposing specific conditions on $\dot{v}_{\mathrm{a}}$. Finally, the explicit form of (1.127) is not unique. Another possible implementation is:

$$\boldsymbol{f} = \begin{bmatrix} (v_h - v_{\mathrm{R}})/T_{\mathrm{R}} \\ (v^{\mathrm{ref}} - v_{\mathrm{R}} - v_{\mathrm{f}} - v_{\mathrm{b}})/T_{\mathrm{b}} \\ (K_{\mathrm{a}} v_{\mathrm{in}} + v_{\mathrm{a}}^{\max} + v_{\mathrm{a}}^{\min} - v_{\mathrm{a}})/T_{\mathrm{a}} \\ \dfrac{K_{\mathrm{f}}}{T_{\mathrm{ef}} T_{\mathrm{f}}}(v_{\mathrm{a}} - (K_{\mathrm{ef}} + \mathrm{sat})\, v_{\mathrm{ef}}) - \dfrac{v_{\mathrm{f}}}{T_{\mathrm{f}}} \\ (v_{\mathrm{a}} - (K_{\mathrm{ef}} + \mathrm{sat})\, v_{\mathrm{ef}})/T_{\mathrm{ef}} \end{bmatrix}, \tag{1.136}$$

$$\boldsymbol{g} = \begin{bmatrix} \dfrac{T_{\mathrm{c}}}{T_{\mathrm{b}}}(v^{\mathrm{ref}} - v_{\mathrm{R}} - v_{\mathrm{f}}) + \left(1 - \dfrac{T_{\mathrm{c}}}{T_{\mathrm{b}}}\right) v_{\mathrm{b}} - v_{\mathrm{in}} \\ \tilde{v}_{\mathrm{ef}} - v_{\mathrm{ef}} \\ v_o^{\mathrm{ref}} - v^{\mathrm{ref}} \end{bmatrix}, \tag{1.137}$$

and the anti-windup limiter of the third element of $\boldsymbol{f}$ in (1.136), say f_3, is handled through an *if-then* clause:

$$\begin{aligned} &\text{if} \quad v_{\mathrm{a}} \geq v_{\mathrm{a}}^{\max} \quad \text{and} \quad \dot{v}_{\mathrm{a}} > 0 \quad : \quad v_{\mathrm{a}} = v_{\mathrm{a}}^{\max} \quad \text{and} \quad f_3 = 0\,, \\ &\text{if} \quad v_{\mathrm{a}} \leq v_{\mathrm{a}}^{\min} \quad \text{and} \quad \dot{v}_{\mathrm{a}} < 0 \quad : \quad v_{\mathrm{a}} = v_{\mathrm{a}}^{\min} \quad \text{and} \quad f_3 = 0\,. \end{aligned} \tag{1.138}$$

The dynamics of the IEEE Type DC1 exciter overlap the time scale of the inertial response of the machine. In some cases, however, faster dynamics can be neglected. The minimal AVR model that retains its basic dynamic behavior is obtained by setting $T_{\mathrm{c}} = T_{\mathrm{b}} = T_{\mathrm{m}} = T_{\mathrm{f}} = K_{\mathrm{f}} = A_{\mathrm{ef}} = 0$ and $K_{\mathrm{ef}} = 1$.

1.3.7 Power System Stabilizer

The PSS is a widely used auxiliary controller of synchronous machines aimed at damping electromechanical oscillations. The PSS takes as input the measurement of the rotor speed of the synchronous machine and utilizes this signal

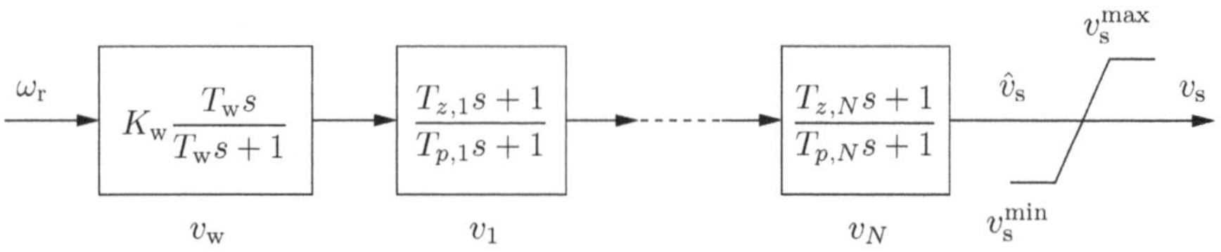

FIGURE 1.12: PSS with wash-out filter (WF) and N lead-lag compensators.

TABLE 1.6: Parameters of the PSS model.

Parameter	Description	Unit
K_w	WF gain	–
$T_{z,k}$	time constant of the k-th lead-lag zero	s
$T_{p,k}$	time constant of the k-th lead-lag pole	s
T_w	WF time constant	s
v_s^{max}	Max. stabilizer output signal	pu(kV)
v_s^{min}	Min. stabilizer output signal	pu(kV)

to modify the reference voltage of the AVR. If properly tuned, the effect of the PSS is to increase the damping of electromechanical oscillations of the rotor, without actually increasing mechanical losses.

PSSs are designed to improve the small-signal angle stability of the system and consist basically in a circuitry that implements a transfer function. Despite being relatively inexpensive (few hundreds of dollars), the impact of PSSs on a system can be significant. It is well known, for example, that the blackout of the WSCC system occurred in 1996 and that led to a total load loss of 35.5 GW and 7.5 million customers experiencing power outage up to 9 hours could have been avoided by simply adding a PSS to a power plant in San Onofre and re-tuning the PSS of the power plant in Palo Verde [82].

These controllers have also inspired a large number of scientific papers, especially related to control theory and small-signal stability analysis. A very good reference on this topic is the book [56], a large part of which is dedicated to the design of PSSs.

A typical PSS scheme with a wash-out filter (WF) and N lead-lag blocks is shown in Figure 1.12, and its parameters are defined in Table 1.6. The set of DAEs that describes such a PSS model is:

$$
\begin{aligned}
T_w \dot{v}_w &= K_w \, \omega_r - v_w \, , \\
T_{p,1} \, \dot{v}_1 - T_w \, \dot{v}_1 &= -v_2 \, , \\
T_{p,k} \, \dot{v}_k - T_{z,k-1} \, \dot{v}_{k-1} &= v_{k-1} - v_k \, , \qquad k = 2, \dots, N \\
-T_{z,N} \, \dot{v}_N &= v_N - \hat{v}_s \, ,
\end{aligned}
\tag{1.139}
$$

where $\hat{v}_s$ is subject to a windup limiter κ_s that constraints the output v_s of the PSS:

$$v_{\rm s} = \kappa_{\rm s}(\hat{v}_{\rm s}) \, , \tag{1.140}$$

where the expression of the limiter is as follows:

$$\kappa_{\rm s}(\hat{v}_{\rm s}) = \begin{cases} v_{\rm s}^{\rm min} & \text{if } \hat{v}_{\rm s} < v_{\rm s}^{\rm min} \, , \\ v_{\rm s}^{\rm max} & \text{if } \hat{v}_{\rm s} > v_{\rm s}^{\rm max} \, , \\ \hat{v}_{\rm s} & \text{otherwise} \, . \end{cases} \tag{1.141}$$

The output of the PSS is added to the reference voltage of the AVR. Recalling the last equation of (1.127), one has:

$$0 = v_o^{\rm ref} \underbrace{+\, v_{\rm s}}_{\text{PSS signal}} - v^{\rm ref} \, , \tag{1.142}$$

and $v_{\rm s}$ has to be included in the vector $\boldsymbol{u}$ of the inputs of the DAE model (1.133)–(1.135) of the AVR.

Using the notation of (1.14), the set (1.139)–(1.141) can be rewritten as:

$$\boldsymbol{x}^{\rm T} = [v_{\rm w}, \, v_1, \, \ldots, \, v_N, \, \omega_{\rm r}] \, , \quad \boldsymbol{y}^{\rm T} = [\hat{v}_{\rm s}, \, v_{\rm s}] \, , \quad \boldsymbol{u} = \emptyset \, , \tag{1.143}$$

$$\boldsymbol{f} = \begin{bmatrix} K_{\rm w}\,\omega_{\rm r} - v_{\rm w} \\ -v_1 \\ v_1 - v_2 \\ \cdots \\ v_{N-1} - v_N \end{bmatrix} , \quad \boldsymbol{g} = \begin{bmatrix} v_N - \hat{v}_{\rm s} \\ \kappa_{\rm s}(\hat{v}_{\rm s}) - v_{\rm s} \end{bmatrix} , \tag{1.144}$$

$$\mathbf{T} = \begin{bmatrix} T_{\rm w} & 0 & 0 & \ldots & 0 \\ -T_{\rm w} & T_{p,1} & 0 & \ldots & 0 \\ 0 & -T_{z,1} & T_{p,2} & \ldots & 0 \\ \vdots & \vdots & \vdots & \vdots & \vdots \\ 0 & \ldots & -T_{z,N-1} & T_{p,N} & 0 \end{bmatrix} , \quad \mathbf{R} = \begin{bmatrix} 0 & \ldots & 0 & -T_{z,N} \\ 0 & \ldots & 0 & 0 \end{bmatrix} , \tag{1.145}$$

or, equivalently, using the explicit form (1.15):

$$\boldsymbol{f} = \begin{bmatrix} (K_{\rm w}\,\omega_{\rm r} - v_{\rm w})/T_{\rm w} \\ (K_{\rm w}\,\omega_{\rm r} - v_{\rm w} - v_1)/T_{p,1} \\ \left[\frac{T_{z,1}}{T_{p,1}}(K_{\rm w}\,\omega_{\rm r} - v_{\rm w}) + \left(\frac{T_{z,1}}{T_{p,1}} - 1\right) v_1 - v_2\right] \frac{1}{T_{p,2}} \\ \cdots \\ \left\{\prod_{k=1}^{N-1} \frac{T_{z,k}}{T_{p,k}}(K_{\rm w}\,\omega_{\rm r} - v_{\rm w}) + \right. \\ \sum_{k=1}^{N-1} \left[\prod_{h=k+1}^{N-1} \frac{T_{z,h}}{T_{p,h}} \left(\frac{T_{z,k}}{T_{p,k}} - 1\right) v_k\right] + \\ \left. \left(\frac{T_{z,N-1}}{T_{p,N-1}} - 1\right) v_{N-1} - v_N \right\} \frac{1}{T_{p,N}} \end{bmatrix} , \tag{1.146}$$

$$
\boldsymbol{g} = \begin{bmatrix} \prod_{k=1}^{N} \frac{T_{z,k}}{T_{p,k}} (K_{\rm w}\,\omega_{\rm r} - v_{\rm w}) + \\ \sum_{k=1}^{N} \left[\prod_{h=k+1}^{N} \frac{T_{z,h}}{T_{p,h}} \left(\frac{T_{z,k}}{T_{p,k}} - 1 \right) v_k \right] + \\ \left(\frac{T_{z,N}}{T_{p,N}} + 1 \right) v_N - \hat{v}_{\rm s} \\ \kappa_{\rm s}(\hat{v}_{\rm s}) - v_{\rm s} \end{bmatrix} . \tag{1.147}
$$

It is relevant to note that the WF is a sort of numerical derivative of the input signal. Thus, the PSS has an effect on the dynamic behavior of the system only during transients for which the frequency if not constant or, equivalently, rate of change of frequency (RoCoF), is not null. The lead-lag compensators that are included in the control diagram of the PSS improve its dynamic response but, in a first approximation, can be neglected. Hence, the minimal PSS model can be defined assuming $T_{p,k} = T_{z,k} = 0$, for $k = 1, \dots, N$.

The signal utilized as input of the PSS does not need to be a local measurement of the machine rotor speed. Frequency signals from remote buses/areas can be utilized as part of a wide-area control system to damp inter-area oscillations [97, 145, 150]. The main difficulty of this kind of controllers is to cope with communication delays, which complicate the small-signal stability analysis and the design of robust controllers. This remark is relevant for some examples discussed in later chapters of the book, in particular Chapters 11 and 12.

1.4 Equilibria

Equilibrium points, or *equilibria*, are special kind of stationarity and are characterized by the condition:

$$
\dot{\boldsymbol{\xi}} = \mathbf{0}_{m,1} \,. \tag{1.148}
$$

Imposing (1.148) in the set of implicit and explicit DAEs defined in Section 1.1, we obtain the following expressions for the equilibrium points.

Definition 1.5 (Equilibrium point)**.** For the implicit model (1.4), $(\boldsymbol{\xi}_o, \boldsymbol{u}_o)$ is an equilibrium point if it satisfies the condition:

$$
\mathbf{0}_{m,1} = \boldsymbol{\varphi}(\boldsymbol{\xi}_o, \boldsymbol{u}_o) \,, \tag{1.149}
$$

where we have removed the dependency on time as the equilibrium point is constant.

Equivalently, splitting the state and algebraic variables as in model (1.14) or (1.15), the point $(\boldsymbol{x}_o, \boldsymbol{y}_o, \boldsymbol{u}_o)$ is an equilibrium if it satisfies the condition:

$$
\begin{aligned} \mathbf{0}_{n,1} &= \boldsymbol{f}(\boldsymbol{x}_o, \boldsymbol{y}_o, \boldsymbol{u}_o) \,, \\ \mathbf{0}_{l,1} &= \boldsymbol{g}(\boldsymbol{x}_o, \boldsymbol{y}_o, \boldsymbol{u}_o) \,. \end{aligned} \tag{1.150}
$$

In (1.150), we have imposed $\dot{\boldsymbol{x}} = \mathbf{0}_{n,1}$, whereas the condition $\dot{\boldsymbol{y}} = \mathbf{0}_{l,1}$ is implicit in the definition of algebraic variables, which can thus be interpreted as states that are always in stationary conditions or in equilibrium.

Similarly, we can define the equilibria for discrete maps. These are often called *fixed points*, which are defined as follows:

Definition 1.6 (Fixed point). A fixed point for the discrete map (1.27) is a point $\boldsymbol{x}_o$ that satisfies the condition:

$$\boldsymbol{\xi}_o = \boldsymbol{\eta}(\boldsymbol{\xi}_o, \boldsymbol{u}_o)\,, \tag{1.151}$$

or, equivalently:

$$\mathbf{0}_{m,1} = \boldsymbol{\eta}(\boldsymbol{\xi}_o, \boldsymbol{u}_o) - \boldsymbol{\xi}_o\,. \tag{1.152}$$

It is important to note that when the set of DAEs describes a power system formulated using the model presented in Section 1.3, an equilibrium represents a stationary condition where all electrical quantities are oscillating at the same frequency, e.g. 50 Hz, and have constant magnitudes and phases. The fact that the models of Section 1.3 show effectively time-invariant quantities at an equilibrium point is because the reference frame is also rotating at the same frequency with such quantities.

In general, one can expect that any system in stationary conditions can actually be described, through an adequate variable transformation, into a set of time-invariant equations. Hence the importance of studying the properties of equilibrium points.

For linear DAEs, the determination of equilibrium points is straightforward. A linear system has no equilibria if its equations are inconsistent or over-determined, one equilibrium if the system is well determined, or infinite equilibria if the system is under-determined or incomplete.

On the other hand, for non-linear DAEs, the determination of equilibrium points is a challenging task. It is, in general, impossible to know how many equilibria there are, if any at all. We illustrate this feature with a simple example.

Example 1.7 (All power flow solutions of the three-bus system). Let us consider the power flow solutions of the three-bus system described in Appendix A. For this example, we neglect the differential equations and consider exclusively the power flow equations of the system as described in Section 1.3.2.4.

The goal is to find all power flow solutions. This can be achieved through a homotopy technique, such as the continuation power flow [113]. With this aim, let us consider a modified version of the power flow equations (1.82), as follows:

$$\varrho\,\bar{\boldsymbol{S}}_o = \bar{\boldsymbol{v}}_{o,\mathrm{dq}} \circ \bar{\boldsymbol{\imath}}^*_{o,\mathrm{dq}}\,, \tag{1.153}$$

or, more in general:

$$\mathbf{0}_l = \boldsymbol{g}(\boldsymbol{y}_o, \boldsymbol{u}_o, \varrho)\,, \tag{1.154}$$

where ϱ is the continuation parameter which represents the loading level of the system. The solutions of the base-case problem are obtained for $\varrho = 1$. A feature of homotopy techniques is that if a solution is known, then all solutions for all values of the continuation parameter can be obtained. The set of solutions is a closed curve in the plain (ϱ, y_i), where y_i is any algebraic variable of (1.154).

Figure 1.13 shows the homotopy curve in the plain (ϱ, v_3), where v_3 is the bus voltage magnitude at the load bus of the three-bus system. Every intersection of the homotopy curve with the vertical dashed line corresponding to $\varrho = 1$ is a solution of (1.153) that satisfies the base-case power flow conditions. We can thus conclude that the three-bus system has at least four solutions. Of these solutions, only the one with $v_3 \approx 1$ pu(kV) is acceptable. Moreover, the solutions with $v_3 < 0$ lack of physical meaning.

We have said that the results of the homotopy technique allow concluding that the three-bus system has *at least* four power flow solutions. Reference [127] discusses an interesting counter-example that explains why one cannot be really sure that the solution shown in Figure 1.13 are actually *all* power flow solutions of the three-bus system. □

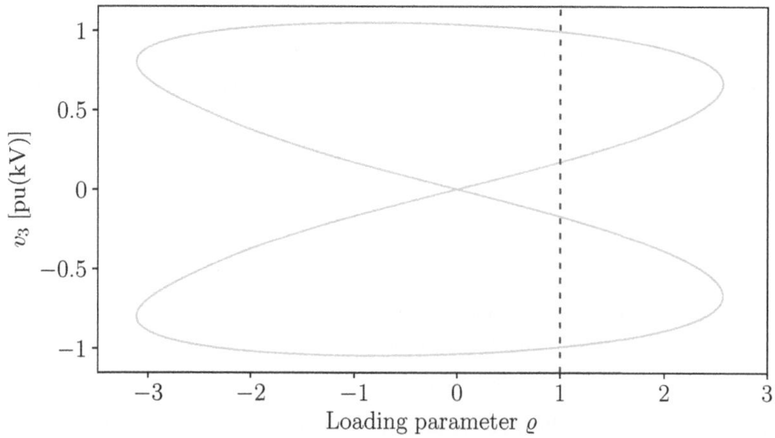

FIGURE 1.13: All power flow solutions for the three-bus system.

1.5 Linearization

The linearization around an equilibrium point is a necessary step for the eigenvalue analysis that is described in this book. In the following, we revisit the set of DAEs presented in Section 1.1 and provide their linearized form.

Consider first the set of implicit DAEs given in (1.14). Its linearization around an equilibrium point $(\boldsymbol{x}_o, \boldsymbol{y}_o, \boldsymbol{u}_o)$ gives:

$$\begin{bmatrix} \mathbf{T} & \mathbf{0}_{n,l} \\ \mathbf{R} & \mathbf{0}_{l,l} \end{bmatrix} \begin{bmatrix} \Delta\dot{\boldsymbol{x}} \\ \mathbf{0}_{l,1} \end{bmatrix} = \begin{bmatrix} \dfrac{\partial \boldsymbol{f}}{\partial \boldsymbol{x}}\Delta\boldsymbol{x} + \dfrac{\partial \boldsymbol{f}}{\partial \boldsymbol{y}}\Delta\boldsymbol{y} + \dfrac{\partial \boldsymbol{f}}{\partial \boldsymbol{u}}\Delta\boldsymbol{u} \\ \\ \dfrac{\partial \boldsymbol{g}}{\partial \boldsymbol{x}}\Delta\boldsymbol{x} + \dfrac{\partial \boldsymbol{g}}{\partial \boldsymbol{y}}\Delta\boldsymbol{y} + \dfrac{\partial \boldsymbol{g}}{\partial \boldsymbol{u}}\Delta\boldsymbol{u} \end{bmatrix}, \tag{1.155}$$

where

$$\Delta\boldsymbol{x} = \boldsymbol{x} - \boldsymbol{x}_o\,, \qquad \Delta\boldsymbol{y} = \boldsymbol{y} - \boldsymbol{y}_o\,, \qquad \Delta\boldsymbol{u} = \boldsymbol{u} - \boldsymbol{u}_o\,,$$

and

$$\begin{aligned} \frac{\partial \boldsymbol{f}}{\partial \boldsymbol{x}} &= \nabla_{\boldsymbol{x}}^{\mathrm{T}}\boldsymbol{f}|_{\boldsymbol{x}_o,\boldsymbol{y}_o,\boldsymbol{u}_o}\,, & \frac{\partial \boldsymbol{f}}{\partial \boldsymbol{y}} &= \nabla_{\boldsymbol{y}}^{\mathrm{T}}\boldsymbol{f}|_{\boldsymbol{x}_o,\boldsymbol{y}_o,\boldsymbol{u}_o}\,, & \frac{\partial \boldsymbol{f}}{\partial \boldsymbol{u}} &= \nabla_{\boldsymbol{u}}^{\mathrm{T}}\boldsymbol{f}|_{\boldsymbol{x}_o,\boldsymbol{y}_o,\boldsymbol{u}_o}\,, \\ \frac{\partial \boldsymbol{g}}{\partial \boldsymbol{x}} &= \nabla_{\boldsymbol{x}}^{\mathrm{T}}\boldsymbol{g}|_{\boldsymbol{x}_o,\boldsymbol{y}_o,\boldsymbol{u}_o}\,, & \frac{\partial \boldsymbol{g}}{\partial \boldsymbol{y}} &= \nabla_{\boldsymbol{y}}^{\mathrm{T}}\boldsymbol{g}|_{\boldsymbol{x}_o,\boldsymbol{y}_o,\boldsymbol{u}_o}\,, & \frac{\partial \boldsymbol{g}}{\partial \boldsymbol{u}} &= \nabla_{\boldsymbol{u}}^{\mathrm{T}}\boldsymbol{g}|_{\boldsymbol{x}_o,\boldsymbol{y}_o,\boldsymbol{u}_o}\,. \end{aligned}$$

Recovering the notation $\boldsymbol{\xi} = (\boldsymbol{x}, \boldsymbol{y})$, (1.155) can be written more compactly as:

$$\mathbf{E}\,\Delta\dot{\boldsymbol{\xi}} = \mathbf{A}\,\Delta\boldsymbol{\xi} + \mathbf{B}\,\Delta\boldsymbol{u}\,, \tag{1.156}$$

where

$$\mathbf{E} = \begin{bmatrix} \mathbf{T} & \mathbf{0}_{n,l} \\ \mathbf{R} & \mathbf{0}_{l,l} \end{bmatrix}, \qquad \mathbf{A} = \begin{bmatrix} \dfrac{\partial \boldsymbol{f}}{\partial \boldsymbol{x}} & \dfrac{\partial \boldsymbol{f}}{\partial \boldsymbol{y}} \\ \\ \dfrac{\partial \boldsymbol{g}}{\partial \boldsymbol{x}} & \dfrac{\partial \boldsymbol{g}}{\partial \boldsymbol{y}} \end{bmatrix}, \qquad \mathbf{B} = \begin{bmatrix} \dfrac{\partial \boldsymbol{f}}{\partial \boldsymbol{u}} \\ \\ \dfrac{\partial \boldsymbol{g}}{\partial \boldsymbol{u}} \end{bmatrix}.$$

Next, we consider the linearization of the set of explicit DAEs given in (1.15). Its linearization around an equilibrium point $(\boldsymbol{x}_o, \boldsymbol{y}_o, \boldsymbol{u}_o)$, with same notation as in (1.155), gives:

$$\begin{bmatrix} \mathbf{I}_n & \mathbf{0}_{n,l} \\ \mathbf{0}_{l,n} & \mathbf{0}_{l,l} \end{bmatrix} \begin{bmatrix} \Delta\dot{\boldsymbol{x}} \\ \mathbf{0}_{l,1} \end{bmatrix} = \begin{bmatrix} \dfrac{\partial \boldsymbol{f}}{\partial \boldsymbol{x}}\Delta\boldsymbol{x} + \dfrac{\partial \boldsymbol{f}}{\partial \boldsymbol{y}}\Delta\boldsymbol{y} + \dfrac{\partial \boldsymbol{f}}{\partial \boldsymbol{u}}\Delta\boldsymbol{u} \\ \\ \dfrac{\partial \boldsymbol{g}}{\partial \boldsymbol{x}}\Delta\boldsymbol{x} + \dfrac{\partial \boldsymbol{g}}{\partial \boldsymbol{y}}\Delta\boldsymbol{y} + \dfrac{\partial \boldsymbol{g}}{\partial \boldsymbol{u}}\Delta\boldsymbol{u} \end{bmatrix}, \tag{1.157}$$

or, in compact notation:

$$\mathbf{E}_{\mathrm{I}}\,\Delta\dot{\boldsymbol{\xi}} = \mathbf{A}\,\Delta\boldsymbol{\xi} + \mathbf{B}\,\Delta\boldsymbol{u}\,, \tag{1.158}$$

where

$$\mathbf{E}_{\mathrm{I}} = \begin{bmatrix} \mathbf{I}_n & \mathbf{0}_{n,l} \\ \mathbf{0}_{l,n} & \mathbf{0}_{l,l} \end{bmatrix}.$$

A special case of (1.158) can be obtained by substituting the identity matrix $\mathbf{I}_n$ with the diagonal matrix $\mathbf{T}_\Delta$ as in (1.10):

$$\mathbf{E}_\Delta \, \Delta\dot{\boldsymbol{\xi}} = \mathbf{A} \, \Delta\boldsymbol{\xi} + \mathbf{B} \, \Delta\boldsymbol{u} \,, \tag{1.159}$$

where

$$\mathbf{E}_\Delta = \begin{bmatrix} \mathbf{T}_\Delta & \mathbf{0}_{n,l} \\ \mathbf{0}_{l,n} & \mathbf{0}_{l,l} \end{bmatrix} .$$

Section 1.1 has discussed how it is generally impossible to calculate $\hat{\boldsymbol{g}}(\boldsymbol{x}, \boldsymbol{u})$ to remove algebraic variables and obtain a set of ODEs. However, this is generally possible for the linearized system (1.157) if the Jacobian matrix $\frac{\partial \boldsymbol{g}}{\partial \boldsymbol{y}}$ is full rank and, hence, invertible.

Definition 1.7 (State matrix)**.** The linearized set of ODEs that can be obtained from (1.157) is:

$$\Delta\dot{\boldsymbol{x}} = \mathbf{A}_\mathrm{S} \, \Delta\boldsymbol{x} + \mathbf{B}_\mathrm{S} \, \Delta\boldsymbol{u} \,, \tag{1.160}$$

where

$$\mathbf{A}_\mathrm{S} = \frac{\partial \boldsymbol{f}}{\partial \boldsymbol{x}} - \frac{\partial \boldsymbol{f}}{\partial \boldsymbol{y}} \left[\frac{\partial \boldsymbol{g}}{\partial \boldsymbol{y}} \right]^{-1} \frac{\partial \boldsymbol{g}}{\partial \boldsymbol{x}} \,, \qquad \mathbf{B}_\mathrm{S} = \frac{\partial \boldsymbol{f}}{\partial \boldsymbol{u}} - \frac{\partial \boldsymbol{f}}{\partial \boldsymbol{y}} \left[\frac{\partial \boldsymbol{g}}{\partial \boldsymbol{y}} \right]^{-1} \frac{\partial \boldsymbol{g}}{\partial \boldsymbol{u}} \,.$$

$\mathbf{A}_\mathrm{S}$ is often called *state matrix.*

The ODE form can be calculated also for the linearized systems (1.156) and (1.159) but, since $\mathbf{T}$ and $\mathbf{T}_\Delta$ are not full rank, their formulation is more involved. In any case, the determination of the linearized ODE system and of the state matrix is not necessary to study the stability of the system through its eigenvalues. How to handle systems with singular $\mathbf{E}$ is thoroughly discussed in Chapters from 2 to 6. For the same reason, we do not provide examples in this section, as plenty of examples illustrating the linear set of equations above are provided in the remainder of the book.

Finally, we provide the linearization of discrete maps in the form of (1.27), which, using the notation utilized in Section 1.2, gives:

$$\boldsymbol{\xi}^{(i+1)} = \mathbf{H}^{(i)} \, \boldsymbol{\xi}^{(i)} \,. \tag{1.161}$$

As discussed in (1.27), in the context of this book, the most relevant linearization for the discrete maps is that obtained as a byproduct of the solution of the $(i+1)$-th step of a numerical integration scheme. In this case, for single step implicit schemes, the Jacobian matrix $\mathbf{H}^{(i+1)}$ takes the form of a matrix pencil, as in equations (1.36) and (1.38).

1.6 Lyapunov Stability Criterion

The well-known Lyapunov stability criterion considers the homogeneous system:

$$\mathbf{E}\dot{\boldsymbol{x}} = \boldsymbol{f}(\boldsymbol{x}) \,, \tag{1.162}$$

with an equilibrium point $\boldsymbol{x}_o = \mathbf{0}_{n,1}$.

Definition 1.8 (Lyapunov function). Assume that we can find a function $\mathcal{V}(\boldsymbol{x}) : \mathbb{R}^{n\times 1} \mapsto \mathbb{R}$ such that:

- $\mathcal{V}(\boldsymbol{x}) = 0$ if and only if $\boldsymbol{x} = \mathbf{0}_{n,1}$;
- $\mathcal{V}(\boldsymbol{x}) > 0$ if and only if $\boldsymbol{x} \neq \mathbf{0}_{n,1}$;
- $\dot{\mathcal{V}}(\boldsymbol{x}) = \dfrac{\partial \mathcal{V}}{\partial \boldsymbol{x}}(\boldsymbol{x})\, \boldsymbol{f}(\boldsymbol{x}) \leq 0$ for all values $\boldsymbol{x} \neq \mathbf{0}_{n,1}$.

If such a function exists, $\mathcal{V}(\boldsymbol{x})$ is called *Lyapunov function* and (1.162) and the equilibrium $\boldsymbol{x}_o$ of such a system is said to be *Lyapunov stable.*

Formally, Lyapunov stability can be defined as follows:

Definition 1.9 (Lyapunov stability). If, for every $\epsilon > 0$, there exists a $\delta = \delta(\epsilon) > 0$ such that, if $\|\boldsymbol{x}(0) - \boldsymbol{x}_o\| < \delta$, then $\|\boldsymbol{x}(t) - \boldsymbol{x}_o\| < \epsilon$, for every $t \geq 0$.

Definition 1.10 (Asymptotic stability). If the condition $\dot{\mathcal{V}}(\boldsymbol{x}) \leq 0$ is substituted for $\dot{\mathcal{V}}(\boldsymbol{x}) < 0$ for all values $\boldsymbol{x} \neq \mathbf{0}_{n,1}$, then system (1.162) is *asymptotically stable.*

Asymptotic stability is stricter than Lyapunov stability and holds if (1.162) is Lyapunov stable and if there exists $\delta > 0$ such that if $\|\boldsymbol{x}(0) - \boldsymbol{x}_o\| < \delta$, then $\lim_{t\to\infty} \|\boldsymbol{x}(t) - \boldsymbol{x}_o\| = 0$.

Lyapunov stability means that, if a dynamic system is "close enough" to a stable equilibrium point, i.e. in a neighborhood with distance δ, the system will remain "close enough" to such an equilibrium point forever, i.e. in a neighborhood with distance δ. On the other hand, asymptotic stability means that the system will also eventually converge to the equilibrium point if it starts close enough from it. Since most physical systems are affected by random noise anyway that prevent them to remain on an equilibrium point, Lyapunov stability is good enough in most engineering applications.

The Lyapunov stability criterion has been extended to non-homogeneous systems, dissipative systems, differential-algebraic systems, discrete systems, delayed systems and stochastic systems. It is hard to overestimate its importance in the study of dynamic systems and electric power systems in particular. However, the Lyapunov stability criterion has a major flaw: it does not provide an algorithm to find the function $\mathcal{V}$.

For many physical systems, this issue can be overcome by observing that the total energy of the system is a good Lyapunov function candidate. In electrical power systems, such an energy function typically accounts for potential and kinetic energy [135]. But this leads to another issue, i.e. such an energy function does not provide a necessary and sufficient stability criterion as losses cannot be taken into account. If losses are not considered, in fact, one cannot conclude that the system is always unstable if $\dot{\mathcal{V}}(\boldsymbol{x}) > 0$.

There are of course attempts to overcome this problem. We cite for example [170] that proposes a "Lyapunov functions family" that allows taking into account losses but is limited to the classical model of the synchronous machine. The main problem, i.e. the lack of a general and systematic algorithm to determine the Lyapunov function, remains unsolved. In power system studies, this insurmountable issue has become even more evident in recent years, due to the fact that several new devices and controllers have drastically changed the dynamic behavior of the grid [121].

While the general problem cannot be solved, there are classes of dynamic systems for which $\mathcal{V}(\boldsymbol{x})$ is well known. This book is concerned with a special class of systems for which the Lyapunov function is known, namely linear(ized) DAEs.

Let us consider again for simplicity but without lack of generality the homogeneous set of ODEs:

$$\mathbf{E}\dot{\boldsymbol{x}} = \boldsymbol{f}(\boldsymbol{x}) \,. \tag{1.163}$$

If $\boldsymbol{x}_o$ is an equilibrium point of (1.163), the linearization at $\boldsymbol{x}_o$ gives:

$$\mathbf{E}\Delta\dot{\boldsymbol{x}} = \frac{\partial \boldsymbol{f}}{\partial \boldsymbol{x}}(\boldsymbol{x}_o)\,\Delta\boldsymbol{x} = \mathbf{A}\,\Delta\boldsymbol{x} \,, \tag{1.164}$$

where $\Delta\boldsymbol{x} = \boldsymbol{x} - \boldsymbol{x}_o$.

The Lyapunov function of (1.164) is:

$$\mathcal{V}(\boldsymbol{x}) = \boldsymbol{x}^{\mathrm{T}}\mathbf{E}^{T}\mathbf{M}\,\boldsymbol{x} \,, \tag{1.165}$$

with $\mathbf{E}^{T}\mathbf{M}$ symmetric, i.e. $\mathbf{E}^{T}\mathbf{M} = \mathbf{M}^{\mathrm{T}}\mathbf{E}$, and positive definite and:

$$\mathbf{A}^{\mathrm{T}}\,\mathbf{M} + \mathbf{M}\,\mathbf{A} \tag{1.166}$$

negative definite. If such a matrix $\mathbf{M}$ exists, then (1.164) is locally asymptotically stable and, hence, also Lyapunov stable.

It is possible to show that the condition of (1.166) being negative definite is equivalent to say that every solution $s = \lambda$ of the equations:

$$\begin{aligned} \mathbf{A}\,\mathbf{u} &= s\,\mathbf{E}\,\mathbf{u} \,, \\ \mathbf{w}\,\mathbf{A} &= s\,\mathbf{w}\,\mathbf{E}, \end{aligned} \tag{1.167}$$

has $\mathrm{Re}(\lambda) < 0$, where λ, $\lambda \in \mathbb{C}$, is called *eigenvalue*; $\mathbf{u} \in \mathbb{C}^{n\times 1}$ is the *right eigenvector*; and $\mathbf{w} \in \mathbb{C}^{1\times n}$ is the *left eigenvector*. These quantities are defined in a rigorous mathematical form in Chapter 2 (see Definition 2.4).

Definition 1.11 (Characteristic equation). From linear system theory, it is also well-known that the solutions of (1.167) also satisfy the condition:

$$\det(\mathbf{A} - s\,\mathbf{E}) = 0\,, \tag{1.168}$$

or, equivalently:

$$a_{n'} s^{n'} + a_{n'-1} s^{n'-1} + \cdots + a_1 s + a_0 = 0\,, \tag{1.169}$$

where the coefficients a_h, $h = 0, \ldots, n'$ are obtained through the Laplace expansion of a determinant by complementary minors. In (1.169), $n' \leq n$ in general and $n' = n$ if $\mathbf{E}$ is non-singular. Equation (1.169) is called *characteristic equation* of the system (1.164) and its roots are eigenvalues of the problem (1.167).

We note since now that the determination of the eigenvalues based on the determination of the roots of (1.169) is not viable if n is high. All numerical methods discussed on Part IV are based on matrices.

For a linear system, i.e. $\boldsymbol{f}(\boldsymbol{x}) = \mathbf{A}\boldsymbol{x}$, the Lyapunov stability is a global property. However, for the linearized system (1.164), the stability condition holds only locally, i.e. sufficiently close to the equilibrium point $\boldsymbol{x}_o$ at which the state matrix $\frac{\partial \boldsymbol{f}}{\partial \boldsymbol{x}}$ is calculated. How "close" is sufficiently close to preserve the stability characteristics of the linearized system is generally not known *a priori*. For practical applications, this is the main limitation of the Lyapunov stability criterion for linearized systems.

Definition 1.12 (Lyapunov stability criterion for linear systems). In summary, the Lyapunov stability criterion for linear systems can be expressed as follows:

- The system is locally stable about $\boldsymbol{x}_o$ if all the eigenvalues of $s\mathbf{E} - \mathbf{A}$ are on the left-half of the complex plane.
- The system is locally unstable about $\boldsymbol{x}_o$ if at least one eigenvalue of $s\mathbf{E} - \mathbf{A}$ is on the right-half of the complex plane.

Remark 1.1. A square matrix is called a stable matrix (or Hurwitz matrix) if every eigenvalue of the matrix has a strictly negative real part.

The conditions above leave out the important case for which at least one eigenvalue λ is on the imaginary axis of the complex plane. Then the Lyapunov stability criterion is inconclusive. In practice, if the real part of an eigenvalue is null, the pencil $s\mathbf{E} - \mathbf{A}$ is not informative on the stability of the equilibrium point $\boldsymbol{x}_o$. We illustrate the applications of the Lyapunov stability criterion to linear(ized) systems below and relevant cases of systems with null eigenvalues in the following examples.

Example 1.8 (Scalar linear time-invariant ODE)**.** We first illustrate the Lyapunov stability criterion for the simplest linear dynamic system possible: a scalar time-invariant ODE:

$$\dot{x} = a\,x\,. \tag{1.170}$$

In this simple case, the state matrix of the system, $\mathbf{A}_{\mathrm{S}} = [a]$, is a 1×1 matrix and, thus, a is also the only eigenvalue of the system.

The equilibrium point is $x_o = 0$. Let assume that $a < 0$. If $x < 0$, then $\dot{x} > 0$ and, hence, x increases. If $x > 0$, then $\dot{x} < 0$ and, hence, x decreases. The equilibrium point is thus attractive and stable, as predicted by the Lyapunov stability criterion. A similar reasoning can be utilized to show that the equilibrium point is unstable for $a > 0$ and we leave this exercise to the reader.

More subtle is the case for which $a = 0$. The Lyapunov stability criterion is not informative in this case. For $a = 0$, however, $\dot{x} \equiv 0$ independently from the value of x. The system has no dynamic at all and it is thus immaterial to discuss its stability. □

Example 1.9 (Stability of the equilibrium points of the OMIB)**.** Consider the OMIB system shown in Figure 1.14. The synchronous machine is modeled with the classical model described in Section 1.3.4.7, whereas the infinite bus is a constant voltage source with infinite inertia. The reactance X_{tot} accounts for the series of the internal reactance of the machine, the reactance of the line that connects the machine with the infinite bus and the Thevenin reactance of the rest of the grid.

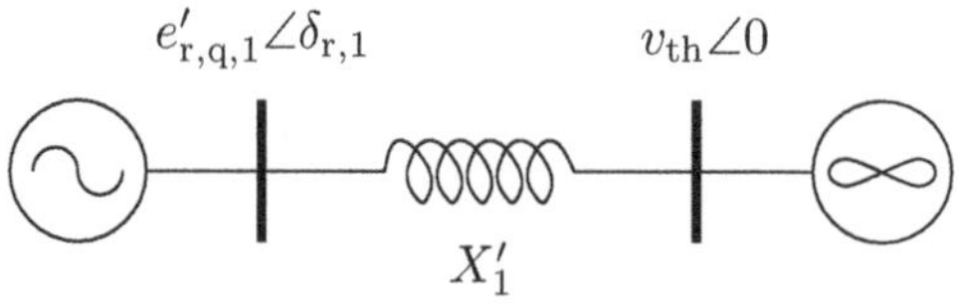

FIGURE 1.14: One-machine infinite-bus system (OMIB).

Let assume the parameters of the OMIB given in Section A.4 of the Appendix and assuming the parameters referred to system bases, namely: $P_{\mathrm{m}} = 6.67$ pu(MW), $X'_1 = 0.1$ pu(Ω), $M_1 = 13\,S_{\mathrm{n,1}}/S_{\mathrm{b}} = 13 \cdot 18 = 234$ s and $\omega_o = 377$ rad/s. Let also assume $e'_{\mathrm{r,q,1}} = 1.05$ pu(kV), $\upsilon_{\mathrm{th}} = 1$ pu(kV), $R_{\mathrm{tot}} = 0$, and $\theta_{\mathrm{th}} = 0$. Then, the differential equations of the system are:

$$\begin{aligned} 377^{-1}\,\dot{\delta}_{\mathrm{r,1}} &= \omega_{\mathrm{r,1}} - 1\,, \\ 234\,\dot{\omega}_{\mathrm{r,1}} &= 6.67 - P_{\mathrm{e}}^{\mathrm{max}}\sin(\delta_{\mathrm{r,1}})\,, \end{aligned} \tag{1.171}$$

where $P_{\mathrm{e}}^{\mathrm{max}} = e'_{\mathrm{r,q,1}}\upsilon_{\mathrm{th}}/X'_1 = 10.5$ pu(MW).

Figure 1.15 shows the electric power P_{e} and the mechanical power P_{m} as functions of the rotor angle $\delta_{\mathrm{r,1}}$. Following the same reasoning of the previous Example 1.8, one can easily find that of the equilibrium points $\boldsymbol{x}_{\mathrm{a}} = (\delta_{\mathrm{r,1,a}}, 1)$

and $\boldsymbol{x}_{\mathrm{b}} = (\delta_{\mathrm{r},1,\mathrm{b}}, 1)$, where $\delta_{\mathrm{r},1,\mathrm{a}} = 0.688$ rad and $\delta_{\mathrm{r},1,\mathrm{b}} = \pi - 0.688 = 2.453$ rad, $\boldsymbol{x}_{\mathrm{a}}$ is stable and $\boldsymbol{x}_{\mathrm{b}}$ is unstable.

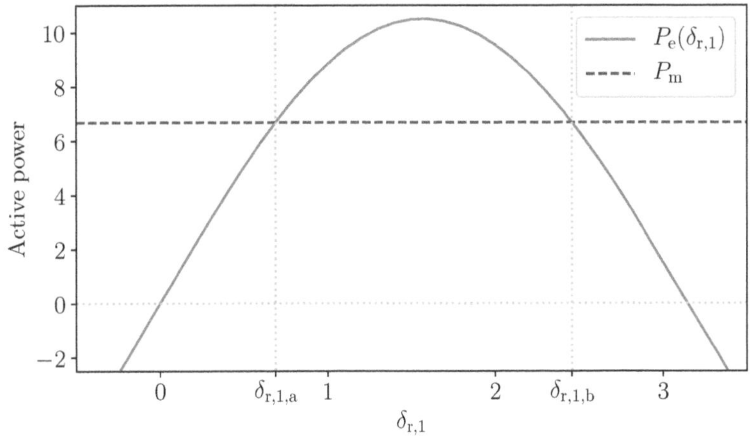

FIGURE 1.15: Active power of the OMIB system.

We discuss qualitatively only the stability of point $\boldsymbol{x}_{\mathrm{a}}$. Assume that $\delta_{\mathrm{r},1}$ is perturbed with a small virtual displacement, say $\delta_{\mathrm{r},1} = \delta_{\mathrm{r},1,\mathrm{a}} + \epsilon$, with $1 \gg \epsilon > 0$. Then P_{e} increases and $\dot{\omega}_{\mathrm{r},1}$ becomes negative. But then $\omega_{\mathrm{r},1}$ decreases and $\omega_{\mathrm{r},1} - 1$ and, hence, $\dot{\delta}_{\mathrm{r},1}$ become negative, which, in turn, make $\delta_{\mathrm{r},1}$ decrease. Similarly, if $-1 \ll \epsilon < 0$, the system reacts so that $\delta_{\mathrm{r},1}$ increases. The equilibrium point $\boldsymbol{x}_{\mathrm{a}}$ is thus attractive and able to compensate small perturbations.

According to the notation of (1.156), the linearization of (1.171) yields:

$$\mathbf{E} = \begin{bmatrix} \omega_o^{-1} & 0 \\ 0 & M_1 \end{bmatrix}, \qquad \mathbf{A} = \begin{bmatrix} 0 & 1 \\ -P_{\mathrm{e}}^{\max} \cos(\delta_{\mathrm{r},1,o}) & 0 \end{bmatrix}.$$

Since in this case $\mathbf{E}$ is full rank, the state matrix can be obtained as

$$\mathbf{A}_{\mathrm{S}} = \mathbf{E}^{-1}\mathbf{A} = \begin{bmatrix} 0 & 377 \\ -0.0449 \cos(\delta_{\mathrm{r},1,o}) & 0 \end{bmatrix}.$$

The eigenvalues of $\mathbf{A}_{\mathrm{S}}$ are the roots of the following characteristic equation:

$$\begin{aligned} 0 = \det(\mathbf{A}_{\mathrm{S}} - s\mathbf{I}_2) &= s^2 + \omega_o \frac{P_{\mathrm{e}}^{\max}}{M_1} \cos(\delta_{\mathrm{r},1,o}) \\ &= s^2 + 16.92 \cos(\delta_{\mathrm{r},1,o}) . \end{aligned}$$

These roots are:

$$\begin{aligned} \lambda_{\mathrm{a},(1,2)} &= \pm 3.615\,\jmath , \\ \lambda_{\mathrm{b},(1,2)} &= \pm 3.615 . \end{aligned}$$

Point $\boldsymbol{x}_{\mathrm{b}}$ is unstable, as expected, as one of the eigenvalues is positive. Point $\boldsymbol{x}_{\mathrm{a}}$, however, is on the imaginary axis and, thus, the Lyapunov stability criterion is inconclusive in this case. Nevertheless, because of the discussion above, we expect that $\boldsymbol{x}_{\mathrm{a}}$ is not unstable.

This is indeed a modeling issue. Since the damping of the machine is null, then, after a small disturbance, the rotor angle and speed of the machine oscillate forever around the equilibrium point $\boldsymbol{x}_{\mathrm{a}}$. The energy given to the machine by the disturbance, in fact, is transformed from kinetic to potential and *vice versa*, without being dissipated. This situation is often referred to as *weak stability*. However, weak stability has little practical interest as real-world machines always have some damping due to the friction of the rotor shaft on the bearings. In particular, the value of the damping depends on the machine model. $D_1 > 1$ (on machine bases) is adequate in low dynamic order, i.e. fourth order or lower. This allows accounting for the damping effect for unmodeled amortisseurs windings. For high-order machine models, i.e. sixth order or higher, $0 \leq D_1 < 1$ (on machine basis) is generally adequate.

Assume $D_1 = 2\,S_{\mathrm{n},1}/S_{\mathrm{b}} = 2{\cdot}18 = 36$. The equations of the system become:

$$\begin{aligned} 377^{-1}\,\dot{\delta}_{\mathrm{r},1} &= \omega_{\mathrm{r},1} - 1\,, \\ 234\,\dot{\omega}_{\mathrm{r},1} &= 6.67 - 10.5\sin(\delta_{\mathrm{r},1}) - 36\,(\omega_{\mathrm{r},1} - 1)\,. \end{aligned} \tag{1.172}$$

With the same equilibrium points as (1.171), state matrix:

$$\mathbf{A}_{\mathrm{S}} = \begin{bmatrix} 0 & 377 \\ -0.0449\cos(\delta_{\mathrm{r},1,o}) & -0.1538 \end{bmatrix},$$

characteristic equation:

$$0 = s^2 + 0.1538\,s + 16.92\,\cos(\delta_{\mathrm{r},1,o})\,,$$

and eigenvalues:

$$\begin{aligned} \lambda_{\mathrm{a},(1,2)} &= -0.077\,\pm\,3.614\,\jmath\,, \\ \lambda_{\mathrm{b},1} &= -3.538\,, \\ \lambda_{\mathrm{b},2} &= 3.692\,, \end{aligned}$$

we find again that point $\boldsymbol{x}_{\mathrm{b}}$ is unstable. Point $\boldsymbol{x}_{\mathrm{a}}$, on the other hand, is now characterized by a pair of complex eigenvalues with negative real part and, thus, is stable according to the Lyapunov stability criterion. After a small disturbance, in fact, the damping dissipates the energy accumulated in the machine and the oscillations of the rotor angle and speed eventually stop. □

Example 1.10 (Normal form of the saddle-node bifurcation). *Limit points* are a relevant class of equilibrium points characterized by a state matrix with one or more eigenvalues on the imaginary axis. These are a special subset of bifurcation points, i.e. points at which the structure of the state flow $\boldsymbol{f}$

changes, thus changing the dynamic behavior of the system itself. It is important to note that limit points are neither stable nor unstable. They have special features that makes them unique and, in fact, each limit point can be characterized by a well defined set of *transversality conditions* [155].

There is a vast literature on bifurcation theory for power system stability analysis. Two limit points are particularly relevant in power systems: saddle-node bifurcation and Hopf bifurcation. In this example, we discuss the *normal form*, i.e. the simplest dynamical system that shows a saddle-node bifurcation.

Consider the following scalar differential equation:

$$\dot{x} = c - x^2 \,. \tag{1.173}$$

The behavior of the system and the properties of its equilibrium points change as c varies:

- Two equilibrium points ($x_o = \pm\sqrt{c}$) for $c > 0$. One point is stable and the other unstable.
- No equilibrium points for $c < 0$.
- One equilibrium point ($x_o = 0$) for $c = 0$.

Figure 1.16 shows the state space $(x, \dot{x})$ for $c = 0$. Intuitively, one can appreciate that the point $(x_o, c_o) = (0, 0)$ defines a structural change of the system: $\dot{x} < 0$ for all values of $x \neq 0$ and $\dot{x} = 0$ only at the bifurcation point. This means that if $x > 0$, the trajectories of x will eventually converge to $x_o = 0$, whereas for $x < 0$, the state variable will further decrease and tend to $-\infty$ for $t \to \infty$.

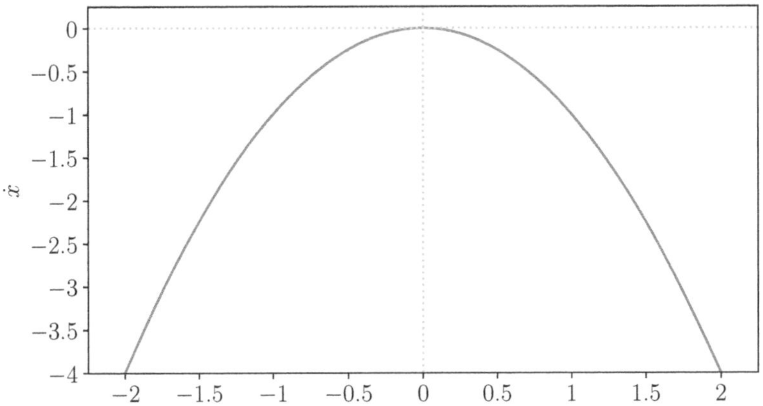

FIGURE 1.16: Saddle-node bifurcation point of the normal form $\dot{x} = -x^2$.

The saddle-node bifurcation is a limit point that occurs when two equilibria, typically one stable and one unstable, merge and disappear as the parameter c changes. The saddle-node bifurcation is also called *fold* bifurcation or *tangent* bifurcation.

For completeness, the transversality conditions of the saddle-node bifurcation at the equilibrium $(\boldsymbol{x}_o, c_o)$ for a system in the form (1.162) are:

$$\begin{aligned} \frac{\partial \boldsymbol{f}}{\partial \boldsymbol{x}}\mathbf{u} = \mathbf{w}\frac{\partial \boldsymbol{f}}{\partial \boldsymbol{x}} &= \mathbf{0}_{n,1}\,, \\ \mathbf{w}\frac{\partial \boldsymbol{f}}{\partial c} &\neq 0\,, \\ \mathbf{w}\frac{\partial^2 \boldsymbol{f}}{\partial \boldsymbol{x}^2}\mathbf{u}\mathbf{u} &\neq 0\,, \end{aligned} \tag{1.174}$$

where $\mathbf{u}$ and $\mathbf{w}$ are the normalized right and left eigenvectors of the Jacobian matrix $\frac{\partial \boldsymbol{f}}{\partial \boldsymbol{x}}(\boldsymbol{x}_o, c_o)$. The reader can easily verify that (1.173) satisfies the conditions (1.174) at $(x, c) = (0, 0)$.

Saddle-node bifurcations are particularly important in power system analysis as the quadratic nature of power flow equations implies that every transmission system has a maximum loading condition which, in turn, are saddle-node bifurcations. This can be observed in Figure 1.13: the continuation curve bends four times. At this turning points the Jacobian matrix of the power flow equations is singular and one can show that these are effectively saddle-node bifurcation points.

Another relevant example is given by the swing equation of the synchronous machine in the OMIB system in Example 1.9. The point $(\delta_{\mathrm{r},o}, \omega_{\mathrm{r},o}, P_{\mathrm{m},o}) = (\pi/2, 1, P_{\mathrm{e}}^{\max})$ is also a saddle-node bifurcation that separates the region for which the machine has two equilibria ($P_{\mathrm{m}} < P_{\mathrm{e}}^{\max}$), one stable and one unstable, from the region with no equilibrium point ($P_{\mathrm{m}} > P_{\mathrm{e}}^{\max}$). □

Example 1.11 (Reference angle of power systems)**.** This example discusses a case where the state matrix shows a zero eigenvalue and, yet, the associated equilibrium point is not a bifurcation. This case is particularly relevant to power system models.

Consider the three-bus system described in Appendix A and assume that its synchronous machines are modeled using the classical second-order model described in Section 1.3.4.7, loads are converted to constant impedances and transmission losses are neglected. With these assumptions, the model of the system can be written, in per unit, as follows:

$$\begin{aligned} \omega_o^{-1}\,\dot{\delta}_{\mathrm{r},1} &= \omega_{\mathrm{r},1} - 1\,, \\ M_1\dot{\omega}_{\mathrm{r},1} &= P_{\mathrm{m},1} - D_1(\omega_{\mathrm{r},1} - 1) - \frac{e'_{\mathrm{r,q},1}e'_{\mathrm{r,q},2}}{X_{12}}\sin(\delta_{\mathrm{r},1} - \delta_{\mathrm{r},2})\,, \\ \omega_o^{-1}\,\dot{\delta}_{\mathrm{r},2} &= \omega_{\mathrm{r},2} - 1\,, \\ M_2\dot{\omega}_{\mathrm{r},2} &= P_{\mathrm{m},2} - D_2(\omega_{\mathrm{r},2} - 1) - \frac{e'_{\mathrm{r,q},1}e'_{\mathrm{r,q},2}}{X_{21}}\sin(\delta_{\mathrm{r},2} - \delta_{\mathrm{r},1})\,, \end{aligned} \tag{1.175}$$

where X_{12} and X_{21} are equivalent reactances that depend on the reactances of transmission lines and loads and the internal reactances of the synchronous machines.

The system (1.175) is in the form (1.4), but since $\mathbf{E}$ is diagonal, constant and full rank, we can focus exclusively on the right-hand side of (1.175).

At the equilibrium point, the matrix $\mathbf{A}$ of (1.175) has the following structure:

$$\mathbf{A} = \begin{bmatrix} 0 & 1 & 0 & 0 \\ K_1 & -D_1 & -K_1 & 0 \\ 0 & 0 & 0 & 1 \\ -K_2 & 0 & K_2 & -D_2 \end{bmatrix}, \tag{1.176}$$

where we have assumed that the equilibrium point is $\boldsymbol{x}_o = [\delta_{\mathrm{r},1,o}, 1, \delta_{\mathrm{r},2,o}, 1]$ and

$$\begin{aligned} K_1 &= \frac{e'_{\mathrm{r,q},1} e'_{\mathrm{r,q},2}}{X_{12}} \cos(\delta_{\mathrm{r},1,o} - \delta_{\mathrm{r},2,o}) \,, \\ K_2 &= \frac{e'_{\mathrm{r,q},1} e'_{\mathrm{r,q},2}}{X_{21}} \cos(\delta_{\mathrm{r},2,o} - \delta_{\mathrm{r},1,o}) \,. \end{aligned}$$

The first and third columns of (1.176), say $\boldsymbol{c}_1$ and $\boldsymbol{c}_3$ are linear dependent, i.e. $\boldsymbol{c}_1 = -\boldsymbol{c}_3$, and, hence, $\mathbf{A}$ is not full rank. In fact, $\mathrm{rank}(\mathbf{A}) = 3 < n = 4$, which is sufficient condition to have a zero eigenvalue.

The result obtained above is more general than the simple system considered in this example. As a matter of fact, all power system models where no reference angle is imposed, show a zero eigenvalue. Such an eigenvalue, however, is not associated with a bifurcation point or any particular dynamic behavior of the system. It is due exclusively to the fact that one equation of the system is redundant. Or, more formally, the equilibrium point of (1.175) is a family of infinite points:

$$\boldsymbol{x}_o(\alpha) = [\delta_{\mathrm{r},1,o} + \alpha, 1, \delta_{\mathrm{r},2,o} + \alpha, 1] \,, \tag{1.177}$$

with $\alpha \in \mathbb{R}$. If we select a rotor angle as the reference, e.g. $\delta_{\mathrm{r},1,o} = 0$, the set of differential equations becomes:

$$\begin{aligned} M_1 \dot{\omega}_{\mathrm{r},1} &= P_{\mathrm{m},1} - D_1(\omega_{\mathrm{r},1} - 1) + \frac{e'_{\mathrm{r,q},1} e'_{\mathrm{r,q},2}}{X_{12}} \sin(\delta_{\mathrm{r},2}) \,, \\ \omega_o^{-1} \, \dot{\delta}_{\mathrm{r},2} &= \omega_{\mathrm{r},2} - 1 \,, \\ M_2 \dot{\omega}_{\mathrm{r},2} &= P_{\mathrm{m},2} - D_2(\omega_{\mathrm{r},2} - 1) + \frac{e'_{\mathrm{r,q},1} e'_{\mathrm{r,q},2}}{X_{21}} \sin(\delta_{\mathrm{r},2}) \,, \end{aligned} \tag{1.178}$$

and, the state matrix at the equilibrium point $\boldsymbol{x}_o = [1, \delta_{\mathrm{r},2,o}, 1]$ is:

$$\mathbf{A} = \begin{bmatrix} -D_1 & -K_1 & 0 \\ 0 & 0 & 1 \\ 0 & K_2 & -D_2 \end{bmatrix}. \tag{1.179}$$

A is full rank except for the cases with null damping, which is unrealistic, and where a machine generates its maximum electrical power, which indicates that the equilibrium is a saddle-node bifurcation. □

A relevant feature of the Lyapunov stability criterion is that it is a quantitative criterion not just a qualitative one. For a qualitative analysis, one can use a numerical time domain integration of the DAEs and deduce the (in)stability of the system by observing the trajectories of the variables. The eigenvalues and eigenvectors of the state matrix of the system, on the other hand, also provide information on "how stable" an equilibrium point is. The "closer" the eigenvalues are to the imaginary axis, in fact, the more critical they are for the dynamic behavior of the system. For this reason, the rightmost eigenvalues are often called *dominant modes.*

This feature has a relevant consequence on the calculation of the eigenvalues. If the analysis is concerned only with the stability of the system, in fact, it is not necessary to calculate *all* of eigenvalues, but only the subset of rightmost ones. Since the calculation of the eigenvalues is cumbersome and time consuming for large systems, the option of calculating a small subset has particularly relevant practical consequences. This aspect of the eigenvalue analysis is so important that we dedicate to numerical methods the entire Part IV.

The effect of dominant modes on the system can be quantitatively evaluated by defining the following quantities: the *damping ratio*, the *natural frequency*, the *damped frequency.*

Definition 1.13 (Damping ratio of an eigenvalue). Let $\lambda = a + \jmath b$ be a complex eigenvalue. The damping ratio is defined as:

$$\zeta = -\frac{a}{|\lambda|} = -\frac{a}{\sqrt{a^2 + b^2}} \,. \tag{1.180}$$

Definition 1.14 (Natural frequency of an eigenvalue). The natural frequency of λ is defined as:

$$f_n = \frac{|\lambda|}{2\pi} = \frac{\sqrt{a^2 + b^2}}{2\pi} \,. \tag{1.181}$$

Definition 1.15 (Damped frequency of an eigenvalue). The damped frequency of λ is defined as:

$$f_d = \left(\sqrt{1 - \zeta^2}\right) f_n \leq f_n \,. \tag{1.182}$$

The definitions of natural and damped frequency indicate that, dimensionally, the eigenvalues have the dimension of an angular frequency. The characteristic equation (1.169) stems, in fact, from the Laplace transform of the linearized ODEs that describe the system. Thus the eigenvalue analysis belongs to the family of frequency-domain methods.

More importantly, the eigenvalues associated with differential equations that describe physical systems carry an information on the *energy* of the system itself. This is a direct consequence of the Lyapunov stability theory and function, which, as we said above, for physical systems, can be formulated in terms of the total energy of the system itself.

The farther the eigenvalues from the imaginary axis on the left half of the complex plane, the lower the energy and the faster the oscillations and dynamics of the associated mode vanish. Positive eigenvalues, on the other hand, are associated with high-energy ever-increasing oscillations and self-amplifying dynamics. In a stable system, the dominant modes are thus the ones with the highest energy, and for this reason they are associated with the most relevant dynamics.

This observation makes also relatively easy to spot inconsistencies in the numerical solution of the eigenvalue problem. It is is not uncommon to obtain very large positive eigenvalues if the state matrix is ill-conditioned.[2] But, such eigenvalues cannot have a physical meaning as they would require more energy than is actually present in the system. For this reason, the positive eigenvalues, of an unstable system are typically "close" to the imaginary axis.

The definitions above originated and are generally utilized in the context of mechanical vibrations but they are also relevant in power system analysis, where the "vibrations" are the oscillations due to the electromechanical modes of the synchronous machines. We illustrate this concept in the following example.

Example 1.12 (Effect of damping on the OMIB). Let us consider again the OMIB system of Example 1.171. The eigenvalues obtained for the equilibrium point $\boldsymbol{x}_{\rm a}$ and for (1.171), i.e. for $D_1 = 0$, have:

$$\zeta = 0 \, , \qquad f_n = f_d = \frac{3.615}{2\,\pi} = 0.5753 \text{ Hz} \, ,$$

whereas the eigenvalues obtained for the equilibrium point $\boldsymbol{x}_{\rm a}$ and for (1.172), i.e. for $D_1 = 2$ on machine bases, have:

$$\zeta = 0.0212 = 2.12\% \, , \qquad f_n = 0.5753 \text{ Hz} \, , \qquad f_d = 0.5751 \text{ Hz} \, .$$

In power system analysis, $\zeta < 5\%$ is considered to indicate a *poorly* damped mode. Such a situation has to be fixed by including additional damping to the system. This is generally achieved with a PSS rather than over-sizing amortisseur windings or, worse, increasing friction. Examples including PSS devices are discussed in Chapters 4, 11 and 12. In this example, let us just assume that the damping coefficient can be increased to $D_1 = 4$, on machine bases. In this case, the eigenvalues obtained for the equilibrium point $\boldsymbol{x}_{\rm a}$ and for (1.172), i.e. for $D_1 = 10$ on machine bases, are:

$$\lambda_{\rm a} = -0.385 \pm 3.594\,\jmath \, ,$$

[2]Note that non-symmetric eigenvalue problems tend to be ill-conditioned.

with

$$\zeta = 0.106 = 10.6\%\,, \qquad f_n = 0.5753\ \text{Hz}\,, \qquad f_d = 0.5720\ \text{Hz}\,.$$

Figure 1.17 shows the effect of the damping on the transient response of the OMIB system following a three-phase short-circuit occurring at $X_1'/2$ at $t = 1$ s and cleared after 60 ms. □

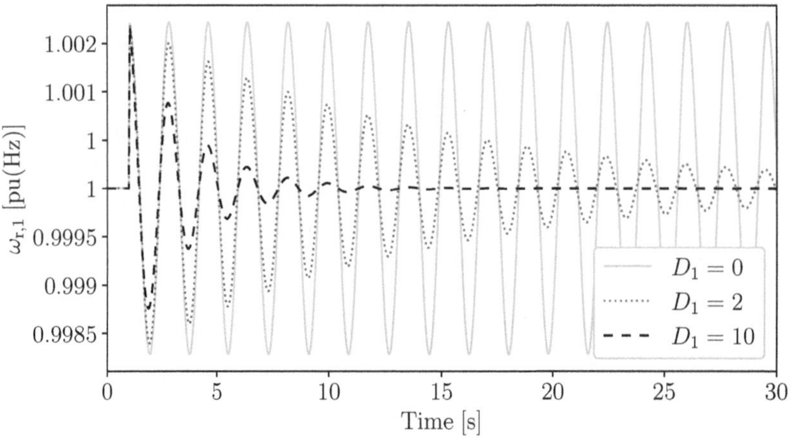

FIGURE 1.17: Effect of damping on the OMIB system following the occurrence and clearance after 60 ms of a short-circuit.

The Lyapunov stability criterion can be extended to discrete maps. This is very relevant for power system analysis as the time domain integration is always carried out through the discretization of the continuous DAEs that model the system.

Let us consider the discrete map:

$$\boldsymbol{x}^{(i+1)} = \boldsymbol{h}(\boldsymbol{x}^{(i)})\,. \tag{1.183}$$

Similarly to the continuous case, (1.183) can be linearized around a fixed point:

$$\Delta\boldsymbol{x}^{(i+1)} = \frac{\partial \boldsymbol{h}}{\partial \boldsymbol{x}}(\boldsymbol{x}_o)\,\Delta\boldsymbol{x}^{(i)} = \mathbf{A}\,\Delta\boldsymbol{x}^{(i)}\,, \tag{1.184}$$

and, in this case, if $\mathbf{A}$ is full rank, the fixed point is $\boldsymbol{x}_o = \mathbf{0}_{n,1}$.

Definition 1.16 (Lyapunov stability criterion for linear maps). The Lyapunov stability criterion for linear discrete maps of the type of (1.184) can be expressed as follows:

- The system is locally stable about $\boldsymbol{x}_o$ if the magnitude of all eigenvalues of $\mathbf{A}$ are lower than 1.

- The system is locally unstable about $\boldsymbol{x}_o$ if at least one eigenvalue of $\mathbf{A}$ has magnitude greater than 1.

Also, the case for which one or more eigenvalues of $\mathbf{A}$ have magnitude equal to 1 is not covered by the Lyapunov criterion. Such fixed points are limit points with special features that need to be considered case by case.

Example 1.13 (Stability of a scalar linear discrete map). We consider the simplest discrete map:

$$x^{(i+1)} = h\,x^{(i)}\,, \tag{1.185}$$

where $h \in \mathbb{R}$. It is easy to observe that if $|h| < 1$, then independently from the initial point $x^{(0)}$, $|x^{(i+1)}| < |x^{(i)}|$, which leads eventually to the fixed point $x_o = 0$. On the other hand, if $|h| > 1$, $|x^{(i+1)}| > |x^{(i)}|$, $\forall i = 1, 2, \dots,$ and eventually the map diverges. For $h = 1$, we have the identity $x^{(i+1)} = \boldsymbol{x}^{(0)}$, while for $h = -1$, $x^{(i+1)} = (-1)^i x^{(i)}$, which is not a fixed point but is nevertheless a stationary condition. It is interesting to note that if h is complex, the region of limit points is the unit circle. □

Example 1.14 (Stability of the Newton method). Consider the set of non-linear algebraic equations:

$$\mathbf{0}_{l,1} = \boldsymbol{g}(\boldsymbol{y})\,. \tag{1.186}$$

A widely used method to find the solution of (1.186) is the Newton method. As already discussed in Section 1.2, this is a discrete map, as follows:

$$\boldsymbol{y}^{(i+1)} = \boldsymbol{y}^{(i)} - \left(\frac{\partial \boldsymbol{g}}{\partial \boldsymbol{y}}(\boldsymbol{y}^{(i)})\right)^{-1} \boldsymbol{g}(\boldsymbol{y}^{(i)})\,, \tag{1.187}$$

with given initial condition $\boldsymbol{y}^{(0)}$. We are interested in studying the stability of the fixed points $\boldsymbol{y}_o$ of (1.187), which are in turn the solutions of (1.186).[3]

We observe that (1.187) can be interpreted as the FEM (see equation (1.29)) with time step $\Delta t = 1$ of a continuous set of implicit ODEs of the form:

$$\mathbf{E}(\boldsymbol{y})\,\dot{\boldsymbol{y}} = -\boldsymbol{g}(\boldsymbol{y})\,, \tag{1.188}$$

where

$$\mathbf{E}(\boldsymbol{y}) = \frac{\partial \boldsymbol{g}}{\partial \boldsymbol{y}}(\boldsymbol{y})\,. \tag{1.189}$$

If $\mathbf{E}$ is not singular, an equilibrium point of (1.188) is thus clearly a solution of (1.186). Moreover, the eigenvalues of the Jacobian matrix $\mathbf{A}(\boldsymbol{y}_o) = -\frac{\partial \boldsymbol{g}}{\partial \boldsymbol{y}}(\boldsymbol{y}_o) = -\mathbf{E}(\boldsymbol{y}_o)$ allows deciding whether the Newton method effectively converges to the fixed point if the initial condition is "sufficiently" close to it. A sufficient condition for convergence is that the eigenvalues of $\mathbf{E}(\boldsymbol{y}_o)$ are all positive.

[3]We do not consider here the case where $\left(\frac{\partial \boldsymbol{g}}{\partial \boldsymbol{y}}(\boldsymbol{y}_o)\right)^{-1} \boldsymbol{g}(\boldsymbol{y}_o) = \mathbf{0}_{l,1}$ with $\boldsymbol{g}(\boldsymbol{y}_o) \neq \mathbf{0}_{l,1}$, i.e. the case for which $\boldsymbol{g}(\boldsymbol{y}_o)$ is an eigenvector of $\left(\frac{\partial \boldsymbol{g}}{\partial \boldsymbol{y}}(\boldsymbol{y}_o)\right)^{-1}$.

If (1.186) represents the power flow equations of a power system (see Section 1.3.2.4), the condition above on the Jacobian matrix is satisfied most of the times. This is also illustrated with Example 8.2 in Section 8.1.2. It is important to note, however, that the fact that an equilibrium point of (1.188) is stable does not mean that it can be reached. This depends exclusively on the choice of the initial guess $\boldsymbol{y}^{(0)}$. The only conclusion that we can draw is that, if the initial guess is within the region of attraction of an equilibrium point of (1.188), then this is stable and can be reached with the Newton method. Unfortunately, there is no algorithm to define the region of attraction or to define a suitable initial guess. The interested reader can find a comprehensive discussion on the stability of the Newton method and its continuous variants in [112] and [117]. □

1.7 Definition of Small-Signal Stability

The conventional taxonomy of power system stability is given in the [76] and illustrated in Table 1.7. This is based, mostly on the phenomenon that originates the instability rather than the mathematical aspects that drive a set of DAEs to instability.

TABLE 1.7: Conventional taxonomy of power system stability

Type of stability	Stability subtype	Short term	Long term
Rotor Angle	Small-disturbance	✓	
	Transient	✓	
Voltage	Large-disturbance	✓	✓
	Small-disturbance	✓	✓
Frequency	–	✓	✓

This phenomenological approach is justified by the practical nature of this paper and by the fact that, at the time of its writing, power system dynamics were dominated by the synchronous machine and its primary regulators. Two out of three types of stability indicated in Table 1.7, namely rotor angle and frequency, are a direct consequence of the swing equation of the synchronous generators and/or their interactions with their AVRs. Voltage stability depends both on the ability of synchronous generators to regulate the voltage and on the transfer capability of the transmission system.

Since this book is concerned with eigenvalue analysis, we focus in this section exclusively on *small disturbance* or *small signal* stability definitions given in [76]. These are as follows:

Definition 1.17 (Small-disturbance rotor angle stability). *Small-disturbance rotor angle stability is concerned with the ability of the power system to maintain synchronism under small disturbances.*

Definition 1.18 (Small-disturbance voltage stability). *Small-disturbance voltage stability refers to the system's ability to maintain steady voltages when subjected to small perturbations such as incremental changes in system load.*

Both definitions above assume that the disturbance that affects the system is *small enough* that the linearization of the model of the system, and hence, its eigenvalues and eigenvectors, are adequate to study its dynamic response. This is, of course, an assumption that can be verified only *a posteriori*, e.g. by solving a time domain simulation and verifying that no non-linearity, such as a hard limit of a controller, modifies the structure of the DAEs and their local stability properties.

It is also interesting to note that the two definitions above only differ in the variables that are considered, namely rotor speed and angles in the former and voltage magnitudes in the latter. The principle, i.e. the ability to recover an acceptable equilibrium point following a disturbance is essentially the same. We elaborate more on this point below.

In recent years, electrical systems have undergone several changes, among which the move from large conventional synchronous power plants to distributed energy resources mostly based on renewables and connected to the grid through power electronics devices. Emerging technologies such as converter-interfaced energy storage systems [122], micro-grids [66] and, more in general, the concept of smart grids [128] are revolutionizing the way the system is planned, operated and, more importantly for the scope of this book, are changing and "enriching" its dynamic behavior.

As a consequence, also the types of stability have increased and there have been some attempts to expand the classical taxonomy shown in Table 1.7 to include new phenomena and devices, such as the effect of converter controllers and micro-grids. The lack of definitive models and standard control schemes for such new devices, however, creates a reasonable doubt on the archival value of these kinds of definitions and classifications. The flexible and ever-changing nature of the architectures and controllers of power electronics converters, in fact, makes very difficult to do a durable classification of the dynamic phenomena that can occur in modern power systems.

For this reasons, we prefer to provide in this section a custom taxonomy of stability, which is based on the properties of the equilibria and of the dynamic behavior of the DAEs. In this way, the classification is not device or power-system dependent. In fact, it can be applied to any dynamic system. We believe that such an approach is more likely to endure the technological changes that will certainly characterize the following decades.

The proposed taxonomy is given in Figure 1.18. In the upper row, three instability types are identified based on the properties of the equilibrium point: i.e. the equilibrium does not exists; exists but is unstable; and exists, is stable but is not reachable. These three cases include as special cases all kinds of instabilities discussed in [76]. Then, in the lower row of Figure 1.18, the resulting dynamic behavior of the DAEs is classified as collapse or any other new stationary condition. These include new equilibrium points, limit cycles, tori, deterministic, and stochastic chaos, etc.

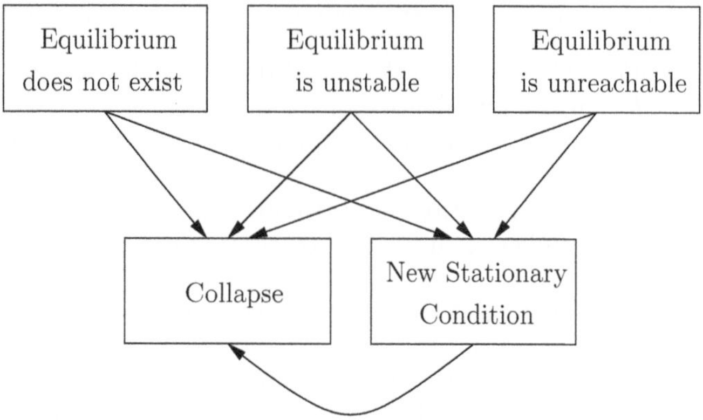

FIGURE 1.18: Types of instability classified based on the properties and reachability of the equilibrium point of a set of DAEs.

Relevant remarks on the three instability cases shown in Figure 1.18 are as follows:

- The case of non-existence (or disappearance after a contingency) of the equilibrium is typical for example of voltage instabilities following a large disturbance and the frequency instability, which, ultimately, is a consequence of a severe power imbalance between generation and consumption.
- The case of a(n) (un)stable equilibrium point can be studied through the small-signal stability analysis. This analysis includes as special cases the small-signal rotor angle and voltage (in)stability defined in [76] but also includes any possible small-signal instability that may be caused by power electronic converters or any other device connected to the grid.
- The case of an equilibrium point that is stable but not reachable is the typical case of transient instability, which results in the synchronous machine to go out of step and, finally, in the loss of synchronism of the grid. This case is not specific of angle stability. The Chilean blackout occurred in May 1997 is an example where the system had a stable equilibrium point but the remedial action was taken too late to avoid the voltage instability and finally the collapse [168].

The collapse is nothing else than the system evolving towards the *trivial solution*, i.e. all variables being null. In power systems this corresponds to a system-wide blackout where all currents and voltages are null. The trivial solution is generally useless from a practical point of view. The blackout is in fact an operating condition for which the system cannot deliver any power to the loads. On the other hand, a blackout, as most trivial solutions of physical systems, is an "extremely" stable equilibrium point. Anyone that has dealt with a blackout knows well how challenging is to restore the normal operating conditions after a transmission network-wide collapse.

Example 1.15 (Lotka-Volterra equations). An interesting and well-known counter-example of trivial solution that is not stable is given by the Lotka-Volterra equations, also known as the predator–prey equations. These are:

$$\begin{aligned} \dot{x}_1 &= a\,x_1 - b\,x_1 x_2\,, \\ \dot{x}_2 &= d\,x_1 x_2 - c\,x_2\,, \end{aligned} \tag{1.190}$$

where x_1 is the number of prey, x_2 is the number of predators and a, b, c and d are positive constants.

At the origin $(x_{1,o}, x_{2,o}) = (0, 0)$ the state matrix of (1.190) is:

$$\mathbf{A} = \begin{bmatrix} a - bx_{1,o} & -bx_{1,o} \\ dx_{2,o} & dx_{1,o} - c \end{bmatrix} = \begin{bmatrix} a & 0 \\ 0 & -c \end{bmatrix}, \tag{1.191}$$

with eigenvalues $\lambda_1 = a > 0$ and $\lambda_2 = -c < 0$. The extinction of the two species is thus an unstable equilibrium point that is not reachable starting with a non-null initial condition ($x_1(0) \neq 0$ and $x_2(0) \neq 0$. This is clearly just an approximation as, unfortunately, species do extinguish. To begin with, the numbers of predator and prey are integer values and thus the model should not be continuous. Then, delays should be taken into account to model the response of a population to the variation of the other. Still, the Lotka–Volterra model has proved to be a very valuable model to study the dynamics of ecological systems. □

The other kind of dynamic response accounts for the situations where the system does not collapse and its trajectories move to a new stationary condition. In power systems, such a new condition is more often not a new feasible operating point but, rather, some form of oscillatory behavior that is not acceptable in practice as it involves losses, stress of the shafts of the generators and possible triggering of relays and the possibility to initiate a cascading phenomenon, which, eventually, may end up in the system collapse.

As it should already be clear to the reader from the discussions above, this book focuses exclusively on the case where the equilibrium exists and provides the theoretical tools to decide whether such a point is stable or unstable according to the Lyapunov stability criterion. Whenever relevant the transient response of the system converging to a stable equilibrium point or moving away from an unstable one is also discussed.

2

Mathematical Outlines

2.1 Matrix Pencils

We introduce the term *polynomial matrix* first.

A polynomial matrix is a matrix $\mathbf{A}(s) \in \mathbb{C}^{r \times m}$, $s \in \mathbb{C}$ with elements $[a_{hk}(s)]_{h=1,2,\ldots,n}^{k=1,2,\ldots,m}$ that are polynomials of s. If l is the highest order of polynomial that appears as an element of $\mathbf{A}(s)$, then there exist matrices $\mathbf{A}_0$, $\mathbf{A}_1$, $\ldots$, $\mathbf{A}_l$ such that:

$$\mathbf{A}(s) = \mathbf{A}_l\, s^l + \mathbf{A}_{l-1}\, s^{l-1} + \cdots + \mathbf{A}_1\, s + \mathbf{A}_0\,.$$

Example 2.1 (Polynomial matrix). Let us consider the following example of a polynomial matrix:

$$\mathbf{A}(s) = \begin{bmatrix} s^3 + 2\,s^2 - 2\,s + 4 & s^3 + s^2 + 3\,s - 2 \\ s - 1 & s - 1 \end{bmatrix}.$$

The polynomial matrix $\mathbf{A}(s)$ can be also written in the form:

$$\mathbf{A}(s) = \begin{bmatrix} 1 & 1 \\ 0 & 0 \end{bmatrix} s^3 + \begin{bmatrix} 2 & 1 \\ 0 & 0 \end{bmatrix} s^2 + \begin{bmatrix} -2 & 3 \\ 1 & 1 \end{bmatrix} s \begin{bmatrix} 4 & -2 \\ -1 & -1 \end{bmatrix}.$$

The order of the polynomial matrix is $l = 3$ and $r = m = 2$. Thus:

$$\mathbf{A}_3 = \begin{bmatrix} 1 & 1 \\ 0 & 0 \end{bmatrix}, \quad \mathbf{A}_2 = \begin{bmatrix} 2 & 1 \\ 0 & 0 \end{bmatrix}, \quad \mathbf{A}_1 = \begin{bmatrix} -2 & 3 \\ 1 & 1 \end{bmatrix}, \quad \mathbf{A}_0 = \begin{bmatrix} 4 & -2 \\ -1 & -1 \end{bmatrix}.$$

□

A matrix pencil is a polynomial matrix – possibly of infinite order –. Matrix pencil theory has been used many times in studying linear dynamical systems such as: linear systems of differential equations, see [24, 34, 37], linear discrete time systems, see [35], and linear systems of fractional operators, see [38]. Before we define the matrix pencil let us understand why the need to introduce this concept. We consider the following system of differential equations:

$$\mathbf{E}\,\dot{\boldsymbol{x}}(t) = \mathbf{A}\boldsymbol{x}(t)\,, \tag{2.1}$$

where $\mathbf{E}, \mathbf{A} \in \mathbb{C}^{r \times m}$, $\boldsymbol{x} : [0, +\infty] \mapsto \mathbb{R}^{m \times 1}$. The matrices $\mathbf{E}$ and $\mathbf{A}$ can be non-square ($r \neq m$) or square ($r = m$) with $\mathbf{E}$ singular, i.e. $\det(\mathbf{E}) = 0$. With $\dot{\boldsymbol{x}}(t)$ we denote the first order derivative of $\boldsymbol{x}(t)$.

Definition 2.1 (Laplace transform). The Laplace transform of a continuous function $f(t)$, defined for all real numbers $t \geq 0$, is the function $\mathcal{L}\{f(t)\} = F(s)$, and defined as

$$\mathcal{L}\{f(t)\} = F(s) = \int_0^{\infty} \exp(-s\,t)\, f(t)\, dt .$$

Let $\mathcal{L}\{\boldsymbol{x}(t)\} = \boldsymbol{z}(s)$ be the Laplace transform of $\boldsymbol{x}(t)$ respectively. By applying the Laplace transform $\mathcal{L}$ into (2.1), we get:

$$\mathbf{E}\mathcal{L}\{\dot{\boldsymbol{x}}(t)\} = \mathbf{A}\mathcal{L}\{\boldsymbol{x}(t)\} ,$$

or, equivalently,

$$\mathbf{E}\big(s\boldsymbol{z}(s) - \boldsymbol{x}_o\big) = \mathbf{A}\,\boldsymbol{z}(s) ,$$

where $\boldsymbol{x}_o = \boldsymbol{x}(0)$, i.e. the initial condition of (2.1). Since we assume that $\boldsymbol{x}_o$ is unknown we can use an unknown constant vector $\boldsymbol{C} \in \mathbb{R}^{m \times 1}$ and give to the above expression the following form:

$$(s\mathbf{E} - \mathbf{A})\boldsymbol{z}(s) = \mathbf{E}\boldsymbol{C} .$$

From the above equation, it is obvious that the polynomial matrix $s\mathbf{E} - \mathbf{A}$ plays an important role in the study of (2.1), especially regarding the existence of solutions and its stability properties.

We will refer to $s\mathbf{E} - \mathbf{A}$ as the pencil of system (2.1). Hence, a matrix pencil is a family of matrices parametrized by a complex number s, see [51].

Notice that we have two cases. The first is (a) $r = m$ and $\det(s\mathbf{E} - \mathbf{A})$ to be equal to a polynomial with order equal or less than m (regular pencil). The second case is (b) $r \neq m$, or $r = m$ with $\det(s\mathbf{E} - \mathbf{A}) \equiv 0$, $\forall$ arbitrary $s \in \mathbb{C}$ (singular pencil). In the case of (a), we have that $\det(s\mathbf{E} - \mathbf{A}) \not\equiv 0$, and hence $s\boldsymbol{E} - \mathbf{A}$ is invertible. Then $\boldsymbol{z}(s)$ can be defined and consequently $\boldsymbol{x}(t)$ always exists and is given by $\boldsymbol{x}(t) = \mathcal{L}^{-1}\{(s\mathbf{E} - \mathbf{A})^{-1}\mathbf{E}\mathbf{C}\}$. Hence in the case of a regular pencil, the solution of (2.1) always exists. In the case of (b), if $r \leq m$ there are at least $m - r$ unknown functions, m equations, and hence $\boldsymbol{z}(s)$ can not be defined; while if $r > m$, $z(s)$ can be defined under specific conditions. We will study these properties in detail and more extensively in Chapter 3.

Definition 2.2 (Regular and singular matrices). A square matrix $\mathbf{E} \in \mathbb{C}^{m \times m}$ is called:

- Regular when $\det(\mathbf{E}) \neq 0$. In this case the matrix $\mathbf{E}$ is invertible;
- Singular when $\det(\mathbf{E}) = 0$. In this case the matrix $\mathbf{E}$ is not invertible.

Definition 2.3 (Regular and singular matrix pencils). Given $\mathbf{E}, \mathbf{A} \in \mathbb{C}^{r \times m}$, and an arbitrary $s \in \mathbb{C}$, the matrix pencil $s\mathbf{E} - \mathbf{A}$ is called:

- Regular when $r = m$ and $\det(s\mathbf{E} - \mathbf{A}) = \varpi(s) \not\equiv 0$;
- Singular when $r \neq m$, or $r = m$ and $\det(s\mathbf{E} - \mathbf{A}) \equiv 0$;

where $\varpi(s)$ is a polynomial of s of degree $\deg(\varpi(s)) \leq m$.

Here are some examples of regular and singular pencils.

Example 2.2 (Singular pencil). Let

$$\mathbf{E} = \begin{bmatrix} 2 & 1 & 1 & 0 & 0 & 0 & 0 \\ 1 & 3 & 1 & 1 & 0 & 0 & 0 \\ 1 & 1 & 2 & 1 & 0 & 0 & 0 \\ 0 & 1 & 1 & 1 & 0 & 0 & 0 \\ 0 & 0 & 0 & 0 & 0 & 0 & 0 \\ 0 & 0 & 0 & 0 & 1 & 0 & 0 \\ 0 & 1 & 0 & 0 & 0 & 0 & 1 \end{bmatrix}, \quad \mathbf{A} = \begin{bmatrix} 1 & 1 & 1 & 0 & 0 & 0 & 1 \\ 0 & 3 & 2 & 2 & 0 & 1 & 1 \\ 1 & 2 & 3 & 2 & 0 & 0 & 0 \\ 0 & 2 & 2 & 2 & 0 & 0 & 0 \\ 0 & 0 & 0 & 0 & 1 & 0 & 0 \\ 0 & 0 & 0 & 0 & 0 & 0 & 0 \\ 0 & 0 & 0 & 0 & 0 & 1 & 0 \end{bmatrix}.$$

Then

$$s\mathbf{E} - \mathbf{A} = \begin{bmatrix} 2s-1 & s-1 & s-1 & 0 & 0 & 0 & -1 \\ s & 3s-3 & s-2 & s-2 & 0 & -1 & -1 \\ s-1 & s-2 & 2s-3 & s-2 & 0 & 0 & 0 \\ 0 & s-2 & s-2 & s-2 & 0 & 0 & 0 \\ 0 & 0 & 0 & 0 & -1 & 0 & 0 \\ 0 & 0 & 0 & 0 & s & 0 & 0 \\ 0 & s & 0 & 0 & -s & -1 & s \end{bmatrix}.$$

The determinant of this pencil is $\det(s\mathbf{E} - \mathbf{A}) = 0$, and hence the pencil is singular. □

Example 2.3 (Singular pencil). Let

$$\mathbf{E} = \begin{bmatrix} 1 & 1 & 1 \\ 0 & 1 & 1 \\ 1 & 1 & 1 \\ 0 & 1 & 1 \end{bmatrix}, \quad \mathbf{A} = \begin{bmatrix} 1 & 2 & 2 \\ 0 & 2 & 2 \\ 1 & 2 & 2 \\ 0 & 2 & 3 \end{bmatrix}.$$

Then

$$s\mathbf{E} - \mathbf{A} = \begin{bmatrix} s-1 & s-2 & s-2 \\ 0 & s-2 & s-2 \\ s-1 & s-2 & s-2 \\ 0 & s-2 & s-3 \end{bmatrix}.$$

The matrices $\mathbf{E}$, $\mathbf{A}$ are non-square. Hence, we are in the case where $r > m$, i.e. $4 > 3$ and the pencil is singular. □

Example 2.4 (Regular pencil). We consider now the matrices

$$\mathbf{E} = \begin{bmatrix} 1 & 1 \\ 0 & 0 \end{bmatrix}, \quad \mathbf{A} = \begin{bmatrix} 1 & 1 \\ 0 & -1 \end{bmatrix}.$$

Then

$$s\mathbf{E} - \mathbf{A} = \begin{bmatrix} s-1 & s-1 \\ 0 & 1 \end{bmatrix}.$$

The determinant of the pencil is $\det(s\mathbf{E} - \mathbf{A}) = s - 1 \not\equiv 0$, and thus the pencil is regular. □

Remark 2.1. Given $\mathbf{E}, \mathbf{A} \in \mathbb{C}^{r \times m}$, and an arbitrary $s \in \mathbb{C}$, if pencil $s\mathbf{E} - \mathbf{A}$ is:

1. Regular, since $\det(s\mathbf{E} - \mathbf{A}) \not\equiv 0$, there exists a matrix function $\tilde{\mathbf{P}} : \mathbb{C} \mapsto \mathbb{R}^{m \times m}$ (which can be computed via the Gauss-Jordan elimination method, see [146]) such that:
$$\tilde{\mathbf{P}}(s)(s\mathbf{E} - \mathbf{A}) = \tilde{\mathbf{A}}(s),$$
where $\tilde{\mathbf{A}} : \mathbb{C} \mapsto \mathbb{R}^{m \times m}$ is a diagonal matrix with non-zero elements;
2. Singular and $r > m$, then there exists a matrix function $\tilde{\mathbf{P}} : \mathbb{C} \mapsto \mathbb{R}^{r \times r}$ (which can be computed via the Gauss-Jordan elimination method) such that
$$\tilde{\mathbf{P}}(s)(s\mathbf{E} - \mathbf{A}) = \begin{bmatrix} \tilde{\mathbf{A}}(s) \\ \mathbf{0}_{r_1,m} \end{bmatrix}, \quad \text{with} \quad \tilde{\mathbf{P}}(s) = \begin{bmatrix} \tilde{\mathbf{P}}_1(s) \\ \tilde{\mathbf{P}}_2(s) \end{bmatrix},$$
where $\tilde{\mathbf{A}} : \mathbb{C} \mapsto \mathbb{R}^{m_1 \times m}$, with $m_1 + r_1 = r$, is a matrix such that if $[\tilde{a}_{hk}]_{1 \le h \le m_1}^{1 \le k \le m}$ are its elements, for $h = k$ all elements are non-zero and for $h \neq k$ all elements are zero and $\tilde{\mathbf{P}}_1(s) \in \mathbb{R}^{m_1 \times r}$, $\tilde{\mathbf{P}}_2(s) \in \mathbb{R}^{r_1 \times r}$.
3. Singular and $m > r$, then there exists a matrix $\tilde{\mathbf{P}} : \mathbb{C} \mapsto \mathbb{R}^{r \times r}$ such that
$$\tilde{\mathbf{P}}(s)(s\mathbf{E} - \mathbf{A}) = \begin{bmatrix} \tilde{\mathbf{A}}(s) & \mathbf{0}_{r,m_1} \end{bmatrix}, \quad \text{with} \quad \tilde{\mathbf{P}}(s) = \begin{bmatrix} \tilde{\mathbf{P}}_1(s) & \tilde{\mathbf{P}}_2(s) \end{bmatrix},$$
where $\tilde{\mathbf{A}} : \mathbb{C} \mapsto \mathbb{R}^{r \times r_1}$, with $m_1 + r_1 = m$, is a matrix such that if $[\tilde{a}_{hk}]_{1 \le h \le r_1}^{1 \le k \le r}$ are its elements, for $h = k$ all elements are non-zero and for $h \neq k$ all elements are zero and $\tilde{\mathbf{P}}_1(s) \in \mathbb{R}^{m \times r_1}$, $\tilde{\mathbf{P}}_2(s) \in \mathbb{R}^{m \times m_1}$.
4. Singular and $r = m$, then if $\text{rank}(s\mathbf{E} - \mathbf{A}) \le m$, we have the same decomposition as in case 2. If $\text{rank}(s\mathbf{E} - \mathbf{A}) \le r$, we have the same decomposition as in case 3.

Here are some examples.

Example 2.5 (Decomposition of a singular pencil)**.** We assume the singular pencil
$$s\mathbf{E} - \mathbf{A} = \begin{bmatrix} s-1 & s-2 & s-2 \\ 0 & s-2 & s-2 \\ s-1 & s-2 & s-2 \\ 0 & s-2 & s-3 \end{bmatrix}.$$
Then there exists the matrix
$$\mathbf{P}(s) = \begin{bmatrix} 1 & -1 & 0 & 0 \\ 0 & 3-s & 0 & -s+2 \\ 0 & 1 & 0 & -1 \\ 0 & 0 & 1 & 0 \end{bmatrix},$$

such that

$$\mathbf{P}(s)(s\mathbf{E}-\mathbf{A})=\begin{bmatrix}\mathbf{A}(s)\\ \mathbf{0}_{1,1}\end{bmatrix},$$

where

$$\mathbf{A}(s)=\begin{bmatrix}s-1 & 0 & 0\\ 0 & s-2 & 0\\ 0 & 0 & 1\end{bmatrix}.$$

In addition, for

$$\mathbf{P}(s)=\begin{bmatrix}\mathbf{P}_1(s)\\ \mathbf{P}_2(s)\end{bmatrix},$$

with

$$\mathbf{P}_1(s)=\begin{bmatrix}1 & -1 & 0 & 0\\ 0 & 3-s & 0 & -s+2\\ 0 & 1 & 0 & -1\end{bmatrix},\quad \mathbf{P}_2(s)=\begin{bmatrix}0 & 0 & 1 & 0\end{bmatrix},$$

we have

$$\mathbf{P}_2(s)\mathbf{E}=\begin{bmatrix}1 & 1 & 1\end{bmatrix}\neq\mathbf{0}_{1,3}.$$

Example 2.6 (Decomposition of a singular pencil). We assume now the singular pencil with

$$s\mathbf{E}-\mathbf{A}=\begin{bmatrix}s-1 & s-2\\ 0 & s-2\\ s-1 & s-2\end{bmatrix}.$$

Then there exists the matrix

$$\mathbf{P}(s)=\begin{bmatrix}1 & -1 & 0\\ 0 & 1 & 0\\ -1 & 0 & 1\end{bmatrix},$$

such that

$$\mathbf{P}(s)(s\mathbf{E}-\mathbf{A})=\begin{bmatrix}\mathbf{A}(s)\\ \mathbf{0}_{1,2}\end{bmatrix},$$

where

$$\mathbf{A}(s)=\begin{bmatrix}s-1 & 0\\ 0 & s-2\end{bmatrix}.$$

In addition, for

$$\mathbf{P}(s)=\begin{bmatrix}\mathbf{P}_1(s)\\ \mathbf{P}_2(s)\end{bmatrix},$$

with

$$\mathbf{P}_1(s)=\begin{bmatrix}1 & -1 & 0\\ 0 & 1 & 0\end{bmatrix},\quad \mathbf{P}_2(s)=\begin{bmatrix}-1 & 0 & 1\end{bmatrix},$$

we have

$$\mathbf{P}_2(s)\mathbf{E}=\mathbf{0}_{1,2}.$$

Example 2.7 (Decomposition of a regular pencil). Finally, we assume the regular pencil

$$s\mathbf{E} - \mathbf{A} = \begin{bmatrix} s-1 & s-1 \\ 0 & 1 \end{bmatrix}.$$

Then there exists the matrix

$$\mathbf{P}(s) = \begin{bmatrix} 1 & -(s-1) \\ 0 & 1 \end{bmatrix},$$

such that

$$\mathbf{P}(s)(s\mathbf{E} - \mathbf{A}) = \mathbf{A}(s),$$

where

$$\mathbf{A}(s) = \begin{bmatrix} s-1 & 0 \\ 0 & 1 \end{bmatrix}.$$

□

2.2 Taxonomy of Eigenvalue Problems

We assume the pencil $s\mathbf{E} - \mathbf{A}$ as introduced in the previous section. Then we state the following definition:

Definition 2.4 (Eigenvalues and eigenvectors). A scalar λ is called an eigenvalue of the pencil $s\mathbf{E} - \mathbf{A}$ if there is a non-trivial solution $\mathbf{u} \in \mathbb{C}^{m\times 1}$ of $\mathbf{A}\,\mathbf{u} = \lambda\mathbf{E}\,\mathbf{u}$, or, equivalently, if there is a non-trivial solution $\mathbf{w} \in \mathbb{C}^{1\times m}$ of $\mathbf{w}\mathbf{A} = \lambda\mathbf{w}\mathbf{E}$. Such a $\mathbf{w}$ is called a left eigenvector corresponding to the eigenvalue λ, and such a $\mathbf{u}$ is called a right eigenvector corresponding to the eigenvalue λ.

In system (2.1) let $r = m$, and $\mathbf{E} = \mathbf{I}_m$, where $\mathbf{I}_m$ is the identity matrix. Then $\mathbf{A}$ is square, and the zeros of the function $\det(s\mathbf{E}-\mathbf{A})$ are the eigenvalues of $\mathbf{A}$. Consequently, the problem of finding the non-trivial solutions of the equation

$$s\mathbf{E}\,\mathbf{x} = \mathbf{A}\,\mathbf{x}\ ,$$

is called the *generalized eigenvalue problem*, see [146]. Although the generalized eigenvalue problem looks like a simple generalization of the conventional eigenvalue problem, it exhibits some important differences. First, it is possible for $\mathbf{E}$ to be singular in which case the problem has eigenvalues at infinity. To see this write the generalized eigenvalue problem in the reciprocal form:

$$\mathbf{E}\,\mathbf{x} = s^{-1}\mathbf{A}\,\mathbf{x}\ .$$

If $\mathbf{E}$ is singular with a null vector $\mathbf{x}$, then $\mathbf{E}\,\mathbf{x} = \mathbf{0}_{m,1}$, so that $\mathbf{x}$ is an eigenvector of the reciprocal problem corresponding to the eigenvalue $\lambda^{-1} = 0$;

i.e. $\lambda \to \infty$. A second non-trivial case is the determinant $\det(s\mathbf{E} - \mathbf{A})$, when $\mathbf{E}$, $\mathbf{A}$ are square matrices, to be identically zero, independent of s. And finally there is the case for both matrices $\mathbf{E}$, $\mathbf{A}$ to be non-square (for $r \neq m$). To sum up, we should have in mind that we consider the following five categories of eigenvalue problems when studying a matrix pencil:

- If the pencil $s\mathbf{E} - \mathbf{A}$ is regular with $\mathbf{E}$ regular, then the pencil has m finite eigenvalues. These eigenvalues can be found as follows:

  Step 1: We compute the determinant of the pencil: $\det(s\mathbf{E} - \mathbf{A})$. Let the result be a polynomial $\varpi(s)$ of s. The polynomial will always be of order m.

  Step 2: We compute the roots of the polynomial $\varpi(s)$. These roots will be the eigenvalues of the pencil $s\mathbf{E} - \mathbf{A}$.

  Step 3: Let λ_i, $i = 1, 2, \ldots, \nu$ be the roots of the polynomial $\varpi(s)$ and p_i, $i = 1, 2, \ldots, \nu$ be the multiplicity of λ_i. Consequently, $\sum_{i=0}^{\nu} p_i = m$. Then for the pencil $s\mathbf{E} - \mathbf{A}$, the eigenvalue λ_i will have algebraic multiplicity p_i.

  Step 4: For the eigenvalue λ_i with algebraic multiplicity p_i we compute its corresponding right eigenvector $\mathbf{u}_i$ by solving the singular algebraic system $(\lambda_i\mathbf{E} - \mathbf{A})\mathbf{u}_i = \mathbf{0}_{m,1}$ which will always have infinitely many solutions. Then:

  - By solving this algebraic system we arrive at a base, i.e. a vector space of p_i linear independent eigenvectors. The dimension of the vector space of solutions is p_i. This vector space is called eigenspace and contains the complete set of the corresponding eigenvectors of λ_i; p_i' is the geometric multiplicity of the eigenvalue λ_i.
  - If by solving this algebraic system we arrive at a number of linear independent eigenvectors that are less than p_i, i.e. $p_i' \leq p_i$, then additional linear independent eigenvectors are needed in order to have the complete set of eigenvectors. These additional linear independent eigenvectors will be called generalized eigenvectors, and can be found as follows. If $p_i' = p_i - 1$, we will need one generalized eigenvector $\mathbf{u}_i^1$, which satisfies $(\lambda_i\mathbf{E} - \mathbf{A})\mathbf{u}_i^1 = \mathbf{u}_i$. If $p_i' = p_i - 2$, we will need two generalized eigenvectors $\mathbf{u}_i^1$, $\mathbf{u}_i^2$, which satisfy $(\lambda_i\mathbf{E} - \mathbf{A})\mathbf{u}_i^1 = \mathbf{u}_i$, and $(\lambda_i\mathbf{E} - \mathbf{A})\mathbf{u}_i^2 = \mathbf{u}_i^1$. If $p_i' = p_i - 3$, we will need three generalized eigenvectors $\mathbf{u}_i^1$, $\mathbf{u}_i^2$, $\mathbf{u}_i^3$, which satisfy $(\lambda_i\mathbf{E} - \mathbf{A})\mathbf{u}_i^1 = \mathbf{u}_i$, $(\lambda_i\mathbf{E} - \mathbf{A})\mathbf{u}_i^2 = \mathbf{u}_i^1$, and $(\lambda_i\mathbf{E} - \mathbf{A})\mathbf{u}_i^3 = \mathbf{u}_i^2$, etc.

  Step 5: For the eigenvalue λ_i with algebraic multiplicity p_i we will compute its corresponding left eigenvector $\mathbf{w}_i$ by solving the singular algebraic system $\mathbf{w}_i(\lambda_i\mathbf{E} - \mathbf{A}) = \mathbf{0}_{1,m}$.

  - If by solving this algebraic system we arrive at p_i linear independent eigenvectors, i.e. the dimension of the eigenspace of the eigenvector, then we will have found the complete set of the corresponding eigenvectors of λ_i.

- If by solving this algebraic system we arrive at a number of linear independent eigenvectors that are less than p_i, i.e. $p_i' \leq p_i$, then additional linear independent eigenvectors are needed to have the complete set of eigenvectors. These additional linear independent eigenvectors will be called generalized left eigenvectors, and can be found as follows. If $p_i' = p_i - 1$, we will need one generalized eigenvector $\mathbf{w}_i^1$, which satisfies $\mathbf{w}_i^1(\lambda_i\mathbf{E} - \mathbf{A}) = \mathbf{w}_i$. If $p_i' = p_i - 2$, we will need two generalized eigenvectors $\mathbf{w}_i^1$, $\mathbf{w}_i^2$, which satisfy $\mathbf{w}_i^1(\lambda_i\mathbf{E} - \mathbf{A}) = \mathbf{w}_i$, and $\mathbf{w}_i^2(\lambda_i\mathbf{E} - \mathbf{A}) = \mathbf{w}_i^1$. If $p_i' = p_i - 3$, we will need three generalized eigenvectors $\mathbf{w}_i^1$, $\mathbf{w}_i^2$, $\mathbf{w}_i^3$, which satisfy $\mathbf{w}_i^1(\lambda_i\mathbf{E} - \mathbf{A}) = \mathbf{w}_i$, $\mathbf{w}_i^2(\lambda_i\mathbf{E} - \mathbf{A}) = \mathbf{w}_i^1$, and $\mathbf{w}_i^3(\lambda_i\mathbf{E} - \mathbf{A}) = \mathbf{w}_i^2$, etc.

- If the pencil $s\mathbf{E} - \mathbf{A}$ is regular with $\mathbf{E}$ is singular, then the pencil has finite eigenvalues, and an infinite eigenvalue. These eigenvalues can be found as follows:

  Step 1: We compute the determinant of the pencil. Let the result be a polynomial $\varpi(s)$ of s, i.e. $\det(s\mathbf{E} - \mathbf{A}) = \varpi(s)$. The polynomial will always be of order $p < m$.

  Step 2: We compute the roots of the polynomial $\varpi(s)$. These roots will be the finite eigenvalues of the pencil $s\mathbf{E} - \mathbf{A}$.

  Step 3: Let λ_i, $i = 1, 2, \ldots, \nu$ be the roots of polynomial $\varpi(s)$ and p_i, $i = 1, 2, \ldots, \nu$ be the multiplicity of λ_i. Consequently, $\sum_{i=0}^{\nu} p_i = p$. Then for the pencil $s\mathbf{E} - \mathbf{A}$, the eigenvalue λ_i will have algebraic multiplicity p_i.

  Step 4: From the result of step 3, there exist an infinite eigenvalue of algebraic multiplicity q.

  Step 5: For the eigenvalue λ_i with algebraic multiplicity p_i we can compute its corresponding eigenvectors by following the steps 4, 5 in case 1.

  Step 6: For the infinite eigenvalue with algebraic multiplicity q we will compute its corresponding right eigenvector $\mathbf{u}^q$ by solving the singular algebraic system $\mathbf{E}\mathbf{u}^q = \mathbf{0}_{q,1}$. Then:

  - If by solving this algebraic system we arrive at q linear independent eigenvectors, i.e. the dimension of the eigenspace of the eigenvector then we have found the complete set of the corresponding eigenvectors of the infinite eigenvalue.
  - If by solving this algebraic system we arrive at a number of linear independent eigenvectors that are less than q, i.e. $q' \leq q$, where q' is

the geometric multiplicity of the infinite eigenvalue, then additional linear independent eigenvectors are needed to have the complete set of eigenvectors. These additional linear independent eigenvectors will be called generalized eigenvectors, and can be found as follows. If $q' = q - 1$, we will need one generalized eigenvector $\mathbf{u}^{q(1)}$, which is the solution of $\mathbf{E}^2\mathbf{u}^{q(1)} = \mathbf{0}_{q,1}$. If $q' = q - 2$, we will need two generalized eigenvectors $\mathbf{u}^{q(1)}$, $\mathbf{u}^{q(2)}$, which are the solutions of $\mathbf{E}^2\mathbf{u}^{q(1)} = \mathbf{0}_{q,1}$, and $\mathbf{E}^3\mathbf{u}^{q(2)} = \mathbf{0}_{q,1}$, etc.

Step 7: For the infinite eigenvalue with algebraic multiplicity q we will compute its corresponding left eigenvector $\mathbf{w}^q$ by solving the singular algebraic system $\mathbf{w}^q\mathbf{E} = \mathbf{0}_{1,q}$. Then:

- If by solving this algebraic system we arrive at q linear independent eigenvectors, i.e. the dimension of the eigenspace of the eigenvector then we have found the complete set of the corresponding eigenvectors of the infinite eigenvalue.
- If by solving this algebraic system we arrive at a number of linear independent eigenvectors that are less than q, i.e. $q' \leq q$, then additional linear independent eigenvectors are needed to have the complete set of eigenvectors. These additional linear independent eigenvectors will be called generalized eigenvectors, and can be found as follows. If $q' = q - 1$, we will need one generalized eigenvector $\mathbf{w}^{q(1)}$ which is the solution of $\mathbf{w}^{q(1)}\mathbf{E}^2 = \mathbf{0}_{1,q}$. If $q' = q - 2$, we will need two generalized eigenvectors $\mathbf{w}^{q(1)}$, $\mathbf{w}^{q(2)}$, which are the solutions of $\mathbf{w}^{q(1)}\mathbf{E}^2 = \mathbf{0}_{1,q}$, and $\mathbf{w}^{q(2)}\mathbf{E}^2 = \mathbf{0}_{1,q}$, etc.

- If $\mathbf{E}$, $\mathbf{A}$ are square and the pencil $s\mathbf{E} - \mathbf{A}$ is singular, the determinant of the pencil is equal to zero, i.e. $\det(s\mathbf{E} - \mathbf{A}) \equiv 0$. Unlike the case of the regular pencil, the determinant of the pencil is not equal to a polynomial that through its roots we can have information on the eigenvalues of the pencil. In this case we may use Remark 2.1 to obtain information on the eigenvalues. Since $s\mathbf{E} - \mathbf{A}$ is singular, it is obvious that $\text{rank}(s\mathbf{E} - \mathbf{A}) < m$. Then there exists a matrix function $\tilde{\mathbf{P}} : \mathbb{C} \mapsto \mathbb{R}^{r \times r}$ such that

$$\tilde{\mathbf{P}}(s)(s\mathbf{E} - \mathbf{A}) = \begin{bmatrix} \tilde{\mathbf{A}}(s) & \mathbf{0}_{r,m_1} \end{bmatrix},$$

with

$$\tilde{\mathbf{P}}(s) = \begin{bmatrix} \tilde{\mathbf{P}}_1(s) & \tilde{\mathbf{P}}_2(s) \end{bmatrix},$$

where $\tilde{\mathbf{A}} : \mathbb{C} \mapsto \mathbb{R}^{r \times r_1}$, with $m_1 + r_1 = m$, is a matrix such that if $[\tilde{a}_{ij}]_{1 \leq i \leq r_1}^{1 \leq j \leq r}$ are its elements, for $i = j$ all elements are non-zero and for $i \neq j$ all elements are zero and $\tilde{\mathbf{P}}_1(s) \in \mathbb{R}^{m \times r_1}$, $\tilde{\mathbf{P}}_2(s) \in \mathbb{R}^{m \times m_1}$. The finite eigenvalues will then lie in the diagonal of $\mathbf{A}(s)$. For the infinite eigenvalue, if it exists, we do the same procedure but for the pencil $\mathbf{E} - s\mathbf{A}$. Then the appearance of the eigenvalue 0 for the pencil $\mathbf{E} - s\mathbf{A}$ will mean that there exists an infinite eigenvalue for $s\mathbf{E} - \mathbf{A}$ with algebraic multiplicity same as the algebraic multiplicity of 0 for $\mathbf{E} - s\mathbf{A}$. The corresponding left, and right eigenvectors are then computed similarly to cases 1, 2.

- If $\mathbf{E}$, $\mathbf{A}$ are non-square with $r > m$, then the pencil $s\mathbf{E} - \mathbf{A}$ is singular. Then by using Remark 2.1, there exists a matrix $\tilde{\mathbf{P}} : \mathbb{C} \mapsto \mathbb{R}^{r \times r}$ such that

$$\tilde{\mathbf{P}}(s)(s\mathbf{E} - \mathbf{A}) = \begin{bmatrix} \tilde{\mathbf{A}}(s) \\ \mathbf{0}_{r_1,m} \end{bmatrix},$$

with

$$\tilde{\mathbf{P}}(s) = \begin{bmatrix} \tilde{\mathbf{P}}_1(s) \\ \tilde{\mathbf{P}}_2(s) \end{bmatrix},$$

where $\tilde{\mathbf{A}} : \mathbb{C} \mapsto \mathbb{R}^{m_1 \times m}$, with $m_1 + r_1 = r$, is a matrix such that if $[\tilde{a}_{ij}]_{1 \leq i \leq m_1}^{1 \leq j \leq m}$ are its elements, for $i = j$ all elements are non-zero and for $i \neq j$ all elements are zero and $\tilde{\mathbf{P}}_1(s) \in \mathbb{R}^{m_1 \times r}$, $\tilde{\mathbf{P}}_2(s) \in \mathbb{R}^{r_1 \times r}$. The finite eigenvalues will then lie in the diagonal of $\mathbf{A}(s)$. For the infinite eigenvalue, if it exists, we do the same procedure but for the pencil $\mathbf{E} - s\mathbf{A}$. Then the appearance of the eigenvalue 0 for the pencil $\mathbf{E} - s\mathbf{A}$ will mean that there exists an infinite eigenvalue for $s\mathbf{E} - \mathbf{A}$ with algebraic multiplicity same as the algebraic multiplicity of 0 for $\mathbf{E} - s\mathbf{A}$. The corresponding left, and right eigenvectors are then computed similarly to cases 1, 2.

- If $\mathbf{E}$, $\mathbf{A}$ are non-square with $m > r$, the pencil $s\mathbf{E} - \mathbf{A}$ is singular. Then by using Remark 2.1, there exists a matrix $\tilde{\mathbf{P}} : \mathbb{C} \mapsto \mathbb{R}^{r \times r}$ such that

$$\tilde{\mathbf{P}}(s)(s\mathbf{E} - \mathbf{A}) = \begin{bmatrix} \tilde{\mathbf{A}}(s) & \mathbf{0}_{r,m_1} \end{bmatrix},$$

with

$$\tilde{\mathbf{P}}(s) = \begin{bmatrix} \tilde{\mathbf{P}}_1(s) & \tilde{\mathbf{P}}_2(s) \end{bmatrix},$$

where $\tilde{\mathbf{A}} : \mathbb{C} \mapsto \mathbb{R}^{r \times r_1}$, with $m_1 + r_1 = m$, is a matrix such that if $[\tilde{a}_{ij}]_{1 \leq i \leq r_1}^{1 \leq j \leq r}$ are its elements, for $i = j$ all elements are non-zero and for $i \neq j$ all elements are zero and $\tilde{\mathbf{P}}_1(s) \in \mathbb{R}^{m \times r_1}$, $\tilde{\mathbf{P}}_2(s) \in \mathbb{R}^{m \times m_1}$. The finite eigenvalues will then lie in the diagonal of $\mathbf{A}(s)$. For the infinite eigenvalue, if it exists, we do the same procedure but for the pencil $\mathbf{E} - s\mathbf{A}$. Then the appearance of the eigenvalue 0 for the pencil $\mathbf{E} - s\mathbf{A}$ will mean that there exist an infinite eigenvalue for $s\mathbf{E} - \mathbf{A}$ with algebraic multiplicity same as the algebraic multiplicity of 0 for $\mathbf{E} - s\mathbf{A}$. The corresponding left, and right eigenvalues are then computed similarly to cases 1 and 2.

Here are some examples.

Example 2.8 (Eigenvalues of a regular pencil). Let

$$\mathbf{E} = \begin{bmatrix} 1 & 0 & -1 & -1 & 1 \\ 0 & -2 & 0 & 1 & 1 \\ -2 & -2 & 2 & 2 & 0 \\ 0 & 2 & 1 & 0 & -2 \\ 0 & 0 & 0 & 0 & 0 \end{bmatrix}, \quad \mathbf{A} = \begin{bmatrix} -1 & 0 & 1 & 1 & 0 \\ -4 & -2 & 5 & 5 & -2 \\ -2 & -2 & 2 & 2 & 0 \\ 4 & 2 & -4 & -4 & 1 \\ -1 & 0 & 2 & 1 & -1 \end{bmatrix}.$$

Then the pencil is regular, and $\det(s\mathbf{E} - \mathbf{A}) = (s-2)(s-1)$. The finite eigenvalues are $\lambda_1 = 1$, $\lambda_2 = 2$, and there is also the infinite eigenvalue of algebraic multiplicity 3. By using the method described in case 2, we obtain the right and left eigenvectors. The matrix that contains the left eigenvectors has the form:

$$\mathbf{P} = \begin{bmatrix} 0 & 1 & 0 & 1 & 0 \\ 0 & 0 & -1 & 0 & 0 \\ 1 & 0 & 0 & 0 & -1 \\ 1 & 0 & 1 & 1 & 0 \\ 0 & 0 & 0 & 0 & 1 \end{bmatrix},$$

while the matrix that contains the right eigenvectors has the form:

$$\mathbf{Q} = \begin{bmatrix} 1 & 0 & 0 & 1 & 0 \\ 0 & 1 & 1 & 0 & 1 \\ 0 & 0 & 0 & 1 & 1 \\ 1 & 0 & 1 & 0 & 0 \\ 0 & 0 & 1 & 1 & 1 \end{bmatrix}.$$

□

Example 2.9 (Eigenvalues of a regular pencil). Let

$$\mathbf{E} = \begin{bmatrix} 0 & 0 & -1 & 0 & 1 \\ 0 & -1 & -1 & 1 & 1 \\ -1 & -1 & 1 & 1 & 0 \\ 0 & 1 & 2 & 0 & -2 \\ 0 & 0 & 0 & 0 & 0 \end{bmatrix}, \quad \mathbf{A} = \frac{1}{5}\begin{bmatrix} -5 & 0 & 8 & 5 & -3 \\ -11 & -1 & 14 & 11 & -8 \\ -2 & -2 & 2 & 2 & 0 \\ 11 & 2 & -14 & -11 & 8 \\ -5 & 0 & 10 & 5 & -5 \end{bmatrix}.$$

Then $\det(s\mathbf{E} - \mathbf{A}) = s(s - \frac{1}{5})(s - \frac{2}{5})$ and the pencil is regular. The finite eigenvalues will be $\lambda_1 = 0$, $\lambda_2 = \frac{1}{5}$, $\lambda_3 = \frac{2}{5}$, and there is also the infinite eigenvalue of algebraic multiplicity 2. By using the method described in case 2, we obtain the right and left eigenvectors. The matrix that contains the right eigenvectors related to the finite eigenvalues has the form:

$$\mathbf{Q}_p = \begin{bmatrix} 1 & 0 & 0 \\ 0 & 1 & 1 \\ 0 & 0 & 0 \\ 1 & 0 & 1 \\ 0 & 0 & 1 \end{bmatrix},$$

while the matrix that contains the right eigenvectors related to the infinite eigenvalue has the form:

$$\mathbf{Q}_q = \begin{bmatrix} 1 & 0 \\ 0 & 1 \\ 1 & 1 \\ 0 & 0 \\ 1 & 1 \end{bmatrix}.$$

The matrix that contains the left eigenvectors related to the finite eigenvalues has the form:

$$\mathbf{P}_1 = \begin{bmatrix} 0 & 1 & 0 & 1 & 0 \\ 0 & 0 & -1 & 0 & 0 \\ 1 & 0 & 0 & 0 & -1 \end{bmatrix}$$

while the matrix that contains the right eigenvectors related to the infinite eigenvalue has the form:

$$\mathbf{P}_2 = \begin{bmatrix} 1 & 0 & 1 & 1 & 0 \\ 0 & 0 & 0 & 0 & 1 \end{bmatrix}.$$

□

Example 2.10 (Eigenvalues of a singular pencil). Let

$$\mathbf{E} = \begin{bmatrix} 1 & 1 & 1 & 1 & 1 \\ 0 & 1 & 1 & 0 & 1 \\ 1 & 1 & 1 & 1 & 1 \\ 0 & 1 & 1 & 0 & 1 \\ 1 & 0 & 1 & 0 & 0 \\ 0 & 0 & 1 & 1 & 1 \end{bmatrix}, \quad \mathbf{A} = \begin{bmatrix} 1 & 2 & 2 & 1 & 2 \\ 0 & 2 & 2 & 0 & 2 \\ 1 & 2 & 2 & 2 & 3 \\ 0 & 2 & 3 & 1 & 3 \\ 0 & 0 & 0 & 0 & 0 \\ 1 & 0 & 1 & 0 & 0 \end{bmatrix}.$$

Since the matrices **E**, **A** are non square, the matrix pencil $s\mathbf{E} - \mathbf{A}$ is singular and has the finite eigenvalues $\lambda_1 = 1$, $\lambda_2 = 2$, and the row minimal indices ζ_1=0, ζ_2=1. Let

$$\mathbf{E} = \begin{bmatrix} 1 & 0 \\ 0 & 1 \\ 0 & 0 \end{bmatrix}, \quad \mathbf{A} = \begin{bmatrix} \frac{3}{4} & 0 \\ 0 & 0 \\ 0 & 1 \end{bmatrix}.$$

Since the matrices **E**, **A** are non square, the matrix pencil $s\mathbf{E} - \mathbf{A}$ is singular and has the finite eigenvalues $\lambda_1 = \frac{3}{4}$ and the row minimal indices ζ_1=0, ζ_2=1. A corresponding eigenvector of the finite eigenvalue is

$$\begin{bmatrix} 1 \\ 0 \end{bmatrix}.$$

□

Example 2.11 (Eigenvalue multiplicity of a regular pencil). Let

$$\mathbf{E} = \begin{bmatrix} 1 & 0 & 0 & 0 & 0 & 0 \\ 0 & 1 & 0 & 0 & 0 & 0 \\ 0 & 0 & 1 & 0 & 0 & 0 \\ 0 & 0 & 0 & 1 & 0 & 0 \\ 0 & 0 & 0 & 0 & 1 & 1 \\ 0 & 0 & 0 & 0 & 0 & 0 \end{bmatrix}, \quad \mathbf{A} = \begin{bmatrix} 0 & 0 & 1 & 0 & 0 & 0 \\ 0 & 0 & 0 & 1 & 0 & 0 \\ 0 & 0 & 0 & 0 & 1 & 0 \\ 0 & 0 & 0 & 0 & 0 & 1 \\ -4 & 2 & 2 & -3 & -2 & -1 \\ 1 & 1 & -1 & -1 & 0 & 0 \end{bmatrix}.$$

The pencil $s\mathbf{E} - \mathbf{A}$ has three finite eigenvalues $\lambda_1 = 3$, $\lambda_2 = 2$, $\lambda_3 = 1$, of algebraic multiplicity $p_1 = p_2 = p_3 = 1$, and an infinite eigenvalue of algebraic multiplicity $q = 3$. The eigenspaces of $s\mathbf{E} - \mathbf{A}$ associated with the eigenvalues 3, 2, 1, are :

$$\left\langle \mathbf{u}_1 \right\rangle = \left\langle \begin{bmatrix} -1 \\ 1 \\ -3 \\ 3 \\ -9 \\ 9 \end{bmatrix} \right\rangle, \quad \left\langle \mathbf{u}_2 \right\rangle = \left\langle \begin{bmatrix} -1 \\ 1 \\ -2 \\ 2 \\ -4 \\ 4 \end{bmatrix} \right\rangle, \quad \left\langle \mathbf{u}_3 \right\rangle = \left\langle \begin{bmatrix} -3 \\ 5 \\ -3 \\ 5 \\ -3 \\ 5 \end{bmatrix} \right\rangle,$$

while the eigenspace of $s\mathbf{E} - \mathbf{A}$ associated with the infinite eigenvalue, including the generalized eigenvectors, is:

$$\left\langle \mathbf{u}_4, \quad \mathbf{u}_5, \quad \mathbf{u}_6 \right\rangle = \left\langle \begin{bmatrix} 0 \\ 0 \\ 0 \\ 0 \\ -1 \\ 1 \end{bmatrix}, \begin{bmatrix} 0 \\ 0 \\ -1 \\ 1 \\ 0 \\ 1 \end{bmatrix}, \begin{bmatrix} -1 \\ 1 \\ 0 \\ 1 \\ 42 \\ -48 \end{bmatrix} \right\rangle.$$

□

Example 2.12 (Eigenvalue multiplicity of a regular pencil). Let

$$\mathbf{E} = \begin{bmatrix} 0 & -3 & 0 & 1 & 1 & 8 & 2 \\ 12 & 9 & -5 & -2 & -4 & -3 & 4 \\ 0 & -4 & -5 & 13 & 3 & 9 & 6 \\ 6 & -2 & -3 & 13 & 3 & -7 & 0 \\ 0 & 0 & 0 & 0 & 0 & 0 & 0 \\ 0 & -1 & -11 & 26 & 4 & -1 & 8 \\ -3 & -3 & 2 & -1 & 1 & 3 & -1 \end{bmatrix},$$

and

$$\mathbf{A} = \begin{bmatrix} 7 & 23 & 13 & 34 & 7 & -8 & 17 \\ -12 & -8 & 8 & 3 & 7 & 6 & -3 \\ -1 & 6 & -5 & -2 & 5 & -6 & -6 \\ 5 & 13 & 1 & -9 & -1 & 38 & 3 \\ 8 & 22 & 22 & 42 & 1 & 16 & 29 \\ 11 & 19 & 31 & 20 & 13 & 14 & 21 \\ 7 & 16 & 5 & 13 & 5 & -6 & 7 \end{bmatrix}.$$

The pencil $s\mathbf{E} - \mathbf{A}$ has three finite eigenvalues $\lambda_1 = -1$, $\lambda_2 = 0$, $\lambda_3 = -2$ of algebraic multiplicity $p_1 = p_2 = p_3 = 1$, and an infinite eigenvalue of algebraic multiplicity $q = 4$. The eigenspaces of $s\mathbf{E} - \mathbf{A}$ associated with the eigenvalues

-1, 0, -2, are :

$$\langle \mathbf{u}_1 \rangle = \left\langle \begin{bmatrix} 0.5981 \\ -0.1773 \\ 0.2936 \\ 0.4628 \\ -0.2070 \\ 0.1516 \\ -1 \end{bmatrix} \right\rangle, \quad \langle \mathbf{u}_2 \rangle = \left\langle \begin{bmatrix} -0.6466 \\ 0.1937 \\ -0.2696 \\ -0.4712 \\ 0.1885 \\ -0.1597 \\ 1 \end{bmatrix} \right\rangle, \quad \langle \mathbf{u}_3 \rangle = \left\langle \begin{bmatrix} -0.6675 \\ 0.2118 \\ -0.3128 \\ -0.4433 \\ 0.1773 \\ -0.1872 \\ 1 \end{bmatrix} \right\rangle,$$

while the eigenspace of $s\mathbf{E} - \mathbf{A}$ associated with the infinite eigenvalue, including the generalized eigenvectors, is:

$$\langle \mathbf{u}_4, \mathbf{u}_5, \mathbf{u}_6, \mathbf{u}_7 \rangle = \left\langle \begin{bmatrix} 0.6283 \\ -0.1760 \\ 0.2854 \\ 0.4603 \\ -0.2130 \\ 0.1531 \\ -1 \end{bmatrix}, \begin{bmatrix} -0.6498 \\ 0.1802 \\ -0.3133 \\ -0.4687 \\ 0.1929 \\ -0.1480 \\ 1 \end{bmatrix}, \begin{bmatrix} -0.5969 \\ -0.1143 \\ 0.2738 \\ 0.4507 \\ -0.1623 \\ -0.1711 \\ -1 \end{bmatrix}, \begin{bmatrix} 0.6240 \\ -0.1652 \\ 0.2997 \\ 0.4692 \\ -0.2273 \\ 0.1578 \\ -1 \end{bmatrix} \right\rangle.$$

□

Example 2.13 (Eigenvalues of a singular pencil). Let

$$\mathbf{E} = \begin{bmatrix} 1 & 1 & 1 \\ 0 & 1 & 1 \\ 1 & 1 & 1 \\ 0 & 1 & 1 \end{bmatrix}, \quad \mathbf{A} = \begin{bmatrix} 1 & 2 & 2 \\ 0 & 2 & 2 \\ 1 & 2 & 2 \\ 0 & 2 & 3 \end{bmatrix}.$$

The matrices **E**, **A** are non-square. Hence, we are in the case where $r > m$, i.e. $4 > 3$ and the pencil

$$s\mathbf{E} - \mathbf{A} = \begin{bmatrix} s-1 & s-2 & s-2 \\ 0 & s-2 & s-2 \\ s-1 & s-2 & s-2 \\ 0 & s-2 & s-3 \end{bmatrix},$$

is singular. From Remark 2.1, there exists the matrix function

$$\mathbf{P}(s) = \begin{bmatrix} 1 & -1 & 0 & 0 \\ 0 & 3-s & 0 & -s+2 \\ 0 & 1 & 0 & -1 \\ 0 & 0 & 1 & 0 \end{bmatrix},$$

such that

$$\mathbf{P}(s)(s\mathbf{E} - \mathbf{A}) = \begin{bmatrix} \mathbf{A}(s) \\ \mathbf{0}_{1,1} \end{bmatrix},$$

where

$$\mathbf{A}(s) = \begin{bmatrix} s-1 & 0 & 0 \\ 0 & s-2 & 0 \\ 0 & 0 & 1 \end{bmatrix}.$$

In addition, for

$$\mathbf{P}(s) = \begin{bmatrix} \mathbf{P}_1(s) \\ \mathbf{P}_2(s) \end{bmatrix},$$

with

$$\mathbf{P}_1(s) = \begin{bmatrix} 1 & -1 & 0 & 0 \\ 0 & 3-s & 0 & -s+2 \\ 0 & 1 & 0 & -1 \end{bmatrix}, \quad \mathbf{P}_2(s) = \begin{bmatrix} 0 & 0 & 1 & 0 \end{bmatrix},$$

we have

$$\mathbf{P}_2(s)\mathbf{E} = \begin{bmatrix} 1 & 1 & 1 \end{bmatrix} \neq \mathbf{0}_{1,3}.$$

Hence, the pencil has the finite eigenvalues $\lambda_1 = 1$, $\lambda_2 = 2$, and an infinite eigenvalue.

Example 2.14 (Eigenvalues of a singular pencil). Let

$$\mathbf{E} = \begin{bmatrix} 1 & 1 \\ 0 & 1 \\ 1 & 1 \end{bmatrix}, \quad \mathbf{A} = \begin{bmatrix} 1 & 2 \\ 0 & 2 \\ 1 & 2 \end{bmatrix}.$$

The matrices **E**, **A** are non-square. Hence, we are in the case where $r > m$, i.e. $3 > 2$ and the pencil

$$s\mathbf{E} - \mathbf{A} = \begin{bmatrix} s-1 & s-2 \\ 0 & s-2 \\ s-1 & s-2 \end{bmatrix},$$

is singular. Then

$$\mathbf{P}(s) = \begin{bmatrix} 1 & -1 & 0 \\ 0 & 1 & 0 \\ -1 & 0 & 1 \end{bmatrix},$$

such that

$$\mathbf{P}(s)(s\mathbf{F} - \mathbf{G}) = \begin{bmatrix} \mathbf{A}(s) \\ \mathbf{0}_{1,2} \end{bmatrix},$$

where

$$\mathbf{A}(s) = \begin{bmatrix} s-1 & 0 \\ 0 & s-2 \end{bmatrix}.$$

In addition, for

$$\mathbf{P}(s) = \begin{bmatrix} \mathbf{P}_1(s) \\ \mathbf{P}_2(s) \end{bmatrix},$$

with

$$\mathbf{P}_1(s) = \begin{bmatrix} 1 & -1 & 0 \\ 0 & 1 & 0 \end{bmatrix}, \quad \mathbf{P}_2(s) = \begin{bmatrix} -1 & 0 & 1 \end{bmatrix},$$

we have

$$\mathbf{P}_2(s)\mathbf{E} = \mathbf{0}_{1,2}.$$

Hence, the pencil has the finite eigenvalues $\lambda_1 = 1$, $\lambda_2 = 2$, and an infinite eigenvalue. The eigenspace of $s\mathbf{E} - \mathbf{A}$ associated with the finite eigenvalues 1, 2, are:

$$\langle \mathbf{u}_1 \rangle = \left\langle \begin{bmatrix} 1 \\ 0 \end{bmatrix} \right\rangle, \quad \langle \mathbf{u}_2 \rangle = \left\langle \begin{bmatrix} 0 \\ 1 \end{bmatrix} \right\rangle.$$

□

2.3 Canonical Forms of Matrix Pencils

We will begin this section with the following definition:

Definition 2.5 (Strictly equivalent matrix pencil). The pencil $s\mathbf{E} - \mathbf{A}$ is said to be *strictly equivalent* to the pencil $s\tilde{\mathbf{E}} - \tilde{\mathbf{A}}$ if and only if there exist non-singular $\mathbf{P} \in \mathbb{C}^{r \times r}$ and $\mathbf{Q} \in \mathbb{C}^{m \times m}$ such that:

$$s\mathbf{E} - \mathbf{A} = \mathbf{P}(s\tilde{\mathbf{E}} - \tilde{\mathbf{A}})\mathbf{Q}\,.$$

Let $\mathbf{B}_{n_1} \in \mathbb{C}^{n_1 \times n_1}$, $\mathbf{B}_{n_2} \in \mathbb{C}^{n_2 \times n_2}$, ..., $\mathbf{B}_{n_r} \in \mathbb{C}^{n_r \times n_r}$. With the direct sum

$$\mathbf{B}_{n_1} \oplus \mathbf{B}_{n_2} \oplus \cdots \oplus \mathbf{B}_{n_r}\,,$$

we will denote the block diagonal matrix:

$$\text{blockdiag} \begin{bmatrix} \mathbf{B}_{n_1} & \mathbf{B}_{n_2} & \dots & \mathbf{B}_{n_r} \end{bmatrix}.$$

We have the following cases:

- The pencil $s\mathbf{E} - \mathbf{A}$ is regular, and $\mathbf{E}$ is regular. Then $s\mathbf{E} - \mathbf{A}$ is characterized by a uniquely defined element, known as complex Jordan canonical form, $s\mathbf{E}_J - \mathbf{A}_J$, see [51], specified by the complete set of eigenvalues which are all finite in this case. If λ_i, $i = 1, 2, \dots, \nu$ is an eigenvalue of the pencil with corresponding algebraic multiplicity p_i then from the regularity of $s\mathbf{E} - \mathbf{A}$, there exist non-singular matrices $\mathbf{P}, \mathbf{Q} \in \mathbb{C}^{m \times m}$ such that

$$\mathbf{PEQ} = \mathbf{E}_J = \mathbf{I}_m, \quad \mathbf{PAQ} = \mathbf{A}_J = \mathbf{J}_m\,,$$

  where $s\mathbf{E}_J - \mathbf{A}_J$ is the complex Jordan form of the regular pencil $s\mathbf{E} - \mathbf{A}$ and is defined by

$$s\mathbf{E}_J - \mathbf{A}_J := s\mathbf{I}_m - \mathbf{J}_m\,,$$

where the Jordan type element is uniquely defined by the set of the eigenvalues,

$$(s-\lambda_1)^{p_1},\ldots,(s-\lambda_\nu)^{p_\nu}$$

of $s\mathbf{E}-\mathbf{A}$ and has the form

$$s\mathbf{I}_m-\mathbf{J}_m := s\mathbf{I}_{p_1}-\mathbf{J}_{p_1}(\lambda_1)\oplus\cdots\oplus s\mathbf{I}_{p_\nu}-\mathbf{J}_{p_\nu}(\lambda_\nu)\,.$$

The matrices $\mathbf{I}_{p_i},\mathbf{J}_{p_i}(\lambda_i)$ are defined as

$$\mathbf{I}_{p_i}=\begin{bmatrix} 1 & 0 & \ldots & 0 & 0\\ 0 & 1 & \ldots & 0 & 0\\ \vdots & \vdots & \ddots & \vdots & \vdots\\ 0 & 0 & \ldots & 0 & 1\end{bmatrix}\in\mathbb{C}^{p_i\times p_i},$$

and

$$\mathbf{J}_{p_i}(\lambda_i)=\begin{bmatrix} \lambda_i & 1 & \ldots & 0 & 0\\ 0 & \lambda_i & \ldots & 0 & 0\\ \vdots & \vdots & \ddots & \vdots & \vdots\\ 0 & 0 & \ldots & \lambda_i & 1\\ 0 & 0 & \ldots & 0 & \lambda_i\end{bmatrix}\in\mathbb{C}^{p_i\times p_i}.$$

- The class of $s\mathbf{E}-\mathbf{A}$ is characterized by a uniquely defined element, known as complex Weierstrass canonical form, $s\mathbf{E}_w-\mathbf{A}_w$, see [35, 51], specified by the complete set of invariants of $s\mathbf{E}-\mathbf{A}$. This is the set of finite and infinite eigenvalues. If λ_i, $i=1,2,\ldots,\nu$, is a finite eigenvalue of the pencil, the corresponding algebraic multiplicity is p_i. Also, q is the multiplicity of the infinite eigenvalue. Note that $\sum_{i=1}^{\nu}p_i=p$, and $p+q=m$. Then from the regularity of $s\mathbf{E}-\mathbf{A}$, there exist non-singular matrices $\mathbf{P},\mathbf{Q}\in\mathbb{C}^{m\times m}$ such that

$$\mathbf{PEQ}=\mathbf{E}_w=\mathbf{I}_p\oplus\mathbf{H}_q\,,\quad \mathbf{PAQ}=\mathbf{A}_w=\mathbf{J}_p\oplus\mathbf{I}_q\,,$$

where $s\mathbf{E}_w-\mathbf{A}_w$ is the complex Weierstrass form of the regular pencil $s\mathbf{E}-\mathbf{A}$ and is defined by

$$s\mathbf{E}_w-\mathbf{A}_w := s\mathbf{I}_p-\mathbf{J}_p\oplus s\mathbf{H}_q-\mathbf{I}_q\,,$$

where the first normal Jordan type element is uniquely defined by the set of the finite eigenvalues, where the Jordan type element is uniquely defined by the set of the eigenvalues

$$(s-\lambda_1)^{p_1},\ldots,(s-\lambda_\nu)^{p_\nu}$$

of $s\mathbf{E}-\mathbf{A}$ and has the form

$$s\mathbf{I}_m-\mathbf{J}_m := s\mathbf{I}_{p_1}-\mathbf{J}_{p_1}(\lambda_1)\oplus\cdots\oplus s\mathbf{I}_{p_\nu}-\mathbf{J}_{p_\nu}(\lambda_\nu)\,.$$

The second uniquely defined block $s\mathbf{H}_q - \mathbf{I}_q$ corresponds to the infinite eigenvalue of $s\mathbf{E} - \mathbf{A}$. Thus, $\mathbf{H}_q$ is a nilpotent element of $\mathbb{C}^{q\times q}$ with index $\tilde{q} = \max\{q_i : j = 1, 2, \ldots, \sigma\}$, where

$$\mathbf{H}_q^{\tilde{q}} = \mathbf{0}_{q,q} \,.$$

The matrix $\mathbf{H}_q$ is defined as:

$$\mathbf{H}_q = \begin{bmatrix} 0 & 1 & \ldots & 0 & 0 \\ 0 & 0 & \ldots & 0 & 0 \\ \vdots & \vdots & \ddots & \vdots & \vdots \\ 0 & 0 & \ldots & 0 & 1 \\ 0 & 0 & \ldots & 0 & 0 \end{bmatrix} \in \mathbb{C}^{q\times q} \,.$$

- If $s\mathbf{E} - \mathbf{A}$ is singular with $\mathbf{E}$, $\mathbf{A}$ square, and $\det(s\mathbf{E} - \mathbf{A}) \equiv 0$, then the class of $s\mathbf{E} - \mathbf{A}$ is characterized by a uniquely defined element, known as complex Kronecker canonical form, $s\mathbf{E}_K - \mathbf{A}_K$, see [37, 51], specified by the complete set of invariants of the singular matrix pencil $s\mathbf{E} - \mathbf{A}$.

  These invariants are the *elementary divisors* (e.d.) and the *minimal indices* (m.i.). The set of e.d. is obtained by factorizing the invariant polynomials into powers of homogeneous polynomials irreducible over $\mathbb{C}$. There are two different types of e.d., the set of type $(s - \lambda_i)^{p_i}$, called *finite elementary divisors* (f.e.d.), where λ_i finite eigenvalue of algebraic multiplicity p_i $(1 \le i \le \nu)$ with $\sum_{i=1}^{\nu} p_i = p$, and the set of type $\hat{s}^q = \frac{1}{s^q}$, called *infinite elementary divisors* (i.e.d.), where q is the algebraic multiplicity of the infinite eigenvalue. The set of the m.i. is defined as follows. The distinguishing feature of a singular pencil $s\mathbf{E} - \mathbf{A}$ is that either $r \ne m$ or $r = m$ and $\det(s\mathbf{E} - \mathbf{A}) \equiv 0$. Let $\mathcal{N}_r$, $\mathcal{N}_l$ be the right, left null space of a matrix respectively. Then the equations:

$$(s\mathbf{E} - \mathbf{A})\mathbf{U}(s) = \mathbf{0}_{r,1},$$

  and

$$\mathbf{V}^{\mathrm{T}}(s)(s\mathbf{E} - \mathbf{A}) = \mathbf{0}_{1,r},$$

  have solutions in $\mathbf{U}(s)$, $\mathbf{V}(s)$, which are vectors in the rational vector spaces $\mathcal{N}_r(s\mathbf{E}-\mathbf{A})$ and $\mathcal{N}_l(s\mathbf{E}-\mathbf{A})$ respectively. The binary vectors $\mathbf{U}(s)$ and $\mathbf{V}^{\mathrm{T}}(s)$ express dependence relationships among the columns or rows of $s\mathbf{E} - \mathbf{A}$ respectively. $\mathbf{U}(s)$, $\mathbf{V}(s)$ are polynomial vectors. Let d=dim$\mathcal{N}_r(s\mathbf{E} - \mathbf{A})$ and t=dim$\mathcal{N}_l(s\mathbf{E} - \mathbf{A})$. The spaces $\mathcal{N}_r(s\mathbf{E} - \mathbf{A})$ and $\mathcal{N}_l(s\mathbf{E} - \mathbf{A})$ as rational vector spaces are spanned by minimal polynomial bases of minimal degrees $\epsilon_1 = \epsilon_2 = \cdots = \epsilon_g = 0 < \epsilon_{g+1} \le \cdots \le \epsilon_d$ and $\zeta_1 = \zeta_2 = \cdots = \zeta_h = 0 < \zeta_{h+1} \le \cdots \le \zeta_\beta$ respectively. The set of minimal indices $\epsilon_1, \epsilon_2, \ldots, \epsilon_d$ and $\zeta_1, \zeta_2, \ldots, \zeta_\beta$ are known as *column minimal indices* (c.m.i.) and *row minimal indices* (r.m.i.) of $s\mathbf{E} - \mathbf{A}$ respectively. To sum up, in the case of a singular matrix pencil with square matrices we have invariants of the following type:

  - e.d. of type $(s - \lambda_i)^{p_i}$, called *finite elementary divisors*;

- e.d. of type $\hat{s}^q = \frac{1}{s^q}$, called *infinite elementary divisors*;
- m.i. of type $\epsilon_1 = \epsilon_2 = \cdots = \epsilon_g = 0 < \epsilon_{g+1} \leq \cdots \leq \epsilon_d$, called *column minimal indices*; and
- m.i. of type $\zeta_1 = \zeta_2 = \cdots = \zeta_h = 0 < \zeta_{h+1} \leq \cdots \leq \zeta_\beta$, called *row minimal indices.*

Then from the singularity of $s\mathbf{E}-\mathbf{A}$, there exist non-singular matrices $\mathbf{P}, \mathbf{Q} \in \mathbb{C}^{m\times m}$ such that

$$\mathbf{PEQ} = \mathbf{E}_K = \mathbf{I}_p \oplus \mathbf{H}_q \oplus \mathbf{E}_\epsilon \oplus \mathbf{E}_\zeta \oplus \mathbf{0}_{h,g},$$

and

$$\mathbf{PAQ} = \mathbf{A}_K = \mathbf{J}_p \oplus \mathbf{I}_q \oplus \mathbf{A}_\epsilon \oplus \mathbf{A}_\zeta \oplus \mathbf{0}_{h,g}\,,$$

where $s\mathbf{E}_K - \mathbf{A}_K$ is the complex Kronecker form of the singular pencil $s\mathbf{E}-\mathbf{A}$ and is defined by

$$s\mathbf{E}_K - \mathbf{A}_K := s\mathbf{I}_p - \mathbf{J}_p \oplus s\mathbf{H}_q - \mathbf{I}_q \oplus s\mathbf{E}_\epsilon - \mathbf{A}_\epsilon \oplus s\mathbf{E}_\zeta - \mathbf{A}_\zeta \oplus \mathbf{0}_{h,g},$$

where $p+q+\epsilon+\zeta+h=r$ and $p+q+\epsilon+\zeta+g=m$. The block $s\mathbf{I}_p - \mathbf{J}_p$ is uniquely defined by the set of f.e.d. of $s\mathbf{E}-\mathbf{A}$:

$$(s-\lambda_1)^{p_1}, \ldots, (s-\lambda_\nu)^{p_\nu}, \quad \sum_{i=1}^{\nu} p_i = p$$

and has the form

$$s\mathbf{I}_p - \mathbf{J}_p := s\mathbf{I}_{p_1} - \mathbf{J}_{p_1}(\lambda_1) \oplus \cdots \oplus s\mathbf{I}_{p_\nu} - \mathbf{J}_{p_\nu}(\lambda_\nu).$$

The q blocks of the second uniquely defined block $s\mathbf{H}_q - \mathbf{I}_q$ correspond to the i.e.d. of $s\mathbf{E}-\mathbf{A}$

$$\hat{s}^{q_1}, \ldots, \hat{s}^{q_\sigma}, \quad \sum_{j=1}^{\sigma} q_i = q$$

and has the form

$$s\mathbf{H}_q - \mathbf{I}_q := s\mathbf{H}_{q_1} - \mathbf{I}_{q_1} \oplus \cdots \oplus s\mathbf{H}_{q_\sigma} - \mathbf{I}_{q_\sigma}.$$

The matrix $\mathbf{H}_q$ is a nilpotent element of $\mathbb{C}^{q\times q}$ with index $q_* = \max\{q_i : j = 1, 2, \ldots, \sigma\}$, i.e.

$$\mathbf{H}_q^{q_*} = \mathbf{0}_{q,q}.$$

In the above notations, the matrices $\mathbf{I}_{p_i}, \mathbf{J}_{p_i}(\lambda_i), \mathbf{H}_{q_i}$ are defined in the previous cases. The rest of the diagonal blocks, i.e. the blocks $s\mathbf{E}_\epsilon - \mathbf{A}_\epsilon$ and $s\mathbf{E}_\zeta - \mathbf{A}_\zeta$, are defined as follows. The matrices $\mathbf{E}_\epsilon$, $\mathbf{A}_\epsilon$ have the form

$$\mathbf{E}_\epsilon = \text{blockdiag}\left\{\mathbf{L}_{\epsilon_{g+1}}, \mathbf{L}_{\epsilon_{g+2}}, \ldots, \mathbf{L}_{\epsilon_d}\right\}\,,$$

and

$$\mathbf{A}_\epsilon = \text{blockdiag}\left\{\bar{\mathbf{L}}_{\epsilon_{g+1}}, \bar{\mathbf{L}}_{\epsilon_{g+2}}, \ldots, \bar{\mathbf{L}}_{\epsilon_d}\right\},$$

where $\mathbf{L}_{\epsilon_i} = \left[\begin{array}{ccc} \mathbf{I}_{\epsilon_i} & \vdots & \mathbf{0}_{\epsilon_i,1} \end{array}\right]$, and $\bar{\mathbf{L}}_{\epsilon_i} = \left[\begin{array}{ccc} \mathbf{0}_{\epsilon_i,1} & \vdots & \mathbf{I}_{\epsilon_i} \end{array}\right]$, for $i = g+1, g+2, \ldots, d$. The matrices $\mathbf{E}_\zeta$, $\mathbf{A}_\zeta$ have the form

$$\mathbf{E}_\zeta = \text{blockdiag}\left\{\mathbf{L}_{\zeta_{h+1}}, \mathbf{L}_{\zeta_{h+2}}, \ldots, \mathbf{L}_{\zeta_\beta}\right\},$$

and

$$\mathbf{A}_\zeta = \text{blockdiag}\left\{\bar{\mathbf{L}}_{\zeta_{h+1}}, \bar{\mathbf{L}}_{\zeta_{h+2}}, \ldots, \bar{\mathbf{L}}_{\zeta_\beta}\right\},$$

where $\mathbf{L}_{\zeta_i} = \begin{bmatrix} \mathbf{I}_{\zeta_i} \\ \mathbf{0}_{1,\zeta_i} \end{bmatrix}$, $\bar{\mathbf{L}}_{\zeta_i} = \begin{bmatrix} \mathbf{0}_{1,\zeta_i} \\ I_{\zeta_i} \end{bmatrix}$, for $i = h+1, h+2, \ldots, t$.

- The *singular pencil* with $r > m$ is characterized by the set of the finite-infinite eigenvalues, and the minimal row indices, see [37, 51]. Let $\mathcal{N}_r$ be the right null space of a matrix respectively. Then the equations $\mathbf{V}^{\mathrm{T}}(s)(s\mathbf{E} - \mathbf{A}) = \mathbf{0}_{1,m}$ have solutions in $\mathbf{V}(s)$, which are vectors in the rational vector space $\mathcal{N}_l(s\mathbf{E} - \mathbf{A})$. The binary vectors $\mathbf{V}^{\mathrm{T}}(s)$ express dependence relationships among the rows of $s\mathbf{E} - \mathbf{A}$. Note that $\mathbf{V}(s) \in \mathbb{C}^{r\times 1}$ are polynomial vectors. Let t=dim$[\mathcal{N}_l(s\mathbf{E} - \mathbf{A})]$. It is known that $\mathcal{N}_l(s\mathbf{E} - \mathbf{A})$ as rational vector spaces, are spanned by minimal polynomial bases of minimal degrees
$$\zeta_1 = \zeta_2 = \cdots = \zeta_h = 0 < \zeta_{h+1} \leq \cdots \leq \zeta_{h+k=\beta},$$
which is the set of *row minimal indices* of $s\mathbf{E} - \mathbf{A}$. This means there are β row minimal indices, but $\beta - h = k$ non-zero row minimal indices. We are interested only in the k non-zero minimal indices. To sum up, the invariants of a singular pencil with $r > m$ are the finite-infinite eigenvalues of the pencil and the minimal row indices as described above. Following the above given analysis, there exist non-singular matrices $\mathbf{P}$, $\mathbf{Q}$ with $\mathbf{P} \in \mathbb{C}^{r\times r}$, $\mathbf{Q} \in \mathbb{C}^{m\times m}$, such that:
$$\mathbf{PEQ} = \mathbf{E}_K = \mathbf{I}_p \oplus \mathbf{H}_q \oplus \mathbf{E}_\zeta,$$
$$\mathbf{PAQ} = \mathbf{A}_K = \mathbf{J}_p \oplus \mathbf{I}_q \oplus \mathbf{A}_\zeta.$$
The matrices $\mathbf{P}$, $\mathbf{Q}$ can be written as:
$$\mathbf{P} = \begin{bmatrix} \mathbf{P}_1 \\ \mathbf{P}_2 \\ \mathbf{P}_3 \end{bmatrix}, \quad \mathbf{Q} = \begin{bmatrix} \mathbf{Q}_p & \mathbf{Q}_q & \mathbf{Q}_\zeta \end{bmatrix},$$
with $\mathbf{P}_1 \in \mathbb{C}^{p\times r}$, $\mathbf{P}_2 \in \mathbb{C}^{q\times r}$, $\mathbf{P}_3 \in \mathbb{C}^{\tilde{\zeta}_1\times r}$, $\tilde{\zeta}_1 = k + \sum_{i=1}^{k}[\zeta_{h+i}]$, and $\mathbf{Q}_p \in \mathbb{C}^{m\times p}$, $\mathbf{Q}_q \in \mathbb{C}^{m\times q}$, $\mathbf{Q}_\zeta \in \mathbb{C}^{m\times\tilde{\zeta}_2}$, and $\tilde{\zeta}_2 = \sum_{i=1}^{k}[\zeta_{h+i}]$, where $\mathbf{J}_p$ is the Jordan matrix for the finite eigenvalues, $\mathbf{H}_q$ a nilpotent matrix with index q_* which is actually the Jordan matrix of the zero eigenvalue of the pencil $s\mathbf{A} - \mathbf{E}$. The matrices $\mathbf{E}_\zeta$, $\mathbf{A}_\zeta$ are defined as
$$\mathbf{E}_\zeta = \begin{bmatrix} \mathbf{I}_{\zeta_{h+1}} \\ \mathbf{0}_{1,\zeta_{h+1}} \end{bmatrix} \oplus \begin{bmatrix} \mathbf{I}_{\zeta_{h+2}} \\ \mathbf{0}_{1,\zeta_{h+2}} \end{bmatrix} \oplus \cdots \oplus \begin{bmatrix} \mathbf{I}_{\zeta_{h+k}} \\ \mathbf{0}_{1,\zeta_{h+k}} \end{bmatrix},$$

and

$$\mathbf{A}_\zeta = \begin{bmatrix} \mathbf{0}_{1,\zeta_{h+1}} \\ \mathbf{I}_{\zeta_{h+1}} \end{bmatrix} \oplus \begin{bmatrix} \mathbf{0}_{1,\zeta_{h+2}} \\ \mathbf{I}_{\zeta_{h+2}} \end{bmatrix} \oplus \cdots \oplus \begin{bmatrix} \mathbf{0}_{1,\zeta_{h+k}} \\ \mathbf{I}_{\zeta_{h+k}} \end{bmatrix},$$

with $p + q + \sum_{i=1}^{k}[\zeta_{h+i}] + k = r$, $p + q + \sum_{i=1}^{k}[\zeta_{h+i}] = m$.

- The *singular pencil* with $r < m$ is characterized by the set of the finite-infinite eigenvalues, and the minimal column indices, see [37, 51]. Let $\mathcal{N}_r$ be the right null space of a matrix. Then the equations $(s\mathbf{E} - \mathbf{A})\mathbf{U}(s) = \mathbf{0}_{r,1}$ have solutions in $\mathbf{U}(s)$, which are vectors in the rational vector space $\mathcal{N}_r(s\mathbf{E} - \mathbf{A})$. The binary vectors $\mathbf{U}(s)$ express dependence relationships among the columns of $s\mathbf{E} - \mathbf{A}$. Note that $\mathbf{U}(s) \in \mathbb{C}^{m\times 1}$ are polynomial vectors. Let d=dim$[\mathcal{N}_r(s\mathbf{E} - \mathbf{A})]$. It is known that $\mathcal{N}_r(s\mathbf{E} - \mathbf{A})$ as rational vector spaces, are spanned by minimal polynomial bases of minimal degrees

$$\epsilon_1 = \epsilon_2 = \cdots = \epsilon_g = 0 < \epsilon_{g+1} \leq \cdots \leq \epsilon_d \ ,$$

which is the set of *column minimal indices* of $s\mathbf{E} - \mathbf{A}$. This means there are d column minimal indices, but $d - g$ non-zero column minimal indices. We are interested only in the $d - g$ non-zero minimal indices. To sum up, the invariants of a singular pencil with $r < m$ are the finite-infinite eigenvalues of the pencil and the minimal column indices as described above. Following the above given analysis, there exist non-singular matrices $\mathbf{P}$, $\mathbf{Q}$ with $\mathbf{P} \in \mathbb{C}^{r\times r}$, $\mathbf{Q} \in \mathbb{C}^{m\times m}$, such that:

$$\mathbf{PEQ} = \mathbf{E}_K = \mathbf{I}_p \oplus \mathbf{H}_q \oplus \mathbf{E}_\epsilon \ ,$$

$$\mathbf{PAQ} = \mathbf{A}_K = \mathbf{J}_p \oplus \mathbf{I}_q \oplus \mathbf{A}_\epsilon \ .$$

The matrices $\mathbf{P}$, $\mathbf{Q}$ can be written as:

$$\mathbf{P} = \begin{bmatrix} \mathbf{P}_1 \\ \mathbf{P}_2 \\ \mathbf{P}_3 \end{bmatrix}, \quad \mathbf{Q} = \begin{bmatrix} \mathbf{Q}_p & \mathbf{Q}_q & \mathbf{Q}_\epsilon \end{bmatrix},$$

with $\mathbf{P}_1 \in \mathbb{C}^{p\times r}$, $\mathbf{P}_2 \in \mathbb{C}^{q\times r}$, $\mathbf{P}_3 \in \mathbb{C}^{\tilde{\epsilon}_1\times r}$, $\tilde{\epsilon}_1 = k + \sum_{i=1}^{k}[\epsilon_{g+i}]$ and $\mathbf{Q}_p \in \mathbb{C}^{m\times p}$, $\mathbf{Q}_q \in \mathbb{C}^{m\times q}$, $\mathbf{Q}_\epsilon \in \mathbb{C}^{m\times\tilde{\epsilon}_2}$ and $\tilde{\epsilon}_2 = \sum_{i=1}^{k}[\epsilon_{g+i}]$, where $\mathbf{J}_p$ is the Jordan matrix for the finite eigenvalues, $\mathbf{H}_q$ a nilpotent matrix with index q_* which is actually the Jordan matrix of the zero eigenvalue of the pencil $s\mathbf{A} - \mathbf{E}$. The matrices $\mathbf{E}_\epsilon$, $\mathbf{A}_\epsilon$ are defined as

$$\mathbf{E}_\epsilon = \begin{bmatrix} \mathbf{I}_{\epsilon_{g+1}} & \mathbf{0}_{1,\epsilon_{g+1}} \end{bmatrix} \oplus \begin{bmatrix} \mathbf{I}_{\epsilon_{g+2}} & \mathbf{0}_{1,\epsilon_{g+2}} \end{bmatrix} \oplus \cdots \oplus \begin{bmatrix} \mathbf{I}_{\epsilon_d} & \mathbf{0}_{1,\epsilon_d} \end{bmatrix},$$

and

$$\mathbf{A}_\epsilon = \begin{bmatrix} \mathbf{0}_{1,\epsilon_{g+1}} & \mathbf{I}_{\epsilon_{g+1}} \end{bmatrix} \oplus \begin{bmatrix} \mathbf{0}_{1,\epsilon_{g+2}} & \mathbf{I}_{\epsilon_{g+2}} \end{bmatrix} \oplus \cdots \oplus \begin{bmatrix} \mathbf{0}_{1,\epsilon_d} & \mathbf{I}_{\epsilon_d} \end{bmatrix},$$

with $p + q + \sum_{i=1}^{k}[\epsilon_{g+i}] + k = r$, $p + q + \sum_{i=1}^{k}[\epsilon_{g+i}] = m$.

Here are some examples.

Example 2.15 (Canonical form of a regular pencil). Let

$$\mathbf{E} = \begin{bmatrix} 1 & 0 & -1 & -1 & 1 \\ 0 & -2 & 0 & 1 & 1 \\ -2 & -2 & 2 & 2 & 0 \\ 0 & 2 & 1 & 0 & -2 \\ 0 & 0 & 0 & 0 & 0 \end{bmatrix},$$

and

$$\mathbf{A} = \begin{bmatrix} -1 & 0 & 1 & 1 & 0 \\ -4 & -2 & 5 & 5 & -2 \\ -2 & -2 & 2 & 2 & 0 \\ 4 & 2 & -4 & -4 & 1 \\ -1 & 0 & 2 & 1 & -1 \end{bmatrix}.$$

Then the pencil is regular, and $\det(s\mathbf{E} - \mathbf{A}) = (s-2)(s-1)$. The finite eigenvalues are $\lambda_1 = 1$, $\lambda_2 = 2$, and there is one infinite eigenvalue of algebraic multiplicity 3. Then:

$$\mathbf{J}_p = \begin{bmatrix} 1 & 0 \\ 0 & 2 \end{bmatrix}, \quad \mathbf{H}_q = \begin{bmatrix} 0 & 0 & 0 \\ 1 & 0 & 0 \\ 0 & 1 & 0 \end{bmatrix},$$

and

$$\mathbf{E}_w = \begin{bmatrix} 1 & 0 & 0 & 0 & 0 \\ 0 & 1 & 0 & 0 & 0 \\ 0 & 0 & 0 & 1 & 0 \\ 0 & 0 & 0 & 0 & 1 \\ 0 & 0 & 0 & 0 & 0 \end{bmatrix}, \quad \mathbf{A}_w = \begin{bmatrix} 1 & 0 & 0 & 0 & 0 \\ 0 & 2 & 0 & 0 & 0 \\ 0 & 0 & 1 & 0 & 0 \\ 0 & 0 & 0 & 1 & 0 \\ 0 & 0 & 0 & 0 & 1 \end{bmatrix}.$$

□

Example 2.16 (Canonical form of a regular pencil). Let

$$\mathbf{E} = \begin{bmatrix} 0 & 0 & -1 & 0 & 1 \\ 0 & -1 & -1 & 1 & 1 \\ -1 & -1 & 1 & 1 & 0 \\ 0 & 1 & 2 & 0 & -2 \\ 0 & 0 & 0 & 0 & 0 \end{bmatrix},$$

and

$$\mathbf{A} = \frac{1}{5}\begin{bmatrix} -5 & 0 & 8 & 5 & -3 \\ -11 & -1 & 14 & 11 & -8 \\ -2 & -2 & 2 & 2 & 0 \\ 11 & 2 & -14 & -11 & 8 \\ -5 & 0 & 10 & 5 & -5 \end{bmatrix}.$$

Then $\det(s\mathbf{E} - \mathbf{A}) = s(s - \frac{1}{5})(s - \frac{2}{5})$ and the pencil is regular. The finite eigenvalues will be $\lambda_1 = 0$, $\lambda_2 = \frac{1}{5}$, $\lambda_3 = \frac{2}{5}$, and there is one infinite eigenvalue of algebraic multiplicity 2. Then:

$$\mathbf{J}_p = \begin{bmatrix} 0 & 0 & 0 \\ 0 & \frac{1}{5} & 0 \\ 0 & 0 & \frac{2}{5} \end{bmatrix}, \quad \mathbf{H}_q = \begin{bmatrix} 0 & 1 \\ 0 & 0 \end{bmatrix},$$

and

$$\mathbf{E}_w = \begin{bmatrix} 1 & 0 & 0 & 0 & 0 \\ 0 & 1 & 0 & 0 & 0 \\ 0 & 0 & 1 & 0 & 0 \\ 0 & 0 & 0 & 0 & 1 \\ 0 & 0 & 0 & 0 & 0 \end{bmatrix}, \quad \mathbf{A}_w = \begin{bmatrix} 0 & 0 & 0 & 0 & 0 \\ 0 & \frac{1}{5} & 0 & 0 & 0 \\ 0 & 0 & \frac{2}{5} & 0 & 0 \\ 0 & 0 & 0 & 1 & 0 \\ 0 & 0 & 0 & 0 & 1 \end{bmatrix}.$$

□

Example 2.17 (Canonical form of a singular pencil). Let

$$\mathbf{E} = \begin{bmatrix} 0 & 1 & 1 & 0 \\ 0 & 0 & 0 & -1 \\ 1 & 1 & 0 & 0 \\ 1 & 1 & 1 & 1 \\ 0 & 0 & 0 & 0 \end{bmatrix}, \quad \mathbf{A} = \begin{bmatrix} 0 & 2 & 0 & 1 \\ 0 & 0 & 0 & 0 \\ 1 & 1 & -1 & 0 \\ 1 & 1 & 0 & 1 \\ 0 & 0 & 1 & 0 \end{bmatrix}.$$

Since the matrices $\mathbf{E}$, $\mathbf{A}$ are non square, the matrix pencil $s\mathbf{E} - \mathbf{A}$ is singular and has the finite eigenvalues $\lambda_1 = 1$, $\lambda_2 = 2$, and the row minimal indices ζ_1=0, ζ_2=2. Then:

$$\mathbf{J}_p = \begin{bmatrix} 1 & 0 \\ 0 & 2 \end{bmatrix}, \quad \mathbf{E}_\zeta = \begin{bmatrix} \mathbf{I}_{\zeta_{h+1}} \\ \mathbf{0}_{1,\zeta_{h+1}} \end{bmatrix} = \begin{bmatrix} 1 & 0 \\ 0 & 1 \\ 0 & 0 \end{bmatrix},$$

and

$$\mathbf{A}_\zeta = \begin{bmatrix} \mathbf{0}_{1,\zeta_{h+1}} \\ \mathbf{I}_{\zeta_{h+1}} \end{bmatrix} = \begin{bmatrix} 0 & 0 \\ 1 & 0 \\ 0 & 1 \end{bmatrix}.$$

Hence

$$\mathbf{E}_K = \begin{bmatrix} 1 & 0 & 0 & 0 \\ 0 & 1 & 0 & 0 \\ 0 & 0 & 1 & 0 \\ 0 & 0 & 0 & 1 \\ 0 & 0 & 0 & 0 \end{bmatrix}, \quad \mathbf{A}_K = \begin{bmatrix} 1 & 0 & 0 & 0 \\ 0 & 2 & 0 & 0 \\ 0 & 0 & 0 & 0 \\ 0 & 0 & 1 & 0 \\ 0 & 0 & 0 & 1 \end{bmatrix}.$$

□

Example 2.18 (Canonical form of a singular pencil). Let

$$\mathbf{E} = \begin{bmatrix} 1 & 1 & 0 & 1 \\ 0 & 0 & 0 & 0 \\ 0 & 0 & 0 & 1 \end{bmatrix}, \quad \mathbf{A} = \begin{bmatrix} 2 & 2 & 1 & 0 \\ 0 & 1 & 0 & 0 \\ 0 & 0 & 1 & 0 \end{bmatrix}.$$

Since the matrices **E**, **A** are non square, the matrix pencil $s\mathbf{E} - \mathbf{A}$ is singular and has invariants the finite eigenvalue $\lambda_1 = 2$, an infinite eigenvalue of algebraic multiplicity one, and the column minimal indices ϵ_1=0, ϵ_2=1. Then $\mathrm{J}_p = 2$, $\mathrm{H}_q = 0$, and:

$$\mathbf{E}_\epsilon = \begin{bmatrix} \mathbf{I}_{\epsilon_{g+1}} & \mathbf{0}_{1,\epsilon_{g+1}} \end{bmatrix} = \begin{bmatrix} 1 & 0 \end{bmatrix},$$

and

$$\mathbf{A}_\epsilon = \begin{bmatrix} \mathbf{0}_{1,\epsilon_{g+1}} & \mathbf{I}_{\epsilon_{g+1}} \end{bmatrix} = \begin{bmatrix} 0 & 1 \end{bmatrix}.$$

Hence

$$\mathbf{E}_K = \begin{bmatrix} 1 & 0 & 0 & 0 \\ 0 & 0 & 0 & 0 \\ 0 & 0 & 1 & 0 \end{bmatrix}, \quad \mathbf{A}_K = \begin{bmatrix} 2 & 0 & 0 & 0 \\ 0 & 1 & 0 & 0 \\ 0 & 0 & 0 & 1 \end{bmatrix}.$$

□

Part II

Linear Eigenvalue Problems

3

Differential Equations with Regular Pencil

3.1 Formulation

In this chapter we consider the following system of differential equations:

$$\mathbf{E}\,\dot{\boldsymbol{x}}(t) = \mathbf{A}\,\boldsymbol{x}(t) + \boldsymbol{V}(t)\,, \tag{3.1}$$

where $\mathbf{E}, \mathbf{A} \in \mathbb{C}^{m\times m}$, $\boldsymbol{x} : [0, +\infty] \mapsto \mathbb{C}^{m\times 1}$, and $\boldsymbol{V} : [0, +\infty] \mapsto \mathbb{C}^{m\times 1}$. The matrices $\mathbf{E}$, $\mathbf{A}$ are square and $\mathbf{E}$ can be regular or singular ($\det(\mathbf{E}) = 0$).

Since the matrices $\mathbf{E}, \mathbf{A} \in \mathbb{C}^{m\times m}$ are square, for an arbitrary $s \in \mathbb{C}$, the matrix pencil $s\mathbf{E} - \mathbf{A}$ can be regular if $\det(s\mathbf{E} - \mathbf{A}) = \varpi(s) \not\equiv 0$; or singular if $\det(s\mathbf{E} - \mathbf{A}) \equiv 0$, where $\varpi(s)$ is a polynomial of s of degree $\deg\varpi(s) \leq m$. For these two type of pencils, we will firstly study the existence of solutions of system (3.1). We state the following Theorem:

Theorem 3.1 (Existence of the solution of differential equations). There exist solutions for (3.1) if and only if the pencil of the system is regular.

Proof. Let $\mathcal{L}\{\boldsymbol{x}(t)\} = \boldsymbol{z}(s)$, $\mathcal{L}\{\boldsymbol{V}(t)\} = \boldsymbol{U}(s)$ be the Laplace transform of $\boldsymbol{x}(t)$, $\boldsymbol{V}(t)$ respectively. By applying the Laplace transform $\mathcal{L}$ into (3.1), we get:

$$\mathbf{E}\mathcal{L}\{\dot{\boldsymbol{x}}(t)\} = \mathbf{A}\mathcal{L}\{\boldsymbol{x}(t)\} + \mathcal{L}\{\boldsymbol{V}(t)\}\,,$$

or, equivalently,

$$\mathbf{E}\big(s\boldsymbol{z}(s) - \boldsymbol{x}_o\big) = \mathbf{A}\,\boldsymbol{z}(s) + \boldsymbol{U}(s)\,,$$

where $\boldsymbol{x}_o = \boldsymbol{x}(0)$, i.e. the initial condition of (3.1). Since we assume that $\boldsymbol{x}_o$ is unknown we can use an unknown constant vector $\boldsymbol{C} \in \mathbb{C}^{m\times 1}$ and give to the above expression the following form:

$$(s\mathbf{E} - \mathbf{A})\,\boldsymbol{z}(s) = \mathbf{E}\,\boldsymbol{C} + \boldsymbol{U}(s)\,. \tag{3.2}$$

We have two cases. The first is (a) the pencil $s\mathbf{E} - \mathbf{A}$ to be regular with $\det(s\mathbf{E} - \mathbf{A}) = \varpi(s) \not\equiv 0$, i.e. the determinant of the pencil to be equal to a polynomial with order less than m. The second case is (b) the pencil $s\mathbf{E} - \mathbf{A}$ to be singular with $\det(s\mathbf{E} - \mathbf{A}) \equiv 0$, $\forall$ arbitrary $s \in \mathbb{C}$.

In the case of (a), since the pencil is assumed regular, we have that $\det(s\mathbf{E} - \mathbf{A}) \not\equiv 0$. Hence $\boldsymbol{z}(s)$ in (3.2) can be defined. Consequently $\boldsymbol{x}(t)$ always exists and is given by $\boldsymbol{x}(t) = \mathcal{L}^{-1}\{(s\mathbf{E} - \mathbf{A})^{-1}\mathbf{E}\boldsymbol{C} + (s\mathbf{E} - \mathbf{A})^{-1}\boldsymbol{U}(s)\}$. Hence in the case of a regular pencil, the solution of (3.1) always exists. In the case of (b), if the pencil is singular then $\text{rank}(s\mathbf{E} - \mathbf{A}) < m$, the matrix pencil $s\mathbf{E} - \mathbf{A}$ is not invertible, and there exists a matrix $\tilde{\mathbf{P}} : \mathbb{C} \mapsto \mathbb{C}^{m\times m}$ (which can be computed via the Gauss-Jordan elimination method, see [146]) such that:

$$\tilde{\mathbf{P}}(s)(s\mathbf{E} - \mathbf{A}) = \tilde{\mathbf{A}}(s) \oplus \mathbf{0}_{r_2,m_2}\,,$$

where $\tilde{\mathbf{A}} : \mathbb{C} \mapsto \mathbb{R}^{r_1\times m_1}$ with $r_1 \leq m_1$. All elements of $\tilde{\mathbf{A}}(s)$ are zero except the ones in the diagonal which are all non-zero. Also, $r_1+r_2 = m_1+m_2 = m$. Then system (3.1) can have solutions if and only if $r_2 = m_2 = 0$, i.e. $r_1 = m_1 = m$; In any other case in (3.2), we have more unknown functions than equations or no solutions. But since we are in the case where $r = m$ and the pencil is singular, i.e. $\det(s\mathbf{E} - \mathbf{A}) \equiv 0$, this assumption can never hold. The proof is completed. ∎

From Remark 2.1, given $\mathbf{E}, \mathbf{A} \in \mathbb{C}^{m\times m}$, and an arbitrary $s \in \mathbb{C}$, if the pencil $s\mathbf{E} - \mathbf{A}$ is regular, and since $\det(s\mathbf{E} - \mathbf{A}) \not\equiv 0$, there exists a matrix $\mathbf{P} : \mathbb{C} \mapsto \mathbb{C}^{m\times m}$ such that

$$\mathbf{P}(s)(s\mathbf{E} - \mathbf{A}) = \mathbf{A}(s)\,, \tag{3.3}$$

where $\mathbf{A} : \mathbb{C} \mapsto \mathbb{C}^{m\times m}$ is a diagonal matrix with non-zero elements in its main diagonal. Since we proved that there exist solutions for (3.1) if the pencil of the system is regular, in this chapter we are interested in the case of system (3.1) with a regular pencil. We state the following Theorem:

Theorem 3.2 (Solution of a regular pencil)**.** Consider system (3.1) with a regular pencil. Then the general solution is given by:

$$\boldsymbol{x}(t) = \boldsymbol{\Phi}_o(t)\boldsymbol{C} + \boldsymbol{\Phi}(t) \tag{3.4}$$

where $\boldsymbol{\Phi}_o(t) = \mathcal{L}^{-1}\{\mathbf{A}^{-1}(s)\mathbf{P}(s)\mathbf{E}\}$, $\boldsymbol{\Phi}(t) = \mathcal{L}^{-1}\{\mathbf{A}^{-1}(s)\mathbf{P}(s)\boldsymbol{U}(s)\}$, $\mathbf{A}(s)$, $\mathbf{P}(s)$ are defined in (3.3) and $\boldsymbol{C} \in \mathbb{R}^{m\times 1}$ is an unknown constant vector. If

$$\mathcal{L}^{-1}\{\mathbf{A}^{-1}(s)\mathbf{P}(s)\} = \boldsymbol{\Phi}_1(t)\,,$$

by using the convolution theorem an alternative formula is:

$$\boldsymbol{x}(t) = \boldsymbol{\Phi}_o(t)\boldsymbol{C} + \int_0^\infty \boldsymbol{\Phi}_1(t-\tau)\boldsymbol{V}(\tau)d\tau\,.$$

Proof. By applying the Laplace transform $\mathcal{L}$ into (3.1), we get

$$\mathbf{E}\mathcal{L}\{\dot{\boldsymbol{x}}(t)\} = \mathbf{A}\mathcal{L}\{\boldsymbol{x}(t)\} + \mathcal{L}\{\boldsymbol{V}(t)\}\,,$$

or, equivalently,

$$\mathbf{E}(s\boldsymbol{z}(s) - \boldsymbol{x}_o) = \mathbf{A}\boldsymbol{z}(s) + \boldsymbol{U}(s)\,,$$

where $\boldsymbol{x}_o = \boldsymbol{x}(0)$, i.e. the initial condition of (3.1). Since we assume that $\boldsymbol{x}_o$ is unknown we can use an unknown constant vector $\boldsymbol{C} \in \mathbb{C}^{m\times 1}$ and give to the above expression the form (3.2):

$$(s\mathbf{E} - \mathbf{A})\boldsymbol{z}(s) = \mathbf{E}\boldsymbol{C} + \boldsymbol{U}(s)\,.$$

Since the pencil is assumed regular and $\det(s\mathbf{E} - \mathbf{A}) \not\equiv 0$, there exists a matrix $\mathbf{P} : \mathbb{C} \mapsto \mathbb{C}^{m\times m}$ such that

$$\mathbf{P}(s)(s\mathbf{E} - \mathbf{A}) = \mathbf{\Lambda}(s)\,,$$

where $\mathbf{\Lambda} : \mathbb{C} \mapsto \mathbb{C}^{m\times m}$ is a diagonal matrix with non-zero elements in its diagonal. Then by multiplying (3.2) with $\mathbf{P}(s)$ we get

$$\mathbf{P}(s)(s\mathbf{E} - \mathbf{A})\boldsymbol{z}(s) = \mathbf{P}(s)\mathbf{E}\boldsymbol{C} + \mathbf{P}(s)\boldsymbol{U}(s),$$

or, equivalently,

$$\mathbf{\Lambda}(s)\boldsymbol{z}(s) = \mathbf{P}(s)\mathbf{E}\boldsymbol{C} + \mathbf{P}(s)\boldsymbol{U}(s),$$

or, equivalently,

$$\boldsymbol{z}(s) = \mathbf{\Lambda}^{-1}(s)\mathbf{P}(s)\mathbf{E}\boldsymbol{C} + \mathbf{\Lambda}^{-1}(s)\mathbf{P}(s)\boldsymbol{U}(s).$$

The inverse Laplace transform of the matrix $\mathbf{\Lambda}^{-1}(s)\mathbf{P}(s)\mathbf{E} = (s\mathbf{E} - \mathbf{A})^{-1}\mathbf{E}$ always exists because its elements are fractions of fractional polynomials with the order of the polynomial in the denominator always being higher than the order of the polynomial in the numerator. Let $\mathcal{L}^{-1}\{\mathbf{\Lambda}^{-1}(s)\mathbf{P}(s)\mathbf{E}\} = \mathbf{\Phi}_o(t)$ and $\mathcal{L}^{-1}\{\mathbf{\Lambda}^{-1}(s)\mathbf{P}(s)\boldsymbol{U}(s)\} = \mathbf{\Phi}(t)$. Then $\boldsymbol{x}(t)$ is given by (3.4). Or, alternatively, if $\mathcal{L}^{-1}\{\mathbf{\Lambda}^{-1}(s)\mathbf{P}(s)\} = \mathbf{\Phi}_1(t)$ then by using the convolution theorem we have

$$\boldsymbol{x}(t) = \mathbf{\Phi}_o(t)\boldsymbol{C} + \int_0^{\infty} \mathbf{\Phi}_1(t-\tau)\boldsymbol{V}(\tau)d\tau.$$

The proof is completed. ■

For system (3.1), in the case that the pencil of the system is regular, the general solution is given by (3.4). The constant column $\boldsymbol{C}$ in the general solution is an unknown constant vector related to the initial conditions of the system since we used the Laplace transform.

Remark 3.1. We have two types of initial conditions, the consistent ones which lead the system to a unique solution, and the non-consistent, that if given, they lead the system to infinite solutions.

Remark 3.2. If the pencil of system (3.1) is regular, and $\mathbf{E}$ is a regular matrix, then for given initial conditions we will have a unique solution. However, if $\mathbf{E}$ is a singular matrix if there exist solutions for system (3.1), it is not guaranteed that for given initial conditions we will have a unique solution. If the given initial conditions are consistent, and there exist solutions for (3.1), then in the formula of the general solution (3.4) we replace $\boldsymbol{C} = \boldsymbol{x}_o$. However, if the given initial conditions are non-consistent but there exist solutions for (3.1), then the general solution (3.4) holds for $t > 0$ and the system is impulsive. In the end of the next subsection we provide a criterion on how to identify if the given initial conditions are consistent or non-consistent.

Remark 3.3. For the case that the initial conditions are consistent, the matrix functions $\boldsymbol{\Phi}_o(t)$, $\boldsymbol{\Psi}_o(t)$ can have elements defined for $t > 0$. But the columns $\boldsymbol{\Phi}_o(t)\boldsymbol{x}_o$, $\boldsymbol{\Psi}_o(t)\boldsymbol{x}_o$ have all their elements always defined for $t \geq 0$.

Example 3.1 (Solution of a set of linear differential equations)**.** We consider system (3.1) with $\boldsymbol{V}(t) = \mathbf{0}_{2,1}$ and

$$\mathbf{E} = \begin{bmatrix} 1 & 1 \\ 0 & 0 \end{bmatrix}, \ \mathbf{A} = \begin{bmatrix} 1 & 1 \\ 0 & -1 \end{bmatrix}.$$

Then

$$s\mathbf{E} - \mathbf{A} = \begin{bmatrix} s-1 & s-1 \\ 0 & 1 \end{bmatrix},$$

with $\det(s\mathbf{E} - \mathbf{A}) \not\equiv 0$, and thus the pencil is regular. From Theorem 3.2 there exists the matrix

$$\mathbf{P}(s) = \begin{bmatrix} 1 & -(s-1) \\ 0 & 1 \end{bmatrix},$$

such that

$$\mathbf{P}(s)(s\mathbf{E} - \mathbf{A}) = \mathbf{A}(s),$$

where

$$\mathbf{A}(s) = \begin{bmatrix} s-1 & 0 \\ 0 & 1 \end{bmatrix}.$$

Hence from Theorem 3.2 there exist solutions for the system given by (3.4):

$$\boldsymbol{\Phi}_o(t) = \mathcal{L}^{-1}\{\mathbf{A}^{-1}(s)\mathbf{P}(s)\mathbf{E}\} = \mathcal{L}^{-1}\Big\{\begin{bmatrix} \frac{1}{s-1} & \frac{1}{s-1} \\ 0 & 0 \end{bmatrix}\Big\},$$

or, equivalently,

$$\boldsymbol{\Phi}_o(t) = \begin{bmatrix} \mathrm{e}^t & \mathrm{e}^t \\ 0 & 0 \end{bmatrix}.$$

Then we have the general solution:

$$\boldsymbol{x}(t) = \begin{bmatrix} \mathrm{e}^t & \mathrm{e}^t \\ 0 & 0 \end{bmatrix}\boldsymbol{C},$$

or, equivalently,

$$\boldsymbol{x}(t) = \begin{bmatrix} \mathrm{e}^t c \\ 0 \end{bmatrix}, \quad c \in \mathbb{R}.$$

□

Example 3.2 (Solution of a set of non-linear differential equations). Consider the following system of implicit non-linear differential equations:

$$\boldsymbol{f}(t, \boldsymbol{x}, \dot{\boldsymbol{x}}) = \mathbf{0}_{m,1},$$

with known initial conditions

$$\boldsymbol{x}(0) = \boldsymbol{x}_o, \quad \dot{\boldsymbol{x}}(0) = \dot{\boldsymbol{x}}_o,$$

where $\boldsymbol{f} : [0, +\infty] \times \mathbb{R}^{m\times 1} \times \mathbb{R}^{m\times 1} \mapsto \mathbb{R}^{m\times 1}$ has continuous partial derivatives of second order. Let $\boldsymbol{f}_o = \boldsymbol{f}(0, \boldsymbol{x}(0), \dot{\boldsymbol{x}}(0)) = \boldsymbol{f}(0, \boldsymbol{x}_o, \dot{\boldsymbol{x}}_o)$. We may use Theorem 3.2 to solve approximately, for small values of t, this non-linear system. If we linearize locally the function $\boldsymbol{f}$ at $(t, \boldsymbol{x}(t), \dot{\boldsymbol{x}}(t)) = (0, \boldsymbol{x}_o, \dot{\boldsymbol{x}}_o)$ we get:

$$\boldsymbol{f}(0, \boldsymbol{x}_o, \dot{\boldsymbol{x}}_o) + \frac{\partial \boldsymbol{f}}{\partial t} t + \frac{\partial \boldsymbol{f}}{\partial \boldsymbol{x}}(\boldsymbol{x} - \boldsymbol{x}_o) + \frac{\partial \boldsymbol{f}}{\partial \dot{\boldsymbol{x}}}(\dot{\boldsymbol{x}} - \dot{\boldsymbol{x}}_o) = \mathbf{0}_{m,1},$$

or, equivalently,

$$-\frac{\partial \boldsymbol{f}}{\partial \dot{\boldsymbol{x}}}\dot{\boldsymbol{x}} = \frac{\partial \boldsymbol{f}}{\partial \boldsymbol{x}}\boldsymbol{x} + \left[\boldsymbol{f}_o + \frac{\partial \boldsymbol{f}}{\partial t} t - \frac{\partial \boldsymbol{f}}{\partial \boldsymbol{x}} \boldsymbol{x}_o - \frac{\partial \boldsymbol{f}}{\partial \dot{\boldsymbol{x}}} \dot{\boldsymbol{x}}_o\right],$$

where the Jacobian matrices $\frac{\partial \boldsymbol{f}}{\partial t}$, $\frac{\partial \boldsymbol{f}}{\partial \boldsymbol{x}}$ and $\frac{\partial \boldsymbol{f}}{\partial \dot{\boldsymbol{x}}}$ are calculated at the initial condition $(0, \boldsymbol{x}_o, \dot{\boldsymbol{x}}_o)$. Note also that $\frac{\partial \boldsymbol{f}}{\partial \dot{\boldsymbol{x}}}, \frac{\partial \boldsymbol{f}}{\partial \boldsymbol{x}} \in \mathbb{C}^{m\times m}$.

We adopt the notation

$$\mathbf{E} = -\frac{\partial \boldsymbol{f}}{\partial \dot{\boldsymbol{x}}}, \qquad \mathbf{A} = \frac{\partial \boldsymbol{f}}{\partial \boldsymbol{x}}, \qquad \boldsymbol{V}(t) = \boldsymbol{f}_o + \frac{\partial \boldsymbol{f}}{\partial t} t - \frac{\partial \boldsymbol{f}}{\partial \boldsymbol{x}}\boldsymbol{x}_o - \frac{\partial \boldsymbol{f}}{\partial \dot{\boldsymbol{x}}}\dot{\boldsymbol{x}}_o,$$

where $\mathbf{E}, \mathbf{A} \in \mathbb{C}^{m\times m}$ and $\boldsymbol{V} : [0, +\infty] \mapsto \mathbb{R}^{m\times 1}$. Then, the above expression is system (3.1). Hence, if the pencil is regular, i.e. $\det(s\mathbf{E} - \mathbf{A}) \not\equiv 0$, the solution of (3.1) with initial conditions $\boldsymbol{x}_o$, $\dot{\boldsymbol{x}}_o$, is an approximate solution of the non-linear system for small values of t.

Let $\boldsymbol{x} : [0, +\infty] \mapsto \mathbb{R}^{3\times 1}$ with $\boldsymbol{x} = \begin{bmatrix} x_1 & x_2 & x_3 \end{bmatrix}^{\mathrm{T}}$, and assume the following non-linear system:

$$\begin{aligned} \dot{x}_1 + \mathrm{e}^{x_1} - x_2 &= 3t, \\ \dot{x}_2\,\dot{x}_3 + \mathrm{e}^{x_2} &= 2\,\mathrm{e}^t, \\ \dot{x}_2 + \dot{x}_3 + x_1\,x_3 - 2\,\mathrm{e}^{x_2} &= 5t, \end{aligned}$$

with initial conditions

$$\boldsymbol{x}(0) = \begin{bmatrix} 0 \\ 0 \\ 0 \end{bmatrix}, \quad \dot{\boldsymbol{x}}(0) = \begin{bmatrix} -1 \\ 1 \\ 1 \end{bmatrix}.$$

We compute the following values:

$$\boldsymbol{f}_o = \begin{bmatrix} 0 \\ 0 \\ 0 \end{bmatrix}, \quad \frac{\partial \boldsymbol{f}}{\partial t} = \begin{bmatrix} -3 \\ -2 \\ -5 \end{bmatrix},$$

and

$$\frac{\partial \boldsymbol{f}}{\partial \boldsymbol{x}} = \begin{bmatrix} 1 & 0 & 0 \\ 0 & 1 & 0 \\ 0 & -2 & 0 \end{bmatrix}, \quad \frac{\partial \boldsymbol{f}}{\partial \dot{\boldsymbol{x}}} = \begin{bmatrix} 1 & 0 & 0 \\ 0 & 1 & 1 \\ 0 & 1 & 1 \end{bmatrix}.$$

We consider the matrices:

$$\mathbf{E} = -\begin{bmatrix} 1 & 0 & 0 \\ 0 & 1 & 1 \\ 0 & 1 & 1 \end{bmatrix}, \quad \mathbf{A} = \begin{bmatrix} 1 & 0 & 0 \\ 0 & 1 & 0 \\ 0 & -2 & 0 \end{bmatrix}, \quad \boldsymbol{V}(t) = \begin{bmatrix} 3t-1 \\ 2t+2 \\ 5t+2 \end{bmatrix}.$$

Since

$$s\mathbf{E} - \mathbf{A} = -\begin{bmatrix} s+1 & 0 & 0 \\ 0 & s+1 & s \\ 0 & s-2 & s \end{bmatrix},$$

by using Theorem 3.2 there exists the matrix

$$\mathbf{P}(s) = -\frac{1}{3}\begin{bmatrix} 3 & 0 & 0 \\ 0 & s+1 & -s-1 \\ 0 & -s+2 & s+1 \end{bmatrix},$$

such that

$$\mathbf{P}(s)(s\mathbf{E} - \mathbf{A}) = \mathbf{A}(s),$$

where

$$\mathbf{A}(s) = \begin{bmatrix} s+1 & 0 & 0 \\ 0 & s+1 & 0 \\ 0 & 0 & s \end{bmatrix}.$$

Furthermore,

$$\boldsymbol{\Phi}_o(t) = \mathcal{L}^{-1}\{\mathbf{A}^{-1}(s)\mathbf{P}(s)\mathbf{E}\} = \mathcal{L}^{-1}\{\begin{bmatrix} \frac{1}{s+1} & 0 & 0 \\ 0 & 0 & 0 \\ 0 & \frac{1}{s} & \frac{1}{s} \end{bmatrix}\},$$

or, equivalently,

$$\boldsymbol{\Phi}_o(t) = \begin{bmatrix} \mathrm{e}^{-t} & 0 & 0 \\ 0 & 0 & 0 \\ 0 & 1 & 1 \end{bmatrix}.$$

In addition,

$$\boldsymbol{\Phi}(t) = \mathcal{L}^{-1}\{\mathbf{A}^{-1}(s)\mathbf{P}(s)\boldsymbol{U}(s)\},$$

where

$$\boldsymbol{U}(s) = \mathcal{L}\{\boldsymbol{V}(t)\} = \mathcal{L}\{\begin{bmatrix} 3t-1 \\ 2t+2 \\ 5t+2 \end{bmatrix}\} = \begin{bmatrix} \frac{3}{s^2} - \frac{1}{s} \\ \frac{2}{s^2} + \frac{2}{s} \\ \frac{5}{s^2} + \frac{2}{s} \end{bmatrix}.$$

Hence

$$\boldsymbol{\Phi}(t) = \mathcal{L}^{-1}\{\begin{bmatrix} \frac{1}{s^2} \\ -\frac{1}{s^2} \\ \frac{5}{s^3} + \frac{2}{s^2} \end{bmatrix}\}.$$

Equivalently,

$$\mathbf{\Phi}(t) = \begin{bmatrix} t \\ -t \\ \frac{5}{2}t^2 + 2t \end{bmatrix}.$$

Then an approximate solution of the non-linear system for small values of t is given by

$$\boldsymbol{x} = \begin{bmatrix} x_1 \\ x_2 \\ x_3 \end{bmatrix} \cong \begin{bmatrix} t \\ -t \\ \frac{5}{2}t^2 + 2t \end{bmatrix}.$$

□

3.2 Solutions and Eigenvalue Analysis

As discussed in Chapter 2, a regular pencil has only finite eigenvalues if $\mathbf{E}$ is regular, and finite/infinite eigenvalues if $\mathbf{E}$ is singular. In this section, we study the solutions of (3.1) by using matrix pencil theory.

From Theorem 3.1 there exist solutions for system (3.1) if the pencil is regular. By focusing on this case, and having in mind to simplify things, we identify two cases:

Case I: The pencil $s\mathbf{E} - \mathbf{A}$ is regular, and $\mathbf{E}$ is a regular matrix. In this case, see [51], there exist non-singular matrices $\mathbf{P}$, $\mathbf{Q} \in \mathbb{C}^{m\times m}$ such that:

$$\begin{aligned} \mathbf{PEQ} &= \mathbf{I}_m \,, \\ \mathbf{PAQ} &= \mathbf{J}_m \,, \end{aligned} \tag{3.5}$$

where $\mathbf{J}_m \in \mathbb{C}^{m\times m}$ is a Jordan matrix constructed by the finite eigenvalues of the pencil and their algebraic multiplicity. $\mathbf{P}$ is a matrix with rows m linear independent (including the generalized) left eigenvectors of the finite eigenvalues of $s\mathbf{E} - \mathbf{A}$; $\mathbf{Q}$ is a matrix with columns m linear independent (including the generalized) right eigenvectors of the finite eigenvalues of $s\mathbf{E} - \mathbf{A}$. We state the following Proposition:

Proposition 3.1 (Solution of a regular pencil)**.** We consider system (3.1) with a regular pencil, and $\mathbf{E}$ regular. Let $\mathbf{P}$ be a matrix with rows m linear independent left eigenvectors of the finite eigenvalues of $s\mathbf{E} - \mathbf{A}$, and $\mathbf{Q}$ be a matrix with columns m linear independent right eigenvectors of the finite eigenvalues of $s\mathbf{E} - \mathbf{A}$. In addition, let $\mathbf{J}_m$ be the Jordan matrix of the finite eigenvalues, as defined in (3.5). Then there exists a solution for (3.1) and it is given by:

$$\boldsymbol{x}(t) = \mathbf{Q}\, e^{\mathbf{J}_m t}\boldsymbol{C} + \int_0^\infty \mathbf{Q}\, e^{\mathbf{J}_m (t-\tau)}\mathbf{P}\boldsymbol{V}(\tau)d\tau \,. \tag{3.6}$$

Proof. If we consider the transformation $\boldsymbol{x}(t) = \mathbf{Q}\boldsymbol{z}(t)$ and substitute it into (3.1) we obtain

$$\mathbf{E}\,\dot{\boldsymbol{x}}(t)\mathbf{Q}\boldsymbol{z}(t) = \mathbf{A}\mathbf{Q}\boldsymbol{z}(t) + \boldsymbol{V}(t)\,,$$

whereby, multiplying by $\mathbf{P}$ and using (3.5) we get

$$\dot{\boldsymbol{z}}(t) = \mathbf{J}_m\boldsymbol{z}(t) + \mathbf{P}\boldsymbol{V}(t)\,.$$

By applying the Laplace transform $\mathcal{L}$ in the above equation we obtain:

$$\mathcal{L}\{\dot{\boldsymbol{z}}(t)\} = \mathbf{J}_m\mathcal{L}\{\boldsymbol{z}(t)\} + \mathbf{P}\mathcal{L}\{\boldsymbol{V}(t)\}\,,$$

or, equivalently,

$$s\boldsymbol{W}(s) - \boldsymbol{z}_0 = \mathbf{J}_m\boldsymbol{W}(s) + \mathbf{P}\boldsymbol{U}(s)\,,$$

where $\mathcal{L}\{\boldsymbol{z}(t)\} = \boldsymbol{W}(s)$, $\mathcal{L}\{\boldsymbol{V}(t)\} = \boldsymbol{U}(s)$ and $\boldsymbol{z}_0 = \boldsymbol{z}(0)$, i.e. the initial condition of the linear system before applying the Laplace transform. Since we assume that $\boldsymbol{z}_0$ is unknown, we set $\boldsymbol{z}_0 = \boldsymbol{C}$, where $\boldsymbol{C}$ unknown column, and give to the above expression the following form:

$$(s\mathbf{I}_m - \mathbf{J}_m)\boldsymbol{W}(s) = \boldsymbol{C} + \mathbf{P}\boldsymbol{U}(s)\,,$$

or, equivalently,

$$\boldsymbol{W}(s) = (s\mathbf{I}_m - \mathbf{J}_m)^{-1}\boldsymbol{C} + (s\mathbf{I}_m - \mathbf{J}_m)^{-1}\mathbf{P}\boldsymbol{U}(s)\,,$$

or, equivalently,

$$\boldsymbol{z}(t) = \mathcal{L}^{-1}\{(s\mathbf{I}_m - \mathbf{J}_m)^{-1}\boldsymbol{C}\} + \mathcal{L}^{-1}\{(s\mathbf{I}_m - \mathbf{J}_m)^{-1}\mathbf{P}\boldsymbol{U}(s)\}\,,$$

or, equivalently,

$$\boldsymbol{z}(t) = \mathrm{e}^{\mathbf{J}_m t}\boldsymbol{C} + \int_0^\infty \mathrm{e}^{\mathbf{J}_m(t-\tau)}\mathbf{P}\boldsymbol{V}(\tau)d\tau\,,$$

and consequently,

$$\boldsymbol{x}(t) = \mathbf{Q}\boldsymbol{z}(t) = \mathbf{Q}\,\mathrm{e}^{\mathbf{J}_m t}\boldsymbol{C} + \int_0^\infty \mathbf{Q}\,\mathrm{e}^{\mathbf{J}_m(t-\tau)}\mathbf{P}\boldsymbol{V}(\tau)d\tau\,.$$

The proof is completed. ■

We can now present the following Corollary:

Corollary 3.1 (Uniqueness of the solution of a regular pencil). If system (3.1) has a regular pencil, and $\mathbf{E}$ is regular, then for given initial conditions $\boldsymbol{x}(0) = \boldsymbol{x}_o$ the solution is always unique.

The general solution is then given by (3.6) and $\boldsymbol{C}$ is the unique solution of the linear system

$$\mathbf{Q}\boldsymbol{C} = \boldsymbol{x}_o\,.$$

Proof. This is a direct result from Proposition 3.1. If we use the formula (3.6) for $t = 0$ we get:

$$\boldsymbol{x}(0) = \mathbf{Q}\boldsymbol{C}\,.$$

Since the columns of the matrix $\mathbf{Q}$ are linear independent, the above system in respect of $\boldsymbol{C}$ has always a unique solution. The proof is completed. ■

Case II: The pencil $s\mathbf{E} - \mathbf{A}$ is regular, and $\mathbf{E}$ is a singular matrix. In this case the pencil has finite eigenvalues and an infinite eigenvalue. Let λ_i be a finite eigenvalue of algebraic multiplicity p_i, $i = 1, 2, \ldots, \nu$, and $\sum_{i=0}^{\nu} p_i = p$; Also let q be the algebraic multiplicity of the infinite eigenvalue. Obviously $p + q = m$.

From the regularity of the pencil, see [35, 51], there exist non-singular matrices $\mathbf{P}$, $\mathbf{Q} \in \mathbb{C}^{m\times m}$ such that:

$$\begin{aligned} \mathbf{PEQ} &= \mathbf{I}_p \oplus \mathbf{H}_q\,, \\ \mathbf{PAQ} &= \mathbf{J}_p \oplus \mathbf{I}_q\,, \end{aligned} \tag{3.7}$$

where

$$\mathbf{P} = \begin{bmatrix} \mathbf{P}_1 \\ \mathbf{P}_2 \end{bmatrix}, \qquad \mathbf{Q} = \begin{bmatrix} \mathbf{Q}_p & \mathbf{Q}_q \end{bmatrix},$$

with $\mathbf{P}_1 \in \mathbb{C}^{p\times m}$, $\mathbf{P}_2 \in \mathbb{C}^{q\times m}$ and $\mathbf{Q}_p \in \mathbb{C}^{m\times p}$, $\mathbf{Q}_q \in \mathbb{C}^{m\times q}$. Furthermore, $\mathbf{H}_q \in \mathbb{C}^{q\times q}$ is a nilpotent matrix with index q_*, constructed by using the algebraic multiplicity of the infinite eigenvalue, and $\mathbf{J}_p \in \mathbb{C}^{p\times p}$ is a Jordan matrix constructed by the finite eigenvalues of the pencil and their algebraic multiplicity. $\mathbf{P}_1$ is a matrix with rows the p linear independent left eigenvectors of the finite eigenvalues of $s\mathbf{E} - \mathbf{A}$; $\mathbf{P}_2$ is a matrix with columns the q linear independent left eigenvectors of the infinite eigenvalue of $s\mathbf{E} - \mathbf{A}$; $\mathbf{Q}_p$ is a matrix with columns the p linear independent right eigenvectors of the finite eigenvalues of $s\mathbf{E} - \mathbf{A}$; and $\mathbf{Q}_q$ is a matrix with columns the q linear independent right eigenvectors of the infinite eigenvalue of $s\mathbf{E} - \mathbf{A}$. By applying the above expressions into (3.7), we get the following eight equalities:

$$\begin{aligned} \mathbf{P}_1\mathbf{AQ}_p &= \mathbf{J}_p\,, & \mathbf{P}_1\mathbf{EQ}_p &= \mathbf{I}_p \\ \mathbf{P}_1\mathbf{AQ}_q &= \mathbf{0}_{p,q}\,, & \mathbf{P}_1\mathbf{EQ}_q &= \mathbf{0}_{p,q} \\ \mathbf{P}_2\mathbf{AQ}_p &= \mathbf{0}_{q,p}\,, & \mathbf{P}_2\mathbf{EQ}_p &= \mathbf{0}_{q,p} \\ \mathbf{P}_2\mathbf{AQ}_q &= \mathbf{I}_q\,, & \mathbf{P}_2\mathbf{EQ}_q &= \mathbf{H}_q\,. \end{aligned}$$

Proposition 3.2 (Solution of a regular pencil)**.** We consider system (3.1) with a regular pencil, and $\mathbf{E}$ regular. Let $\mathbf{J}_m$ be the Jordan matrix of the finite eigenvalues of $s\mathbf{E} - \mathbf{A}$, and $\mathbf{Q}$ the matrix that contains all linear independent eigenvectors as defined in (3.7). Then there always exists a solution and it is given by:

$$\boldsymbol{x}(t) = \mathbf{Q}_p\left[\mathrm{e}^{\mathbf{J}_p t}\boldsymbol{C} + \int_0^\infty \mathrm{e}^{\mathbf{J}_p(t-\tau)}\mathbf{P}_1\boldsymbol{V}(\tau)d\tau\right] - \mathbf{Q}_q\sum_{i=0}^{q_*-1}\mathbf{H}_q^i\mathbf{P}_2\boldsymbol{V}^{[i]}(t)\,, \tag{3.8}$$

where $\boldsymbol{C} \in \mathbb{C}^{m \times m}$ is constant vector and $^{[i]}$ indicates the i-th derivative with respect to time.

Proof. If we consider the transformation $\boldsymbol{x}(t) = \mathbf{Q}\boldsymbol{z}(t)$ and substitute it into (3.1) we obtain

$$\mathbf{E}\,\dot{\boldsymbol{x}}(t)\mathbf{Q}\,\boldsymbol{z}(t) = \mathbf{AQ}\,\boldsymbol{z}(t) + \boldsymbol{V}(t),$$

whereby, multiplying by $\mathbf{P}$ and using (3.5), we get

$$\begin{bmatrix} \mathbf{I}_p & \mathbf{0}_{p,q} \\ \mathbf{0}_{q,p} & \mathbf{H}_q \end{bmatrix} \begin{bmatrix} \dot{\boldsymbol{z}}_p(t) \\ \dot{\boldsymbol{z}}_q(t) \end{bmatrix} = \begin{bmatrix} \mathbf{J}_p & \mathbf{0}_{p,q} \\ \mathbf{0}_{q,p} & \mathbf{I}_q \end{bmatrix} \begin{bmatrix} \boldsymbol{z}_p(t) \\ \boldsymbol{z}_q(t) \end{bmatrix} + \begin{bmatrix} \mathbf{P}_1 \\ \mathbf{P}_2 \end{bmatrix} \boldsymbol{V}(t),$$

where

$$\boldsymbol{z}(t) = \begin{bmatrix} \boldsymbol{z}_p(t) \\ \boldsymbol{z}_q(t) \end{bmatrix},$$

with $\boldsymbol{z}_p(t) \in \mathbb{C}^{p\times 1}$, $\boldsymbol{z}_q(t) \in \mathbb{C}^{q\times 1}$. From the above expressions we arrive easily at the subsystems:

$$\dot{\boldsymbol{z}}_p(t) = \mathbf{J}_p \boldsymbol{z}_p(t) + \mathbf{P}_1 \boldsymbol{V}(t),$$

and

$$\mathbf{H}_q \dot{\boldsymbol{z}}_q(t) = \boldsymbol{z}_q(t) + \mathbf{P}_2 \boldsymbol{V}(t).$$

The first subsystem is of similar type to the one studied in the first case. Its general solution is equal to:

$$\boldsymbol{z}_p(t) = \mathrm{e}^{\mathbf{J}_p t}\boldsymbol{C} + \int_0^\infty \mathrm{e}^{\mathbf{J}_p(t-\tau)}\mathbf{P}_1\boldsymbol{V}(\tau)d\tau.$$

Let q_* be the index of the nilpotent matrix $\mathbf{H}_q$, i.e. $\mathbf{H}_q^{q_*} = \mathbf{0}_{q,q}$. Then if we obtain the following matrix equations

$$\begin{aligned}
\mathbf{H}_q \dot{\boldsymbol{z}}_q(t) &= \boldsymbol{z}_q(t) + \mathbf{P}_2 \boldsymbol{V}(t), \\
\mathbf{H}_q^2 \boldsymbol{z}_q^{[2]}(t) &= \mathbf{H}_q \dot{\boldsymbol{z}}_q(t) + \mathbf{H}_q \mathbf{P}_2 \dot{\boldsymbol{V}}(t), \\
\mathbf{H}_q^3 \boldsymbol{z}_q^{[3]}(t) &= \mathbf{H}_q^2 \boldsymbol{z}_q^{[2]}(t) + \mathbf{H}_q^2 \mathbf{P}_2 \boldsymbol{V}^{[2]}(t), \\
\mathbf{H}_q^4 \boldsymbol{z}_q^{[4]}(t) &= \mathbf{H}_q^3 \boldsymbol{z}_q^{[3]}(t) + \mathbf{H}_q^3 \mathbf{P}_2 \boldsymbol{V}^{[3]}(t), \\
&\vdots \\
\mathbf{H}_q^{q_*-1} \boldsymbol{z}_q^{[q_*-1]}(t) &= \mathbf{H}_q^{q_*-2} \boldsymbol{z}_q^{[q_*-2]}(t) + \mathbf{H}_q^{q_*-2} \mathbf{P}_2 \boldsymbol{V}^{[q_*-2]}(t), \\
\mathbf{H}_q^{q_*} \boldsymbol{z}_q^{[q_*]}(t) &= \mathbf{H}_q^{q_*-1} \boldsymbol{z}_q^{[q_*-1]}(t) + \mathbf{H}_q^{q_*-1} \mathbf{P}_2 \boldsymbol{V}^{[q_*-1]}(t),
\end{aligned}$$

by taking the sum of the above equations and using the fact that $\mathbf{H}_q^{q_*} = \mathbf{0}_{q,q}$, we arrive easily at the solution of the second subsystem which has the unique solution

$$\boldsymbol{z}_q(t) = -\sum_{i=0}^{q_*-1} \mathbf{H}_q^i \mathbf{P}_2 \boldsymbol{V}^{[i]}(t).$$

To conclude, by combining the solutions of the two subsystems, for the case of a regular pencil with $\mathbf{E}$ being singular, system (3.1) has the general solution

$$\boldsymbol{x}(t) = \mathbf{Q}\boldsymbol{z}(t) = \begin{bmatrix} \mathbf{Q}_p & \mathbf{Q}_q \end{bmatrix} \begin{bmatrix} e^{\mathbf{J}_p t}\boldsymbol{C} + \int_0^\infty e^{\mathbf{J}_p(t-\tau)}\mathbf{P}_1\boldsymbol{V}(\tau)d\tau \\ -\sum_{i=0}^{q_*-1}\mathbf{H}_q^i\mathbf{P}_2\boldsymbol{V}^{[i]}(t) \end{bmatrix},$$

or, equivalently,

$$\boldsymbol{x}(t) = \mathbf{Q}_p\left[e^{\mathbf{J}_p t}\boldsymbol{C} + \int_0^\infty e^{\mathbf{J}_p(t-\tau)}\mathbf{P}_1\boldsymbol{V}(\tau)d\tau\right] - \mathbf{Q}_q\sum_{i=0}^{q_*-1}\mathbf{H}_q^i\mathbf{P}_2\boldsymbol{V}^{[i]}(t).$$

The proof is completed. ■

We can now present the following Corollary:

Corollary 3.2 (Uniqueness of the solution of a regular pencil)**.** If system (3.1) has a regular pencil with $\mathbf{E}$ singular, then for given initial conditions $\boldsymbol{x}(0) = \boldsymbol{x}_o$ the solution is unique if and only if:

$$\boldsymbol{x}_o \in \text{colspan}\mathbf{Q}_p - \mathbf{Q}_q\sum_{i=0}^{q_*-1}\mathbf{H}_q^i\mathbf{P}_2\boldsymbol{V}^{[i]}(0).$$

In case that the above relation holds, the unique solution is given by (3.8) and $\boldsymbol{C}$ is the unique solution of the linear system

$$\mathbf{Q}_p\boldsymbol{C} = \left[\boldsymbol{x}_o + \mathbf{Q}_q\sum_{i=0}^{q_*-1}\mathbf{H}_q^i\mathbf{P}_2\boldsymbol{V}^{[i]}(0)\right].$$

Proof. In the transformation $\boldsymbol{x}(t) = \mathbf{Q}\boldsymbol{z}(t)$ used in the proof of Proposition 3.2 we use the initial condition $\boldsymbol{x}(0) = \boldsymbol{x}_o$:

$$\boldsymbol{x}_o = \mathbf{Q}\boldsymbol{z}_o = \mathbf{Q}\begin{bmatrix} \boldsymbol{z}_o^p \\ \boldsymbol{z}_o^q \end{bmatrix}.$$

We observe from the solutions of the two subsystems that appear in the proof of Proposition 3.2 after applying the transformation $\boldsymbol{x}(t) = \mathbf{Q}\boldsymbol{z}(t)$ into (3.1), that although $\boldsymbol{z}_p(0)$ can be chosen arbitrary, $\boldsymbol{z}_q(0)$ must be satisfying the relation

$$\boldsymbol{z}_q(0) = -\sum_{i=0}^{q_*-1}\mathbf{H}_q^i\mathbf{P}_2\boldsymbol{V}^{[i]}(0).$$

Hence

$$\boldsymbol{x}_o = \begin{bmatrix} \mathbf{Q}_p & \mathbf{Q}_q \end{bmatrix}\begin{bmatrix} \boldsymbol{z}_p(0) \\ -\sum_{i=0}^{q_*-1}\mathbf{H}_q^i\mathbf{P}_2\boldsymbol{V}^{[i]}(0) \end{bmatrix},$$

or, equivalently,

$$\boldsymbol{x}_o = \mathbf{Q}_p\boldsymbol{z}_p(0) - \mathbf{Q}_q\sum_{i=0}^{q_*-1}\mathbf{H}_q^i\mathbf{P}_2\boldsymbol{V}^{[i]}(0),$$

or, equivalently,

$$\boldsymbol{x}_o \in \text{colspan}\mathbf{Q}_p - \mathbf{Q}_q \sum_{i=0}^{q_*-1} \mathbf{H}_q^i \mathbf{P}_2 \boldsymbol{V}^{[i]}(0) .$$

If the above relation holds, we may use the formula (3.8) for $t = 0$ and get:

$$\boldsymbol{x}(0) = \mathbf{Q}_p \boldsymbol{C} - \mathbf{Q}_q \sum_{i=0}^{q_*-1} \mathbf{H}_q^i \mathbf{P}_2 \boldsymbol{V}^{[i]}(0) ,$$

or, equivalently,

$$\mathbf{Q}_p \boldsymbol{C} = [\boldsymbol{x}_o + \mathbf{Q}_q \sum_{i=0}^{q_*-1} \mathbf{H}_q^i \mathbf{P}_2 \boldsymbol{V}^{[i]}(0)] .$$

The above system of linear equations in respect to $\boldsymbol{C}$ has always a unique solution since the matrix $[\boldsymbol{x}_o + \mathbf{Q}_q \sum_{i=0}^{q_*-1} \mathbf{H}_q^i \mathbf{P}_2 \boldsymbol{V}^{[i]}(0)] \in \text{colspan}\mathbf{Q}_p$. The proof is completed. ■

Remark 3.4. In Corollary 3.2 if $\boldsymbol{x}_o \in \text{colspan}\mathbf{Q}_p - \mathbf{Q}_q \sum_{i=0}^{q_*-1} \mathbf{H}_q^i \mathbf{P}_2 \boldsymbol{V}^{[i]}(0)$ holds then the initial conditions (IC) $\boldsymbol{x}(0) = \boldsymbol{x}_o$ will be called consistent IC. If it does not hold, then the IC will be called non-consistent IC because in this case it would be not possible for $\boldsymbol{C}$ to be identified uniquely in the system $\mathbf{Q}_p \boldsymbol{C} = [\boldsymbol{x}_o + \mathbf{Q}_q \sum_{i=0}^{q_*-1} \mathbf{H}_q^i \mathbf{P}_2 \boldsymbol{V}^{[i]}(0)]$.

Remark 3.5. To sum up, here are the steps we have to undertake when we need to provide the solutions of (3.1):

- Step 1. Identify that the pencil is regular, i.e. $\det(s\mathbf{E} - \mathbf{A}) = \varpi(s) \neq 0$, and if $\mathbf{E}$ is a regular or singular matrix.
- Step 2. A regular pencil has only finite eigenvalues if $\mathbf{E}$ is regular, and finite/infinite eigenvalues if $\mathbf{E}$ is singular. Compute the finite eigenvalues which are the roots of $\varpi(s)$.
- Step 3. Form the Jordan matrices $\mathbf{J}_p$, $\mathbf{H}_q$ in the case that $\mathbf{E}$ is singular, or $\mathbf{J}_m$ in the case that $\mathbf{E}$ is regular.
- Step 4. Compute the left, and right eigenvectors of the eigenvalues of $s\mathbf{E} - \mathbf{A}$ in order to form the matrices $\mathbf{P}$, $\mathbf{Q}$, respectively.
- Step 5. If $\mathbf{E}$ is regular, use the formula (3.6) to provide the solution of (3.1), or the formula (3.8) if $\mathbf{E}$ is singular.
- Step 6. For given initial conditions, if $\mathbf{E}$ is regular, use Corollary 3.1 to provide the unique solution of the system. If $\mathbf{E}$ is singular, use Corollary 3.2 to identify if the initial conditions are consistent or non-consistent. In the case that they are consistent provide the unique solution by using the formula in Corollary 3.2.

Example 3.3 (Regular pencil with multiple solutions). We consider system (3.1) with $\boldsymbol{V}(t) = \mathbf{0}_{2,1}$ and

$$\mathbf{E} = \begin{bmatrix} 1 & 1 \\ 0 & 0 \end{bmatrix}, \quad \mathbf{A} = \begin{bmatrix} 1 & 1 \\ 0 & -1 \end{bmatrix}.$$

Then

$$s\mathbf{E} - \mathbf{A} = \begin{bmatrix} s-1 & s-1 \\ 0 & 1 \end{bmatrix},$$

with $\det(s\mathbf{E}-\mathbf{A}) = s-1 \not\equiv 0$, and thus the pencil is regular while $\mathbf{E}$ is singular. Hence the pencil has a finite eigenvalue $\lambda_1 = 1$, and an infinite eigenvalue. We form $\mathbf{Q}_p$ from the corresponding eigenvector to the finite eigenvalue:

$$\mathbf{Q}_p = \begin{bmatrix} 1 \\ 0 \end{bmatrix}.$$

Hence from Proposition 3.2, there exist solutions for the system given by (3.8):

$$\boldsymbol{x}(t) = \begin{bmatrix} 1 \\ 0 \end{bmatrix} \mathrm{e}^t \boldsymbol{C},$$

or, equivalently,

$$\boldsymbol{x}(t) = \begin{bmatrix} \mathrm{e}^t c \\ 0 \end{bmatrix}, \quad c \in \mathbb{R}.$$

We consider now system (3.1) with $\boldsymbol{V}(t) = 0_{3,1}$, and:

$$\mathbf{E} = -\begin{bmatrix} 1 & 0 & 0 \\ 0 & 1 & 1 \\ 0 & 1 & 1 \end{bmatrix}, \quad \mathbf{A} = \begin{bmatrix} 1 & 0 & 0 \\ 0 & 1 & 0 \\ 0 & -2 & 0 \end{bmatrix}, \quad \boldsymbol{V}(t) = \begin{bmatrix} 3t-1 \\ 2t+2 \\ 5t+2 \end{bmatrix}.$$

Since

$$s\mathbf{E} - \mathbf{A} = -\begin{bmatrix} s+1 & 0 & 0 \\ 0 & s+1 & s \\ 0 & s-2 & s \end{bmatrix},$$

we have $\det(s\mathbf{E} - \mathbf{A}) = 4s(s+1) \not\equiv 0$, and thus the pencil is regular while $\mathbf{E}$ is singular. Hence the pencil has finite eigenvalues $\lambda_1 = -1$, $\lambda_2 = 0$, and an infinite eigenvalue. We form $\mathbf{Q}_p$ from the corresponding eigenvector to the finite eigenvalues:

$$\mathbf{Q}_p = \begin{bmatrix} 1 & 0 \\ 0 & 0 \\ 0 & 1 \end{bmatrix}.$$

Hence from Proposition 3.2, there exist solutions for the system given by (3.8):

$$\boldsymbol{x}(t) = \begin{bmatrix} 1 & 0 \\ 0 & 0 \\ 0 & 1 \end{bmatrix} \begin{bmatrix} \mathrm{e}^{-t} & 0 \\ 0 & 1 \end{bmatrix} \boldsymbol{C},$$

or, equivalently,

$$\boldsymbol{x}(t) = \begin{bmatrix} e^{-t}c_1 \\ 0 \\ c_2 \end{bmatrix}, \quad c_1, c_2 \in \mathbb{R}.$$

If

$$\boldsymbol{x}(0) = \boldsymbol{x}_o = \begin{bmatrix} 1 \\ 0 \\ 2 \end{bmatrix}$$

are initial conditions, then $\boldsymbol{x}_o \in \text{colspan}\mathbf{Q}_p$, and $\boldsymbol{x}_o$ are consistent. Hence the unique solution is:

$$\boldsymbol{x}(t) = \begin{bmatrix} e^{-t} \\ 0 \\ 2 \end{bmatrix}.$$

If

$$\boldsymbol{x}(0) = \boldsymbol{x}_o = \begin{bmatrix} 0 \\ 1 \\ 1 \end{bmatrix}$$

are initial conditions, then $\boldsymbol{x}_o \notin \text{colspan}\mathbf{Q}_p$, and $\boldsymbol{x}_o$ are non-consistent. Hence there does not exist a unique solution for the system in this case. □

Example 3.4 (Solution of a regular pencil with infinite eigenvalues). We consider now the system (3.1) with $\boldsymbol{V}(t) = \mathbf{0}_{6,1}$, and

$$\mathbf{E} = \begin{bmatrix} 1 & 0 & 0 & 0 & 0 & 0 \\ 0 & 1 & 0 & 0 & 0 & 0 \\ 0 & 0 & 1 & 0 & 0 & 0 \\ 0 & 0 & 0 & 1 & 0 & 0 \\ 0 & 0 & 0 & 0 & 1 & 1 \\ 0 & 0 & 0 & 0 & 0 & 0 \end{bmatrix}, \quad \mathbf{A} = \begin{bmatrix} 0 & 0 & 1 & 0 & 0 & 0 \\ 0 & 0 & 0 & 1 & 0 & 0 \\ 0 & 0 & 0 & 0 & 1 & 0 \\ 0 & 0 & 0 & 0 & 0 & 1 \\ -4 & 2 & 2 & -3 & -2 & -1 \\ 1 & 1 & -1 & -1 & 0 & 0 \end{bmatrix}.$$

The pencil $s\mathbf{E} - \mathbf{A}$ has three finite eigenvalues $\lambda_1 = 3$, $\lambda_2 = 2$, $\lambda_3 = 1$, of algebraic multiplicity $p_1 = p_2 = p_3 = 1$, and an infinite eigenvalue of algebraic multiplicity $q = 3$. The matrix $\mathbf{Q}_p$ associated with the finite eigenvalues is:

$$\mathbf{Q}_p = \begin{bmatrix} -1 & -1 & -3 \\ 1 & 1 & 5 \\ -3 & -2 & -3 \\ 3 & 2 & 5 \\ -9 & -4 & -3 \\ 9 & 4 & 5 \end{bmatrix}.$$

The Jordan matrix related to the finite eigenvalues is given by:

$$\mathbf{J}_p = \begin{bmatrix} 3 & 0 & 0 \\ 0 & 2 & 0 \\ 0 & 0 & 1 \end{bmatrix}.$$

Then

$$x(t) = \begin{bmatrix} -1 & -1 & -3 \\ 1 & 1 & 5 \\ -3 & -2 & -3 \\ 3 & 2 & 5 \\ -9 & -4 & -3 \\ 9 & 4 & 5 \end{bmatrix} \begin{bmatrix} e^{3t} & 0 & 0 \\ 0 & e^{2t} & 0 \\ 0 & 0 & e^{t} \end{bmatrix} \boldsymbol{C},$$

or, equivalently,

$$x(t) = \begin{bmatrix} -e^{3t}c_1 - e^{2t}c_2 - 3e^{t}c_3 \\ e^{3t}c_1 + e^{2t}c_2 + 5e^{t}c_3 \\ -3e^{3t}c_1 - 2e^{2t}c_2 - 3e^{t}c_3 \\ 3e^{3t}c_1 + 2e^{2t}c_2 + 5e^{t}c_3 \\ -9e^{3t}c_1 - 4e^{2t}c_2 - 3e^{t}c_3 \\ 9e^{3t}c_1 + 4e^{2t}c_2 + 5e^{t}c_3 \end{bmatrix}, \quad c_1, c_2, c_3 \in \mathbb{R}.$$

□

3.3 Applications to Electrical Circuits and Systems

This section applies the theory presented in this chapter on a classical linear circuit. The reader is also invited to revisit Examples 1.9 and 1.12 that discuss the OMIB system for examples of non-linear ODEs.

Example 3.5 (*RLC* circuit)**.** Let us consider the *RLC* circuit shown in Figure 3.1. The capacitor C is charged before the time $t = 0$, when the switch turns

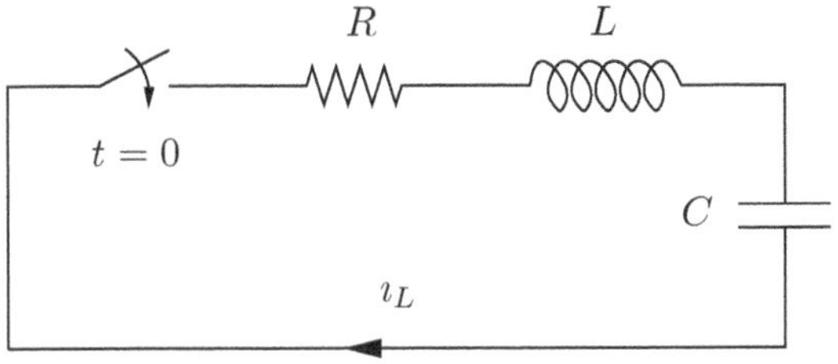

FIGURE 3.1: *RLC* circuit.

on to start the discharge of the capacitor. For $t \geq 0$, according to Kirchhoff Voltage Law (KVL), it must hold:

$$v_C + v_R + v_L = 0\,. \tag{3.9}$$

Since

$$\imath_L = C\,\dot{v}_C\,, \tag{3.10}$$

we can deduce:

$$v_R = R\,\imath_L = RC\,\dot{v}_C\,, \tag{3.11}$$

and

$$v_L = L\,\dot{\imath}_L\,. \tag{3.12}$$

Assuming $\boldsymbol{x} = [\imath_L \;\; v_C]^{\mathrm{T}}$, equations (3.9)–(3.12) lead to the pencil $s\mathbf{E} - \mathbf{A}$, where:

$$\mathbf{E} = \begin{bmatrix} L & 0 \\ 0 & C \end{bmatrix}, \; \mathbf{A} = \begin{bmatrix} -R & -1 \\ 1 & 0 \end{bmatrix}, \tag{3.13}$$

which is regular, being both $\mathbf{E}$ and $\mathbf{A}$ regular. It is worth noting that physical systems are rarely described by singular pencils, except for very specific situations, e.g. limit points and bifurcations.

Also, it is uncommon that physical systems show multiple solutions such as the case discussed in Example 3.3. To obtain a similar case as Example 3.3, either C or L in the circuit of Figure 3.1 should be null. In that case, however, it simply means that the circuit element is a short-circuit and no differential equation is required (see also the discussion of the case for $a = 0$ in Example 1.8).

Assuming that $C \neq 0$ and $L \neq 0$, one can define the state matrix of the system as:

$$\mathbf{A}_{\mathrm{S}} = \mathbf{E}^{-1}\mathbf{A} = \begin{bmatrix} -R/L & -1/L \\ 1/C & 0 \end{bmatrix}. \tag{3.14}$$

In analogy with Example 1.9, the eigenvalues of (3.14) are found by calculating the roots of its characteristic equation. Since the system is linear, we can actually rewrite the differential equation as a function of v_C by observing that:

$$L\,\dot{\imath}_L = LC\,\ddot{v}_C\,,$$

and substituting this expression and (3.11) into (3.9):

$$LC\,\ddot{v}_C + RC\,\dot{v}_C + v_C = 0\,. \tag{3.15}$$

The characteristic equation of (3.15) can be deduced from the Laplace transform:

$$LC\,s^2 + RC\,s + 1 = 0\,. \tag{3.16}$$

As a second-order ODE, the equation (3.15) has two eigenvalues, which numerically equal to the roots of the characteristic equation (3.16), which can be computed as follows:

$$\lambda_{1,2} = -\frac{R}{2L} \pm \sqrt{\left(\frac{R}{2L}\right)^2 - \frac{1}{LC}}\,.$$

With $R, C, L \in \mathbb{R}^+$, it must hold $\mathrm{Re}(\lambda_{1,2}) < 0$. Therefore, the RLC circuit is stable and can always reach a steady state at the end of the discharge, namely $\lim_{t\to\infty} v_c(t) = 0$.

The transient behavior of v_C during the discharge can be described as:

$$v_C(t) = K_1\,e^{s_1 t} + K_2\,e^{s_2 t}\,, \tag{3.17}$$

where $K_{1,2} \in \mathbb{R}$ and can be deduced through a given initial condition $v_C(0) = v_o$ and $i(0) = 0$:

$$\begin{aligned} v_o &= K_1 + K_2\,, \\ 0 &= s_1 K_1 + s_2 K_2\,. \end{aligned}$$

The second-order ODE (3.15) can be implemented as the following set of two linear first-order ODEs:

$$\begin{aligned} \dot{v}_C &= -\frac{\imath_L}{C}\,, \\ L\,\dot{\imath}_L &= v_C - R\,\imath_L\,. \end{aligned} \tag{3.18}$$

Assume $v_o = 5$ V, $C = 2.2$ mF, $L = 22$ mH. Let us consider the following two scenarios:

1. $R = 10\ \Omega$: In this scenario, the circuit has a pair of real eigenvalues at $\lambda_1 = -51.23$ and $\lambda_2 = -403.32$.
2. $R = 1\ \Omega$: In this scenario, the circuit has a pair of conjugate complex eigenvalues at $\lambda_{1,2} = -22.73 \pm \jmath\,141.93$.

The trajectories of v_C during the discharge of C with $R = 10$ and $1\ \Omega$ are shown in Figure 3.2. The trajectories are obtained from the time-domain simulation of the ODEs (3.18) with a fixed time step of 0.1 ms.

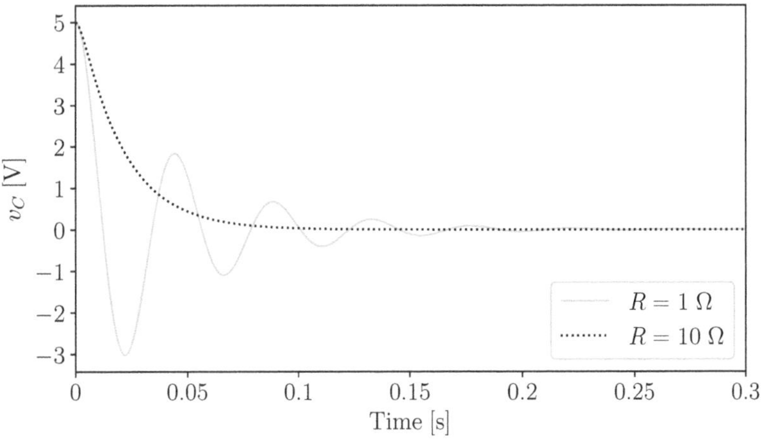

FIGURE 3.2: Trajectories of v_C during the discharge of C for different values of R.

Figure 3.2 shows that, if the resistance of the circuit is sufficiently high, the voltage (and similarly other variables of the circuit) do not oscillate. R in this example plays the same role of the damping coefficient D in Example 1.12 presented in Chapter 1. As the resistance decreases, the capacitor voltage v_C oscillates. The linearity of the circuit allows drawing the same conclusion by simply looking at the eigenvalues, as for linear differential equations, the properties of the equilibrium point are *global.* The same conclusion can be also deduced through the general solution (3.17). □

Example 3.6 (Dommel method)**.** In this example, we consider again the RLC circuit of Figure 3.1. If we apply the ITM to the set of differential equations of the system and recalling the expressions (1.37), at the i-th step of the integration one needs to solve the equation:

$$\mathbf{0}_{2,1} = (0.5\,\Delta t\mathbf{A} + \mathbf{E})\,\boldsymbol{x}^{(i)} + (0.5\,\Delta t\,\mathbf{A} - \mathbf{E})\,\boldsymbol{x}^{(i+1)}\,, \tag{3.19}$$

where $\boldsymbol{x}^{(i+1)}$ are the unknowns and $\mathbf{E}$ and $\mathbf{A}$ are given in (3.13), or equivalently:

$$\mathbf{0}_{2,1} = (0.5\,\Delta t\mathbf{A}_{\mathrm{S}} + \mathbf{I}_2)\,\boldsymbol{x}^{(i)} + (0.5\,\Delta t\,\mathbf{A}_{\mathrm{S}} - \mathbf{I}_2)\,\boldsymbol{x}^{(i+1)}\,, \tag{3.20}$$

where $\mathbf{A}_{\mathrm{S}}$ is defined in (3.14).

Equations (3.19) and (3.20) are linear. If one utilizes a constant time step Δt, then, matrix $0.5\,\Delta t\,\mathbf{A} - \mathbf{E}$ or matrix $0.5\,\Delta t\,\mathbf{A}_{\mathrm{S}} - \mathbf{I}_2$ have to be factorized only once, or until the structure of the system changes, e.g. the switch toggles.

The solution of linear circuits through the ITM is, as it appears, nothing else than the well-known method invented by Dommel, which is the work-horse of electronic circuit simulators such a SPICE [164]. In its classical formulation, Dommel introduced virtual current generators for the elements L and C, which accounted for the "memory" of these components in the previous step of the integration. This, of course, is not really needed as it is a consequence of the ITM, but the physical interpretation of the known term in (3.19) and (3.20) is nevertheless, interesting. More importantly, we observe that (3.19) is a special form of the Möbius transform, which will be introduced in Chapter 7 and, in particular, in Section 7.2. □

4

Explicit Differential-Algebraic Equations

4.1 Formulation

In mathematics, differential-algebraic equations (DAEs) are a system of equations that either contains differential equations and algebraic equations, or is equivalent to such a system. Such systems occur as the general form of systems of differential equations for vector-valued functions $\boldsymbol{x}$ in one independent variable t. Overall DAEs have the following singular non-linear form:

$$\mathbf{E}_{\mathrm{I}}\,\dot{\boldsymbol{\xi}} = \boldsymbol{\varphi}(\boldsymbol{\xi})\,, \tag{4.1}$$

where

$$\mathbf{E}_{\mathrm{I}} = \begin{bmatrix} \mathbf{I}_n & \mathbf{0}_{n,l} \\ \mathbf{0}_{l,n} & \mathbf{0}_{l,l} \end{bmatrix}, \quad \boldsymbol{\xi} = \begin{bmatrix} \boldsymbol{x} \\ \boldsymbol{y} \end{bmatrix}, \quad \boldsymbol{\varphi}(\boldsymbol{\xi}) = \begin{bmatrix} \boldsymbol{f}(\boldsymbol{x},\boldsymbol{y}) \\ \boldsymbol{g}(\boldsymbol{x},\boldsymbol{y}) \end{bmatrix},$$

and $\boldsymbol{x} = \boldsymbol{x}(t) \in \mathbb{C}^{n\times 1}$, $\boldsymbol{y} = \boldsymbol{y}(t) \in \mathbb{C}^{l\times 1}$, $\boldsymbol{f}(\boldsymbol{x},\boldsymbol{y}) \in \mathbb{C}^{n\times 1}$, $\boldsymbol{g}(\boldsymbol{x},\boldsymbol{y}) \in \mathbb{C}^{l\times 1}$. The functions $\boldsymbol{f},\boldsymbol{g}$ are at least C^1, i.e. differentiable functions whose derivative is continuous. If we substitute $\mathbf{E}_{\mathrm{I}}$, $\boldsymbol{\varphi}(\boldsymbol{\xi})$ into the system we get the following form of dynamic equations:

$$\begin{bmatrix} \mathbf{I}_n & \mathbf{0}_{n,l} \\ \mathbf{0}_{l,n} & \mathbf{0}_{l,l} \end{bmatrix} \begin{bmatrix} \dot{\boldsymbol{x}} \\ \dot{\boldsymbol{y}} \end{bmatrix} = \begin{bmatrix} \boldsymbol{f}(\boldsymbol{x},\boldsymbol{y}) \\ \boldsymbol{g}(\boldsymbol{x},\boldsymbol{y}) \end{bmatrix},$$

or, equivalently,

$$\begin{aligned} \dot{\boldsymbol{x}} &= \boldsymbol{f}(\boldsymbol{x},\boldsymbol{y})\,, \\ \mathbf{0}_{l,1} &= \boldsymbol{g}(\boldsymbol{x},\boldsymbol{y})\,. \end{aligned} \tag{4.2}$$

System (4.2) is an explicit DAE system involving the state variables vector $\boldsymbol{x}$ and the algebraic variables vector $\boldsymbol{y}$ changing during the time t.

Lemma 4.1 (Local differentiable map of algebraic variables)**.** Let $\boldsymbol{g}(\boldsymbol{x},\boldsymbol{y}) : \mathbb{C}^{l\times 1} \times \mathbb{C}^{l\times 1} \mapsto \mathbb{C}^{l\times 1}$ be a differentiable function as defined in equation (4.2). Then we can locally express the surface $\boldsymbol{g}(\boldsymbol{x},\boldsymbol{y}) = \mathbf{0}_{l,1}$ by $\boldsymbol{x} = \boldsymbol{x}(\boldsymbol{y})$ for some differentiable map $\boldsymbol{x}(\boldsymbol{y})$ in the neighborhood of $(\boldsymbol{x},\boldsymbol{y})$.

Proof. From equation (4.2), we have

$$\dot{\boldsymbol{g}} = \mathbf{0}_{l,1}\,,$$

or, equivalently,

$$\frac{\partial \boldsymbol{g}}{\partial \boldsymbol{x}}\dot{\boldsymbol{x}} + \frac{\partial \boldsymbol{g}}{\partial \boldsymbol{y}}\dot{\boldsymbol{y}} = \mathbf{0}_{l,1}\,,$$

or, equivalently,

$$\frac{\partial \boldsymbol{g}}{\partial \boldsymbol{x}}\dot{\boldsymbol{x}} = -\frac{\partial \boldsymbol{g}}{\partial \boldsymbol{y}}\dot{\boldsymbol{y}}\,.$$

From equation (4.2) we have

$$\frac{\partial \boldsymbol{g}}{\partial \boldsymbol{x}}\boldsymbol{f}(\boldsymbol{x},\boldsymbol{y}) = -\frac{\partial \boldsymbol{g}}{\partial \boldsymbol{y}}\dot{\boldsymbol{y}}\,.$$

Note that $\frac{\partial \boldsymbol{g}}{\partial \boldsymbol{y}}$ is a square matrix $l \times l$ and since $\boldsymbol{f}(\boldsymbol{x},\boldsymbol{y})$ is uniquely defined, the inverse of the matrix $\frac{\partial \boldsymbol{g}}{\partial \boldsymbol{y}}$ exists. If this inverse did not exist, then the function $\boldsymbol{f}$ would be either not unique or would not exist. Hence:

$$\det\left(\frac{\partial \boldsymbol{g}}{\partial \boldsymbol{y}}\right) \neq 0\,.$$

Then from (4.2), the above expression and the implicit function Theorem, we can locally express the surface $\boldsymbol{g}(\boldsymbol{x},\boldsymbol{y}) = \mathbf{0}_{l,1}$ by $\boldsymbol{x} = \boldsymbol{x}(\boldsymbol{y})$ for some differentiable map $\boldsymbol{x}(\boldsymbol{y})$ in the neighborhood of $(\boldsymbol{x},\boldsymbol{y})$. The proof is completed. ■

If the conditions in Lemma 4.1 hold, there exists the function $\boldsymbol{x} = \boldsymbol{x}(\boldsymbol{y})$ defined by

$$\boldsymbol{g}\big(\boldsymbol{x}(\boldsymbol{y}),\boldsymbol{y}\big) = \mathbf{0}_{l,1}\,,$$

such that a solution can be obtained of the form

$$\boldsymbol{y} = \hat{\boldsymbol{g}}(\boldsymbol{x})\,.$$

Then by replacing the above expression into (4.2) we reduce the DAEs into a system of regular differential equations of the form:

$$\dot{\boldsymbol{x}} = \boldsymbol{f}\big(\boldsymbol{x},\hat{\boldsymbol{g}}(\boldsymbol{x})\big) = \hat{\boldsymbol{f}}(\boldsymbol{x})\,.$$

We consider now the linear DAEs:

$$\mathbf{E}_{\text{I}}\,\dot{\boldsymbol{\xi}} = \mathbf{A}\,\boldsymbol{\xi}\,,$$

with

$$\mathbf{E}_{\text{I}} = \begin{bmatrix} \mathbf{I}_n & \mathbf{0}_{n,l} \\ \mathbf{0}_{l,n} & \mathbf{0}_{l,l} \end{bmatrix}, \quad \boldsymbol{\xi} = \begin{bmatrix} \boldsymbol{x} \\ \boldsymbol{y} \end{bmatrix}, \quad \mathbf{A} = \begin{bmatrix} \mathbf{A}_{11} & \mathbf{A}_{12} \\ \mathbf{A}_{21} & \mathbf{A}_{22} \end{bmatrix},$$

and $\mathbf{A}_{11} \in \mathbb{C}^{n\times n}$, $\mathbf{A}_{12} \in \mathbb{C}^{n\times l}$, $\mathbf{A}_{21} \in \mathbb{C}^{l\times n}$, $\mathbf{A}_{22} \in \mathbb{C}^{l\times l}$, and $\boldsymbol{x} : [0,+\infty] \mapsto \mathbb{C}^{n\times 1}$, $\boldsymbol{y} : [0,+\infty] \mapsto \mathbb{C}^{l\times 1}$. Then, the linear DAEs take the following form of dynamic equations:

$$\begin{bmatrix} \mathbf{I}_n & \mathbf{0}_{n,l} \\ \mathbf{0}_{l,n} & \mathbf{0}_{l,l} \end{bmatrix}\begin{bmatrix} \dot{\boldsymbol{x}} \\ \dot{\boldsymbol{y}} \end{bmatrix} = \begin{bmatrix} \mathbf{A}_{11} & \mathbf{A}_{12} \\ \mathbf{A}_{21} & \mathbf{A}_{22} \end{bmatrix}\begin{bmatrix} \boldsymbol{x} \\ \boldsymbol{y} \end{bmatrix},$$

or, equivalently,

$$\begin{aligned}\dot{\boldsymbol{x}} &= \mathbf{A}_{11}\boldsymbol{x} + \mathbf{A}_{12}\boldsymbol{y}\,,\\ \mathbf{0}_{l,1} &= \mathbf{A}_{21}\boldsymbol{x} + \mathbf{A}_{22}\boldsymbol{y}\,.\end{aligned} \tag{4.3}$$

The pencil of system (4.3) is $s\mathbf{E}_{\mathrm{I}} - \mathbf{A}$, or, equivalently,

$$s\begin{bmatrix} \mathbf{I}_n & \mathbf{0}_{n,l} \\ \mathbf{0}_{l,n} & \mathbf{0}_{l,l} \end{bmatrix} - \begin{bmatrix} \mathbf{A}_{11} & \mathbf{A}_{12} \\ \mathbf{A}_{21} & \mathbf{A}_{22} \end{bmatrix}, \tag{4.4}$$

or equivalently,

$$\begin{bmatrix} s\mathbf{I}_n - \mathbf{A}_{11} & -\mathbf{A}_{12} \\ -\mathbf{A}_{21} & -\mathbf{A}_{22} \end{bmatrix},$$

whereby taking its determinant, and since $s\mathbf{I}_n - \mathbf{A}_{11}$ is a regular pencil, see [51], we have that

$$\det(s\mathbf{E}_{\mathrm{I}} - \mathbf{A}) = \det(s\mathbf{I}_n - \mathbf{A}_{11})\det(\mathbf{A}_{22} + \mathbf{A}_{21}(s\mathbf{I}_n - \mathbf{A}_{11})^{-1}\mathbf{A}_{12}).$$

Lemma 4.2 (Reduction of explicit differential-algebraic equations)**.** Consider the DAEs (4.3). If the matrix $\mathbf{A}_{22}$ is regular then (4.3) can be reduced to the regular system of differential equations

$$\dot{\boldsymbol{x}} = (\mathbf{A}_{11} - \mathbf{A}_{12}\mathbf{A}_{22}^{-1}\mathbf{A}_{21})\,\boldsymbol{x} = \mathbf{A}_{\mathrm{S}}\,\boldsymbol{x}\,.$$

In addition if λ_i, $i = 1, 2, \ldots, \nu$ is a finite eigenvalue of (4.4) with algebraic multiplicity p_i then $\sum_{i=1}^{\nu} p_i = n$.

Proof. If the matrix $\mathbf{A}_{22}$ is a regular matrix then it is invertible and

$$\mathbf{0}_{l,1} = \mathbf{A}_{21}\boldsymbol{x} + \mathbf{A}_{22}\boldsymbol{y}\,,$$

can be written as

$$\boldsymbol{y} = -\mathbf{A}_{22}^{-1}\mathbf{A}_{21}\boldsymbol{x}\,.$$

Then (4.3) can be reduced to

$$\dot{\boldsymbol{x}} = \mathbf{A}_{11}\boldsymbol{x} + \mathbf{A}_{12}(-\mathbf{A}_{22}^{-1}\mathbf{A}_{21}\boldsymbol{x})\,,$$

or, equivalently,

$$\dot{\boldsymbol{x}} = (\mathbf{A}_{11} - \mathbf{A}_{12}\mathbf{A}_{22}^{-1}\mathbf{A}_{21})\boldsymbol{x} = \mathbf{A}_{\mathrm{S}}\,\boldsymbol{x}\,.$$

This is a system of regular type with $\mathbf{A}_{\mathrm{S}} = (\mathbf{A}_{11} - \mathbf{A}_{12}\mathbf{A}_{22}^{-1}\mathbf{A}_{21}) \in \mathbb{C}^{n\times n}$. Hence if λ_i, $i = 1, 2, \ldots, \nu$ is an eigenvalue of this matrix with algebraic multiplicity p_i then $\sum_{i=1}^{\nu} p_i = n$. Consequently, λ_i is a finite eigenvalue of the pencil (4.4) with the same properties. The proof is completed. ∎

If $\mathbf{A}_{22}$ is a singular matrix then we may use Proposition 3.2 to obtain the solutions of (4.3) and Corollary 3.2 to study uniqueness of solutions if the initial conditions are given. To sum up in the case of a singular $\mathbf{A}_{22}$ here are the steps we have to undertake when we need to provide the solutions of (4.3):

- Step 1. By using the pencil (4.4) of (4.3) we compute the polynomial

$$\det(s\mathbf{I}_n - \mathbf{A}_{11})\det(\mathbf{A}_{22} + \mathbf{A}_{21}(s\mathbf{I}_n - \mathbf{A}_{11})^{-1}\mathbf{A}_{12}) = \varpi(s)\,.$$

- Step 2. The pencil has finite/infinite eigenvalues since

$$\mathbf{E}_{\rm I} = \begin{bmatrix} \mathbf{I}_n & \mathbf{0}_{n,l} \\ \mathbf{0}_{l,n} & \mathbf{0}_{l,l} \end{bmatrix}$$

  is singular. We compute the finite eigenvalues which are the roots of $\varpi(s)$.
- Step 3. We form the Jordan matrices $\mathbf{J}_p$, $\mathbf{H}_q$ as described in Proposition 3.2.
- Step 4. We compute the left, and right eigenvectors of the eigenvalues of $s\mathbf{E}_{\rm I} - \mathbf{A}$ in order to form the matrices $\mathbf{P}$, $\mathbf{Q}$, respectively as described in Proposition 3.2.
- Step 5. We use the formula (3.8) to provide the solution of (4.3) since $\mathbf{E}_{\rm I}$ is singular.
- Step 6. For given initial conditions, since $\mathbf{E}_{\rm I}$ is singular we use Corollary 3.2 to identify if the initial conditions are consistent or non-consistent. In the case that they are consistent we provide the unique solution by using the formula in Corollary 3.2.

4.2 Power Systems Modeled as Explicit DAEs

DAEs are the conventional form of the model of power system for transient stability analysis. This is the model that is utilized in the practical totality of the books on the topic and the one that is described in Chapter 1. This section is mainly dedicated to the comprehensive example on the three-bus system described in Appendix A.

Before entering into the details of the three-bus system, it is worth noticing that, while the number of states in the system is well determined by the dynamics of interest that are modeled in the system itself, the number of algebraic variables depends on the specific formulation (for example, current injection as opposed to power injection model of the transmission lines) and on the "auxiliary" quantities that are defined as algebraic variables.

This concept is best illustrated with a simple example.

Example 4.1 (Lorenz system). Consider the set of ODEs of the well-known Lorenz system:

$$\begin{aligned} \dot{x}_1 &= a\,(x_2 - x_1)\,, \\ \dot{x}_2 &= x_1\,(b - x_3) - x_2\,, \\ \dot{x}_3 &= x_1\,x_2 - c\,x_3\,. \end{aligned} \tag{4.5}$$

The system can be rewritten as a set of DAEs, for example:

$$
\begin{aligned}
\dot{x}_1 &= a\,(x_2 - x_1)\,, \\
\dot{x}_2 &= y_1 - x_2\,, \\
\dot{x}_3 &= y_2 - c\,x_3\,, \\
0 &= x_1\,(b - x_3) - y_1\,, \\
0 &= x_1\,x_2 - y_2\,.
\end{aligned}
\tag{4.6}
$$

The sets of equations (4.5) and (4.6) have exactly same stability properties and three state variables. The number of algebraic variables, on the other hand, is 0 for (4.5) and 2 for (4.6).

One can, in fact, reformulate the equations including an arbitrary number of algebraic variables. One may also argue that the introduction of the "auxiliary" variables y_1 and y_2 is not necessary and thus the only meaningful form of the system is (4.5). This is however, not necessarily true if one considers computational aspects.

We have already discussed in Chapter 1 that the elimination of the algebraic variables is not possible, in general, for non-linear DAEs. Power systems are a very relevant case. On the other hand, it is also not always true that keeping the algebraic variables to a "minimum" number is useful. Increasing the number of algebraic equations and variables certainly increases the size of the system. However, increasing the number of algebraic variables generally also increases the sparsity of the resulting Jacobian matrices of differential and algebraic equations.

In practical applications, it is the sparsity of the Jacobian matrices, more than their size, that contributes most to speed up numerical analyses. Of course, there is a trade-off between the size of the vector of algebraic variables and speed-up. Unfortunately, there is currently no algorithm to determine such a trade-off.

The discussion above leads to the following conclusion: the ODE formulation is the "most dense" form of a set of differential equations. Recalling the notation utilized above and in Section 1.5, even if $\mathbf{A}$ and $\mathbf{E}_{\mathrm{I}}$ are sparse, or even very sparse, matrices, the state matrix $\mathbf{A}_{\mathrm{S}}$ tends to be a dense, generally full, matrix.

In the example above, $\mathbf{E}_{\mathrm{I}}$ and $\mathbf{A}$ are 5×5 matrices, with 3 and 12 non-zero elements, namely 12% and 48% of the total number of elements, respectively. On the other hand, $\mathbf{A}_{\mathrm{S}}$ is 3×3 and has 8 non-zero elements, namely 89% of the total. Of course, this is a tiny system, but the dramatic increase of sparsity as the number of algebraic variables increases is evident. The determination of $\mathbf{A}_{\mathrm{S}}$ should thus be limited to relatively small systems. In our experience a good trade-off is when the order of the state matrix is about 2,000. □

The next two examples illustrate the small-signal stability analysis of the three-bus test system. The compete set of static and dynamic data is given in Appendix A. Two scenarios are considered: base-case (Example 4.2); and with inclusion of a PSS (Example 4.3).

Example 4.2 (LEP for the three-bus system). The examined (base-case) equilibrium point $(\boldsymbol{x}_o, \boldsymbol{y}_o)$ is obtained with the solution of the power flow problem and the initialization of dynamic devices. The system dynamic model has in total $n = 19$ state variables and $l = 37$ algebraic variables. The system state vector is as follows:

$$\boldsymbol{x} = [\, \delta_{\mathrm{r},1} \; \omega_{\mathrm{r},1} \; e'_{\mathrm{r,q},1} \; \psi_{\mathrm{s,d},1} \; \psi_{\mathrm{s,q},1} \; \delta_{\mathrm{r},2} \; \omega_{\mathrm{r},2} \; e'_{\mathrm{r,d},2} \; e'_{\mathrm{r,q},2} \; e''_{\mathrm{r,d},2} \; e''_{\mathrm{r,q},2} \\ v_{\mathrm{R},1} \; v_{\mathrm{R},2} \; v_{\mathrm{ef},1} \; v_{\mathrm{ef},2} \; v_{\mathrm{a},1} \; v_{\mathrm{a},2} \; v_{\mathrm{f},1} \; v_{\mathrm{f},2} \,]^{\mathrm{T}} \, .$$

Linearization of the system around $(\boldsymbol{x}_o, \boldsymbol{y}_o)$ and elimination of the algebraic variables leads to a system of ODEs, which is in the form of (1.160). The state matrix of this system is $\mathbf{A}_{\mathrm{S}}$, $\mathbf{A}_{\mathrm{S}} \in \mathbb{R}^{19\times 19}$ and we are interested in the solution of the corresponding conventional Linear Eigenvalue Problem (LEP). The pencil of the system is $s\mathbf{I}_{19} - \mathbf{A}_{\mathrm{S}}$ and has 19 finite eigenvalues. The eigenvalues, along with their damping ratios, their natural and damped frequencies, are summarized in Table 4.1. Notice that the system has a zero eigenvalue, due to one rotor angle being redundant.

TABLE 4.1: Three-bus system: eigenvalues of the base-case operating point.

Mode	ζ (%)	f_n (Hz)	f_d (Hz)
$\lambda_1 = -35.2064$	100.00	5.60	0
$\lambda_2 = -33.7327$	100.00	5.37	0
$\lambda_3 = -23.7196$	100.00	3.78	0
$\lambda_4 = -13.6561$	100.00	2.17	0
$\lambda_{5,6} = -20.5133 \pm \jmath 5.9284$	96.07	3.39	0.94
$\lambda_{7,8} = -19.6722 \pm \jmath 4.1469$	97.85	3.20	0.66
$\lambda_9 = -3.6091$	100.00	0.57	0
$\lambda_{10,11} = -0.2516 \pm \jmath 4.4309$	5.67	0.71	0.71
$\lambda_{12} = -1.0147$	100.00	0.16	0
$\lambda_{13,14} = -0.8211 \pm \jmath 3.4132$	23.39	0.56	0.54
$\lambda_{15,16} = -1.1003 \pm \jmath 2.5390$	39.76	0.44	0.40
$\lambda_{17} = -1.0096$	100.00	0.16	0
$\lambda_{18,19} = 0.0000$ $(p = 2)$	–	–	–

Since the real parts of all (not redundant) eigenvalues are negative, the system is stable. The most poorly damped pair of eigenvalues is $\lambda_{10,11} = -0.2516 \pm \jmath 4.4309$, which represents the electromechanical oscillatory mode of the system. This is clear when one calculates and observes the modal participation factors of the system. We will discuss in detail the participation factors of power systems in Chapter 8.

Given an eigenvalue that represents an electromechanical mode, then identifying which machines contribute to this mode is possible by observing the *shape* of the mode. The shape of an electromechanical mode is given by the associated right eigenvector, and in particular by those elements that correspond to the rotor angles or speeds of the synchronous machines.

For example, the right eigenvector $\mathbf{u}_{10} = \mathbf{Q}(:, 10)$ is associated with $\lambda_{10} = -0.2516 + \jmath 4.4309$. Then, the elements $\mathbf{u}_{10}(1)$, $\mathbf{u}_{10}(6)$, which correspond to $\delta_{r,1}$, $\delta_{r,2}$, respectively, provide the mode shape, as shown in Figure 4.1.

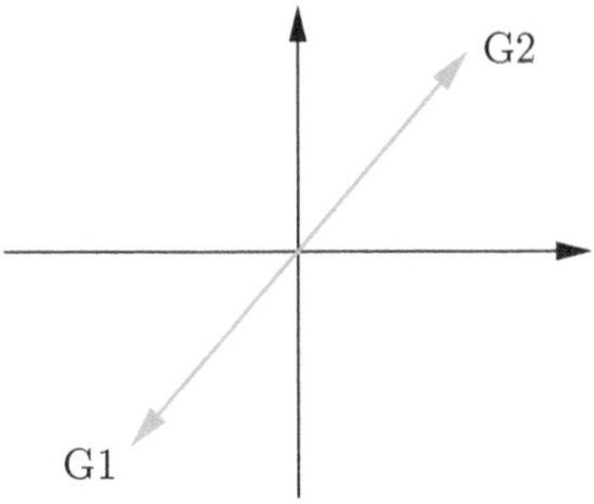

FIGURE 4.1: Mode shape of λ_{10} for the three-bus test system.

The vectors have almost equal magnitudes, which indicates that the mode represents an inter-area oscillation between the two machines. In addition, the vectors have a phase difference about 180 degrees, which implies that the two machines oscillate in counter-phase.

Figure 4.2 shows the transient response of the synchronous machine rotor speeds following a three-phase fault. The fault is applied at $t = 1$ s and is cleared after 60 ms by tripping the line that connects buses 1 and 3.

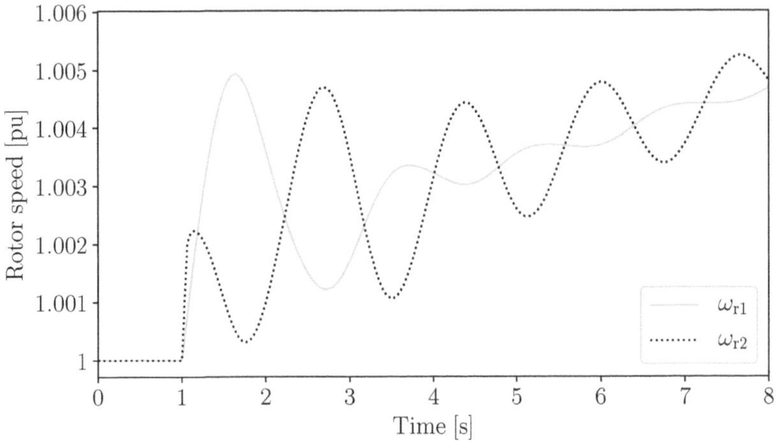

FIGURE 4.2: Transient response of rotor speeds following a fault for the three-bus system.

For completeness, the state matrix $\mathbf{A}_S$, as well as the modal matrices $\mathbf{Q}$ and $\mathbf{P}$ are given in Appendix B. □

Example 4.3 (LEP for the three-bus system with PSS). In this example we design a PSS to increase the damping of the poorly damped electromechanical oscillatory mode of the three-bus system, i.e. $\lambda_{10,11} = -0.2516 \pm \jmath 4.4309$ with damping ratio 5.67%. To this aim, we consider a well-known PSS design technique, i.e. the method of residues, see for example, [1]. The PSS is assumed to consist of a wash-out filter and a product of multiple lead-lag blocks and thus to have a transfer function:

$$G_{\rm PSS}(s) = K_{\rm w}\frac{T_{\rm w}s}{1+T_{\rm w}s}\left(\frac{1+T_{z,1}s}{1+T_{p,1}s}\right)^N ,$$

where $K_{\rm w}$ is the PSS gain; $T_{\rm w}$ is the time-constant of the wash-out filter; $T_{z,1}$, $T_{p,1}$ are the time constants of the lead-lags; and N is the number of lead-lag blocks.

Since an oscillatory mode defines a pair of complex conjugate eigenvalues, it suffices to apply the method for one of them, say $\lambda_{11} = -0.2516 - \jmath 4.4309$.

- Step 1: Choose signal and input control placement:

  Let the rotor speed of machine 1 $\omega_{\rm r,1}$ be a candidate signal to be input to the PSS. We define the 1×19 output vector $\mathbf{C}$ as follows:

$$\mathbf{C} = \begin{bmatrix} 0 & 1 & 0 & 0 & 0 & 0 & 0 & 0 & \cdots & 0 \end{bmatrix} ,$$

  where the position of the non-zero element of $\mathbf{C}$ corresponds to the position of the rotor speed $\omega_{\rm r,1}$ in the system state vector. If $\mathbf{u}_{11}$ is the right eigenvector associated to λ_{11}, the magnitude of the complex number

$$\frac{\mathbf{C}\mathbf{u}_{11}}{||\mathbf{C}||\ ||\mathbf{u}_{11}||} = 7.986 \cdot 10^{-4} + \jmath 6.314 \cdot 10^{-4} ,$$

  is a measure for the observability of λ_{11} from $\omega_{\rm r,1}$; $||\cdot||$ is the Euclidean norm.

  Let also the reference voltage $v_1^{\rm ref}$ of the AVR of machine 1 be a candidate control placement for the PSS output. The reference voltage impacts the system dynamics via the differential equation of $v_{\rm a,1}$:

$$\dot{v}_{\rm a,1} = \frac{K_{\rm a,1}}{T_{\rm a,1}}(v_1^{\rm ref} - v_{\rm R,1} - v_{\rm f,1} - \frac{K_{\rm f,1}}{T_{\rm f,1}}v_{\rm ef,1}) - \frac{v_{\rm a,1}}{T_{\rm a,1}} .$$

  The Jacobian of $\dot{v}_{\rm a,1}$ with respect to $v_1^{\rm ref}$ is $\frac{K_{\rm a,1}}{T_{\rm a,1}} = 727.27$. Thus, the 19×1 input matrix $\mathbf{B}$ is defined as:

$$\mathbf{B} = \begin{bmatrix} 0 & 0 & \cdots & 727.27 & 0 & 0 & 0 \end{bmatrix}^{\rm T} . \tag{4.7}$$

If $\mathbf{w}_{11}$ is the left eigenvector associated to λ_{11}, the magnitude of the complex number

$$\frac{\mathbf{w}_{11}\mathbf{B}}{||\mathbf{w}_{11}||\ ||\mathbf{B}||} = 4.745 \cdot 10^{-5} + \jmath 1.360 \cdot 10^{-5} \ ,$$

is a measure for the controllability of λ_{11} from the voltage reference of the AVR of machine 1.

The associated (normalized) residue of λ_{11} is then defined as:

$$\mathrm{R}_{11} = \frac{\mathbf{C}\mathbf{u}_{11}}{||\mathbf{C}||\ ||\mathbf{u}_{11}||} \frac{\mathbf{w}_{11}\mathbf{B}}{||\mathbf{w}_{11}||\ ||\mathbf{B}||} = 2.931 \cdot 10^{-8} + \jmath 4.082 \cdot 10^{-8} \ . \tag{4.8}$$

The magnitude $|\mathrm{R}_{11}|$ provides a measure for the joint observability-controllability of the mode from the specific set of input-output. Then, from given multiple sets of inputs-outputs, we select the one that provides the highest joint observability-controllability index. In (4.8), one can conveniently compare different kinds of signals and input placements. For example, the effectiveness of a rotor speed can be compared to that of an active power flow signal.

Finally, we mention that, in general, inter-area modes are likely to be better observable from a composite signal that includes information from both areas, e.g. the differential rotor speed $\omega_{\mathrm{r},1} - \omega_{\mathrm{r},2}$. In this case, a part of the desired signal is necessarily remote and thus, the impact of communication phenomena such as time varying delays and noise has to be taken into account. We will extensively consider the effect of time delays on the small-signal stability of power systems in Chapters 10–12.

We proceed in the design of the PSS using the signal $\omega_{\mathrm{r},1}$ and the voltage reference of AVR of machine 1 as control placement. The procedure is the same for any selected set.

- Step 2: Calculate the phase compensation that must be provided by the PSS, as follows:

$$\phi = \begin{cases} 180^\circ - \arg(\mathrm{R}_{11}), & \text{if } \arg(\mathrm{R}_{11}) \geq 0^\circ \ , \\ -180^\circ - \arg(\mathrm{R}_{11}), & \text{if } \arg(\mathrm{R}_{11}) < 0^\circ \ . \end{cases}$$

Since $\arg(\mathrm{R}_{11}) = 54.32^\circ$, we get $\phi = 125.68^\circ$.

- Step 3: Design the lead-lag block, as follows:

The number of the lead-lag blocks required is:

$$N = \begin{cases} 1, & \text{if } 0^\circ < |\phi| \leq 60^\circ \ , \\ 2, & \text{if } 60^\circ < |\phi| \leq 120^\circ \ , \\ 3, & \text{if } 120^\circ < |\phi| \leq 180^\circ \ . \end{cases}$$

Since $120^\circ < |\phi| \leq 180^\circ$, we need $N = 3$ lead-lag blocks.

The time constant values of the lead-lag blocks are calculated as follows:

$$T_{z,1} = \alpha T \,,$$
$$T_{p,1} = T \,,$$

where

$$\alpha = \frac{1+\sin(\frac{\phi}{N})}{1-\sin(\frac{\phi}{N})} \,,$$
$$T = \frac{1}{2\pi f_{o,k}\sqrt{\alpha}} \,,$$

where $f_{o,k}$ is the frequency of oscillation of λ_k, in our case of λ_{11}. We have $\omega_{n,11} = 4.4309$ rad/s. Hence:

$$\alpha = \frac{1+\sin(\frac{-125.68}{3})}{1-\sin(\frac{-125.68}{3})} = 5.0193 \,,$$
$$T = \frac{1}{2\pi f_{o,11}\sqrt{\alpha}} = 0.1007 \,.$$

Finally, $T_{z,1} = 0.5056$ s and $T_{p,1} = 0.1007$ s.

- Step 4: Determine the time-constant of the wash-out block and the PSS gain:

  The wash-out time-constant T_{w} is tuned so that the PSS contributes in the damping of the slowest electromechanical mode of the system (possibly an inter-area mode). To this aim, we first find the frequency of oscillation of the slowest electromechanical mode, let's say $f_{o,\min}$. Then, a rule of thumb is to choose a T_{w} roughly ten times slower, i.e.:

$$T_w \approx 10\frac{1}{f_{o,\min}} \,.$$

  In our case, $f_{o,\min} = f_{o,11} = \frac{4.4309}{2\pi} = 0.7052$ Hz and hence,

$$10\frac{1}{0.7052} = 14.2 \text{ s} \,,$$

  and we choose $T_{\mathrm{w}} = 14$ s.

  Finally, the PSS gain K_{w} can be easily tuned using trial and error, to achieve the highest damping possible for all electromechanical modes of the system. With the PSS installed to the system, Figure 4.3 shows the minimum damping ratio $\zeta_{\min}$ of the system eigenvalues as K_{w} varies. Finally we select the gain that achieves maximum $\zeta_{\min}$. This is $K_{\mathrm{w}} = 11.1$ which gives $\zeta_{\min} = 21.39$ %. □

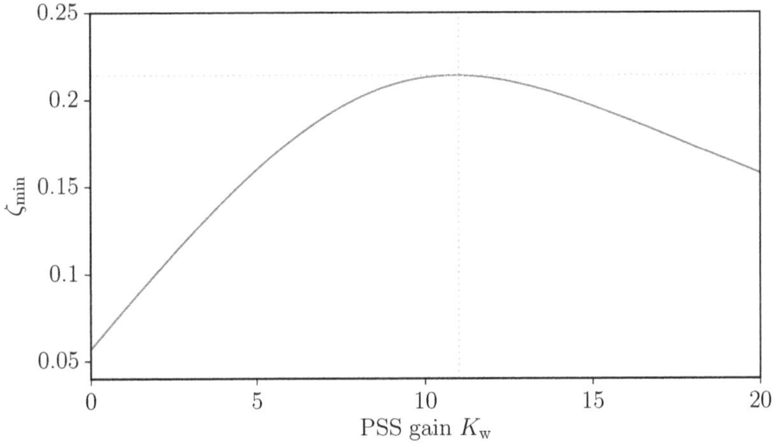

FIGURE 4.3: PSS gain K_{w} vs minimum damping ratio $\zeta_{\min}$ for the three-bus system.

5

Implicit Differential-Algebraic Equations

5.1 Formulation

This chapter focuses on implicit differential-algebraic equations (DAEs) formulated as the following singular non-linear system:

$$\begin{bmatrix} \mathbf{T} & \mathbf{0}_{n,l} \\ \mathbf{R} & \mathbf{0}_{l,l} \end{bmatrix} \begin{bmatrix} \dot{\boldsymbol{x}} \\ \dot{\boldsymbol{y}} \end{bmatrix} = \begin{bmatrix} \boldsymbol{f}(\boldsymbol{x}, \boldsymbol{y}) \\ \boldsymbol{g}(\boldsymbol{x}, \boldsymbol{y}) \end{bmatrix},$$

or, equivalently,

$$\begin{aligned} \mathbf{T}\,\dot{\boldsymbol{x}} &= \boldsymbol{f}(\boldsymbol{x}, \boldsymbol{y}), \\ \mathbf{R}\,\dot{\boldsymbol{x}} &= \boldsymbol{g}(\boldsymbol{x}, \boldsymbol{y}), \end{aligned} \tag{5.1}$$

where $\boldsymbol{x} = \boldsymbol{x} \in \mathbb{C}^{n\times 1}$, $\boldsymbol{y} = \boldsymbol{y} \in \mathbb{C}^{l\times 1}$, $\boldsymbol{f}(\boldsymbol{x}, \boldsymbol{y}) \in \mathbb{C}^{n\times 1}$, $\boldsymbol{g}(\boldsymbol{x}, \boldsymbol{y}) \in \mathbb{C}^{l\times 1}$, and $\mathbf{T}, \mathbf{R} \in \mathbb{C}^{n\times n}$. The functions $\boldsymbol{f}, \boldsymbol{g}$ are at least C^1, i.e. differentiable functions whose derivative is continuous. If we set

$$\mathbf{E} = \begin{bmatrix} \mathbf{T} & \mathbf{0}_{n,l} \\ \mathbf{R} & \mathbf{0}_{l,l} \end{bmatrix}, \quad \boldsymbol{\xi} = \begin{bmatrix} \boldsymbol{x} \\ \boldsymbol{y} \end{bmatrix}, \quad \boldsymbol{\varphi}(\boldsymbol{\xi}) = \begin{bmatrix} \boldsymbol{f}(\boldsymbol{x}, \boldsymbol{y}) \\ \boldsymbol{g}(\boldsymbol{x}, \boldsymbol{y}) \end{bmatrix},$$

then it can be observed that system (5.1) is a case of a singular non-linear system with general form:

$$\mathbf{E}\,\dot{\boldsymbol{\xi}} = \boldsymbol{\varphi}(\boldsymbol{\xi}). \tag{5.2}$$

If we set

$$\boldsymbol{\varphi}(\boldsymbol{\xi}) = \mathbf{A}\,\boldsymbol{\xi},$$

then we arrive at the linear singular system

$$\mathbf{E}\,\dot{\boldsymbol{\xi}} = \mathbf{A}\,\boldsymbol{\xi}.$$

Furthermore if

$$\mathbf{E} = \begin{bmatrix} \mathbf{T} & \mathbf{0}_{n,l} \\ \mathbf{R} & \mathbf{0}_{l,l} \end{bmatrix}, \quad \boldsymbol{\xi} = \begin{bmatrix} \boldsymbol{x} \\ \boldsymbol{y} \end{bmatrix}, \quad \mathbf{A} = \begin{bmatrix} \mathbf{A}_{11} & \mathbf{A}_{12} \\ \mathbf{A}_{21} & \mathbf{A}_{22} \end{bmatrix},$$

with $\mathbf{A}_{11} \in \mathbb{C}^{n\times n}$, $\mathbf{A}_{12} \in \mathbb{C}^{n\times l}$, $\mathbf{A}_{21} \in \mathbb{C}^{l\times n}$, $\mathbf{A}_{22} \in \mathbb{C}^{l\times l}$, and $\boldsymbol{x} : [0, +\infty] \mapsto \mathbb{C}^{n\times 1}$, $\boldsymbol{y} : [0, +\infty] \mapsto \mathbb{C}^{l\times 1}$, then we arrive at the following implicit DAEs:

$$\begin{bmatrix} \mathbf{T} & \mathbf{0}_{n,l} \\ \mathbf{R} & \mathbf{0}_{l,l} \end{bmatrix} \begin{bmatrix} \dot{\boldsymbol{x}} \\ \dot{\boldsymbol{y}} \end{bmatrix} = \begin{bmatrix} \mathbf{A}_{11} & \mathbf{A}_{12} \\ \mathbf{A}_{21} & \mathbf{A}_{22} \end{bmatrix} \begin{bmatrix} \boldsymbol{x} \\ \boldsymbol{y} \end{bmatrix},$$

or, equivalently,

$$\begin{aligned} \mathbf{T}\,\dot{\boldsymbol{x}} &= \mathbf{A}_{11}\,\boldsymbol{x} + \mathbf{A}_{12}\,\boldsymbol{y}\,, \\ \mathbf{R}\,\dot{\boldsymbol{x}} &= \mathbf{A}_{21}\,\boldsymbol{x} + \mathbf{A}_{22}\,\boldsymbol{y}\,. \end{aligned} \tag{5.3}$$

The pencil of system (5.3) is $s\mathbf{E} - \mathbf{A}$, or, equivalently,

$$s\begin{bmatrix} \mathbf{T} & \mathbf{0}_{n,l} \\ \mathbf{R} & \mathbf{0}_{l,l} \end{bmatrix} - \begin{bmatrix} \mathbf{A}_{11} & \mathbf{A}_{12} \\ \mathbf{A}_{21} & \mathbf{A}_{22} \end{bmatrix}. \tag{5.4}$$

Lemma 5.1 (Reduction of implicit differential-algebraic equations)**.** Consider the DAEs (5.3). If the matrix $\mathbf{A}_{22}$ is regular then (5.3) can be reduced to the generalized system of differential equations

$$\tilde{\mathbf{E}}\,\dot{\boldsymbol{x}} = \tilde{\mathbf{A}}\,\boldsymbol{x}\,,$$

where

$$\tilde{\mathbf{E}} = \mathbf{R} - \mathbf{A}_{12}\mathbf{A}_{22}^{-1}\mathbf{T}, \quad \tilde{\mathbf{A}} = \mathbf{A}_{11} - \mathbf{A}_{12}\mathbf{A}_{22}^{-1}\mathbf{A}_{21}\,.$$

In addition if $\tilde{\mathbf{E}}$ is invertible, and λ_i, $i = 1, 2, \dots, \nu$ is a finite eigenvalue of (5.4) with algebraic multiplicity p_i then $\sum_{i=1}^{\nu} p_i = n$.

Proof. If the matrix $\mathbf{A}_{22}$ is a regular matrix then it is invertible and

$$\mathbf{T}\,\dot{\boldsymbol{x}} = \mathbf{A}_{21}\,\boldsymbol{x} + \mathbf{A}_{22}\,\boldsymbol{y}\,,$$

can be written as

$$\boldsymbol{y} = \mathbf{A}_{22}^{-1}\mathbf{T}\,\dot{\boldsymbol{x}} - \mathbf{A}_{22}^{-1}\mathbf{A}_{21}\,\boldsymbol{x}\,.$$

Then (5.3) can be reduced to

$$\mathbf{R}\,\dot{\boldsymbol{x}} = \mathbf{A}_{11}\,\boldsymbol{x} + \mathbf{A}_{12}(\mathbf{A}_{22}^{-1}\mathbf{T}\,\dot{\boldsymbol{x}} - \mathbf{A}_{22}^{-1}\mathbf{A}_{21}\,\boldsymbol{x})\,,$$

or, equivalently,

$$\mathbf{R}\,\dot{\boldsymbol{x}} = \mathbf{A}_{11}\,\boldsymbol{x} + \mathbf{A}_{12}\mathbf{A}_{22}^{-1}\mathbf{T}\,\dot{\boldsymbol{x}} - \mathbf{A}_{12}\mathbf{A}_{22}^{-1}\mathbf{A}_{21}\,\boldsymbol{x}\,,$$

or, equivalently,

$$(\mathbf{R} - \mathbf{A}_{12}\mathbf{A}_{22}^{-1}\mathbf{T})\,\dot{\boldsymbol{x}} = (\mathbf{A}_{11} - \mathbf{A}_{12}\mathbf{A}_{22}^{-1}\mathbf{A}_{21})\,\boldsymbol{x}\,.$$

This is a generalized system with matrices $\tilde{\mathbf{E}}, \tilde{\mathbf{A}} \in \mathbb{C}^{n\times n}$. Hence, if $\tilde{\mathbf{E}}$ is invertible and λ_i, $i = 1, 2, \dots, \nu$ is a finite eigenvalue of the pencil $s\tilde{\mathbf{E}} - \tilde{\mathbf{A}}$ with algebraic multiplicity p_i then $\sum_{i=1}^{\nu} p_i = n$. Consequently, λ_i is a finite eigenvalue of the pencil (5.4) with the same properties. The proof is completed. ∎

If $\mathbf{A}_{22}$ is a singular matrix then we may use Proposition 3.2 to obtain the solutions of (5.3) and Corollary 3.2 to study uniqueness of solutions if the initial conditions are given. To sum up in the case of a singular $\mathbf{A}_{22}$ here are the steps we have to undertake when we need to provide the solutions of (5.3):

- Step 1. By using the pencil (5.4) of (5.3) we compute the matrix polynomial $s\mathbf{E} - \mathbf{A} = \varpi(s)$, where

$$\mathbf{E} = \begin{bmatrix} \mathbf{T} & \mathbf{0}_{n,l} \\ \mathbf{R} & \mathbf{0}_{l,l} \end{bmatrix}, \quad \mathbf{A} = \begin{bmatrix} \mathbf{A}_{11} & \mathbf{A}_{12} \\ \mathbf{A}_{21} & \mathbf{A}_{22} \end{bmatrix}.$$

- Step 2. The pencil has finite/infinite eigenvalues since $\mathbf{E}$ is singular. We compute the finite eigenvalues which are the roots of $\varpi(s)$.
- Step 3. We form the Jordan matrices $\mathbf{J}_p$, $\mathbf{H}_q$ as described in Proposition 3.2.
- Step 4. We compute the left, and right eigenvectors of the eigenvalues of $s\mathbf{E} - \mathbf{A}$ in order to form the matrices $\mathbf{P}$, $\mathbf{Q}$, respectively as described in Proposition 3.2.
- Step 5. We use the formula (3.8) to provide the solution of (5.3) since $\mathbf{E}$ is singular.
- Step 6. For given initial conditions, since $\mathbf{E}$ is singular we use Corollary 3.2 to identify if the initial conditions are consistent or non-consistent. In the case that they are consistent we provide the unique solution by using the formula in Corollary 3.2.

5.2 Power Systems Modeled as Implicit DAEs

Implicit DAEs have been presented only very recently in power system analysis [115]. The observation that motivated [115] is the following. There are infinite equivalent ways to formulate a set of DAEs. Which, among these infinite ways, is the one that leads to the highest sparsity of the Jacobian matrices of the system?

The answer to this question appears to be to give up the conventional explicit formulation and utilize a (semi)implicit one. Section 1.3.7 shows a relevant example of implicit formulation that results in a much more compact form than the explicit one, namely the model of the PSS which includes a series of N lead-lag blocks – compare equations (1.144) and (1.146). Reference [115] also shows that $\mathbf{A}$ is 10 to 15% sparser than $\mathbf{A}$ for large power systems and that the non-null elements of $\mathbf{E}$ are a subset of $\mathbf{A}$. These properties lead to reduced memory requirements and faster matrix-vector and factorization operations for the implicit DAEs than for the explicit DAEs.

In the following example, we revisit the small-signal stability analysis for the three-bus system described in Appendix A and provide the General Eigenvalue Problem (GEP) formulation. The reader is invited to compare this example with Example 4.2 presented in Chapter 4, that presents the LEP for the same system. The finite eigenvalues are obviously the same, however the structure and the sparsity of the Jacobian matrices for the two formulations is different.

Example 5.1 (GEP of the three-bus system). We study the GEP of the three-bus system. The system is linearized around the base-case equilibrium point. The full vector of the $n = 19$ state variables and $l = 37$ algebraic variables is as follows:

$$\begin{aligned}\boldsymbol{\xi} = [\, & \delta_{\mathrm{r},1}\ \omega_{\mathrm{r},1}\ e'_{\mathrm{r,d},1}\ e'_{\mathrm{r,q},1}\ e''_{\mathrm{r,d},1}\ e''_{\mathrm{r,q},1}\ \psi_{\mathrm{s,d},1}\ \psi_{\mathrm{s,q},1}\ \delta_{\mathrm{r},2}\ \omega_{\mathrm{r},2}\ e'_{\mathrm{r,d},2}\ e'_{\mathrm{r,q},2} \\ & e''_{\mathrm{r,d},2}\ e''_{\mathrm{r,q},2}\ \psi_{\mathrm{s,d},2}\ \psi_{\mathrm{s,q},2}\ v_{\mathrm{R},1}\ v_{\mathrm{R},2}\ v_{\mathrm{ef},1}\ v_{\mathrm{ef},2}\ v_{\mathrm{a},1}\ v_{\mathrm{a},2}\ v_{\mathrm{f},1}\ v_{\mathrm{f},2}\ v_{\mathrm{B1,d}} \\ & v_{\mathrm{B2,d}}\ v_{\mathrm{B3,d}}\ v_{\mathrm{B1,q}}\ v_{\mathrm{B2,q}}\ v_{\mathrm{B3,q}}\ \imath_{\mathrm{D1,d}}\ \imath_{\mathrm{D2,d}}\ \imath_{\mathrm{D3,d}}\ \imath_{\mathrm{D1,q}}\ \imath_{\mathrm{D2,q}}\ \imath_{\mathrm{D3,q}}\ P_{\mathrm{m},1} \\ & \tilde{v}_{\mathrm{ef},1}\ v_{1,\mathrm{d}}\ v_{1,\mathrm{q}}\ \imath_{1,\mathrm{d}}\ \imath_{1,\mathrm{q}}\ P_{\mathrm{e},1}\ Q_{\mathrm{e},1}\ P_{\mathrm{m},2}\ \tilde{v}_{\mathrm{ef},2}\ v_{2,\mathrm{d}}\ v_{2,\mathrm{q}}\ \imath_{2,\mathrm{d}}\ \imath_{2,\mathrm{q}}\ P_{\mathrm{e},2} \\ & Q_{\mathrm{e},2}\ v_1\ v_2\ v_1^{\mathrm{ref}}\ v_2^{\mathrm{ref}}\,]^{\mathrm{T}}\,,\end{aligned}$$

where the subscript B denotes variables of the system buses; the subscript D denotes variables of the system loads.

The linearized system can be written in the form of (1.156), where in this example $\mathbf{E}$, $\mathbf{A}$ are 56×56 matrices. The matrix pencil $s\mathbf{E} - \mathbf{A}$ of the system has in total 19 finite eigenvalues, plus the infinite eigenvalue with multiplicity 37. The rightmost eigenvalues of the system are shown in Figure 5.1.

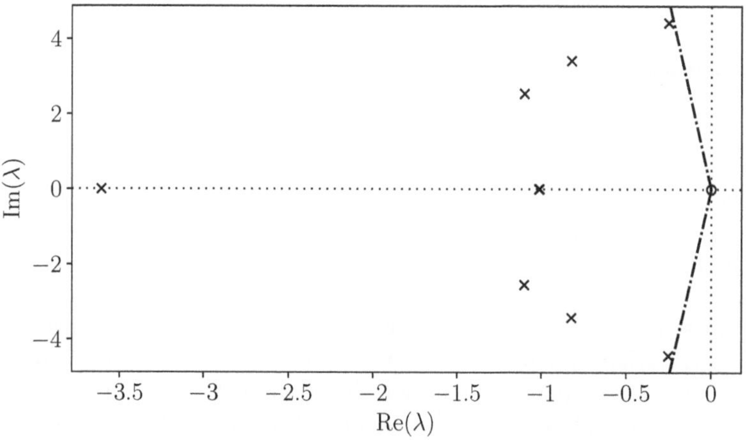

FIGURE 5.1: Relevant eigenvalues for the three-bus system.

The matrices $\mathbf{A}$, $\mathbf{E}$ are given in Appendix C. Matrix $\mathbf{E}$ is very sparse. In particular, it has in total $56 \cdot 56 = 3,136$ entries but only 23 of them are non-zero, which yields a sparsity degree of 99.27%. In fact, the larger a power system model is, the larger the sparsity degree of both $\mathbf{E}$, $\mathbf{A}$. This is a feature of the formulation in (1.156), as opposed to (1.160), in which the state matrix $\mathbf{A}_{\mathrm{S}}$ is dense. □

5.3 Floquet Multipliers

In this section, we are concerned with periodic solutions of hybrid (5.2), i.e. implicit DAEs with discontinuous variables $\boldsymbol{u}$:

$$\mathbf{E}(t)\,\dot{\boldsymbol{\xi}}(t) = \boldsymbol{\varphi}(\boldsymbol{\xi}(t), \boldsymbol{u}(t)) \,, \tag{5.5}$$

i.e. solutions that satisfy the condition:

$$\boldsymbol{\xi}(t+T) = \boldsymbol{\xi}(t) \,, \tag{5.6}$$

where T is the period of the solution. Periodic solutions can be forced oscillations, of which ac power systems in stationary conditions are probably the most common and well-known practical application, or limit cycles originated by some sort of resonance, e.g. the electromechanical oscillations of synchronous machines with no damping described in Examples 1.9 and 1.12.

Whenever (5.5) admits a T-periodic solution, say γ_T, its stability can be studied through the Floquet multipliers, i.e. the eigenvalues of its monodromy matrix, which we define below.

The goal of this section is not to provide a complete treatise on the stability analysis of periodic solutions of implicit DAEs. The interested reader can refer to [89, 16] for a more rigorous discussion.

Definition 5.1 (Fundamental matrix solution). If $\boldsymbol{\varphi}$ is smooth, i.e. continuous and differentiable, the fundamental solution matrix of (5.5) with initial condition $\boldsymbol{\xi}(t_o) = \boldsymbol{\xi}_o$, and its $\boldsymbol{\Xi}(t, t_o)$ fundamental solution matrix is the solution of the *variational equation*:

$$\mathbf{E}(t)\,\dot{\boldsymbol{\Xi}}(t, t_o) = \mathbf{A}(t)\,\boldsymbol{\Xi}(t, t_o) \,, \tag{5.7}$$

with the initial condition $\boldsymbol{\Xi}(t_o, t_o) = \mathbf{I}_m$ and where $\mathbf{A}(t) = \frac{\partial \boldsymbol{\varphi}}{\partial \boldsymbol{\xi}}(t)$.

Remark 5.1. $\boldsymbol{\Xi}(t, t_o)$ provides the sensitivities of the solution of $\mathbf{E}\,\dot{\boldsymbol{\xi}} = \boldsymbol{\varphi}(\boldsymbol{\xi}, \boldsymbol{u})$ with respect to $\boldsymbol{\xi}_o$. This property has found many applications to electronic circuits, e.g. oscillators. There are also some studies based on the fundamental solution matrix in power system analysis, in particular the works by Hiskens, e.g. [71] and, more recently [55], where second-order sensitivities are considered.

Definition 5.2 (Monodromy matrix). If $\boldsymbol{\xi}(t_o) \in \gamma_T$, then $\boldsymbol{\xi}(t_o) = \boldsymbol{\xi}(t_o + T)$ and $\boldsymbol{\mathcal{M}} \equiv \boldsymbol{\Xi}(T + t_o, t_o)$ is called *monodromy matrix*.

Definition 5.3 (Floquet multipliers). The eigenvalues of the monodromy matrix are the λ_k $(k = 1, ..., n)$ *Floquet multipliers* associated to γ_T. If the condition $|\lambda_k| \leq 1\ \forall k$ holds, then γ_T is a stable limit cycle [47]. Floquet multipliers are also called Floquet exponents or characteristic exponents.

Since the variational equation (5.7) can be solved in parallel with the DAEs in (5.5), and since in most cases, the numerical integration of the system is obtained through an implicit method, the Jacobian matrix $\mathbf{A}(t)$ does not need to be calculated on purpose for the integration of the fundamental matrix equation. Then, the computation of the fundamental solution matrix can be achieved using a forward sensitivity analysis [103].

Unfortunately, $\Xi(t, t_o)$, while is very sparse at t_o, quickly becomes full, which makes its calculation and storage both time- and memory-consuming for large systems. To overcome this issue, the solution of the variational equation can be conveniently obtained as a by-product of the integration of the original DAEs [6]. A computationally efficient approach based on the implicit DAE formulation is given in [67].

The basic definition above can be generalized in various ways, e.g. by taking into account discontinuities and delays. For the sake of example, we consider the case for which φ in (5.5) is not smooth, e.g. because variables $\boldsymbol{u}$ jump. These conditions lead to hybrid DAEs with discontinuous right-hand side, which makes the definition of the monodromy matrix $\boldsymbol{\mathcal{M}}$ more involved. The main difficulty is that $\mathbf{A}(t)$ is not defined at the points where φ is not continuous.

Let us assume that an event occurs at $t = t_1$. Such an event can be monitored by a proper switching manifold:

$$h(\boldsymbol{\xi}) = 0 \, ,$$

where $h(\boldsymbol{\xi}) : \mathbb{R}^{m \times 1} \mapsto \mathbb{R}$. Relevant examples of manifolds are the windup limiter (1.120) of the TG described in Section 1.3.5 and the anti-windup limiters (1.130) and (1.131) of the AVR described in Section 1.3.6. The case of multiple events can be easily deduced by defining a vector of manifolds $\boldsymbol{h}$. To model the aforementioned manifolds, one can usually adopt *if-then* rules as in hybrid automata [70].

We are now ready to provide the following definition.

Definition 5.4 (Saltation matrix). If (5.5) admits a limit cycle γ_T with a single discontinuity of φ at, say, $\boldsymbol{\xi}_1$, occurring at $t_1 \in (t_o, t_o + T)$, $\boldsymbol{\mathcal{M}}$ is computed as $\Xi(T + t_o, t_1)\, \mathbf{S}\, \Xi(t_1, t_0)$ where $\mathbf{S}$ is the *saltation matrix* operator, i.e. a proper correction factor to be inserted when the trajectory satisfies the condition $h(\boldsymbol{\xi}(t_1)) = 0$ [45]. The saltation matrix is computed as:

$$\mathbf{S} = \mathbf{I}_n + \frac{\varphi(\boldsymbol{\xi}_1, \boldsymbol{u}_1^+) - \varphi(\boldsymbol{\xi}_1, \boldsymbol{u}_1^-)}{\dfrac{\partial h}{\partial \boldsymbol{\xi}} \varphi(\boldsymbol{\xi}_1, \boldsymbol{u}_1^-)} \left(\frac{\partial h}{\partial \boldsymbol{\xi}} \right)^{\mathrm{T}} , \tag{5.8}$$

where $\boldsymbol{u}_1^-$ and $\boldsymbol{u}_1^+$ are the values of the input variable right before and after, respectively, the event. It is easy to observe that, if the event does not change the structure of the differential equations, i.e. $\varphi(\boldsymbol{\xi}_1, \boldsymbol{u}_1^+) = \varphi(\boldsymbol{\xi}_1, \boldsymbol{u}_1^-)$, which is the case of smooth φ, then $\mathbf{S} = \mathbf{I}_n$ and the event has no effect on the monodromy matrix. Note that more general formulations of the saltation matrix involve the definition of reset functions that define the values of the state variable before and after the event.

Applications of the saltation matrix for the determination of the Floquet multipliers in power systems can be found in [96] where the discontinuity is caused by a dead-band in the input signal of the PSS and in Chapter 12 of [124] where the discontinuity is caused by the variations of the tap ratio of an ULTC. In practice, however, the Floquet theory has marginal applications to power systems. This is due to two reasons. On one hand, in transient stability models, the fundamental frequency oscillation at ω_o is removed from the system as Park vectors project all ac quantities onto a reference frame rotating at ω_o. The fundamental frequency oscillation with period $2\pi\omega_o^{-1}$ thus disappears from the model. On the other hand, limit cycles originated by resonant devices and controllers are either stable, and thus visible in the time domain simulations, or unstable and thus have no effect on the stationary conditions of the system.

It is nevertheless useful to have a tool to evaluate the properties of limit cycles as these, similarly to bifurcation points, are generally the boundary between different dynamic behaviors of the system. The Floquet multipliers provide precisely this kind of information.

Proposition 5.1 (Properties of the Floquet multipliers)**.** Assuming that the rank of $s\mathbf{E} - \mathbf{A}$ is n, $n \leq m$, then $l = m - n$ eigenvalues of the monodromy matrix are null and one eigenvalue is always $\lambda_1 = 1$. The remaining $n - 1$ eigenvalues λ_i, $i = 2, \dots, n$ of the monodromy matrix provide the following information:

- $|\lambda_i| < 1$, the system is asymptotically stable.
- $|\lambda_i| = 1$, the system has a pseudo-periodic solution.
- $\lambda_i = \pm 1$, the system has a periodic solution $\boldsymbol{\gamma}_T$.
- $|\lambda_i| > 1$, the trajectories diverge.

The unit circle is thus the stability boundary for the Floquet multipliers, in the same way it is for discrete maps. This comes with no surprise as the monodromy matrix satisfies the condition:

$$\boldsymbol{\Xi}(t + T, t_o) = \boldsymbol{\Xi}(t, t_o)\,\boldsymbol{\mathcal{M}}\,, \tag{5.9}$$

which descends from the fact that $\boldsymbol{\mathcal{M}}$ is time-independent and formally resembles the structure of the fixed point of a linear discrete map where t corresponds to step (i) and $t + T$ corresponds to step $(i + 1)$ – see Section 1.2 and, in particular, equation (1.27).

Example 5.2 (Floquet multipliers of the three-bus system). Section 1.7 provides a taxonomy of the stability of dynamic systems and indicates that stationary solutions such as limit cycles, i.e. undamped oscillations beside the fundamental frequency, are a possible behavior of power systems.

Limit cycles can be triggered by a sufficiently high loading level and/or high gains of the AVRs and PSSs. In this example, we consider the three-bus system with inclusion of a PSS as in Example 4.3 and optimal parameters except for the wash-out filter (WF) gain of the PSS, which is set to $K_{\rm w} = 50$. This leads to an unstable operating point at the base-case loading conditions.

Figure 5.2 shows the limit cycle obtained by perturbing the system with $\Delta\omega_{\rm r,1} = 10^{-3}$ pu(Hz) and including TGs and CoI to avoid the drift of rotor angles and speeds of the synchronous machines. The parameters of the TGs are $\mathcal{R} = 0.05$ pu(MW) on machine bases, $T_{\rm g} = 0.1$ s, $T_{\rm sm} = 0.45$ s, $T_{\rm t} = 0$ s, $T_{\rm rh} = 50$ s, and $\alpha_{\rm rh} = 0.25$.

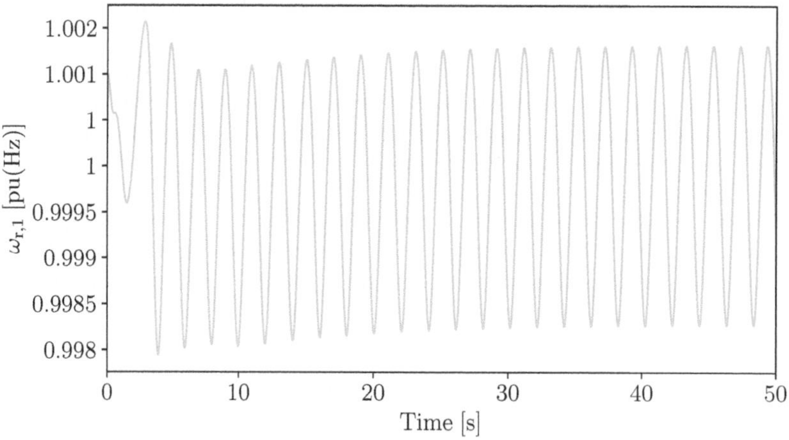

FIGURE 5.2: Limit cycle of the three-bus system obtained with a PSS gain $K_{\rm w} = 50$.

The model is a set of DAEs with $n = 29$ state variables and $l = 38$ algebraic variables. The period of the limit cycle is $T \approx 2.019$ s, and the largest-magnitude Floquet multipliers obtained for $\boldsymbol{\mathcal{M}}(47\,{\rm s} + T)$ and assuming $\boldsymbol{\mathcal{M}}(47\,{\rm s}) = \mathbf{I}_{29}$ are $\lambda_1 = \lambda_2 = 1.00$ and $\lambda_3 = 0.960$. Note also that 38 eigenvalues of the monodromy matrix are null.

As stated in Proposition 5.1, one Floquet multiplier is always 1, and is due to the fact that the DAEs that model the system are autonomous, i.e. t does not appear explicitly in the equations. The second unitary Floquet multiplier is counterpart of the null eigenvalue of the state matrix of a power system where no synchronous machine angle is taken as phase reference, as discussed in Example 1.11. Finally, the third eigenvalue $\lambda_3 \approx 1$ confirms the existence of a stable limit cycle. □

5.4 Lyapunov Exponents

Lyapunov characteristic exponents or, simply, Lyapunov exponents (LEs) are measures that capture the sensitivity of the trajectories of a dynamic system to its initial conditions. Their sign defines whether two orbits that start close together converge (or diverge) as time progresses. More precisely, the definition of LEs is the following.

Definition 5.5 (Lyapunov exponents). Let us consider again the fundamental matrix solution of an implicit set of DAEs given in equation (5.7). The Lyapunov exponents are the eigenvalues of the matrix $\mathbf{\Lambda}$, which is defined as:

$$\mathbf{\Lambda} = \lim_{t \to \infty} \frac{1}{2t} \log \left(\mathbf{\Xi}(t, t_o) \, \mathbf{\Xi}^{\mathrm{T}}(t, t_o) \right) . \tag{5.10}$$

In practical terms, the Lyapunov exponents define whether a small perturbation of a system trajectory, say $\delta\boldsymbol{\xi}(t_o)$, increases or decreases with time:

$$|\delta\boldsymbol{\xi}(t)| \approx \mathrm{e}^{\lambda t} |\delta\boldsymbol{\xi}(t_o)| ,$$

where λ is a LE. From the expression above, it is clear that if $\lambda > 0$ the trajectory diverges, while if $\lambda < 0$ the trajectory converges. The analogy with the eigenvalues of the state matrix of a linearized system at the equilibrium point is apparent. However, it is important to note that "instability" in this case, does not refer to the fact that the system actually collapses.

LEs are utilized principally to identify chaotic orbits and strange attractors, e.g. trajectories that are bounded, aperiodic and infinitely sensitive to initial conditions. Since chaotic orbits are a sort of stationary conditions, very similar in many aspects to stationary stochastic processes, it has to be expected that there exists a set of eigenvalues, in this case, the LEs, that are able to quantitatively define the properties of the flow of the system.

Proposition 5.2 (Properties of Lyapunov exponents). The following are relevant properties of LEs:

- If the trajectory does not converge to an equilibrium point, one LE is always null as this is the eigenvalue associated with the eigenvector in the direction of the flow.
- If at least one LE is positive, then the trajectory is chaotic.
- The sum of all LEs is zero for conservative systems (no losses) and negative for dissipative (lossy) systems.

The main issue with LEs is their calculation, which is cumbersome even for relatively small systems, even if sparsity and implicit formulation are taken into account [67]. However, if one is interested only in the largest Lyapunov exponent (LLE), i.e. only in knowing whether the flow is chaotic or not, then the largest LLE, say λ_1, can then be estimated with:

$$\lambda_1 \approx \frac{1}{t} \ln \|\Xi(t, t_o)\, \mathbf{u}_o\| \,, \tag{5.11}$$

where $\mathbf{u}_o$ is a randomly generated vector of order m and $\|\cdot\|$ is the Euclidean norm. There are three situations to consider for λ_1:

- $\lambda_1 < 0$: The flow is attracted to a stable fixed point or stable periodic orbit.
- $\lambda_1 = 0$: The orbit is a neutral fixed point.
- $\lambda_1 > 0$: The flow is unstable and chaotic.

Note that the rationale behind (5.11) is the well-known power method, which is described in Chapter 13.

Example 5.3 (Lyapunov exponents of the three-bus system)**.** Similarly to limit cycles, chaotic orbits are typically originated by a sufficiently high loading level and/or high gains of the AVRs and PSSs. In this example, we consider again the three-bus system with inclusion of a PSS as in Example 5.2 and same set-up and parameters except for the amplifier gain of the AVR, which is set to $K_{\rm a} = 400$, and the WF gain of the PSS, which is set to $K_{\rm w} = 200$. These values are clearly out of range as it is typical in most examples of power systems that show deterministic chaos [91]. This does not exclude, however, the possibility that chaos occurs in power systems in normal operating conditions and with properly tuned controllers.

Figure 5.3 shows the chaotic orbit obtained by perturbing the system with $\Delta\omega_{\rm r,1} = 10^{-3}$ pu(Hz). The lack of a period is apparent from the state-space trajectory. The confirmation that this is in fact a chaotic orbit is given by Figure 5.4 that shows the LLE for the three-bus system as function of time. Note that the definition of Lyapunov exponents would require to run the simulation for $t \to \infty$. This is obviously impossible. Besides, the calculation of (5.10) for large simulation times tends to be numerically unstable and to cause overflows. The simulation, thus, has to be sufficiently "long" to allow for the Lyapunov exponents to settle but sufficiently "short" to avoid numerical issues. □

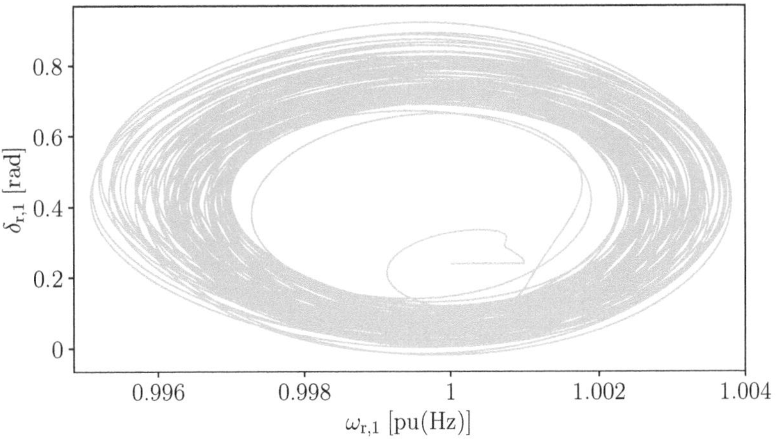

FIGURE 5.3: Chaotic orbit of the three-bus system obtained with an amplifier gain of the AVR $K_a = 400$ and a PSS gain $K_w = 200$.

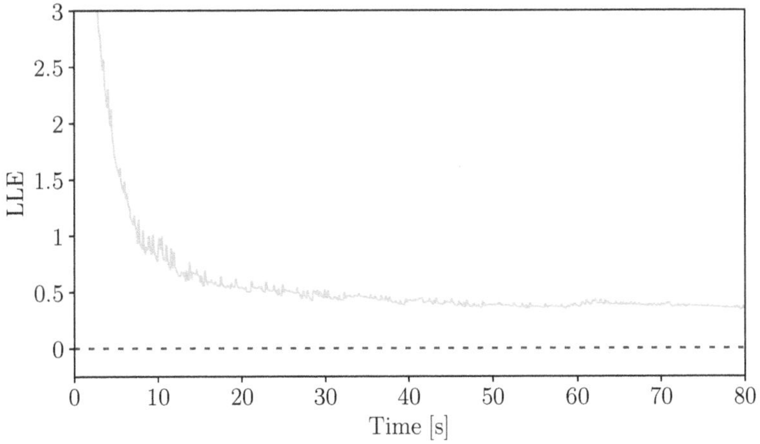

FIGURE 5.4: LLE for the three-bus system obtained with an amplifier gain of the AVR $K_a = 400$ and a PSS gain $K_w = 200$.

6

Differential Equations with Singular Pencil

6.1 Formulation

In this chapter we consider the following system of differential equations:

$$\mathbf{E}\,\dot{\boldsymbol{x}}(t) = \mathbf{A}\,\boldsymbol{x}(t) + \boldsymbol{V}(t)\,, \tag{6.1}$$

where $\mathbf{E}, \mathbf{A} \in \mathbb{C}^{r\times m}$, $\boldsymbol{x} : [0, +\infty] \mapsto \mathbb{C}^{m\times 1}$, and $\boldsymbol{V} : [0, +\infty] \mapsto \mathbb{C}^{r\times 1}$. We will focus on the case that the matrices $\mathbf{E}$ and $\mathbf{A}$ are non-square ($r \neq m$), or square ($r = m$) but with $\det(s\mathbf{E} - \mathbf{A}) \equiv 0$. This means that we are interested in system (6.1) with a singular pencil. For this type of pencil we will firstly study the existence of solutions of system (6.1).

Lemma 6.1 (Singular pencil)**.** Given $\mathbf{E}, \mathbf{A} \in \mathbb{C}^{r\times m}$, and an arbitrary $s \in \mathbb{C}$, if pencil $s\mathbf{E} - \mathbf{A}$ is singular and $r > m$, then there exists a matrix function $\tilde{\mathbf{P}} : \mathbb{C} \mapsto \mathbb{R}^{r\times r}$ (which can be computed via the Gauss-Jordan elimination method) such that

$$\tilde{\mathbf{P}}(s)(s\mathbf{E} - \mathbf{A}) = \begin{bmatrix} \tilde{\mathbf{A}}(s) \\ \mathbf{0}_{r_1,m} \end{bmatrix}, \quad \text{with} \quad \tilde{\mathbf{P}}(s) = \begin{bmatrix} \tilde{\mathbf{P}}_1(s) \\ \tilde{\mathbf{P}}_2(s) \end{bmatrix}, \tag{6.2}$$

where $\tilde{\mathbf{A}} : \mathbb{C} \mapsto \mathbb{R}^{m_1\times m}$, with $m_1 + r_1 = r$, is a matrix such that if $[\tilde{a}_{ij}]_{1\leq i\leq m_1}^{1\leq j\leq m}$ are its elements, for $i = j$ all elements are non-zero and for $i \neq j$ all elements are zero and $\tilde{\mathbf{P}}_1(s) \in \mathbb{R}^{m_1\times r}$, $\tilde{\mathbf{P}}_2(s) \in \mathbb{R}^{r_1\times r}$.

We are now ready to state the following Theorem:

Theorem 6.1 (Existence of the solution)**.** We assume system (6.1) with a singular pencil. Then there exist solutions for (6.1) if and only if $r > m$ and

$$\tilde{\mathbf{P}}_2(s)(\mathbf{E}\boldsymbol{C} + \boldsymbol{U}(s)) = \mathbf{0}_{m_1,1}, \quad \text{and} \quad m_1 = m\,, \tag{6.3}$$

where $\tilde{\mathbf{P}}_2(s)$ is defined in (6.2).

Proof. Let $\mathcal{L}\{\boldsymbol{x}(t)\} = \boldsymbol{z}(s)$, $\mathcal{L}\{\boldsymbol{V}(t)\} = \boldsymbol{U}(s)$ be the Laplace transform of $\boldsymbol{x}(t), \boldsymbol{V}(t)$ respectively. By applying the Laplace transform $\mathcal{L}$ into (6.1), we get:

$$\mathbf{E}\mathcal{L}\{\dot{\boldsymbol{x}}(t)\} = \mathbf{A}\mathcal{L}\{\boldsymbol{x}(t)\} + \mathcal{L}\{\boldsymbol{V}(t)\}\,,$$

or, equivalently,

$$\mathbf{E}\big(s\boldsymbol{z}(s) - \boldsymbol{x}_o\big) = \mathbf{A}\boldsymbol{z}(s) + \boldsymbol{U}(s)\,,$$

where $\boldsymbol{x}_o = \boldsymbol{x}(0)$, i.e. the initial condition of (6.1). Since we assume that $\boldsymbol{x}_o$ is unknown, we can use an unknown constant vector $\boldsymbol{C} \in \mathbb{C}^{m\times 1}$ and give to the above expression the following form:

$$(s\mathbf{E} - \mathbf{A})\boldsymbol{z}(s) = \mathbf{E}\boldsymbol{C} + \boldsymbol{U}(s)\,.$$

Since the pencil of (6.1) is assumed singular, if $r = m$ then $\det(s\mathbf{E} - \mathbf{A}) \equiv 0$ and hence the polynomial matrix $s\mathbf{E} = \mathbf{A}$ is not invertible. Consequently $\boldsymbol{z}(s)$ can not be defined in the above expression which means that there do not exist solutions in this case for (6.1). This leaves us with two cases. The first is $r > m$. The second case is $r < m$. If $r > m$ then there exists a matrix function $\tilde{\mathbf{P}} : \mathbb{C} \mapsto \mathbb{R}^{r\times r}$ such that

$$\tilde{\mathbf{P}}(s)(s\mathbf{E} - \mathbf{A}) = \begin{bmatrix} \tilde{\mathbf{A}}(s) \\ \mathbf{0}_{r_1,m} \end{bmatrix},$$

where $\tilde{\mathbf{A}} : \mathbb{C} \mapsto \mathbb{R}^{m_1\times m}$, with $m_1 + r_1 = r$, is a matrix such that if $[\tilde{a}_{ij}]_{1\leq i\leq m_1}^{1\leq j\leq m}$ are its elements, for $i = j$ all elements are non-zero and for $i \neq j$ all elements are zero. Then by setting

$$\tilde{\mathbf{P}}(s) = \begin{bmatrix} \tilde{\mathbf{P}}_1(s) \\ \tilde{\mathbf{P}}_2(s) \end{bmatrix},$$

where $\tilde{\mathbf{P}}_1(s) \in \mathbb{R}^{m_1\times r}$, $\tilde{\mathbf{P}}_2(s) \in \mathbb{R}^{r_1\times r}$, system (6.1) takes the form:

$$\begin{bmatrix} \tilde{\mathbf{A}}(s) \\ \mathbf{0}_{r_1,m} \end{bmatrix} \boldsymbol{z}(s) = \begin{bmatrix} \tilde{\mathbf{P}}_1(s) \\ \tilde{\mathbf{P}}_2(s) \end{bmatrix} \mathbf{E}\boldsymbol{C},$$

from where we get

$$\tilde{\mathbf{A}}(s)\boldsymbol{z}(s) = \tilde{\mathbf{P}}_1(s)\mathbf{E}\boldsymbol{C}\,, \quad \text{and} \quad \mathbf{0}_{r_1,m}\boldsymbol{z}(s) = \tilde{\mathbf{P}}_2(s)(\mathbf{E}\boldsymbol{C} + \boldsymbol{U}(s))\,.$$

If $\mathbf{0}_{r_1,m}\boldsymbol{z}(s) = \tilde{\mathbf{P}}_2(s)(\mathbf{E}\boldsymbol{C} + \boldsymbol{U}(s))$ holds, then $\boldsymbol{z}(s)$ can be defined in $\tilde{\mathbf{A}}(s)\boldsymbol{z}(s) = \tilde{\mathbf{P}}_1(s)\mathbf{E}\boldsymbol{C}$ if $m_1 = m$. Hence, $\boldsymbol{z}(s)$ can be defined, and consequently $\boldsymbol{x}(t)$ always exists and is given by $\boldsymbol{x}(t) = \mathcal{L}^{-1}\{\tilde{\mathbf{A}}(s)^{-1}\mathbf{E}\boldsymbol{C}\}$ if and only if (6.3) holds. In any other case we have more unknown functions than equations or no solutions. If $r < m$ there are at least $m - r$ unknown functions and m equations. Hence $\boldsymbol{z}(s)$ can not be defined uniquely. To sum up, there exist solutions for the system if $r > m$, $\tilde{\mathbf{A}}(s)$ $m \times m$, and $\tilde{\mathbf{P}}_2(s)(\mathbf{E}\boldsymbol{C} + \boldsymbol{U}(s)) = \mathbf{0}_{m-r,1}$. The proof is completed. ■

Since we proved that there exist solutions for (6.1) for $r > m$ and under the conditions described in Theorem 6.1, we are interested in the case of system (6.1) with $r > m$. We state the following Theorem:

Theorem 6.2 (Existence of the solution). Consider system (6.1) with $r > m$ and

$$\tilde{\mathbf{P}}_2(s)(\mathbf{E}\boldsymbol{C} + \boldsymbol{U}(s)) = \mathbf{0}_{m_1,1}, \quad m_1 = m.$$

In this case the general solution is given by:

$$\boldsymbol{x}(t) = \boldsymbol{\Psi}_0(t)\boldsymbol{C} + \boldsymbol{\Psi}(t), \tag{6.4}$$

where $\boldsymbol{\Psi}_0(t) = \mathcal{L}^{-1}\{\mathbf{A}^{-1}(s)\tilde{\mathbf{P}}_1(s)\mathbf{E}\}$, $\boldsymbol{\Psi}(t) = \mathcal{L}^{-1}\{\mathbf{A}^{-1}(s)\tilde{\mathbf{P}}(s)\boldsymbol{U}(s)\}$, $\mathbf{A}(s)$, $\tilde{\mathbf{P}}(s)$, $\tilde{\mathbf{P}}_1(s)$, $\tilde{\mathbf{P}}_2(s)$ are defined in (6.2), $\boldsymbol{C} \in \mathbb{R}^{m\times 1}$ is an unknown constant vector. If

$$\mathcal{L}^{-1}\{\mathbf{A}^{-1}(s)\tilde{\mathbf{P}}_1(s)\} = \boldsymbol{\Psi}_1(t),$$

by using the convolution theorem, an alternative formula is:

$$\boldsymbol{x}(t) = \boldsymbol{\Psi}_0(t)\boldsymbol{C} + \int_0^\infty \boldsymbol{\Psi}_1(t-\tau)\boldsymbol{V}(\tau)d\tau.$$

Proof. By applying the Laplace transform $\mathcal{L}$ into (6.1), we get

$$\mathbf{E}\mathcal{L}\Big\{\dot{\boldsymbol{x}}(t)\Big\} = \mathbf{A}\mathcal{L}\{\boldsymbol{x}(t)\} + \mathcal{L}\{\boldsymbol{V}(t)\},$$

or, equivalently,

$$\mathbf{E}(s\boldsymbol{z}(s) - \boldsymbol{x}_o) = \mathbf{A}\boldsymbol{z}(s) + \boldsymbol{U}(s),$$

where $\boldsymbol{x}_o = \boldsymbol{x}(0)$, i.e. the initial condition of (6.1). Since we assume that $\boldsymbol{x}_o$ is unknown, we can use an unknown constant vector $\boldsymbol{C} \in \mathbb{C}^{m\times 1}$ and give to the above expression the form:

$$(s\mathbf{E} - \mathbf{A})\boldsymbol{z}(s) = \mathbf{E}\boldsymbol{C} + \boldsymbol{U}(s).$$

If $r > m$ then there exists a matrix function $\tilde{\mathbf{P}} : \mathbb{C} \mapsto \mathbb{R}^{r\times r}$ such that

$$\tilde{\mathbf{P}}(s)(s\mathbf{E} - \mathbf{A}) = \begin{bmatrix} \boldsymbol{A}(s) \\ \mathbf{0}_{r_1,m} \end{bmatrix}.$$

where $\mathbf{A} : \mathbb{C} \mapsto \mathbb{R}^{m_1\times m}$, with $m_1 + r_1 = r$, is a matrix such that if $[a_{ij}]_{1\leq i\leq m_1}^{1\leq j\leq m}$ are its elements, for $i = j$ all elements are non-zero and for $i \neq j$ all elements are zero.

$$\tilde{\mathbf{P}}(s) = \begin{bmatrix} \tilde{\mathbf{P}}_1(s) \\ \tilde{\mathbf{P}}_2(s) \end{bmatrix},$$

where $\tilde{\mathbf{P}}_1(s) \in \mathbb{R}^{m_1\times r}$, $\tilde{\mathbf{P}}_2(s) \in \mathbb{R}^{r_1\times r}$. Then system (6.1) has a unique solution if and only if (6.3) holds. In any other case we have more unknown functions than equations or no solutions. If (6.3) holds then

$$\tilde{\mathbf{P}}(s)(s\mathbf{E} - \mathbf{A}) = \begin{bmatrix} \boldsymbol{A}(s) \\ \mathbf{0}_{r_1,m} \end{bmatrix},$$

and we have

$$\tilde{\mathbf{P}}(s)(s\mathbf{E} - \mathbf{A})\boldsymbol{z}(s) = \tilde{\mathbf{P}}(s)\mathbf{E}\boldsymbol{C} + \tilde{\mathbf{P}}(s)\boldsymbol{U}(s)\,,$$

or, equivalently,

$$\mathbf{A}(s)\boldsymbol{z}(s) = \tilde{\mathbf{P}}_1(s)\mathbf{E}\boldsymbol{C} + \tilde{\mathbf{P}}_1(s)\boldsymbol{U}(s)\,,$$

or, equivalently,

$$\boldsymbol{z}(s) = \boldsymbol{A}^{-1}(s)\tilde{\mathbf{P}}_1(s)\mathbf{E}\boldsymbol{C} + \boldsymbol{A}^{-1}\tilde{\mathbf{P}}_1(s)\boldsymbol{U}(s)\,.$$

The inverse Laplace transform of $\mathbf{A}^{-1}(s)\tilde{\mathbf{P}}_1(s)\mathbf{E}$ always exists because it is a matrix with elements fractions of fractional polynomials and with the order of the polynomial in the denominator always being higher than the order of the polynomial in the numerator. Let $\mathcal{L}^{-1}\{\mathbf{A}^{-1}(s)\tilde{\mathbf{P}}_1(s)\mathbf{E}\} = \boldsymbol{\Psi}_0(t)$ and $\mathcal{L}^{-1}\{\mathbf{A}^{-1}(s)\tilde{\mathbf{P}}(s)\boldsymbol{U}(s)\} = \boldsymbol{\Psi}(t)$. Then $\boldsymbol{x}(t)$ is given (6.4). Or, alternatively, if $\mathcal{L}^{-1}\{\mathbf{A}^{-1}(s)\tilde{\mathbf{P}}_1(s)\} = \boldsymbol{\Psi}_1(t)$, then by using the convolution theorem we have

$$\boldsymbol{x}(t) = \boldsymbol{\Psi}_0(t)\boldsymbol{C} + \int_0^\infty \boldsymbol{\Psi}_1(t-\tau)\boldsymbol{V}(\tau)d\tau\,.$$

The proof is completed. ■

Remark 6.1. For system (6.1), the general solution is given by (6.4). The constant column $\boldsymbol{C}$ in the general solution is an unknown constant vector related to the initial conditions of the system since we used the Laplace transform.

Remark 6.2. If there exist solutions for system (6.1), it is not guaranteed that for given initial conditions we will have a unique solution. If the given initial conditions are consistent, and there exist solutions for (6.1), then in the formula of the general solution (6.4) we replace $\boldsymbol{C} = \boldsymbol{x}_o$. However, if the given initial conditions are non-consistent but there exist solutions for (6.1), then the general solution (6.4) holds for $t > 0$ and the system is impulsive.

Remark 6.3. For the case that the initial conditions are consistent, the matrix functions $\boldsymbol{\Psi}_0(t)$ can have elements defined for $t > 0$. But the columns $\boldsymbol{\Psi}_0(t)\boldsymbol{x}_o$ have all its elements always defined for $t \geq 0$.

Other formulas based on the spectrum of the pencil

We consider system (6.1) with $r > m$, and provide some other formulas by using matrix pencil theory. In general, as already discussed in previous chapters, the class of $s\mathbf{E} - \mathbf{A}$ is then characterized by a uniquely defined element, known as the complex Kronecker canonical form, see [37, 39, 51], specified by the complete set of invariants of the singular pencil $s\mathbf{E} - \mathbf{A}$. This is the set of the finite-infinite eigenvalues and the minimal column-row indices. In the case of $r > m$ there exist only row minimal indices. Let $\mathcal{N}_l$ be the left null space of a matrix respectively. Then the equations $\boldsymbol{V}^{\mathrm{T}}(s)(s\mathbf{E}-\mathbf{A}) = \mathbf{0}_{1,m}$, have solutions in $\boldsymbol{V}(s)$, which are vectors in the rational vector spaces $\mathcal{N}_l(s\mathbf{E}-\mathbf{A})$. The binary vectors $\boldsymbol{V}^{\mathrm{T}}(s)$ express dependence relationships among the rows of $s\mathbf{E}-\mathbf{A}$. Note that $\boldsymbol{V}(s)\in\mathbb{C}^{r\times 1}$ are polynomial vectors. Let $t=\dim\mathcal{N}_l(s\mathbf{E}-\mathbf{A})$. It is known that $\mathcal{N}_l(s\mathbf{E}-\mathbf{A})$ as rational vector spaces are spanned

by minimal polynomial bases of minimal degrees

$$\zeta_1=\zeta_2=\cdots=\zeta_h=0<\zeta_{h+1}\leq\cdots\leq\zeta_{h+k=\beta}\,,$$

which is the set of *row minimal indices* of $s\mathbf{E}-\mathbf{A}$. This means there are β row minimal indices, but $\beta-h=k$ non-zero row minimal indices. We are interested only in the k non zero minimal indices. To sum up the invariants of a singular pencil with $r>m$ are the finite-infinite eigenvalues of the pencil and the minimal row indices as described above. Following the above given analysis, there exist non-singular matrices $\mathbf{P}$, $\mathbf{Q}$ with $\mathbf{P}\in\mathbb{C}^{r\times r}$, $\mathbf{Q}\in\mathbb{C}^{m\times m}$, such that

$$\begin{aligned}\mathbf{PEQ}&=\mathbf{E}_K=\mathbf{I}_p\oplus\mathbf{H}_q\oplus\mathbf{E}_\zeta\,,\\ \mathbf{PAQ}&=\mathbf{A}_K=\mathbf{J}_p\oplus\mathbf{I}_q\oplus\mathbf{A}_\zeta\,,\end{aligned}\tag{6.5}$$

where $\mathbf{J}_p$ is the Jordan matrix for the finite eigenvalues, $\mathbf{H}_q$ a nilpotent matrix with index q_* which is actually the Jordan matrix of the zero eigenvalue of the pencil $s\mathbf{A}-\mathbf{E}$. The matrices $\mathbf{E}_\zeta$, $\mathbf{A}_\zeta$ are defined as

$$\mathbf{E}_\zeta=\begin{bmatrix}\mathbf{I}_{\zeta_{h+1}}\\ \mathbf{0}_{1,\zeta_{h+1}}\end{bmatrix}\oplus\begin{bmatrix}\mathbf{I}_{\zeta_{h+2}}\\ \mathbf{0}_{1,\zeta_{h+2}}\end{bmatrix}\oplus\cdots\oplus\begin{bmatrix}\mathbf{I}_{\zeta_{h+k}}\\ \mathbf{0}_{1,\zeta_{h+k}}\end{bmatrix},$$

$$\mathbf{A}_\zeta=\begin{bmatrix}\mathbf{0}_{1,\zeta_{h+1}}\\ I_{\zeta_{h+1}}\end{bmatrix}\oplus\begin{bmatrix}\mathbf{0}_{1,\zeta_{h+2}}\\ \mathbf{I}_{\zeta_{h+2}}\end{bmatrix}\oplus\cdots\oplus\begin{bmatrix}\mathbf{0}_{1,\zeta_{h+k}}\\ I_{\zeta_{h+k}}\end{bmatrix},$$

with $p+q+\sum_{i=1}^{k}[\zeta_{h+i}]+k=r$, $p+q+\sum_{i=1}^{k}[\zeta_{h+i}]=m$. Finally, the matrices $\mathbf{P}$, $\mathbf{Q}$ can be written as

$$\mathbf{P}=\begin{bmatrix}\mathbf{P}_1\\ \mathbf{P}_2\\ \mathbf{P}_3\end{bmatrix},\quad \mathbf{Q}=\begin{bmatrix}\mathbf{Q}_p & \mathbf{Q}_q & \mathbf{Q}_\zeta\end{bmatrix},$$

with $\mathbf{P}_1\in\mathbb{C}^{p\times r}$, $\mathbf{P}_2\in\mathbb{C}^{q\times r}$, $\mathbf{P}_3\in\mathbb{C}^{\zeta_1\times r}$, $\zeta_1=k+\sum_{i=1}^{k}[\zeta_{h+i}]$ and $\mathbf{Q}_p\in\mathbb{C}^{m\times p}$, $\mathbf{Q}_q\in\mathbb{C}^{m\times q}$, $\mathbf{Q}_\zeta\in\mathbb{C}^{m\times\zeta_2}$ and $\zeta_2=\sum_{i=1}^{k}[\zeta_{h+i}]$.

By substituting the transformation $\boldsymbol{x}(t)=\mathbf{Q}\boldsymbol{z}(t)$ into (6.1) we obtain

$$\mathbf{EQ}\,\dot{\boldsymbol{z}}(t)=\mathbf{AQ}\,\boldsymbol{z}(t)+\boldsymbol{V}(t)\,,$$

whereby, multiplying by $\mathbf{P}$, and setting $\boldsymbol{z}(t)=\begin{bmatrix}\boldsymbol{z}_p(t)\\ \boldsymbol{z}_q(t)\\ \boldsymbol{z}_\zeta(t)\end{bmatrix}$, $\boldsymbol{z}_p(t)\in\mathbb{C}^{p\times 1}$, $\boldsymbol{z}_p(t)\in\mathbb{C}^{q\times 1}$ and $\boldsymbol{z}_\zeta(t)\in\mathbb{C}^{\zeta_2\times 1}$, we arrive at the subsystems

$$\dot{\boldsymbol{z}}_p(t)=\mathbf{J}_p\boldsymbol{z}_p(t)+\mathbf{P}_1\boldsymbol{V}(t)\,,$$

$$\mathbf{H}_q\, \dot{\boldsymbol{z}}_q(t) = \boldsymbol{z}_q(t) + \mathbf{P}_2 \boldsymbol{V}(t)\,,$$

and

$$\mathbf{E}_\zeta \dot{\boldsymbol{z}}_\zeta(t) = \mathbf{A}_\zeta \boldsymbol{z}_\zeta(t) + \mathbf{P}_3 \boldsymbol{V}(t)\,.$$

Recall that the solutions of the first two subsystems are:

$$\boldsymbol{z}_p(t) = \mathrm{e}^{\mathbf{J}_p t}\boldsymbol{C} + \int_0^\infty \mathrm{e}^{\mathbf{J}_p(t-\tau)}\boldsymbol{V}(\tau)d\tau\,,$$

and

$$\boldsymbol{z}_q(t) = -\sum_{i=0}^{q_*-1} \mathbf{H}_q^i \mathbf{P}_2 \boldsymbol{V}^{[i]}(t)\,.$$

For the third subsystem let

$$\boldsymbol{z}_\zeta(t) = \begin{bmatrix} \boldsymbol{z}_{\zeta_{h+1}}(t) \\ \boldsymbol{z}_{\zeta_{h+2}}(t) \\ \vdots \\ \boldsymbol{z}_{\zeta_{h+k}}(t) \end{bmatrix}, \quad \boldsymbol{z}_{\zeta_{h+i}}(t) \in \mathbb{C}^{(\zeta_{h+i})\times 1}, \quad i = 1, 2, \ldots, k\,, \tag{6.6}$$

with

$$\boldsymbol{z}_{\zeta_{h+i}}(t) = \begin{bmatrix} z_{\zeta_{h+i},1}(t) \\ z_{\zeta_{h+i},2}(t) \\ \vdots \\ z_{\zeta_{h+i},\zeta_{h+i}}(t) \end{bmatrix},$$

and

$$\mathbf{P}_3 \boldsymbol{V}(t) = \begin{bmatrix} \boldsymbol{U}_1(t) \\ \boldsymbol{U}_2(t) \\ \vdots \\ \boldsymbol{U}_k(t) \end{bmatrix}, \quad \boldsymbol{U}_i(t) \in \mathbb{C}^{(\zeta_{h+i}+1)\times 1}, \quad i = 1, 2, \ldots, k\,,$$

with

$$\boldsymbol{U}_i(t) = \begin{bmatrix} u_{i0} \\ u_{i1} \\ u_{i2} \\ \vdots \\ u_{i\zeta_{h+i}} \end{bmatrix}, \quad i = 1, 2, \ldots, k\,.$$

By replacing and using these relations we get

$$\begin{bmatrix} \mathbf{I}_{\zeta_{h+i}} \\ \mathbf{0}_{1,\zeta_{h+i}} \end{bmatrix} \dot{\boldsymbol{z}}_{\zeta_{h+i}}(t) = \begin{bmatrix} \mathbf{0}_{1,\zeta_{h+i}} \\ \mathbf{I}_{\zeta_{h+i}} \end{bmatrix} \boldsymbol{z}_{\zeta_{h+i}}(t) + \boldsymbol{U}_i(t)\,,$$

or, equivalently, by using the above expressions

$$\begin{bmatrix} 1 & 0 & \dots & 0 \\ 0 & 1 & \dots & 0 \\ \vdots & \vdots & \dots & \vdots \\ 0 & 0 & \dots & 1 \\ 0 & 0 & \dots & 0 \end{bmatrix} \begin{bmatrix} \dot{z}_{\zeta_{h+i},1}(t) \\ \dot{z}_{\zeta_{h+i},2}(t) \\ \vdots \\ \dot{z}_{\zeta_{h+i},\zeta_{h+i}}(t) \end{bmatrix} = \begin{bmatrix} 0 & 0 & \dots & 0 \\ 1 & 0 & \dots & 0 \\ \vdots & \vdots & \dots & \vdots \\ 0 & 0 & \dots & 0 \\ 0 & 0 & \dots & 1 \end{bmatrix} \begin{bmatrix} z_{\zeta_{h+i},1}(t) \\ z_{\zeta_{h+i},2}(t) \\ \vdots \\ z_{\zeta_{h+i},\zeta_{h+i}}(t) \end{bmatrix} + \begin{bmatrix} u_{i0} \\ u_{i1} \\ u_{i2} \\ \vdots \\ u_{i\zeta_{h+i}} \end{bmatrix},$$

or, equivalently,

$$\begin{aligned} \dot{z}_{\zeta_{h+i},1}(t) &= u_{i0}\,, \\ \dot{z}_{\zeta_{h+i},2}(t) &= z_{\zeta_{h+i},1}(t) + u_{i1}\,, \\ &\vdots \\ \dot{z}_{\zeta_{h+i},\zeta_{h+i}}(t) &= z_{\zeta_{h+i},\zeta_{h+i}-1}(t) + u_{i(\zeta_{h+i}-1)}\,, \\ 0 &= z_{\zeta_{h+i},\zeta_{h+i}}(t) + u_{i\zeta_{h+i}}\,. \end{aligned}$$

We have a system of $\zeta_{h+i}+1$ differential equations and ζ_{h+i} unknown functions. If we denote the n-th order derivative with $^{[n]}$, for $n \geq 4$, starting from the last equation we get the solutions:

$$\begin{aligned} z_{\zeta_{h+i},\zeta_{h+i}}(t) &= -u_{i\zeta_{h+i}}\,, \\ z_{\zeta_{h+i},\zeta_{h+i}-1}(t) &= -u_{i(\zeta_{h+i}-1)} - \dot{u}_{i\zeta_{h+i}}\,, \\ z_{\zeta_{h+i},\zeta_{h+i}-2}(t) &= -u_{i(\zeta_{h+i}-2)} - \dot{u}_{i(\zeta_{h+i}-1)} - \ddot{u}_{i\zeta_{h+i}}\,, \\ &\vdots \\ z_{\zeta_{h+i},1}(t) &= -u_{i1} - \dot{u}_{i2} - \dots - u_{i\zeta_{h+i}}^{[\zeta_{h+i}-1]}\,. \end{aligned}$$

In order to solve the system we used the last ζ_j equations. By applying these results in the first equation we get

$$u_{i0} = -\dot{u}_{i1} - \ddot{u}_{i2} - \dots - u_{i\zeta_{h+i}}^{[\zeta_{h+i}]}\,,$$

or, equivalently,

$$\sum_{\rho=0}^{\zeta_{h+i}} u_{i\rho}^{[\rho]} = 0\,,$$

which is a necessary and sufficient condition for the system to have a solution. If this does not hold, then the system has no solution.

To conclude, in the case of a singular pencil with $r > m$, system (6.1) has the solution

$$\boldsymbol{x}(t) = \mathbf{Q}\boldsymbol{z}(t) = \begin{bmatrix} \mathbf{Q}_p & \mathbf{Q}_q & \mathbf{Q}_\zeta \end{bmatrix} \begin{bmatrix} e^{\mathbf{J}_p t}\boldsymbol{C} + \int_0^\infty e^{\mathbf{J}_p(t-\tau)}\boldsymbol{V}(\tau)d\tau \\ -\sum_{i=0}^{q_*-1} \mathbf{H}_q^i \mathbf{P}_2 \boldsymbol{V}^{[i]}(t) \\ \boldsymbol{z}_\zeta \end{bmatrix},$$

or, equivalently,

$$\boldsymbol{x}(t) = \mathbf{Q}_p \left[e^{\mathbf{J}_p t}\boldsymbol{C} + \int_0^\infty e^{\mathbf{J}_p(t-\tau)}\boldsymbol{V}(\tau)d\tau \right] - \mathbf{Q}_q \sum_{i=0}^{q_*-1} \mathbf{H}_q^i \mathbf{P}_2 \boldsymbol{V}^{[i]}(t) + \mathbf{Q}_\zeta \boldsymbol{z}_\zeta .$$

To sum up, we proved the following Theorem:

Theorem 6.3 (Existence of the solution)**.** There exist solutions for the system (6.1) if and only if $r > m$ and (6.3) holds. Then the general solution is given by

$$\boldsymbol{x}(t) = \mathbf{Q}_p \left[e^{\mathbf{J}_p t}\boldsymbol{C} + \int_0^\infty e^{\mathbf{J}_p(t-\tau)}\boldsymbol{V}(\tau)d\tau \right] - \mathbf{Q}_q \sum_{i=0}^{q_*-1} \mathbf{H}_q^i \mathbf{P}_2 \boldsymbol{V}^{[i]}(t) + \mathbf{Q}_\zeta \boldsymbol{z}_\zeta , \tag{6.7}$$

where $\mathbf{Q}_p$, $\mathbf{J}_p$, $\mathbf{Q}_q$, $\mathbf{H}_q$, $\mathbf{P}_2$, $\mathbf{Q}_\zeta$ are defined in (6.5), and $\boldsymbol{z}_\zeta$ is given by (6.6).

Having identified the conditions under which there exist solutions for systems in the form of (6.1), we can now present the following Corollary:

Corollary 6.1 (Uniqueness of the solution)**.** If there exist solutions for system (6.1), then if $r > m$ and (6.3) holds, for given initial conditions $\boldsymbol{x}(t_o) = \boldsymbol{x}_o$ the solution is unique if and only if:

$$\boldsymbol{x}_o \in \text{colspan}\mathbf{Q}_p - \mathbf{Q}_q \sum_{i=0}^{q_*-1} \mathbf{H}_q^i \mathbf{P}_2 \boldsymbol{V}^{[i]}(0) + \mathbf{Q}_\zeta \boldsymbol{z}_\zeta .$$

Then in the general solution of (6.1), $\boldsymbol{C}$ is the unique solution of the linear system

$$\mathbf{Q}_p \boldsymbol{C} = \left[\boldsymbol{x}_o + \mathbf{Q}_q \sum_{i=0}^{q_*-1} \mathbf{H}_q^i \mathbf{P}_2 \boldsymbol{V}^{[i]}(0) + \mathbf{Q}_\zeta \boldsymbol{z}_\zeta \right] .$$

Proof. This is a direct result from Theorem 6.3. If we use the formula of solutions for $t = 0$, we get:

$$\boldsymbol{x}(0) = \mathbf{Q}_p \boldsymbol{C} - \mathbf{Q}_q \sum_{i=0}^{q_*-1} \mathbf{H}_q^i \mathbf{P}_2 \boldsymbol{V}^{[i]}(0) + \mathbf{Q}_\zeta \boldsymbol{z}_\zeta ,$$

and from here we arrive easily at the desired condition. The proof is completed. ■

Example 6.1 (Singular pencil with no solution). We consider system (6.1) with $\boldsymbol{V}(t) = \mathbf{0}_{4,1}$ and

$$\mathbf{E} = \begin{bmatrix} 1 & 1 & 1 \\ 0 & 1 & 1 \\ 1 & 1 & 1 \\ 0 & 1 & 1 \end{bmatrix}, \qquad \mathbf{A} = \begin{bmatrix} 1 & 2 & 2 \\ 0 & 2 & 2 \\ 1 & 2 & 2 \\ 0 & 2 & 3 \end{bmatrix}.$$

The matrices $\mathbf{E}$, $\mathbf{A}$ are non-square. Hence, we are in the case where $r > m$, i.e. $4 > 3$ and the pencil

$$s\mathbf{E} - \mathbf{A} = \begin{bmatrix} s-1 & s-2 & s-2 \\ 0 & s-2 & s-2 \\ s-1 & s-2 & s-2 \\ 0 & s-2 & s-3 \end{bmatrix}$$

is singular. There exists the matrix

$$\mathbf{P}(s) = \begin{bmatrix} 1 & -1 & 0 & 0 \\ 0 & 3-s & 0 & -s+2 \\ 0 & 1 & 0 & -1 \\ 0 & 0 & 1 & 0 \end{bmatrix},$$

such that

$$\mathbf{P}(s)(s\mathbf{E} - \mathbf{A}) = \begin{bmatrix} \mathbf{A}(s) \\ \mathbf{0}_{1,1} \end{bmatrix},$$

where

$$\mathbf{A}(s) = \begin{bmatrix} s-1 & 0 & 0 \\ 0 & s-2 & 0 \\ 0 & 0 & 1 \end{bmatrix}.$$

In addition, for

$$\mathbf{P}(s) = \begin{bmatrix} \mathbf{P}_1(s) \\ \mathbf{P}_2(s) \end{bmatrix},$$

with

$$\mathbf{P}_1(s) = \begin{bmatrix} 1 & -1 & 0 & 0 \\ 0 & 3-s & 0 & -s+2 \\ 0 & 1 & 0 & -1 \end{bmatrix}, \qquad \mathbf{P}_2(s) = \begin{bmatrix} 0 & 0 & 1 & 0 \end{bmatrix},$$

we have

$$\mathbf{P}_2(s)\mathbf{E} = \begin{bmatrix} 1 & 1 & 1 \end{bmatrix} \neq \mathbf{0}_{1,3}.$$

Hence, (6.3) does not hold and the system (6.1) does not have solutions. □

Example 6.2 (Singular pencil with solutions). We consider again system (6.1) with $\boldsymbol{V}(t) = \mathbf{0}_{3,1}$ and

$$\mathbf{E} = \begin{bmatrix} 1 & 1 \\ 0 & 1 \\ 1 & 1 \end{bmatrix}, \qquad \mathbf{A} = \begin{bmatrix} 1 & 2 \\ 0 & 2 \\ 1 & 2 \end{bmatrix}.$$

The matrices $\mathbf{E}$, $\mathbf{A}$ are non-square. Hence, we are in the case where $r > m$, i.e. $3 > 2$ and the pencil

$$s\mathbf{E} - \mathbf{A} = \begin{bmatrix} s-1 & s-2 \\ 0 & s-2 \\ s-1 & s-2 \end{bmatrix}$$

is singular. There exists the matrix

$$\mathbf{P}(s) = \begin{bmatrix} 1 & -1 & 0 \\ 0 & 1 & 0 \\ -1 & 0 & 1 \end{bmatrix},$$

such that

$$\mathbf{P}(s)(s\mathbf{E} - \mathbf{A}) = \begin{bmatrix} \mathbf{A}(s) \\ \mathbf{0}_{1,2} \end{bmatrix}.$$

where

$$\mathbf{A}(s) = \begin{bmatrix} s-1 & 0 \\ 0 & s-2 \end{bmatrix}.$$

In addition, for

$$\mathbf{P}(s) = \begin{bmatrix} \mathbf{P}_1(s) \\ \mathbf{P}_2(s) \end{bmatrix},$$

with

$$\mathbf{P}_1(s) = \begin{bmatrix} 1 & -1 & 0 \\ 0 & 1 & 0 \end{bmatrix}, \qquad \mathbf{P}_2(s) = \begin{bmatrix} -1 & 0 & 1 \end{bmatrix},$$

we have

$$\mathbf{P}_2(s)\mathbf{E} = \mathbf{0}_{1,2}.$$

Hence, (6.3) holds and there exist solutions for the system given by (6.4):

$$\boldsymbol{\Psi}_0(t) = \mathcal{L}^{-1}\{\mathbf{A}^{-1}\mathbf{P}_1(s)\mathbf{E}\} = \mathcal{L}^{-1}\{\begin{bmatrix} \frac{1}{s-1} & 0 \\ 0 & \frac{1}{s-2} \end{bmatrix}\},$$

or, equivalently,

$$\boldsymbol{\Psi}_0(t) = \begin{bmatrix} \mathrm{e}^t & 0 \\ 0 & \mathrm{e}^{2t} \end{bmatrix}.$$

Then for $\boldsymbol{C} = \begin{bmatrix} c_1 & c_2 \end{bmatrix}^{\mathrm{T}}$ we have:

$$\boldsymbol{x}(t) = \begin{bmatrix} \mathrm{e}^t c_1 \\ \mathrm{e}^{2t} c_2 \end{bmatrix}.$$

□

Example 6.3 (Singular pencil with a solution). We consider now system (6.1) with $\boldsymbol{V}(t) = \mathbf{0}_{5,1}$ and

$$\mathbf{E} = \begin{bmatrix} 0 & 1 & 1 & 0 \\ 0 & 0 & 0 & -1 \\ 1 & 1 & 0 & 0 \\ 1 & 1 & 1 & 1 \\ 0 & 0 & 0 & 0 \end{bmatrix}, \quad \mathbf{A} = \begin{bmatrix} 0 & 2 & 0 & 1 \\ 0 & 0 & 0 & 0 \\ 1 & 1 & -1 & 0 \\ 1 & 1 & 0 & 1 \\ 0 & 0 & 1 & 0 \end{bmatrix}.$$

In this example we will use the spectrum of the pencil $s\mathbf{E} - \mathbf{A}$ to investigate the solutions of the system. The pencil has the finite eigenvalues $\lambda_1 = 1$, $\lambda_2 = 2$, and the row minimal indices $\zeta_1 = 0$, $\zeta_2 = 2$. Then since

$$\mathbf{J}_p = \begin{bmatrix} 1 & 0 \\ 0 & 2 \end{bmatrix}, \quad \mathbf{Q}_p = \begin{bmatrix} 1 & 1 \\ 0 & 1 \\ 0 & 0 \\ 0 & 0 \end{bmatrix},$$

by using Theorem 6.3 the solution of (6.1) is

$$\boldsymbol{x}(t) = \mathbf{Q}_p \, \mathrm{e}^{\mathbf{J}_p t} \boldsymbol{C} = \begin{bmatrix} 1 & 1 \\ 0 & 1 \\ 0 & 0 \\ 0 & 0 \end{bmatrix} \begin{bmatrix} \mathrm{e}^t & 0 \\ 0 & \mathrm{e}^{2t} \end{bmatrix} \begin{bmatrix} c_1 \\ c_2 \end{bmatrix},$$

or, equivalently,

$$\boldsymbol{x}(t) = \begin{bmatrix} c_1 \, \mathrm{e}^t + c_2 \, \mathrm{e}^{2t} \\ c_2 \, \mathrm{e}^{2t} \\ 0 \\ 0 \end{bmatrix}.$$

Furthermore if

$$\boldsymbol{x}(0) = \begin{bmatrix} 2 \\ 1 \\ 0 \\ 0 \end{bmatrix}$$

are the initial conditions of the system, after observing that $\boldsymbol{x}(0) \in \mathrm{colspan}\mathbf{Q}_p$, by using Corollary 6.1 we get the unique solution of (6.1):

$$\boldsymbol{x}(t) = \begin{bmatrix} \mathrm{e}^t + \mathrm{e}^{2t} \\ \mathrm{e}^{2t} \\ 0 \\ 0 \end{bmatrix}.$$

It is worth noting that we can obtain the solution of (6.1) also by using the Kronecker canonical form of the pencil as described in Section 2.3. We have

$$\mathbf{E}_\zeta = \begin{bmatrix} 1 & 0 \\ 0 & 1 \\ 0 & 0 \end{bmatrix}, \quad \mathbf{A}_\zeta = \begin{bmatrix} 0 & 0 \\ 1 & 0 \\ 0 & 1 \end{bmatrix}, \quad \mathbf{Q}_\zeta = \begin{bmatrix} 0 & 0 \\ 0 & 0 \\ 0 & 1 \\ 1 & 0 \end{bmatrix}.$$

Hence,

$$\mathbf{E}_K = \begin{bmatrix} 1 & 0 & 0 & 0 \\ 0 & 1 & 0 & 0 \\ 0 & 0 & 1 & 0 \\ 0 & 0 & 0 & 1 \\ 0 & 0 & 0 & 0 \end{bmatrix}, \qquad \mathbf{A}_K = \begin{bmatrix} 1 & 0 & 0 & 0 \\ 0 & 2 & 0 & 0 \\ 0 & 0 & 0 & 0 \\ 0 & 0 & 1 & 0 \\ 0 & 0 & 0 & 1 \end{bmatrix}.$$

By setting $\boldsymbol{x} = \mathbf{Q}\boldsymbol{z}$, $\mathbf{Q} = \begin{bmatrix} \mathbf{Q}_p & \mathbf{Q}_\zeta \end{bmatrix}$, we have

$$\mathbf{E}_K \dot{\boldsymbol{z}} = \mathbf{A}_K \boldsymbol{z}\,,$$

or, equivalently,

$$\begin{bmatrix} 1 & 0 & 0 & 0 \\ 0 & 1 & 0 & 0 \\ 0 & 0 & 1 & 0 \\ 0 & 0 & 0 & 1 \\ 0 & 0 & 0 & 0 \end{bmatrix} \begin{bmatrix} \dot{z}_{p1} \\ \dot{z}_{p2} \\ \dot{z}_{\zeta 1} \\ \dot{z}_{\zeta 2} \end{bmatrix} = \begin{bmatrix} 1 & 0 & 0 & 0 \\ 0 & 2 & 0 & 0 \\ 0 & 0 & 0 & 0 \\ 0 & 0 & 1 & 0 \\ 0 & 0 & 0 & 1 \end{bmatrix} \begin{bmatrix} z_{p1} \\ z_{p2} \\ z_{\zeta 1} \\ z_{\zeta 2} \end{bmatrix},$$

or, equivalently,

$$\begin{bmatrix} \dot{z}_{p1} \\ \dot{z}_{p2} \\ \dot{z}_{\zeta 1} \\ \dot{z}_{\zeta 2} \\ 0 \end{bmatrix} = \begin{bmatrix} z_{p1} \\ 2z_{p1} \\ 0 \\ z_{\zeta 1} \\ z_{\zeta 2} \end{bmatrix},$$

or, equivalently,

$$\boldsymbol{z} = \begin{bmatrix} c_1\,\mathrm{e}^t \\ c_2\,\mathrm{e}^{2t} \\ 0 \\ 0 \end{bmatrix}.$$

Since $\boldsymbol{x} = \mathbf{Q}\boldsymbol{z}$, we get as previously

$$\boldsymbol{x}(t) = \begin{bmatrix} c_1\,\mathrm{e}^t + c_2\,\mathrm{e}^{2t} \\ c_2\,\mathrm{e}^{2t} \\ 0 \\ 0 \end{bmatrix}.$$

Example 6.4 (Singular system with no solution)**.** We consider now system (6.1) with $\boldsymbol{V}(t) = \mathbf{0}_{4,1}$ and let

$$\mathbf{E} = \begin{bmatrix} 1 & 1 & 0 & 1 \\ 0 & 0 & 0 & 0 \\ 0 & 0 & 0 & 1 \end{bmatrix}, \qquad \mathbf{A} = \begin{bmatrix} 2 & 2 & 1 & 0 \\ 0 & 1 & 0 & 0 \\ 0 & 0 & 1 & 0 \end{bmatrix}.$$

Since the matrices $\mathbf{E}$, $\mathbf{A}$ are non square with $r < m$, from Theorem 6.1 there do not exist solutions for system (6.1). We can also prove this by using the spectrum of the pencil $s\mathbf{E}-\mathbf{A}$ which has invariants the finite eigenvalue $\lambda_1 = 2$, an infinite eigenvalue of algebraic multiplicity one, and the column minimal indices $\epsilon_1=0$, $\epsilon_2=1$. Then $\mathbf{J}_p = 2$, $\mathbf{H}_q = 0$, and:

$$\mathbf{E}_\epsilon = \begin{bmatrix} 1 & 0 \end{bmatrix}, \qquad \mathbf{A}_\epsilon = \begin{bmatrix} 0 & 1 \end{bmatrix}.$$

Hence

$$\mathbf{E}_K = \begin{bmatrix} 1 & 0 & 0 & 0 \\ 0 & 0 & 0 & 0 \\ 0 & 0 & 1 & 0 \end{bmatrix}, \qquad \mathbf{A}_K = \begin{bmatrix} 2 & 0 & 0 & 0 \\ 0 & 1 & 0 & 0 \\ 0 & 0 & 0 & 1 \end{bmatrix}.$$

By setting $\boldsymbol{x} = \mathbf{Q}\boldsymbol{z}$, $\mathbf{Q} = \begin{bmatrix} \mathbf{Q}_p & \mathbf{Q}_q & \mathbf{Q}_\epsilon \end{bmatrix}$, we have

$$\mathbf{E}_K \dot{\boldsymbol{z}} = \mathbf{A}_K \boldsymbol{z} \, ,$$

or, equivalently,

$$\begin{bmatrix} 1 & 0 & 0 & 0 \\ 0 & 0 & 0 & 0 \\ 0 & 0 & 1 & 0 \end{bmatrix} \begin{bmatrix} \dot{z}_p \\ \dot{z}_q \\ \dot{z}_{\epsilon 1} \\ \dot{z}_{\epsilon 2} \end{bmatrix} = \begin{bmatrix} 2 & 0 & 0 & 0 \\ 0 & 1 & 0 & 0 \\ 0 & 0 & 0 & 1 \end{bmatrix} \begin{bmatrix} z_p \\ z_q \\ z_{\epsilon 1} \\ z_{\epsilon 2} \end{bmatrix},$$

or, equivalently,

$$\begin{bmatrix} \dot{z}_p \\ 0 \\ \dot{z}_{\epsilon 1} \end{bmatrix} = \begin{bmatrix} 2z_p \\ z_q \\ z_{\epsilon 2} \end{bmatrix}.$$

It is obvious from the above equations that $\boldsymbol{z}$ cannot be defined and hence there do not exist solutions for system (6.1). □

6.2 Singular Power System Models

The singular systems described in the section above have an interesting application to physical systems. We are, of course, interested only in the case for which a system in the form of (6.1) has solutions.

Reference [30] describes a non-dissipative mechanical pendulum (no friction). This system, if perturbed should oscillate forever with oscillations of constant amplitude. Reference [30] shows, however, that numerical integration of the conventional determined differential equations that describe the pendulum leads to slightly increasing or decreasing oscillations when integrated with the forward Euler method (FEM) and backward Euler method (BEM), respectively (see Section 1.2). On the other hand, when the set of differential equations is augmented with a constraint that imposes that the variation of the total free energy (kinetic + potential) of the pendulum is zero, the numerical integration shows stationary oscillations, as expected.

The inconsistency of the results shown in [30] for the determined set of equations of the mechanical pendulum originates from the integration scheme, not the model of the system. Numerical integration methods, in fact, add an extra "layer" to the stability analysis of a physical system which has to be properly taken into account to avoid inconsistent or erroneous interpretations of the results.

The FEM is known to be the most numerically unstable integration method, meaning that the integration can show unstable trajectories even if the system is actually perfectly stable. For this reason, when integrated with the FEM, the pendulum seems to increase its total free energy and thus create energy out from thin air! On the other hand, the BEM is hyperstable. This means that it can numerically "stabilize" an unstable system. In the case of the pendulum, the hyperstability of the BEM artificially damps the oscillations, thus decreasing the total free energy without the need of friction.

The situation described in [30] is extreme. Since the original system is non-dissipative, it is characterized by undamped oscillations. This is the same situation as the Example 1.12 with $D_1 = 0$ (see also Figure 1.17). From small-signal analysis point of view, it has a pair of complex conjugate eigenvalues on the imaginary axis. Hence, any numerical error that perturbs the real part of these eigenvalues by a small ϵ will show oscillations that either increase ($\epsilon > 0$) or decrease ($\epsilon < 0$) over time.

While it is more difficult to observe such numerical issues with systems that are naturally well-damped or unstable, the issues introduced by numerical methods are always behind the corner.

Example 6.5 (Hyperstability). We consider the OMIB presented in Example 1.12 and assume that $D_1 = 0$. We thus expect that the system oscillates forever after a three-phase short-circuit occurring at $X_1'/2$ at $t = 10$ and cleared after 60 ms.

We are interested in the small-signal stability analysis of the discrete maps of three integration schemes, FEM, BEM, and ITM. Since the OMIB model is non-linear, we consider the maps of the linearized system at the initial (pre-fault) equilibrium point, which is also a fixed point for the maps.

Elaborating on the equations given in Section 1.2, the i-th step of the FEM, BEM and ITM for a linear(ized) set of implicit ODEs gives:

$$\begin{aligned}
\text{FEM} &: \quad \boldsymbol{x}^{(i+1)} = [\mathbf{E}^{-1}(\Delta t\, \mathbf{A} + \mathbf{E})]\, \boldsymbol{x}^{(i)}\,, \\
\text{BEM} &: \quad \boldsymbol{x}^{(i+1)} = -[(\Delta t\, \mathbf{A} - \mathbf{E})^{-1}\mathbf{E}]\, \boldsymbol{x}^{(i)}\,, \\
\text{ITM} &: \quad \boldsymbol{x}^{(i+1)} = -[(0.5\, \Delta t\, \mathbf{A} - \mathbf{E})^{-1}(0.5\, \Delta t\, \mathbf{A} + \mathbf{E})]\, \boldsymbol{x}^{(i)}\,.
\end{aligned} \tag{6.8}$$

With the data of the OMIB as in Example 1.12 and assuming $D_1 = 0$, we obtain that the eigenvalues of (6.8) at the equilibrium point are complex conjugate for all methods with the following magnitudes:

$$\begin{aligned}\text{FEM} &: \quad |\lambda_{1,2}| = 1.0006530376933853\,,\\ \text{BEM} &: \quad |\lambda_{1,2}| = 0.9993473884865322\,,\\ \text{ITM} &: \quad |\lambda_{1,2}| = 1\,.\end{aligned}$$

There is indeed a tiny difference among the three integration schemes, still the eigenvalues obtained with the FEM have magnitude greater than 1, which leads to unstable oscillations, and those obtained with the BEM have magnitude lower than 1, which leads to the damped effect in the resulting trajectories.

In practical applications, nobody really utilizes the FEM. The numerical instability described in [30] is, thus, hardly a problem in the simulation of physical systems. Similarly, the BEM *per se* is never really utilized either, but since the ITM cannot start alone after an event as it needs to know $\boldsymbol{f}^{(i)}$ at the previous integration step, the BEM is often used to initialize the ITM. Both BEM and ITM, thus, are typically included in standard software tools for circuit analysis.

Figure 6.1 shows the simulation results using the BEM and the ITM and fixed time step $\Delta t = 0.01$ s. The ITM returns the expected results, however, the BEM shows damped oscillations that are comparable with those obtained using the ITM and $D_1 = 2$ (see Figure 1.17).

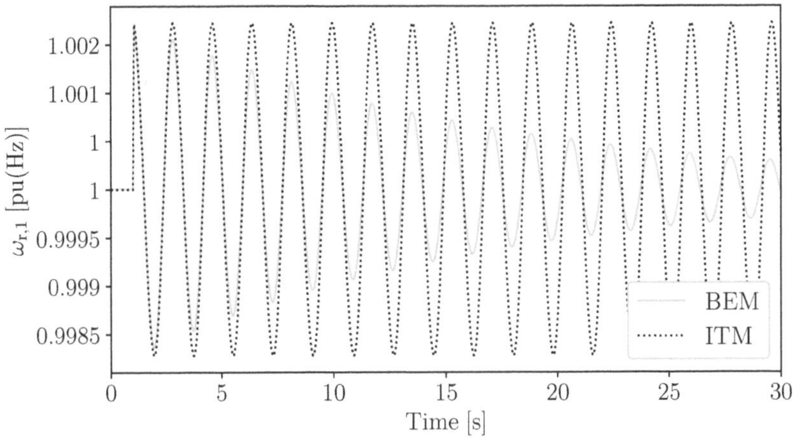

FIGURE 6.1: Comparison of the BEM and ITM for the OMIB and with time step $\Delta t = 0.01s$.

The damping effect of the BEM cannot be eliminated even using a smaller time step Δt. For example:

$$\begin{aligned}\Delta t = 0.001\,\text{s} &: \quad |\lambda_{1,2}| = 0.9999934675547848\,,\\ \Delta t = 0.0001\,\text{s} &: \quad |\lambda_{1,2}| = 0.9999999346749142\,.\end{aligned}$$

The magnitude of the eigenvalues increases as Δt decreases, but is always lower than 1. Figure 6.2 shows the effect of Δt on the time domain integration using the BEM. Note that the simulation takes less than a second for $\Delta t = 0.01$ s, while it takes about 30 s for $\Delta t = 0.0001$ s. Analogously, we can show that the instability of the FEM cannot be removed by reducing the time step.

The issues above are the phenomenon discussed in [30]. But, rather than adding an extra equation to the system, we now know that one simply has to use an adequate integration scheme. The ITM, in fact, is characterized by eigenvalues with magnitude exactly equal to 1, up to the floating point tolerance of the processor that we used for the calculations. □

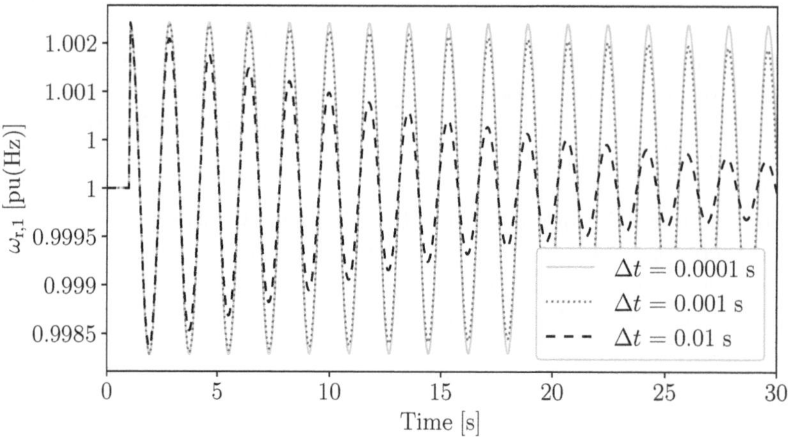

FIGURE 6.2: Comparison of the BEM for the OMIB and with different time steps.

Example 6.6 (Over-determined synchronous machine model)**.** Consider the classical synchronous machine model introduced in Section 1.3.4.7. The machine is assumed to be connected to an infinite bus h with voltage $\bar{v}_h = v_h \angle \theta_h = 1$ pu $\angle 0$ rad. Assuming zero damping as well as constant EMF $e'_{\mathrm{r,q}}$, the swing equations (1.113) can be rewritten as follows:

$$\omega_o^{-1} \dot{\delta}_{\mathrm{r}} = \omega_{\mathrm{r}} - 1 \,, \tag{6.9}$$

$$M \dot{\omega}_{\mathrm{r}} = P_{\mathrm{m}} - P_{\mathrm{e}}^{\max} \sin \delta_{\mathrm{r}} \,, \tag{6.10}$$

where $P_{\mathrm{e}}^{\max} = \frac{e'_{\mathrm{r,q}}}{X'_{\mathrm{d}}}$.

Equation (6.10) can be expressed as [135]:

$$M \dot{\omega}_{\mathrm{r}} = -\frac{\partial \mathcal{V}}{\partial \delta_{\mathrm{r}}} \,, \tag{6.11}$$

where $-\frac{\partial \mathcal{V}}{\partial \delta_{\mathrm{r}}}$ is the negative gradient of the potential energy function:

$$\mathcal{V} = -P_{\mathrm{m}} \delta_{\mathrm{r}} - P_{\mathrm{e}}^{\max} \cos \delta_{\mathrm{r}} \,. \tag{6.12}$$

Then, according to the discussion in the previous example, and following from [30], we can include a constraint to ensure that the there is no variation of the total free energy of system (6.9)-(6.10). This constraint can be expressed as follows:

$$\frac{1}{2} M \omega_{\mathrm{r}}^2 + \mathcal{V} = c\,, \tag{6.13}$$

where c is a constant. Substitution of (6.12) in the last equation yields:

$$\frac{1}{2} M \omega_{\mathrm{r}}^2 - P_{\mathrm{m}} \delta_{\mathrm{r}} - P_{\mathrm{e}}^{\max} \cos \delta_{\mathrm{r}} = c\,. \tag{6.14}$$

The state vector of the DAE system defined by equations (6.9), (6.10) and (6.14) is:

$$\boldsymbol{x} = [\delta_{\mathrm{r}},\, \omega_{\mathrm{r}}]^{\mathrm{T}}\,.$$

Assuming that the mechanical power P_{m} is an input, linearization of the system around an equilibrium point $(\boldsymbol{x}_o, \boldsymbol{V}_o)$, gives:

$$\begin{aligned} \omega_o^{-1} \Delta \dot{\delta}_{\mathrm{r}} &= \Delta \omega_{\mathrm{r}}\,, \\ M \Delta \dot{\omega}_{\mathrm{r}} &= \Delta P_{\mathrm{m}} - P_{\mathrm{e}}^{\max} \cos \delta_{\mathrm{r},o}\, \Delta \delta_{\mathrm{r}}\,, \\ 0 &= M \omega_{\mathrm{r},o} \Delta \omega_{\mathrm{r}} - \Delta \delta_{\mathrm{r}} - \delta_{\mathrm{r},o} \Delta P_{\mathrm{m}} + P_{\mathrm{e}}^{\max} \sin \delta_{\mathrm{r},o}\, \Delta \delta_{\mathrm{r}}\,. \end{aligned} \tag{6.15}$$

In matrix form, system (6.15) reads:

$$\mathbf{E}\, \Delta \dot{\boldsymbol{x}}(t) = \mathbf{A}\, \Delta \boldsymbol{x}(t) + \boldsymbol{V}(t)\,, \tag{6.16}$$

where

$$\mathbf{E} = \begin{bmatrix} \omega_o^{-1} & 0 \\ 0 & M \\ 0 & 0 \end{bmatrix}, \quad \mathbf{A} = \begin{bmatrix} 0 & 1 \\ -P_{\mathrm{e}}^{\max} \cos \delta_{\mathrm{r},o} & 0 \\ -P_{\mathrm{m},o} + P_{\mathrm{e}}^{\max} \sin \delta_{\mathrm{r},o} & M \omega_{\mathrm{r},o} \end{bmatrix},$$

$$\boldsymbol{V}(t) = \begin{bmatrix} 0 \\ \Delta P_{\mathrm{m}} \\ -\delta_{\mathrm{r},o} \Delta P_{\mathrm{m}} \end{bmatrix}.$$

Notice that model (6.16) is over-determined, since $\mathbf{A}, \mathbf{E} \in \mathbb{R}^{3 \times 2}$.

We consider a numerical example. In particular, we assume $\omega_o = 100\pi$ rad/s (50 Hz system), $M = 14$ s, $P_{\mathrm{e}}^{\max} = 2$ pu(MW), $P_{\mathrm{m}} = 1$ pu(MW), Then, the equilibrium of the system state is:

$$\boldsymbol{x}_o = \left[\frac{\pi}{6},\, 1\right]^{\mathrm{T}}.$$

The matrices that describe the system become:

$$\mathbf{E} = \begin{bmatrix} \frac{1}{100\pi} & 0 \\ 0 & 14 \\ 0 & 0 \end{bmatrix}, \quad \mathbf{A} = \begin{bmatrix} 0 & 1 \\ -\sqrt{3} & 0 \\ 0 & 14 \end{bmatrix}, \quad \boldsymbol{V}(t) = \begin{bmatrix} 0 \\ \Delta P_{\mathrm{m}} \\ -\delta_{\mathrm{r},o} \Delta P_{\mathrm{m}} \end{bmatrix}.$$

The pencil of the system will then be

$$s\mathbf{E} - \mathbf{A} = s\begin{bmatrix} \frac{1}{100\pi} & 0 \\ 0 & 14 \\ 0 & 0 \end{bmatrix} - \begin{bmatrix} 0 & 1 \\ -\sqrt{3} & 0 \\ 0 & 14 \end{bmatrix},$$

or, equivalently,

$$s\mathbf{E} - \mathbf{A} = \begin{bmatrix} \frac{1}{100\pi}s & -1 \\ \sqrt{3} & 14s \\ 0 & -14 \end{bmatrix},$$

The pencil has the finite eigenvalues $\lambda_1 = i\sqrt{\frac{\sqrt{3}\pi}{14}}$, $\lambda_2 = -i\sqrt{\frac{\sqrt{3}\pi}{14}}$, and row minimal index $\zeta_1 = 0$.

7

Möbius Transform

7.1 Formulation

We consider the following system:

$$\mathbf{E}\,\dot{\boldsymbol{x}}(t) = \mathbf{A}\,\boldsymbol{x}(t)\,, \tag{7.1}$$

where $\mathbf{E}, \mathbf{A} \in \mathbb{C}^{r\times m}$, $\boldsymbol{x} : [0, +\infty] \mapsto \mathbb{C}^{m\times 1}$. The matrices $\mathbf{E}$, $\mathbf{A}$ can be non-square ($r \neq m$) or square ($r = m$) with $\mathbf{E}$ regular, or singular ($\det(\mathbf{E}) = 0$).

Definition 7.1 (Möbius transformation)**.** The general form of the Möbius transformation is given by

$$s := f(z) = \frac{az+b}{cz+d}, \quad a, b, c, d \in \mathbb{C}, \quad ad - bc \neq 0\,. \tag{7.2}$$

The restriction in Definition 7.1 is necessary because if $ad = bc$ then s is constant which can not be possible; Let the pencil $s\mathbf{E} - \mathbf{A}$ of system (7.1) be regular. Its eigenvalues are then given by solving the following characteristic equation:

$$\det(s\,\mathbf{E} - \mathbf{A}) = 0\,,$$

whereby applying the transform (7.2) we get

$$\det\Big(\frac{az+b}{cz+d}\,\mathbf{E} - \mathbf{A}\Big) = 0\,,$$

or, equivalently, by using determinant properties

$$\det\big((az+b)\,\mathbf{E} - (cz+d)\,\mathbf{A}\big) = 0\,,$$

or, equivalently,

$$\det\big((a\,\mathbf{E} - c\,\mathbf{A})\,z - (d\,\mathbf{A} - b\,\mathbf{E})\big) = 0\,,$$

which is the characteristic equation of a linear dynamical system

$$\tilde{\mathbf{E}}\,\dot{\tilde{\boldsymbol{x}}}(t) = \tilde{\mathbf{A}}\,\tilde{\boldsymbol{x}}(t), \tag{7.3}$$

with pencil

$$z\tilde{\mathbf{E}} - \tilde{\mathbf{A}}\,,$$

where

$$\tilde{\mathbf{E}} = a\,\mathbf{E} - c\,\mathbf{A}, \quad \tilde{\mathbf{A}} = d\,\mathbf{A} - b\,\mathbf{E}\,.$$

Definition 7.2 (Prime system and M-systems)**.** The system (7.1) will be referred as the prime system, and the family of systems (7.3) will be defined as the proper $M(a, b, c, d)$-systems, or simply "M-systems."

The essence of the M-systems is of similar nature to that of projective geometry, and may be stated for autonomous systems. A main result of this chapter is that, if the solution of the prime system is known, then the solution of any of its M-systems can be represented without further computation. The corresponding matrix pencil of (7.1) is $s\mathbf{E} - \mathbf{A}$, and of (7.3) is $z\tilde{\mathbf{E}} - \tilde{\mathbf{A}}$, and their connection may be seen as a consequence of the special transformation. The study of the relationship between (7.1), (7.3) is then reduced to an investigation of the links between their pencils. The notions of their relation may be qualified algebraically in terms of relationships between the strict equivalence of the invariants of the associated pencils. These relationships are summarized below. Let λ_i, $\hat{\lambda}_i$, $i = 1, 2, \ldots, \nu$ be non-zero finite eigenvalues of the prime and M-system, respectively.

- For $a, c, d \neq 0$:
  - If $s \longrightarrow 0$ then $z \longrightarrow -\dfrac{b}{a}$;
  - If $s \longrightarrow \infty$ then $z \longrightarrow -\dfrac{d}{c}$;
  - If $s \longrightarrow \lambda_i$ then $z \longrightarrow \dfrac{-d\lambda_i + b}{c\lambda_i - a}$;
  - If $z \longrightarrow 0$ then $s \longrightarrow \dfrac{b}{d}$;
  - If $z \longrightarrow \infty$ then $s \longrightarrow -\dfrac{a}{c}$;
  - If $z \longrightarrow \hat{\lambda}_i$ then $s \longrightarrow \dfrac{a\hat{\lambda}_i + b}{c\hat{\lambda}_i + d}$;
- For $a = 0$:
  - If $s \longrightarrow 0$ then $z \longrightarrow \infty$;
  - If $s \longrightarrow \infty$ then $z \longrightarrow -\dfrac{d}{c}$;
  - If $s \longrightarrow \lambda_i$ then $z \longrightarrow \dfrac{-d\lambda_i + b}{c\lambda_i}$;
  - If $z \longrightarrow 0$ then $s \longrightarrow \dfrac{b}{d}$;
  - If $z \longrightarrow \infty$ then $s \longrightarrow 0$;
  - If $z \longrightarrow \hat{\lambda}_i, \lambda_i$ then $s \longrightarrow \dfrac{b}{c\hat{\lambda}_i + d}$;
- For $c = 0$:

- If $s \longrightarrow 0$ then $z \longrightarrow -\frac{b}{a}$;
- If $s \longrightarrow \infty$ then $z \longrightarrow \infty$;
- If $s \longrightarrow \lambda_i$ then $z \longrightarrow \frac{-d\lambda_i + b}{-a}$;
- If $z \longrightarrow 0$ then $s \longrightarrow \frac{b}{d}$;
- If $z \longrightarrow \infty$ then $s \longrightarrow \infty$;
- If $z \longrightarrow \hat{\lambda}_i$ then $s \longrightarrow \frac{a\hat{\lambda}_i + b}{d}$;

- For $d = 0$:
  - If $s \longrightarrow 0$ then $z \longrightarrow -\frac{b}{a}$;
  - If $s \longrightarrow \infty$ then $z \longrightarrow 0$;
  - If $s \longrightarrow \lambda_i$ then $z \longrightarrow \frac{b}{c\lambda_i - a}$;
  - If $z \longrightarrow 0$ then $s \longrightarrow \infty$;
  - If $z \longrightarrow \infty$ then $s \longrightarrow -\frac{a}{c}$;
  - If $z \longrightarrow \hat{\lambda}_i$ then $s \longrightarrow \frac{a\hat{\lambda}_i + b}{c\hat{\lambda}_i}$.

In this chapter we consider that the pencil is *regular* with invariants of the following type:

- ν finite eigenvalues of algebraic multiplicity p_i; $i = 1, 2, \ldots, \nu$;
- an infinite eigenvalue of algebraic multiplicity q;

where $\sum_{i=1}^{\nu} p_i = p$ and $p+q = m$. As seen in Chapters 3 and 6, unlike singular pencils, when a system has a regular pencil there always exist solutions. A reminder that this can be seen as follows. Let $\mathcal{L}\{\boldsymbol{x}(t)\} = \boldsymbol{z}(s)$ be the Laplace transform of $\boldsymbol{x}(t)$. By applying the Laplace transform $\mathcal{L}$ into (7.1) we get

$$\mathbf{E}\mathcal{L}\{\dot{\boldsymbol{x}}(t)\} = \mathbf{A}\mathcal{L}\{\boldsymbol{x}(t)\},$$

or, equivalently,

$$\mathbf{E}\big(s\boldsymbol{z}(s) - \boldsymbol{x}_o\big) = \mathbf{A}\,\boldsymbol{z}(s),$$

where $\boldsymbol{x}_o = \boldsymbol{x}(0)$, i.e. the initial condition of (7.1). Since we assume that $\boldsymbol{x}_o$ is unknown, we can use an unknown constant vector $\boldsymbol{C} \in \mathbb{C}^{m\times 1}$ and give to the above expression the following form:

$$(s\mathbf{E} - \mathbf{A})\boldsymbol{z}(s) = \mathbf{E}\boldsymbol{C}.$$

From the above expression it can be seen that when $s\mathbf{E} - \mathbf{A}$ is a regular pencil, i.e. $\det(s\mathbf{E} - \mathbf{A}) \not\equiv 0$, there always exists a solution whereas if the pencil is singular existence is not guaranteed. However, as seen in Chapter 3, even if there exist solutions for a regular pencil, it is not guaranteed that for given initial conditions a singular system will have a unique solution. If the given initial conditions are consistent, and there exist solutions for (7.1), then in the formulas of the general solutions we replace $\boldsymbol{C} = \boldsymbol{x}_o$. However, if the given initial conditions are non-consistent but there exist solutions for (7.1), then the general solution holds for $t > 0$ and the system is impulsive.

From Chapter 3, it is known that:

Lemma 7.1 (Existence of the solution for a prime system)**.** If $s\mathbf{E} - \mathbf{A}$ is regular, then from its regularity there exist non-singular matrices $\mathbf{P}$, $\mathbf{Q} \in \mathbb{C}^{m\times m}$ such that

$$\mathbf{P\,E\,Q} = \mathbf{I}_p \oplus \mathbf{H}_q,$$

$$\mathbf{P\,A\,Q} = \mathbf{J}_p \oplus \mathbf{I}_q\,,$$

where $\mathbf{J}_p \in \mathbb{C}^{p\times p}$ is the Jordan matrix related to the zero and finite eigenvalues, $\mathbf{H}_q \in \mathbb{C}^{q\times q}$ is a nilpotent matrix with index q_*, constructed by using the algebraic multiplicity of the infinite eigenvalue, and $p + q = m$.

Furthermore:

Proposition 7.1 (Uniqueness of the solution for a prime system)**.** Consider the system (7.1) with known initial conditions $\boldsymbol{x}(0) = \boldsymbol{x}_o$ and a regular pencil. Let $\mathbf{J}_p$ be the Jordan matrix of the finite eigenvalues, and $\mathbf{Q}_p$ the matrix that contains all linear independent eigenvalues including the generalized. Then there exists a unique solution if and only if:

$$\boldsymbol{x}_o \in \text{colspan}\mathbf{Q}_p\,.$$

In this case, the unique solution is given by

$$\boldsymbol{x}(t) = \mathbf{Q}_p\, e^{\mathbf{J}_p t}\boldsymbol{z}_p(0)\,,$$

where $\boldsymbol{z}_p(0)$ is the unique solution of the linear system

$$\boldsymbol{x}_o = \mathbf{Q}_p\boldsymbol{z}_p(0)\,.$$

By using only the invariants of the pencil of system (7.1), we will provide insight on the solutions of the family of systems (7.3). We provide the following Theorem:

Theorem 7.1 (Uniqueness of the solution for M-systems)**.** Consider system (7.1) with a regular pencil, and the family of systems (7.3) with known initial conditions $\tilde{\boldsymbol{x}}(0) = \tilde{\boldsymbol{x}}_o$. Then:

- If $a, c \neq 0$, then there exists a unique solution for (7.3) if and only if:

  $$\tilde{\boldsymbol{x}}_o \in \text{colspan}\begin{bmatrix} \mathbf{Q}_p & \mathbf{Q}_q \end{bmatrix},$$

  where:

- $\mathbf{Q}_p \in \mathbb{C}^{m\times p}$ are the linear independent eigenvectors (including the generalized) of all finite eigenvalue of $s\mathbf{E}-\mathbf{A}$ except for the eigenvectors of λ_o, an eigenvalue of the pencil $s\mathbf{E}-\mathbf{A}$ such that $a = c\lambda_o$;
- $\mathbf{Q}_q \in \mathbb{C}^{m\times q}$ are the linear independent eigenvectors (including the generalized) of the infinite eigenvalue of $s\mathbf{E}-\mathbf{A}$.

Then the solution is given by:

$$\tilde{\boldsymbol{x}}(t) = \begin{bmatrix} \mathbf{Q}_p & \mathbf{Q}_q \end{bmatrix} e^{\mathbf{J}_{p+q}t}\tilde{\boldsymbol{z}}_{p+q}(0)\,,$$

where $e^{\mathbf{J}_{p+q}t} = e^{\tilde{\mathbf{J}}_p t} \oplus e^{\tilde{\mathbf{J}}_q t}$, $\tilde{\boldsymbol{z}}_{p+q}(0) = \begin{bmatrix} \tilde{\boldsymbol{z}}_p(0) \\ \tilde{\boldsymbol{z}}_q(0) \end{bmatrix}$, $\tilde{\mathbf{J}}_p = (a\,\mathbf{I}_p - c\,\mathbf{J}_p)^{-1}(d\,\mathbf{J}_p - b\,\mathbf{I}_p)$, $\tilde{\mathbf{J}}_q = (a\,\mathbf{H}_q - c\,\mathbf{I}_q)^{-1}(d\,\mathbf{I}_q - b\,\mathbf{H}_q)$, $\mathbf{J}_p \in \mathbb{C}^{p\times p}$ is the Jordan matrix related to the finite eigenvalues except for λ_o, and $\mathbf{H}_q$ is a nilpotent matrix with index q_*, constructed by using the algebraic multiplicity of the infinite eigenvalue. Finally, $\tilde{\boldsymbol{z}}_{p+q}(0)$ is the unique solution of the algebraic system

$$\tilde{\boldsymbol{x}}_o = \begin{bmatrix} \mathbf{Q}_p & \mathbf{Q}_q \end{bmatrix} \tilde{\boldsymbol{z}}_{p+q}(0)\,.$$

- If $a = 0$, then there exists a unique solution for (7.3) if and only if:

$$\tilde{\boldsymbol{x}}_o \in \operatorname{colspan}\begin{bmatrix} \mathbf{Q}_p & \mathbf{Q}_q \end{bmatrix},$$

where:

- $\mathbf{Q}_p \in \mathbb{C}^{m\times p}$ are the linear independent eigenvectors (including the generalized) of all non-zero finite eigenvalue of $s\mathbf{E}-\mathbf{A}$;
- $\mathbf{Q}_q \in \mathbb{C}^{m\times q}$ are the linear independent eigenvectors (including the generalized) of the infinite eigenvalue of $s\mathbf{E}-\mathbf{A}$.

Then the solution is given by:

$$\tilde{\boldsymbol{x}}(t) = \begin{bmatrix} \mathbf{Q}_p & \mathbf{Q}_q \end{bmatrix} e^{\mathbf{J}_{p+q}t}\tilde{\boldsymbol{z}}_{p+q}(0)\,,$$

where $e^{\mathbf{J}_{p+q}t} = e^{\tilde{\mathbf{J}}_p t} \oplus e^{\tilde{\mathbf{J}}_q t}$, $\tilde{\boldsymbol{z}}_{p+q}(0) = \begin{bmatrix} \tilde{\boldsymbol{z}}_p(0) \\ \tilde{\boldsymbol{z}}_q(0) \end{bmatrix}$, $\tilde{\mathbf{J}}_p = -\frac{1}{c}\mathbf{J}_p^{-1}(d\,\mathbf{J}_p - b\,\mathbf{I}_p)$, $\tilde{\mathbf{J}}_q = -\frac{1}{c}(d\,\mathbf{I}_q - b\,\boldsymbol{H}_q)$, $\mathbf{J}_p \in \mathbb{C}^{p\times p}$ is the Jordan matrix related to the finite non-zero eigenvalues, and $\mathbf{H}_q$ is a nilpotent matrix with index q_*, constructed by using the algebraic multiplicity of the infinite eigenvalue. Finally, $\tilde{\boldsymbol{z}}_{p+q}(0)$ is the unique solution of the algebraic system

$$\tilde{\boldsymbol{x}}_o = \begin{bmatrix} \mathbf{Q}_p & \mathbf{Q}_q \end{bmatrix} \tilde{\boldsymbol{z}}_{p+q}(0)\,.$$

- If $c = 0$, then there exists a unique solution for (7.3) if and only if:

$$\tilde{\boldsymbol{x}}_o \in \operatorname{colspan}\mathbf{Q}_p\,,$$

where:

– $\mathbf{Q}_p \in \mathbb{C}^{m\times p}$ are the linear independent eigenvectors (including the generalized) of all finite eigenvalue of $s\mathbf{E} - \mathbf{A}$;

Then the solution is given by:

$$\tilde{\boldsymbol{x}}(t) = \mathbf{Q}_p \, \mathrm{e}^{\tilde{\mathbf{J}}_p t} \tilde{\boldsymbol{z}}_p(0) \,,$$

where $\tilde{\mathbf{J}}_p = \frac{1}{a}(d\,\mathbf{J}_p - b\,\mathbf{I}_p)$, $\mathbf{J}_p \in \mathbb{C}^{p\times p}$ is the Jordan matrix related to the finite eigenvalues, and $\tilde{\boldsymbol{z}}_p(0)$ is the unique solution of the algebraic system

$$\tilde{\boldsymbol{x}}_o = \mathbf{Q}_p \tilde{\boldsymbol{z}}_p(0) \,.$$

Proof. • Let λ_o be an eigenvalue of the pencil $s\mathbf{E} - \mathbf{A}$ such that $a = c\lambda_o$. Then we can give $\mathbf{Q}$, defined in Lemma 7.1, the following form:

$$\mathbf{Q} = \left[\begin{array}{ccc} \mathbf{Q}_{p_o} & \mathbf{Q}_p & \mathbf{Q}_q \end{array} \right],$$

where:

– $\mathbf{Q}_{p_o} \in \mathbb{C}^{m\times p_o}$ are the linear independent eigenvectors (including the generalized) of the eigenvalue λ_o of $s\mathbf{E} - \mathbf{A}$;
– $\mathbf{Q}_p \in \mathbb{C}^{m\times p}$ are the linear independent eigenvectors (including the generalized) of all finite eigenvalue of $s\mathbf{E} - \mathbf{A}$ except for the eigenvectors of λ_o;
– $\mathbf{Q}_q \in \mathbb{C}^{m\times q}$ are the linear independent eigenvectors (including the generalized) of the infinite eigenvalue of $s\mathbf{E} - \mathbf{A}$.

In this case:

$$\mathbf{P}\,\mathbf{E}\,\mathbf{Q} = \mathbf{I}_{p_o} \oplus \mathbf{I}_p \oplus \mathbf{H}_q,$$

$$\mathbf{P}\,\mathbf{E}\,\mathbf{Q} = \mathbf{J}_{p_o} \oplus \mathbf{J}_p \oplus \mathbf{I}_q \,,$$

where $\mathbf{J}_{p_o} \in \mathbb{C}^{p_o\times p_o}$, $\mathbf{J}_p \in \mathbb{C}^{p\times p}$ are the Jordan matrices related to λ_o, and the finite eigenvalues except for λ_o, respectively, $\mathbf{H}_q \in \mathbb{C}^{q\times q}$ is a nilpotent matrix with index q_*, constructed by using the algebraic multiplicity of the infinite eigenvalue, and $p_o + p + q = m$. We apply the transformation

$$\tilde{\boldsymbol{x}}(t) = \mathbf{Q}\,\tilde{\boldsymbol{z}}(t)$$

into (7.3), and multiply by $\mathbf{P}$:

$$\mathbf{P}\,\tilde{\mathbf{E}}\,\mathbf{Q}\,\dot{\tilde{\boldsymbol{z}}}(t) = \mathbf{P}\,\tilde{\mathbf{A}}\,\mathbf{Q}\,\tilde{\boldsymbol{z}}(t) \,,$$

or, equivalently,

$$\mathbf{P}\,(a\,\mathbf{E} - c\,\mathbf{A})\,\mathbf{Q}\,\dot{\tilde{\boldsymbol{z}}}(t) = \mathbf{P}\,(d\,\mathbf{A} - b\,\mathbf{E})\,\mathbf{Q}\,\tilde{\boldsymbol{z}}(t) \,,$$

or, equivalently,

$$[(a\,\mathbf{I}_{p_o} - c\,\mathbf{J}_{p_o}) \oplus (a\,\mathbf{I}_p - c\,\mathbf{J}_p) \oplus (a\,\mathbf{H}_q - c\,\mathbf{I}_q)]\dot{\tilde{\boldsymbol{z}}}(t) =$$

$$[(d\,\mathbf{J}_{p_o} - b\,\mathbf{I}_{p_o}) \oplus (d\,\mathbf{J}_p - b\,\mathbf{I}_p) \oplus (d\,\mathbf{I}_q - b\,\mathbf{H}_q)]\tilde{\boldsymbol{z}}(t)\,,$$

whereby setting

$$\tilde{\boldsymbol{z}}(t) = \begin{bmatrix} \tilde{\boldsymbol{z}}_{p_o}(t) \\ \tilde{\boldsymbol{z}}_p(t) \\ \tilde{\boldsymbol{z}}_q(t) \end{bmatrix},$$

with $\tilde{\boldsymbol{z}}_{p_o}(t) \in \mathbb{C}^{p_o \times 1}$, $\tilde{\boldsymbol{z}}_p(t) \in \mathbb{C}^{p \times 1}$, $\tilde{\boldsymbol{z}}_q(t) \in \mathbb{C}^{q \times 1}$, and using the above written notations we arrive easily at three subsystems of (7.3):

$$(a\,\mathbf{I}_{p_o} - c\,\mathbf{J}_{p_o})\dot{\tilde{\boldsymbol{z}}}_{p_o}(t) = (d\,\mathbf{J}_{p_o} - b\,\mathbf{I}_{p_o})\tilde{\boldsymbol{z}}_{p_o}(t)\,,$$

$$(a\,\mathbf{I}_p - c\,\mathbf{J}_p)\dot{\tilde{\boldsymbol{z}}}_p(t) = (d\,\mathbf{J}_p - b\,\mathbf{I}_p)\tilde{\boldsymbol{z}}_p(t)\,,$$

$$(a\,\mathbf{H}_q - c\,\mathbf{I}_q)\dot{\tilde{\boldsymbol{z}}}_q(t) = (d\,\mathbf{I}_q - b\,\mathbf{H}_q)\tilde{\boldsymbol{z}}_q(t)\,.$$

Note that the matrix $a\,\mathbf{I}_{p_o} - c\,\mathbf{J}_{p_o}$ has only zeros in its diagonal because $a = c\lambda_o$. Hence, this matrix is equal to a nilpotent matrix, i.e. $a\,\mathbf{I}_{p_o} - c\,\mathbf{J}_{p_o} = \mathbf{H}_{p_o}$. Let p_o^* be the index of the nilpotent matrix $\mathbf{H}_{p_o}$, i.e. $\mathbf{H}_{p_o}^{p_o^*} = \mathbf{0}_{p_o,p_o}$. Then we obtain the following matrix equations:

$$\begin{aligned}
\mathbf{H}_{p_o}\dot{\boldsymbol{z}}_{p_o}(t) &= \boldsymbol{z}_{p_o}(t)\,,\\
\mathbf{H}_{p_o}^2\ddot{\boldsymbol{z}}_{p_o}(t) &= \mathbf{H}_{p_o}\dot{\boldsymbol{z}}_{p_o}(t)\,,\\
\mathbf{H}_{p_o}^3\dddot{\boldsymbol{z}}_{p_o}(t) &= \mathbf{H}_{p_o}^2\ddot{\boldsymbol{z}}_{p_o}(t)\,,\\
\mathbf{H}_{p_o}^4\boldsymbol{z}_{p_o}^{[4]}(t) &= \mathbf{H}_{p_o}^3\dddot{\boldsymbol{z}}_{p_o}(t)\,,\\
&\;\;\vdots\\
\mathbf{H}_{p_o}^{p_o^*-1}\boldsymbol{z}_{p_o}^{[p_o^*-1]}(t) &= \mathbf{H}_{p_o}^{p_o^*-2}\boldsymbol{z}_{p_o}^{[p_o^*-2]}(t)\,,\\
\mathbf{H}_{p_o}^{p_o^*}\boldsymbol{z}_{p_o}^{[p_o^*]}(t) &= \mathbf{H}_{p_o}^{p_o^*-1}\boldsymbol{z}_{p_o}^{[p_o^*-1]}(t)\,.
\end{aligned}$$

By taking the sum of the above equations and using the fact that $\mathbf{H}_{p_o}^{p_o^*} = \mathbf{0}_{p_o,p_o}$ we arrive at the solution of the first subsystem which has the unique solution

$$\boldsymbol{z}_{p_o}(t) = \mathbf{0}_{p_o,1}\,.$$

Furthermore the matrices $a\,\mathbf{I}_p - c\,\mathbf{J}_p$, $a\,\mathbf{H}_q - c\,\mathbf{I}_q$ are both invertible since all elements in their diagonal are non-zero. The two other subsystems have solutions:

$$\tilde{\boldsymbol{z}}_p(t) = \mathrm{e}^{\tilde{\mathbf{J}}_p t}\tilde{\boldsymbol{z}}_p(0), \quad \text{and} \quad \tilde{\boldsymbol{z}}_q(t) = \mathrm{e}^{\tilde{\mathbf{J}}_q t}\tilde{\boldsymbol{z}}_q(0)\,,$$

respectively, where

$$\begin{aligned}
\tilde{\mathbf{J}}_p &= (a\,\mathbf{I}_p - c\,\mathbf{J}_p)^{-1}(d\,\mathbf{J}_p - b\,\mathbf{I}_p)\,,\\
\tilde{\mathbf{J}}_q &= (a\,\mathbf{H}_q - c\,\mathbf{I}_q)^{-1}(d\,\mathbf{I}_q - b\,\mathbf{H}_q)\,.
\end{aligned}$$

By using the solutions of the three subsystems, and the notation for $\mathbf{Q}$ as written in the beginning of the proof, we obtain:

$$\tilde{\boldsymbol{x}}(t) = \mathbf{Q}\,\tilde{\boldsymbol{z}}(t) = \begin{bmatrix} \mathbf{Q}_{p_o} & \mathbf{Q}_p & \mathbf{Q}_q \end{bmatrix} \begin{bmatrix} \mathbf{0}_{p_o,1} \\ \mathrm{e}^{\tilde{\mathbf{J}}_p t}\tilde{\boldsymbol{z}}_p(0) \\ \mathrm{e}^{\tilde{\mathbf{J}}_q t}\tilde{\boldsymbol{z}}_q(0) \end{bmatrix},$$

or, equivalently,

$$\tilde{\boldsymbol{x}}(t) = \mathbf{Q}_p\,\mathrm{e}^{\tilde{\mathbf{J}}_p t}\tilde{\boldsymbol{z}}_p(0) + \mathbf{Q}_q\,\mathrm{e}^{\tilde{\mathbf{J}}_q t}\tilde{\boldsymbol{z}}_q(0)\,,$$

or, equivalently,

$$\tilde{\boldsymbol{x}}(t) = \begin{bmatrix} \mathbf{Q}_p & \mathbf{Q}_q \end{bmatrix} \mathrm{e}^{\mathbf{J}_{p+q} t}\tilde{\boldsymbol{z}}_{p+q}(0)\,,$$

where $\mathrm{e}^{\mathbf{J}_{p+q}t} = \mathrm{e}^{\tilde{\mathbf{J}}_p t} \oplus \mathrm{e}^{\tilde{\mathbf{J}}_q t}$, $\tilde{\boldsymbol{z}}_{p+q}(0) = \begin{bmatrix} \tilde{\boldsymbol{z}}_p(0) \\ \tilde{\boldsymbol{z}}_q(0) \end{bmatrix}$. This solution is unique if and only if:

$$\tilde{\boldsymbol{x}}_o \in \operatorname{colspan} \begin{bmatrix} \mathbf{Q}_p & \mathbf{Q}_q \end{bmatrix}.$$

In this case $\tilde{\boldsymbol{z}}_{p+q}(0)$ is the unique solution of

$$\tilde{\boldsymbol{x}}_o = \begin{bmatrix} \mathbf{Q}_p & \mathbf{Q}_q \end{bmatrix} \tilde{\boldsymbol{z}}_{p+q}(0)\,.$$

- Let 0 be an eigenvalue of the pencil $s\mathbf{E} - \mathbf{A}$. Then we can give $\mathbf{Q}$, defined in Lemma 7.1, the following form:

$$\mathbf{Q} = \begin{bmatrix} \mathbf{Q}_{p_o} & \mathbf{Q}_p & \mathbf{Q}_q \end{bmatrix},$$

where:

  - $\mathbf{Q}_{p_o} \in \mathbb{C}^{m\times p_o}$ are the linear independent eigenvectors (including the generalized) of the zero eigenvalue of $s\mathbf{E} - \mathbf{A}$;
  - $\mathbf{Q}_p \in \mathbb{C}^{m\times p}$ are the linear independent eigenvectors (including the generalized) of all non-zero finite eigenvalue of $s\mathbf{E} - \mathbf{A}$;
  - $\mathbf{Q}_q \in \mathbb{C}^{m\times q}$ are the linear independent eigenvectors (including the generalized) of the infinite eigenvalue of $s\mathbf{E} - \mathbf{A}$.

In this case:

$$\mathbf{P}\,\mathbf{E}\,\mathbf{Q} = \mathbf{I}_{p_o} \oplus \mathbf{I}_p \oplus \mathbf{H}_q\,,$$

$$\mathbf{P}\,\mathbf{E}\,\mathbf{Q} = \mathbf{J}_{p_o} \oplus \mathbf{J}_p \oplus \mathbf{I}_q\,.$$

where $\mathbf{J}_{p_o} \in \mathbb{C}^{p_o\times p_o}$, $\mathbf{J}_p \in \mathbb{C}^{p\times p}$ are the Jordan matrices related to the zero, and the finite non-zero eigenvalues, respectively, $\mathbf{H}_q \in \mathbb{C}^{q\times q}$ is a nilpotent matrix with index q_*, constructed by using the algebraic multiplicity of the infinite eigenvalue, and $p_o + p + q = m$. We apply the transformation

$$\tilde{\boldsymbol{x}}(t) = \mathbf{Q}\,\tilde{\boldsymbol{z}}(t)$$

into (7.3), and multiply by $\mathbf{P}$:

$$\mathbf{P}\,\tilde{\mathbf{E}}\,\mathbf{Q}\,\dot{\tilde{z}}(t) = \mathbf{P}\,\tilde{\mathbf{A}}\,\mathbf{Q}\,\tilde{z}(t)\,,$$

or, equivalently,

$$-\mathbf{P}\,(c\,\mathbf{A})\,\mathbf{Q}\,\dot{\tilde{z}}(t) = \mathbf{P}\,(d\,\mathbf{A} - b\,\mathbf{E})\,\mathbf{Q}\,\tilde{z}(t)\,,$$

or, equivalently,

$$-c[\mathbf{J}_{p_o} \oplus \mathbf{J}_p \oplus \mathbf{I}_q]\dot{\tilde{z}}(t) = [(d\,\mathbf{J}_{p_o} - b\,\mathbf{I}_{p_o}) \oplus (d\,\mathbf{J}_p - b\,\mathbf{I}_p) \oplus (d\,\mathbf{I}_q - b\,\mathbf{H}_q)]\tilde{z}(t)\,,$$

whereby setting

$$\tilde{z}(t) = \begin{bmatrix} \tilde{z}_{p_o}(t) \\ \tilde{z}_p(t) \\ \tilde{z}_q(t) \end{bmatrix},$$

with $\tilde{z}_{p_o}(t) \in \mathbb{C}^{p_o \times 1}$, $\tilde{z}_p(t) \in \mathbb{C}^{p \times 1}$, $\tilde{z}_q(t) \in \mathbb{C}^{q \times 1}$, and using the above written notations we arrive at three subsystems of (7.3):

$$c\,\mathbf{J}_{p_o}\dot{\tilde{z}}_{p_o}(t) = (d\,\mathbf{J}_{p_o} - b\,\mathbf{I}_{p_o})\tilde{z}_{p_o}(t)\,,$$

$$-c\,\mathbf{J}_p\dot{\tilde{z}}_p(t) = (d\,\mathbf{J}_p - b\,\mathbf{I}_p)\tilde{z}_p(t)\,,$$

$$-c\,\mathbf{I}_q\dot{\tilde{z}}_q(t) = (d\,\mathbf{I}_q - b\,\mathbf{H}_q)\tilde{z}_q(t)\,.$$

Note that $c \neq 0$ since $a = 0$ and the relation in (7.2) must hold. Furthermore, the matrix $-c\,\mathbf{J}_{p_o}$ has only zeros in its diagonal because $\mathbf{J}_{p_o}$ is the Jordan matrix of the zero eigenvalue of $s\mathbf{E} - \mathbf{A}$. Finally, the matrices $-c\,\mathbf{J}_p$, $-c\,\mathbf{I}_q$ are both invertible since all elements in their diagonal are non-zero. The solution of the first subsystem is

$$\tilde{z}_{p_o}(t) = \mathbf{0}_{p_o,1}\,.$$

This can be easily proved similarly to the relevant part of the proof in the previous case. The two other subsystems have solutions:

$$\tilde{z}_p(t) = \mathrm{e}^{\tilde{\mathbf{J}}_p t}\tilde{z}_p(0), \quad \text{and} \quad \tilde{z}_q(t) = \mathrm{e}^{\tilde{\mathbf{J}}_q t}\tilde{z}_q(0)\,,$$

respectively, where

$$\tilde{\mathbf{J}}_p = -\frac{1}{c}\mathbf{J}_p^{-1}(d\,\mathbf{J}_p - b\,\mathbf{I}_p)\,, \quad \tilde{\mathbf{J}}_q = -\frac{1}{c}(d\,\mathbf{I}_q - b\,\mathbf{H}_q)\,.$$

By using the solutions of the three subsystems, and the notation for $\mathbf{Q}$ as written in the beginning of the proof, we obtain:

$$\tilde{x}(t) = \mathbf{Q}\tilde{z}(t) = \begin{bmatrix} \mathbf{Q}_{p_o} & \mathbf{Q}_p & \mathbf{Q}_q \end{bmatrix} \begin{bmatrix} \mathbf{0}_{p_o,1} \\ \mathrm{e}^{\tilde{\mathbf{J}}_p t}\tilde{z}_p(0) \\ \mathrm{e}^{\tilde{\mathbf{J}}_q t}\tilde{z}_q(0) \end{bmatrix},$$

or, equivalently,

$$\tilde{\boldsymbol{x}}(t) = \mathbf{Q}_p \, \mathrm{e}^{\tilde{\mathbf{J}}_p t} \tilde{\boldsymbol{z}}_p(0) + \mathbf{Q}_q \, \mathrm{e}^{\tilde{\mathbf{J}}_q t} \tilde{\boldsymbol{z}}_q(0)\,,$$

or, equivalently,

$$\tilde{\boldsymbol{x}}(t) = \begin{bmatrix} \mathbf{Q}_p & \mathbf{Q}_q \end{bmatrix} \mathrm{e}^{\mathbf{J}_{p+q} t} \tilde{\boldsymbol{z}}_{p+q}(0)\,,$$

where $\mathrm{e}^{\mathbf{J}_{p+q} t} = \mathrm{e}^{\tilde{\mathbf{J}}_p t} \oplus \mathrm{e}^{\tilde{\mathbf{J}}_q t}$, $\tilde{\boldsymbol{z}}_{p+q}(0) = \begin{bmatrix} \tilde{z}_p(0) \\ \tilde{z}_q(0) \end{bmatrix}$. This solution is unique if and only if:

$$\tilde{\boldsymbol{x}}_o \in \text{colspan} \begin{bmatrix} \mathbf{Q}_p & \mathbf{Q}_q \end{bmatrix}.$$

In this case, $\tilde{\boldsymbol{z}}_{p+q}(0)$ is the unique solution of

$$\tilde{\boldsymbol{x}}_o = \begin{bmatrix} \mathbf{Q}_p & \mathbf{Q}_q \end{bmatrix} \tilde{\boldsymbol{z}}_{p+q}(0)\,.$$

- We can give $\mathbf{Q}$, defined in Lemma 7.1, the following form:

$$\mathbf{Q} = \begin{bmatrix} \mathbf{Q}_p & \mathbf{Q}_q \end{bmatrix},$$

where:

  - $\mathbf{Q}_p \in \mathbb{C}^{m\times p}$ are the linear independent eigenvectors (including the generalized) of all finite eigenvalue of $s\mathbf{E} - \mathbf{A}$;
  - $\mathbf{Q}_q \in \mathbb{C}^{m\times q}$ are the linear independent eigenvectors (including the generalized) of the infinite eigenvalue of $s\mathbf{E} - \mathbf{A}$.

From Lemma 7.1, we have:

$$\mathbf{P}\,\mathbf{E}\,\mathbf{Q} = \mathbf{I}_p \oplus \mathbf{H}_q\,,$$

$$\mathbf{P}\,\mathbf{E}\,\mathbf{Q} = \mathbf{J}_p \oplus \mathbf{I}_q\,,$$

where $\mathbf{J}_p \in \mathbb{C}^{p\times p}$ is the Jordan matrix related to the finite eigenvalues, $\mathbf{H}_q \in \mathbb{C}^{q\times q}$ is a nilpotent matrix with index q_*, constructed by using the algebraic multiplicity of the infinite eigenvalue, and $p + q = m$. We apply the transformation

$$\tilde{\boldsymbol{x}}(t) = \mathbf{Q}\,\tilde{\boldsymbol{z}}(t)$$

into (7.3), and multiply by $\mathbf{P}$:

$$\mathbf{P}\,\tilde{\mathbf{E}}\,\mathbf{Q}\,\dot{\tilde{\boldsymbol{z}}}(t) = \mathbf{P}\,\tilde{\mathbf{A}}\,\mathbf{Q}\,\tilde{\boldsymbol{z}}(t)\,,$$

or, equivalently,

$$\mathbf{P}\,(a\,\mathbf{E})\,\mathbf{Q}\,\dot{\tilde{\boldsymbol{z}}}(t) = \mathbf{P}\,(d\,\mathbf{A} - b\,\mathbf{E})\,\mathbf{Q}\,\tilde{\boldsymbol{z}}(t)\,,$$

or, equivalently,

$$a[\mathbf{I}_p \oplus \mathbf{H}_q]\dot{\tilde{\boldsymbol{z}}}(t) = [(d\,\mathbf{J}_p - b\,\mathbf{I}_p) \oplus (d\,\mathbf{I}_q - b\,\mathbf{H}_q)]\tilde{\boldsymbol{z}}(t),$$

whereby setting

$$\tilde{\boldsymbol{z}}(t) = \begin{bmatrix} \tilde{\boldsymbol{z}}_p(t) \\ \tilde{\boldsymbol{z}}_q(t) \end{bmatrix},$$

with $\tilde{\boldsymbol{z}}_p(t) \in \mathbb{C}^{p\times 1}$, $\tilde{\boldsymbol{z}}_q(t) \in \mathbb{C}^{q\times 1}$, and using the above written notations we arrive easily at two subsystems of (7.3):

$$a\,\mathbf{I}_p\dot{\tilde{\boldsymbol{z}}}_p(t) = (d\,\mathbf{J}_p - b\,\mathbf{I}_p)\tilde{\boldsymbol{z}}_p(t)\,,$$

$$a\,\mathbf{H}_q\dot{\tilde{\boldsymbol{z}}}_q(t) = (d\,\mathbf{I}_q - b\,\mathbf{H}_q)\tilde{\boldsymbol{z}}_q(t)\,.$$

The solutions of the two subsystems are:

$$\tilde{\boldsymbol{z}}_p(t) = \mathrm{e}^{\tilde{\mathbf{J}}_p t}\tilde{\boldsymbol{z}}_p(0), \quad \text{and} \quad \tilde{\boldsymbol{z}}_q(t) = \mathbf{0}_{q,1},$$

respectively, where

$$\tilde{\mathbf{J}}_p = \frac{1}{a}(d\,\mathbf{J}_p - b\,\mathbf{I}_p).$$

By using the solutions of the two subsystems, and the notation for $\mathbf{Q}$ as written in the beginning of the proof we obtain:

$$\tilde{\boldsymbol{x}}(t) = \mathbf{Q}\tilde{\boldsymbol{z}}(t) = \begin{bmatrix} \mathbf{Q}_p & \mathbf{Q}_q \end{bmatrix}\begin{bmatrix} \mathrm{e}^{\tilde{\mathbf{J}}_p t}\tilde{\boldsymbol{z}}_p(0) \\ \mathbf{0}_{q,1} \end{bmatrix},$$

or, equivalently,

$$\tilde{\boldsymbol{x}}(t) = \mathbf{Q}_p\,\mathrm{e}^{\tilde{\mathbf{J}}_p t}\tilde{\boldsymbol{z}}_p(0)\,.$$

This solution is unique if and only if:

$$\tilde{\boldsymbol{z}}_o \in \operatorname{colspan}\mathbf{Q}_p\,.$$

In this case, $\tilde{\boldsymbol{z}}_p(0)$ is the unique solution of

$$\tilde{\boldsymbol{x}}_o = \mathbf{Q}_p\tilde{\boldsymbol{z}}_p(0)\,.$$

The proof is completed. ■

Example 7.1 (Möbius transform). We consider system (7.1) with

$$\mathbf{E} = \begin{bmatrix} 1 & 0 & 0 & 0 & 0 & 0 \\ 0 & 1 & 0 & 0 & 0 & 0 \\ 0 & 0 & 1 & 0 & 0 & 0 \\ 0 & 0 & 0 & 1 & 0 & 0 \\ 0 & 0 & 0 & 0 & 1 & 1 \\ 0 & 0 & 0 & 0 & 0 & 0 \end{bmatrix}, \quad \mathbf{A} = \begin{bmatrix} 0 & 0 & 1 & 0 & 0 & 0 \\ 0 & 0 & 0 & 1 & 0 & 0 \\ 0 & 0 & 0 & 0 & 1 & 0 \\ 0 & 0 & 0 & 0 & 0 & 1 \\ -4 & 2 & 2 & -3 & -2 & -1 \\ 1 & 1 & -1 & -1 & 0 & 0 \end{bmatrix}.$$

By applying the Möbius transform (7.2) into the pencil of (7.1) we arrive at the family of systems (7.3). Let

$$\tilde{\boldsymbol{x}}_o = \begin{bmatrix} -4 & 6 & -5 & 7 & -7 & 9 \end{bmatrix}^{\mathrm{T}}$$

be the initial conditions of (7.3). We may now use Theorem 7.1, and the invariants of the pencil $s\mathbf{E} - \mathbf{A}$ of (7.1) in order to investigate the solutions of (7.3) $\forall a, b, c, d \in \mathbb{C}$. The pencil $s\mathbf{E} - \mathbf{A}$ has three finite eigenvalues $\lambda_1 = 3$, $\lambda_2 = 2$, $\lambda_3 = 1$, of algebraic multiplicity $p_1 = p_2 = p_3 = 1$, and an infinite eigenvalue of algebraic multiplicity $q = 3$. The eigenspaces of $s\mathbf{E} - \mathbf{A}$ associated with the eigenvalues 3, 2, 1, are:

$$\langle \mathbf{u}_1 \rangle = \left\langle \begin{bmatrix} -1 \\ 1 \\ -3 \\ 3 \\ -9 \\ 9 \end{bmatrix} \right\rangle, \quad \langle \mathbf{u}_2 \rangle = \left\langle \begin{bmatrix} -1 \\ 1 \\ -2 \\ 2 \\ -4 \\ 4 \end{bmatrix} \right\rangle, \quad \langle \mathbf{u}_3 \rangle = \left\langle \begin{bmatrix} -3 \\ 5 \\ -3 \\ 5 \\ -3 \\ 5 \end{bmatrix} \right\rangle,$$

while the eigenspace of $s\mathbf{E} - \mathbf{A}$ associated with the infinite eigenvalue, including the generalized eigenvectors, is:

$$\langle \mathbf{u}_4, \quad \mathbf{u}_5, \quad \mathbf{u}_6 \rangle = \left\langle \begin{bmatrix} 0 \\ 0 \\ 0 \\ 0 \\ -1 \\ 1 \end{bmatrix}, \begin{bmatrix} 0 \\ 0 \\ -1 \\ 1 \\ 0 \\ 1 \end{bmatrix}, \begin{bmatrix} -1 \\ 1 \\ 0 \\ 1 \\ 42 \\ -48 \end{bmatrix} \right\rangle.$$

The Jordan matrix related to the finite eigenvalues, and, the matrix $\mathbf{H}_q$ are given by:

$$\mathbf{J}_p = \begin{bmatrix} 3 & 0 & 0 \\ 0 & 2 & 0 \\ 0 & 0 & 1 \end{bmatrix}, \qquad \mathbf{H}_q = \begin{bmatrix} 0 & 1 & 0 \\ 0 & 0 & 1 \\ 0 & 0 & 0 \end{bmatrix}.$$

The matrix $\mathbf{Q}_q$ is defined as $\mathbf{Q}_q = \begin{bmatrix} \mathbf{u}_4 & \mathbf{u}_5 & \mathbf{u}_6 \end{bmatrix}$. We have the following cases:

- If $a, c \neq 0$, then from Theorem 7.1, the general solution of system (7.3) is given by

$$\tilde{\boldsymbol{x}}(t) = \begin{bmatrix} \mathbf{Q}_p & \mathbf{Q}_q \end{bmatrix} e^{\mathbf{J}_{p+q} t} \tilde{\boldsymbol{z}}_{p+q}(0).$$

The matrices $e^{\mathbf{J}_{p+q} t}$, $\mathbf{Q}_p$ are defined as follows:

(i) If $a \neq 3c$, $a \neq 2c$, $a \neq c$, then $\mathbf{Q}_p = \begin{bmatrix} \mathbf{u}_1 & \mathbf{u}_2 & \mathbf{u}_3 \end{bmatrix}$, and:

$$e^{\mathbf{J}_{p+q} t} = \begin{bmatrix} e^{\frac{-b+3d}{a-3c} t} & 0 & 0 & 0 & 0 & 0 \\ 0 & e^{\frac{-b+2d}{a-2c} t} & 0 & 0 & 0 & 0 \\ 0 & 0 & e^{\frac{-b+d}{a-c} t} & 0 & 0 & 0 \\ 0 & 0 & 0 & e^{-\frac{d}{c} t} & \frac{bc-ad}{c^2} & \frac{abc-ad}{c^3} \\ 0 & 0 & 0 & 0 & e^{-\frac{d}{c} t} & \frac{bc-ad}{c^2} \\ 0 & 0 & 0 & 0 & 0 & e^{-\frac{d}{c} t} \end{bmatrix}. \tag{7.4}$$

(ii) If $a = 3c$, $\mathbf{Q}_p = \begin{bmatrix}\mathbf{u}_2 & \mathbf{u}_3\end{bmatrix}$, and:

$$\mathrm{e}^{\mathbf{J}_{p+q}t} = \begin{bmatrix} \mathrm{e}^{\frac{-b+2d}{a-2c}t} & 0 & 0 & 0 & 0 \\ 0 & \mathrm{e}^{\frac{-b+d}{a-c}t} & 0 & 0 & 0 \\ 0 & 0 & \mathrm{e}^{-\frac{d}{c}t} & \frac{bc-ad}{c^2} & \frac{abc-ad}{c^3} \\ 0 & 0 & 0 & \mathrm{e}^{-\frac{d}{c}t} & \frac{bc-ad}{c^2} \\ 0 & 0 & 0 & 0 & \mathrm{e}^{-\frac{d}{c}t} \end{bmatrix};$$

(iii) If $a = 2c$, then $\mathbf{Q}_p = \begin{bmatrix}\mathbf{u}_1 & \mathbf{u}_3\end{bmatrix}$, and:

$$\mathrm{e}^{\mathbf{J}_{p+q}t} = \begin{bmatrix} \mathrm{e}^{\frac{-b+3d}{a-3c}t} & 0 & 0 & 0 & 0 \\ 0 & \mathrm{e}^{\frac{-b+d}{a-c}t} & 0 & 0 & 0 \\ 0 & 0 & \mathrm{e}^{-\frac{d}{c}t} & \frac{bc-ad}{c^2} & \frac{abc-ad}{c^3} \\ 0 & 0 & 0 & \mathrm{e}^{-\frac{d}{c}t} & \frac{bc-ad}{c^2} \\ 0 & 0 & 0 & 0 & \mathrm{e}^{-\frac{d}{c}t} \end{bmatrix};$$

(iv) If $a = c$, then $\mathbf{Q}_p = \begin{bmatrix}\mathbf{u}_2 & \mathbf{u}_3\end{bmatrix}$, and:

$$\mathrm{e}^{\mathbf{J}_{p+q}t} = \begin{bmatrix} \mathrm{e}^{\frac{-b+3d}{a-3c}t} & 0 & 0 & 0 & 0 \\ 0 & \mathrm{e}^{\frac{-b+2d}{a-2c}t} & 0 & 0 & 0 \\ 0 & 0 & \mathrm{e}^{-\frac{d}{c}t} & \frac{bc-ad}{c^2} & \frac{abc-ad}{c^3} \\ 0 & 0 & 0 & \mathrm{e}^{-\frac{d}{c}t} & \frac{bc-ad}{c^2} \\ 0 & 0 & 0 & 0 & \mathrm{e}^{-\frac{d}{c}t} \end{bmatrix}.$$

It is easy to observe that in the cases of (i) and (ii), we have $\tilde{\boldsymbol{x}}_o \in \operatorname{colspan}\begin{bmatrix}\mathbf{Q}_p & \mathbf{Q}_q\end{bmatrix}$, and hence the solution is unique. For both (i) and (ii), the unique solution is given by:

$$\tilde{\boldsymbol{x}}(t) = \begin{bmatrix} -3\,\mathrm{e}^{\frac{-b+d}{a-c}t} - \mathrm{e}^{\frac{-b+2d}{a-2c}t} \\ 5\,\mathrm{e}^{\frac{-b+d}{a-c}t} + \mathrm{e}^{\frac{-b+2d}{a-2c}t} \\ -3\,\mathrm{e}^{\frac{-b+d}{a-c}t} - 2\,\mathrm{e}^{\frac{-b+2d}{a-2c}t} \\ 5\,\mathrm{e}^{\frac{-b+d}{a-c}t} + 2\,\mathrm{e}^{\frac{-b+2d}{a-2c}t} \\ -3\,\mathrm{e}^{\frac{-b+d}{a-c}t} - 4\,\mathrm{e}^{\frac{-b+2d}{a-2c}t} \\ 5\,\mathrm{e}^{\frac{-b+d}{a-c}t} + 4\,\mathrm{e}^{\frac{-b+2d}{a-2c}t} \end{bmatrix}.$$

It is also easy to observe that for both (iii), (iv), $\tilde{\boldsymbol{x}}_o \notin \operatorname{colspan}\begin{bmatrix}\mathbf{Q}_p & \mathbf{Q}_q\end{bmatrix}$, i.e. from Theorem 7.1 there does not exist a unique solution for these systems. Since $\tilde{\boldsymbol{z}}_{p+q}(0)$ can not be defined uniquely, we set $\tilde{\boldsymbol{z}}_{p+q}(0) = \begin{bmatrix}c_1 & c_3 & c_4 & c_5 & c_6\end{bmatrix}$, and $\tilde{c}_4 = \frac{bc-ad}{c^2}c_5 + \frac{abc-ad}{c^3}c_6$, $\tilde{c}_5 = \frac{bc-ad}{c^2}c_6$. Then the general solution for (iii) is:

$$\tilde{\boldsymbol{x}}(t) = \mathrm{e}^{\frac{-b+3d}{a-3c}t}c_1\mathbf{u}_1 + \mathrm{e}^{\frac{-b+d}{a-c}t}c_3\mathbf{u}_3 + \mathrm{e}^{\frac{-d}{c}t}\sum_{i=4}^{6} c_i\mathbf{u}_i + \tilde{c}_4\mathbf{u}_4 + \tilde{c}_5\mathbf{u}_5\,.$$

For (iv) again $\tilde{\boldsymbol{z}}_{p+q}(0)$ can not be defined uniquely. We set $\tilde{\boldsymbol{z}}_{p+q}(0) = \begin{bmatrix} c_1 & c_2 & c_4 & c_5 & c_6 \end{bmatrix}$, and the general solution is:

$$\tilde{\boldsymbol{x}}(t) = \mathrm{e}^{\frac{-b+3d}{a-3c}t} c_1 \mathbf{u}_1 + \mathrm{e}^{\frac{-b+2d}{a-2c}t} c_2 \mathbf{u}_2 + \mathrm{e}^{\frac{-d}{c}t} \sum_{i=4}^{6} c_i \mathbf{u}_i + \tilde{c}_4 \mathbf{u}_4 + \tilde{c}_5 \mathbf{u}_5 \,.$$

- If $a = 0$, then from Theorem 7.1, the general solution of system (7.3) is given by
$$\tilde{\boldsymbol{x}}(t) = \begin{bmatrix} \mathbf{Q}_p & \mathbf{Q}_q \end{bmatrix} \mathrm{e}^{\mathbf{J}_{p+q} t} \tilde{\boldsymbol{z}}_{p+q}(0) \,,$$
where, $\mathbf{Q}_p = \begin{bmatrix} \mathbf{u}_1 & \mathbf{u}_2 & \mathbf{u}_3 \end{bmatrix}$, and $\mathrm{e}^{\mathbf{J}_{p+q}t}$ is given by (7.4). It is easy to observe that $\tilde{\boldsymbol{x}}_o \in \mathrm{colspan} \begin{bmatrix} \mathbf{Q}_p & \mathbf{Q}_q \end{bmatrix}$, and hence the solution is unique.
- If $c = 0$, then from Theorem 7.1, the general solution of system (7.3) is given by
$$\tilde{\boldsymbol{x}}(t) = \mathbf{Q}_p \, \mathrm{e}^{\tilde{\mathbf{J}}_p t} \tilde{\boldsymbol{z}}_p(0) \,,$$
where $\mathbf{Q}_p = \begin{bmatrix} \mathbf{u}_1 & \mathbf{u}_2 & \mathbf{u}_3 \end{bmatrix}$, and:
$$\mathrm{e}^{\tilde{\mathbf{J}}_p t} = \begin{bmatrix} \mathrm{e}^{\frac{-b+3d}{a}t} & 0 & 0 \\ 0 & \mathrm{e}^{\frac{-b+2d}{a}t} & 0 \\ 0 & 0 & \mathrm{e}^{\frac{-b+d}{a}t} \end{bmatrix} .$$
It is easy to observe that $\tilde{\boldsymbol{x}}_o \in \mathrm{colspan}\mathbf{Q}_p$, and hence the solution is unique.
□

7.2 Special Cases

The utility of the family of systems of type (7.3) has been further emphasized by the features of some particular special cases. The most commonly employed Möbius transforms and the corresponding matrix pencils for the GEP are summarized in Table 7.1. The values of the parameters a, b, c, d that lead to each of these transforms are given in Table 7.2.

We further discuss the definition, through special Möbius transforms, of the important notions of (i) duality; and (ii) the stability analysis of a discrete time system through the spectrum of the pencil of (7.1).

Duality

If we consider the transformation (7.2) for $a = d = 0$, and $b = c = 1$, then

$$s = \frac{1}{z} \,,$$

TABLE 7.1: Common linear spectral transforms.

Name	z	Pencil	s
Prime system	s	$s\mathbf{E} - \mathbf{A}$	z
Invert	$\frac{1}{s}$	$z\mathbf{A} - \mathbf{E}$	$\frac{1}{z}$
Shift & invert	$\frac{1}{s-\sigma}$	$z(\sigma\mathbf{E} - \mathbf{A}) + \mathbf{E}$	$\frac{1}{z} + \sigma$
Cayley[a]	$\frac{s+\sigma}{s-\sigma}$	$z(\sigma\mathbf{E} - \mathbf{A}) - (\mathbf{A} + \sigma\mathbf{E})$	$\sigma\frac{z-1}{z+1}$
Generalized Cayley	$\frac{s+\nu}{s-\sigma}$	$z(\sigma\mathbf{E} - \mathbf{A}) - (\mathbf{A} + \nu\mathbf{E})$	$\frac{\sigma z-\nu}{z+1}$
Möbius	$\frac{-ds+b}{cs-a}$	$z(a\mathbf{E} - c\mathbf{A}) - (d\mathbf{A} - b\mathbf{E})$	$\frac{az+b}{cz+d}$

[a]If $\sigma > 0$, this is equivalent to the bilinear transform $z := \frac{\frac{T}{2}s+1}{\frac{T}{2}s-1}$, where $T = \frac{2}{\sigma}$.

TABLE 7.2: Coefficients of special Möbius transforms.

	a	b	c	d
Prime system	-1	0	0	-1
Dual system	0	1	1	0
Shift & invert	σ	1	1	0
Cayley	σ	$-\sigma$	1	1
Generalized Cayley	σ	$-\nu$	1	1

and by applying it into the pencil of (7.1) we get the pencil

$$\mathbf{A}z - \mathbf{E},$$

which is the pencil of the dual system

$$\mathbf{A}\,\dot{\tilde{\boldsymbol{x}}}(t) = \mathbf{E}\,\tilde{\boldsymbol{x}}(t),$$

of system (7.1). Some studies on the duality of systems can be found in [36, 40, 57, 120].

Linear map between continuous and discrete systems

If we consider the inequality $\mathrm{Re}(s) < 0$, and rewrite it as

$$\frac{s+\bar{s}}{2} < 0\,,$$

or, equivalently,

$$s+\bar{s} < 0,$$

by applying (7.2) for $a = c = d = 1$, and $b = -1$, i.e. the Cayley transform with $\sigma = 1 > 0$, we get

$$\frac{z-1}{z+1} + \frac{\bar{z}-1}{\bar{z}+1} < 0\,,$$

or, equivalently,

$$(\bar{z}+1)(z-1) + (\bar{z}-1)(z+1) < 0\,,$$

or, equivalently, by taking into account that $\bar{z}z = |z|^2$

$$|z| < 1\,.$$

Hence the set $\{Re(w) < 0, \quad \forall w \in \mathbb{C}\}$ maps to the set $\{|z| < 1, \quad \forall z \in \mathbb{C}\}$ through (7.1) for $a = c = d = 1$, and $b = -1$, i.e.

$$s = \frac{z-1}{z+1}\,.$$

Thus if we consider the transformation (7.2), by using it into the pencil of system (7.1), the stability of this continuous time system can be studied through the stability of the discrete time system

$$(\mathbf{E}-\mathbf{A})\,\tilde{\boldsymbol{x}}_{k+1} = (\mathbf{A}+\mathbf{E})\,\tilde{\boldsymbol{x}}_k\,,$$

with pencil

$$(\mathbf{E}-\mathbf{A})\,z - (\mathbf{A}+\mathbf{E})\,,$$

where $\boldsymbol{X} : \mathbb{N} \mapsto \mathbb{C}^{m\times 1}$. For example, if we consider system (7.1) for $\mathbf{E} = \mathbf{I}_m$, i.e.

$$\dot{\boldsymbol{x}}(t) = \mathbf{A}\,\boldsymbol{x}\,,$$

then instead of studying the eigenvalues of $\mathbf{A}$ with $Re(w) < 0$ we can study the eigenvalues of the pencil $(\mathbf{I}_m - \mathbf{A})z - (\mathbf{I}_m + \mathbf{A})$ for $|z| < 1$.

Example 7.2 (Special Möbius transforms)**.** We consider now the system (7.1) with

$$\mathbf{E} = \begin{bmatrix} 0 & -3 & 0 & 1 & 1 & 8 & 2 \\ 12 & 9 & -5 & -2 & -4 & -3 & 4 \\ 0 & -4 & -5 & 13 & 3 & 9 & 6 \\ 6 & -2 & -3 & 13 & 3 & -7 & 0 \\ 0 & 0 & 0 & 0 & 0 & 0 & 0 \\ 0 & -1 & -11 & 26 & 4 & -1 & 8 \\ -3 & -3 & 2 & -1 & 1 & 3 & -1 \end{bmatrix},$$

and

$$\mathbf{A}=\begin{bmatrix} 7 & 23 & 13 & 34 & 7 & -8 & 17 \\ -12 & -8 & 8 & 3 & 7 & 6 & -3 \\ -1 & 6 & -5 & -2 & 5 & -6 & -6 \\ 5 & 13 & 1 & -9 & -1 & 38 & 3 \\ 8 & 22 & 22 & 42 & 1 & 16 & 29 \\ 11 & 19 & 31 & 20 & 13 & 14 & 21 \\ 7 & 16 & 5 & 13 & 5 & -6 & 7 \end{bmatrix}.$$

By applying the Möbius transform (7.2) into the pencil of (7.1) we arrive at the family of systems (7.3). Let

$$\tilde{\boldsymbol{x}}_o=\begin{bmatrix}1 & 0 & 1 & 0 & 1 & 1 & \end{bmatrix}^{\mathrm{T}},$$

be the initial conditions of (7.3). In this example we will focus on two cases:

(i) For $a=d=0$, $b=d=1$, we get the dual system of (7.1):

$$\mathbf{E}\,\boldsymbol{x}=\mathbf{A}\,\dot{\boldsymbol{x}}\,;$$

(ii) For $a=c=d=1$, $b=-1$, we get the discrete time system:

$$(\mathbf{E}-\mathbf{A})\,\boldsymbol{x}_{k+1}=(\mathbf{E}+\mathbf{A})\,\boldsymbol{x}_k,\quad k=0,1,2,\ldots$$

We may now use Theorem 7.1, and the invariants of the pencil $s\mathbf{E}-\mathbf{A}$ of (7.1) in order to investigate the solutions of (i) and (ii). The pencil $s\mathbf{E}-\mathbf{A}$ has three finite eigenvalues $\lambda_1=-1$, $\lambda_2=0$, $\lambda_3=-2$ of algebraic multiplicity $p_1=p_2=p_3=1$, and an infinite eigenvalue of algebraic multiplicity $q=4$. The eigenspaces of $s\mathbf{E}-\mathbf{A}$ associated with the eigenvalues -1, 0, -2, are :

$$\langle\mathbf{u}_1\rangle=\left\langle\begin{bmatrix}0.5981\\-0.1773\\0.2936\\0.4628\\-0.2070\\0.1516\\-1\end{bmatrix}\right\rangle,\quad \langle\mathbf{u}_2\rangle=\left\langle\begin{bmatrix}-0.6466\\0.1937\\-0.2696\\-0.4712\\0.1885\\-0.1597\\1\end{bmatrix}\right\rangle,$$

and

$$\langle\mathbf{u}_3\rangle=\left\langle\begin{bmatrix}-0.6675\\0.2118\\-0.3128\\-0.4433\\0.1773\\-0.1872\\1\end{bmatrix}\right\rangle,$$

while the eigenspace of $s\mathbf{E} - \mathbf{A}$ associated with the infinite eigenvalue, including the generalized eigenvectors, is:

$$\left\langle \mathbf{u}_4, \mathbf{u}_5, \mathbf{u}_6, \mathbf{u}_7 \right\rangle = \left\langle \begin{bmatrix} 0.6283 \\ -0.1760 \\ 0.2854 \\ 0.4603 \\ -0.2130 \\ 0.1531 \\ -1 \end{bmatrix}, \begin{bmatrix} -0.6498 \\ 0.1802 \\ -0.3133 \\ -0.4687 \\ 0.1929 \\ -0.1480 \\ 1 \end{bmatrix}, \begin{bmatrix} -0.5969 \\ -0.1143 \\ 0.2738 \\ 0.4507 \\ -0.1623 \\ -0.1711 \\ -1 \end{bmatrix}, \begin{bmatrix} 0.6240 \\ -0.1652 \\ 0.2997 \\ 0.4692 \\ -0.2273 \\ 0.1578 \\ -1 \end{bmatrix} \right\rangle.$$

The Jordan matrix $\mathbf{J}_p$ related to the finite eigenvalues, and the matrix $\mathbf{H}_q$ are:

$$\mathbf{J}_p = \begin{bmatrix} -1 & 0 & 0 \\ 0 & 0 & 0 \\ 0 & 0 & -2 \end{bmatrix}, \ \mathbf{H}_q = \begin{bmatrix} 0 & 0 & 0 & 0 \\ 0 & 0 & 0 & 0 \\ 0 & 0 & 0 & 0 \\ 0 & 0 & 0 & 0 \end{bmatrix}.$$

The matrix $\mathbf{Q}_q$ is defined as $\mathbf{Q}_q = \begin{bmatrix} \mathbf{u}_4 & \mathbf{u}_5 & \mathbf{u}_6 & \mathbf{u}_7 \end{bmatrix}$. For the dual system (i), since $a = 0$ then from Theorem 7.1, the general solution of system (7.3) is given by

$$\tilde{\boldsymbol{x}}(t) = \begin{bmatrix} \mathbf{Q}_p & \mathbf{Q}_q \end{bmatrix} \mathrm{e}^{\mathbf{J}_{p+q}t} \tilde{\boldsymbol{z}}_{p+q}(0),$$

where $\mathbf{Q}_p = \begin{bmatrix} \mathbf{u}_1 & \mathbf{u}_3 \end{bmatrix}$, and:

$$\mathrm{e}^{\mathbf{J}_{p+q}t} = \begin{bmatrix} \mathrm{e}^{-t} & 0 & 0 & 0 & 0 & 0 \\ 0 & \mathrm{e}^{-\frac{1}{2}t} & 0 & 0 & 0 & 0 \\ 0 & 0 & 1 & 0 & 0 & 0 \\ 0 & 0 & 0 & 1 & 0 & 0 \\ 0 & 0 & 0 & 0 & 1 & 0 \\ 0 & 0 & 0 & 0 & 0 & 1 \end{bmatrix}.$$

It is easy to observe that $\tilde{\boldsymbol{x}}_o \notin \text{colspan} \begin{bmatrix} \mathbf{Q}_p & \mathbf{Q}_q \end{bmatrix}$, and hence there does not exist a unique solution. Since $\tilde{\boldsymbol{z}}_{p+q}(0)$ can not be defined uniquely, we set $\tilde{\boldsymbol{z}}_{p+q}(0) = \begin{bmatrix} c_1 & c_3 & c_4 & c_5 & c_6 & c_7 \end{bmatrix}$, and the general solution is:

$$\tilde{\boldsymbol{x}}(t) = \mathrm{e}^{-t} c_1 \mathbf{u}_1 + \mathrm{e}^{-\frac{1}{2}t} c_3 \mathbf{u}_3 + \sum_{i=4}^{7} c_i \mathbf{u}_i \,.$$

For the discrete time system (ii), it is worth noting that since the steady state of (7.1) is weakly stable, the steady state of the discrete time system will be weakly stable as well. Furthermore, $a, c \neq 0$, and there does not exist an eigenvalue λ_o such that $a = \lambda_o c$. Then from Theorem 7.1, it is easy to observe that $\tilde{\boldsymbol{x}}_o \notin \text{colspan} \begin{bmatrix} \mathbf{Q}_p & \mathbf{Q}_q \end{bmatrix}$, and there does not exist a unique solution.

Hence, since

$$\mathbf{J}_{p+q} = \begin{bmatrix} 0 & 0 & 0 & 0 & 0 & 0 & 0 \\ 0 & 1 & 0 & 0 & 0 & 0 & 0 \\ 0 & 0 & -\frac{1}{3} & 0 & 0 & 0 & 0 \\ 0 & 0 & 0 & -1 & 0 & 0 & 0 \\ 0 & 0 & 0 & 0 & -1 & 0 & 0 \\ 0 & 0 & 0 & 0 & 0 & -1 & 0 \\ 0 & 0 & 0 & 0 & 0 & 0 & -1 \end{bmatrix},$$

if we set $\tilde{\boldsymbol{z}}_{p+q}(0) = \begin{bmatrix} c_2 & c_3 & c_4 & c_5 & c_6 & c_7 \end{bmatrix}$, the general solution is given by:

$$\tilde{\boldsymbol{x}}_k = c_2 \mathbf{u}_2 + \left(-\frac{1}{3}\right)^k c_3 \mathbf{u}_3 + (-1)^k \sum_{i=4}^{7} c_i \mathbf{u}_i \, .$$

□

7.3 Applications of the Möbius Transform

We complete this chapter with an example that illustrates some special cases of the Möbius transform on the three bus system.

Example 7.3 (Applications of Möbius transforms)**.** In this example, we show the image of the spectrum of the three-bus system for a couple of common special Möbius transforms, in particular the shift & invert and the Cayley transform.

The results, presented in Figures 7.1–7.3, refer to the GEP, which has been already discussed in Chapter 5. We denote eigenvalues of the transformed pencil with $\hat{\lambda}$. In each figure, the stable region is shaded, while the stability boundary is indicated with a black solid line. The 5% damping boundary is indicated with a dash-dotted line.

For the shift & invert transform, the stability boundary is defined by the circle with center $c = (1/2\sigma, 0)$ and radius $\rho = 1/2\sigma$. If $\sigma > 0$, that is the case of Figure 7.1, stable eigenvalues are mapped inside the circle. On the other hand, stable eigenvalues are mapped outside the circle, if $\sigma < 0$, that is the case shown in Figure 7.2. If $\sigma = 0$, we obtain the dual pencil with the corresponding invert transform, and the stable region is the full negative right have plane.

Figure 7.3 shows the image of the Cayley transform of the three-bus system for $\sigma = 2$. All stable eigenvalues are located inside the unit circle with center the origin. □

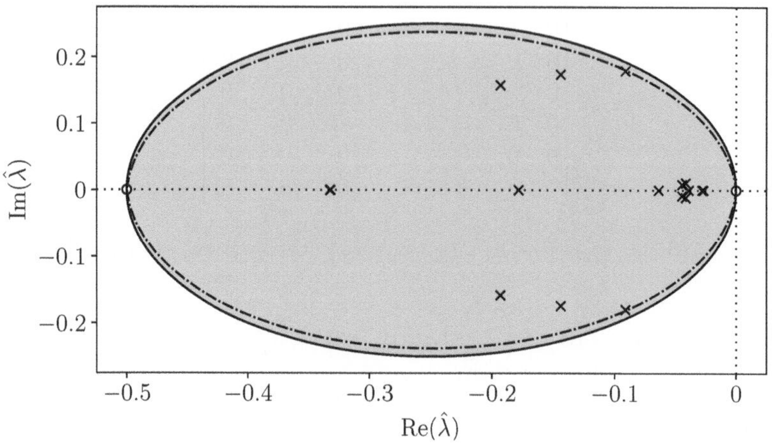

FIGURE 7.1: Shift & invert transform image spectrum with $\sigma = 2$ for the three-bus system.

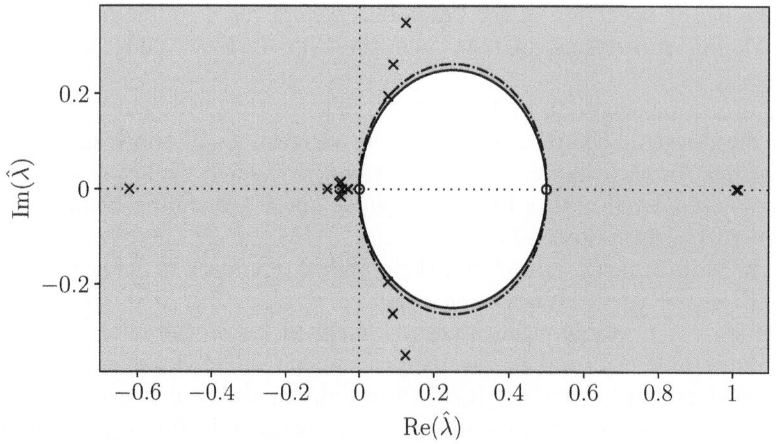

FIGURE 7.2: Shift & invert transform image spectrum with $\sigma = -2$ for the three-bus system.

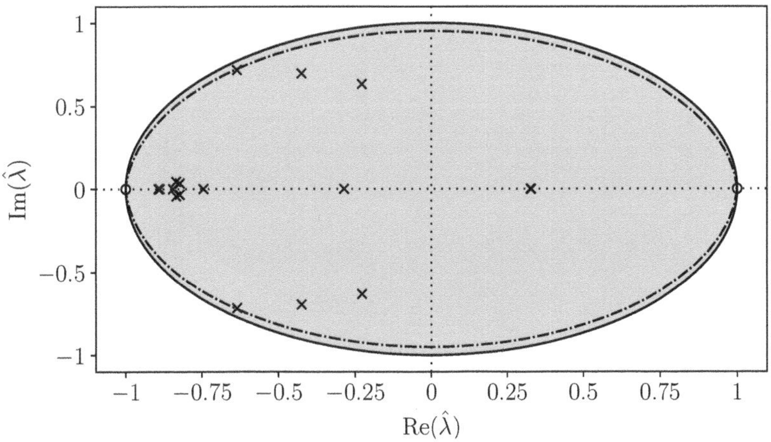

FIGURE 7.3: Cayley transform image of the spectrum with $\sigma = 2$ for the three-bus system.

8

Participation Factors

8.1 Classical Participation Factors

Participation factors (PFs) were firstly introduced by Perez-Arriaga *et al.* in [140] to carry out modal analysis of a linear time-invariant dynamic system of ODEs. Application of appropriate initial conditions to the time response of such a system allows to determine a measure that expresses the relative activity of a state in the structure of an eigenvalue and *vice versa*. This measure is termed PF, see [42, 165].

Definition 8.1 (Participation factor). Consider a linear system of ordinary differential equations in the form:

$$\dot{\boldsymbol{x}} = \mathbf{A}\,\boldsymbol{x} \ ,$$

where $\boldsymbol{x} \in \mathbb{C}^{n\times 1}$, are the state variables and $\mathbf{A} \in \mathbb{C}^{n\times n}$, is the state matrix. Let λ_i be an eigenvalue of $\mathbf{A}$ (or more precisely of $s\mathbf{I}_n - \mathbf{A}$, where $\mathbf{I}_n$ is the $n \times n$ identity matrix) and all the eigenvalues be distinct, i.e. $\lambda_i \neq \lambda_j$, $i \neq j$, and $i, j = 1, 2, \dots, n$. Let also $\mathbf{u}_i$, $\mathbf{u}_i \in \mathbb{C}^{n\times 1}$, and $\mathbf{w}_i$, $\mathbf{w}_i \in \mathbb{C}^{1\times n}$ be right and left eigenvectors associated with λ_i, respectively. If x_k is the k-th state of the system, the PF is defined as:

$$\pi_{k,i} = \mathrm{w}_{i,k}\mathrm{v}_{k,i} \ ,$$

where $\mathrm{v}_{k,i}$ is the k-th row element of $\mathbf{u}_i$ and $\mathrm{w}_{i,k}$ is the k-th column element of $\mathbf{w}_i$.

From Definition 8.1, we can see that the main assumptions of classical modal participation analysis are the following:

- All eigenvalues are distinct.
- The system is modeled as a set of ODEs, i.e. all eigenvalues are finite.

The PFs of a system are typically collected to form the participation matrix $\mathbf{\Pi}_{\mathrm{PF}}$, which is defined as follows:

$$\mathbf{\Pi}_{\mathrm{PF}} = \mathbf{P}^{\mathrm{T}} \circ \mathbf{Q} \ , \tag{8.1}$$

where $\circ$ denotes the Hadamard product, i.e. the element-wise multiplication; and $\mathbf{Q}$, $\mathbf{P}$, are the right and left modal matrices, respectively. That is, the columns of $\mathbf{Q}$ are the right eigenvectors $\mathbf{u}_i$ and the rows of $\mathbf{P}$ are the left eigenvectors $\mathbf{w}_i$.

8.1.1 Residues

The PF $\pi_{k,i}$ basically expresses the relative contribution of x_k in the structure of the eigenvalue λ_i, and *vice versa*, but has also various other interpretations. It is known to represent the sensitivity of an eigenvalue to variations of an element of the state matrix [134]. It has been also viewed as modal energy in the sense described by MacFarlane [64] and utilized in the application of model equivalencing techniques [32].

In the state space representation, PFs can be studied as an important case of residue analysis of the system transfer function and thus, as joint observabilities/controllabilities of the geometric approach, which play an important role during the design of linear control systems [65]. Consider the following single-input single-output system:

$$\begin{aligned} \dot{\boldsymbol{x}} &= \mathbf{A}\,\boldsymbol{x} + \mathbf{b}\,u_1 \\ w_1 &= \mathbf{c}\,\boldsymbol{x} \ , \end{aligned} \tag{8.2}$$

where $\mathbf{b}$ is the column vector of the input u_1; $\mathbf{c}$ is the row vector of the output w_1. Then, the residue of the transfer function of system (8.2) associated with the i-th mode is given by:

$$\mathrm{R}_i = \mathbf{c}\,\mathbf{u}_i\,\mathbf{w}_i\,\mathbf{b} \ . \tag{8.3}$$

That said, the PF of the k-th state in the i-th mode can be viewed as the residue of system (8.2) transfer function associated with the i-th mode, when the input is a perturbation in the differential equation that defines $\dot{\boldsymbol{x}}$ and the output is x_k. Indeed, if

$$\begin{aligned} \mathbf{c} &= \begin{bmatrix} c_1 & \dots & c_k & \dots & c_n \end{bmatrix} = \begin{bmatrix} 0 & \dots & 1 & \dots & 0 \end{bmatrix} , \\ \mathbf{b}^{\mathrm{T}} &= \begin{bmatrix} b_1 & \dots & b_k & \dots & b_n \end{bmatrix}^{\mathrm{T}} = \begin{bmatrix} 0 & \dots & 1 & \dots & 0 \end{bmatrix}^{\mathrm{T}} , \end{aligned}$$

equation (8.3) becomes:

$$\mathrm{R}_i = \mathrm{w}_{i,k}\mathrm{v}_{k,i} = \pi_{k,i} \ . \tag{8.4}$$

In the case of a multiple-input multiple-output system representation, the PFs appear as the diagonal elements of the emerging residue matrix. The ability to calculate only a subset of all residue elements and acquire an approximate but yet accurate measure of the contribution of system states in system modes (and *vice versa*), features the physical importance and the computational efficiency of the PFs.

Example 8.1 (PFs of the state matrix of the three-bus system). We consider the three-bus system. The right and left modal matrices $\mathbf{Q}$ and $\mathbf{P}$ of the LEP, given in Example 4.2, are utilized to compute the participation matrix $\mathbf{\Pi}_{\rm PF}$. The mostly participating (dominant) states in the system eigenvalues and the magnitudes of the corresponding PFs are summarized in Table 8.1.

TABLE 8.1: Three-bus system: dominant states and their PFs.

Mode	Dominant x	$\lvert\pi\rvert_{\rm max}$
$\lambda_1 = -35.2064$	$\psi_{\rm s,d,1}$	0.6559
$\lambda_2 = -33.7327$	$e''_{\rm r,q,2}$	0.5671
$\lambda_3 = -23.7196$	$e''_{\rm r,d,2}$	0.8280
$\lambda_4 = -13.6561$	$\psi_{\rm s,q,1}$	0.8530
$\lambda_{5,6} = -20.5133 \pm \jmath\, 5.9284$	$v_{\rm a,2}$	0.3121
$\lambda_{7,8} = -19.6722 \pm \jmath\, 4.1469$	$v_{\rm R,1}$	0.3694
$\lambda_9 = -3.6091$	$e'_{\rm r,d,2}$	0.8407
$\lambda_{10,11} = -0.2516 \pm \jmath\, 4.4309$	$\delta_{\rm r,2}$, $\omega_{\rm r,2}$	0.2067
$\lambda_{12} = -1.0147$	$v_{\rm f,2}$	0.5396
$\lambda_{13,14} = -0.8211 \pm \jmath\, 3.4132$	$e'_{\rm r,q,2}$	0.2577
$\lambda_{15,16} = -1.1003 \pm \jmath\, 2.5390$	$v_{\rm ef,1}$	0.2123
$\lambda_{17} = -1.0096$	$v_{\rm f,1}$	0.5478
$\lambda_{18,19} = 0.0000$ (2)	$\delta_{\rm r,1}$, $\omega_{\rm r,1}$	0.2525

The states that participate to the most poorly damped pair of eigenvalues $\lambda_{10,11} = -0.2516 \pm \jmath\, 4.4309$ are given in Table 8.2. It is clear that since the mostly participating states are the synchronous machine rotor angles and speeds, $\lambda_{10,11}$ represents an electromechanical oscillation mode. In addition, the variables of the two machines have a similar participation, which implies that the pair represents an inter-area mode.

TABLE 8.2: Three-bus system: dominant states to mode $\lambda_{10,11}$.

State	PF magnitude
$\delta_{\rm r,1}$	0.2023
$\omega_{\rm r,1}$	0.2023
$\delta_{\rm r,2}$	0.2067
$\omega_{\rm r,2}$	0.2067
Rest	0.1820

Figure 8.1 shows matrix $\mathrm{abs}(\mathbf{\Pi}_{\rm PF})$, which includes the magnitudes of all PFs of the three-bus system. □

$$
\begin{bmatrix}
0 & 0 & 0 & 0.01 & 0 & 0 & 0 & 0 & 0.02 & 0.20 & 0.20 & 0 & 0 & 0 & 0.03 & 0.03 & 0 & 0.25 & 0.25 \\
0 & 0 & 0 & 0.01 & 0 & 0 & 0 & 0 & 0.02 & 0.20 & 0.20 & 0 & 0 & 0 & 0.03 & 0.03 & 0 & 0.25 & 0.25 \\
0.02 & 0.01 & 0 & 0 & 0 & 0 & 0.07 & 0.07 & 0 & 0.03 & 0.03 & 0.01 & 0.16 & 0.16 & 0.21 & 0.21 & 0.01 & 0 & 0 \\
0.66 & 0.27 & 0 & 0 & 0.01 & 0.01 & 0.06 & 0.06 & 0 & 0 & 0 & 0 & 0.01 & 0.01 & 0.01 & 0.01 & 0 & 0 & 0 \\
0 & 0 & 0.05 & 0.85 & 0 & 0 & 0.01 & 0.01 & 0.01 & 0.01 & 0.01 & 0 & 0 & 0 & 0.02 & 0.02 & 0 & 0 & 0 \\
0 & 0 & 0 & 0.01 & 0 & 0 & 0 & 0 & 0.02 & 0.21 & 0.21 & 0 & 0 & 0 & 0.03 & 0.03 & 0 & 0.25 & 0.25 \\
0 & 0 & 0 & 0.01 & 0 & 0 & 0 & 0 & 0.02 & 0.21 & 0.21 & 0 & 0 & 0 & 0.03 & 0.03 & 0 & 0.25 & 0.25 \\
0 & 0 & 0.08 & 0.04 & 0 & 0 & 0 & 0 & 0.84 & 0.01 & 0.01 & 0 & 0.01 & 0.01 & 0.03 & 0.03 & 0 & 0 & 0 \\
0 & 0.02 & 0 & 0 & 0.09 & 0.09 & 0 & 0 & 0 & 0.04 & 0.04 & 0.01 & 0.26 & 0.26 & 0.16 & 0.16 & 0 & 0 & 0 \\
0.27 & 0.57 & 0 & 0 & 0.13 & 0.13 & 0 & 0 & 0 & 0 & 0 & 0 & 0.03 & 0.03 & 0.01 & 0.01 & 0 & 0 & 0 \\
0 & 0 & 0.83 & 0.04 & 0.01 & 0.01 & 0 & 0 & 0.05 & 0.01 & 0.01 & 0 & 0 & 0 & 0.01 & 0.01 & 0 & 0 & 0 \\
0.01 & 0.01 & 0 & 0.01 & 0.01 & 0.01 & 0.37 & 0.37 & 0 & 0.01 & 0.01 & 0 & 0.03 & 0.03 & 0.03 & 0.03 & 0 & 0 & 0 \\
0.01 & 0.04 & 0.01 & 0 & 0.31 & 0.31 & 0.02 & 0.02 & 0 & 0.01 & 0.01 & 0 & 0.05 & 0.05 & 0.02 & 0.02 & 0 & 0 & 0 \\
0 & 0 & 0 & 0.01 & 0 & 0 & 0.09 & 0.09 & 0 & 0.02 & 0.02 & 0.01 & 0.15 & 0.15 & 0.18 & 0.18 & 0 & 0 & 0 \\
0 & 0.02 & 0 & 0 & 0.10 & 0.10 & 0 & 0 & 0.03 & 0.03 & 0.03 & 0.01 & 0.20 & 0.20 & 0.12 & 0.12 & 0 & 0 & 0 \\
0.01 & 0.01 & 0 & 0.01 & 0.01 & 0.01 & 0.35 & 0.35 & 0 & 0.01 & 0.01 & 0 & 0.03 & 0.03 & 0.03 & 0.03 & 0 & 0 & 0 \\
0.01 & 0.04 & 0.01 & 0 & 0.31 & 0.31 & 0.02 & 0.02 & 0 & 0.01 & 0.01 & 0 & 0.05 & 0.05 & 0.02 & 0.02 & 0 & 0 & 0 \\
0 & 0 & 0 & 0 & 0 & 0 & 0 & 0 & 0 & 0 & 0 & 0.42 & 0.01 & 0.01 & 0.01 & 0.01 & 0.55 & 0 & 0 \\
0 & 0 & 0 & 0 & 0 & 0 & 0 & 0 & 0 & 0 & 0 & 0.54 & 0.01 & 0.01 & 0.01 & 0.01 & 0.43 & 0 & 0
\end{bmatrix}
$$

FIGURE 8.1: Participation factor matrix abs($\mathbf{\Pi}_{\rm PF}$) for the base-case operating condition of the three-bus system.

8.1.2 Power Flow Modal Analysis

A classical application of participation factors is the power flow modal analysis [52]. The basic idea, which has later been further developed with several variants to include dynamics, e.g. [53], is as follows:

Let recall the power flow equations of (1.82), written using polar coordinates and assuming that $\mathbb{B}$ is the set of b network buses:

$$
\begin{aligned}
P_h &= v_h \sum_{k \in \mathbb{B}} v_k \big(G_{hk} \cos(\theta_{hk}) + B_{hk} \sin(\theta_{hk})\big), \quad h \in \mathbb{B} \,, \\
Q_h &= v_h \sum_{k \in \mathbb{B}} v_k \big(G_{hk} \sin(\theta_{hk}) - B_{hk} \cos(\theta_{hk})\big), \quad h \in \mathbb{B} \,,
\end{aligned}
\tag{8.5}
$$

or, equivalently, in compact vector form:

$$
\begin{aligned}
\boldsymbol{P} &= \boldsymbol{g}_P(\boldsymbol{v}, \boldsymbol{\theta}) \,, \\
\boldsymbol{Q} &= \boldsymbol{g}_Q(\boldsymbol{v}, \boldsymbol{\theta}) \,.
\end{aligned}
\tag{8.6}
$$

The linearization of (8.6) at a solution of the power flow problem, gives:

$$
\Delta \boldsymbol{P} = \frac{\partial \boldsymbol{g}_P}{\partial \boldsymbol{\theta}} \Delta \boldsymbol{\theta} + \frac{\partial \boldsymbol{g}_P}{\partial \boldsymbol{v}} \Delta \boldsymbol{v} \,, \tag{8.7}
$$

$$
\Delta \boldsymbol{Q} = \frac{\partial \boldsymbol{g}_Q}{\partial \boldsymbol{\theta}} \Delta \boldsymbol{\theta} + \frac{\partial \boldsymbol{g}_Q}{\partial \boldsymbol{v}} \Delta \boldsymbol{v} \,, \tag{8.8}
$$

where, in general all Jacobian matrices are not null $b \times b$ matrices. The assumption made in [52] is that, for the purpose of the voltage stability analysis,

$\Delta \boldsymbol{P} \approx \mathbf{0}_{b,1}$. This allows obtaining an expression of the variations of the reactive powers with respect to the variation of bus voltage magnitudes:

$$\Delta \boldsymbol{Q} = \left[\frac{\partial \boldsymbol{g}_Q}{\partial \boldsymbol{v}} - \frac{\partial \boldsymbol{g}_Q}{\partial \boldsymbol{\theta}} \left(\frac{\partial \boldsymbol{g}_P}{\partial \boldsymbol{\theta}} \right)^{-1} \frac{\partial \boldsymbol{g}_P}{\partial \boldsymbol{v}} \right] \Delta \boldsymbol{v} = \mathbf{G}_{\mathrm{R}} \, \Delta \boldsymbol{v} \,, \tag{8.9}$$

where $\Delta \boldsymbol{\theta}$ in (8.8) has been substituted with the expression:

$$\Delta \boldsymbol{\theta} = - \left(\frac{\partial \boldsymbol{g}_P}{\partial \boldsymbol{\theta}} \right)^{-1} \frac{\partial \boldsymbol{g}_P}{\partial \boldsymbol{v}} \Delta \boldsymbol{v} \,, \tag{8.10}$$

obtained from imposing $\Delta \boldsymbol{P} = \mathbf{0}_{b,1}$ in (8.7).

$\mathbf{G}_{\mathrm{R}}$ is a "reduced" power flow Jacobian matrix that can be utilized to infer the sensitivity of bus voltages to the injection of reactive power at buses.

Equation (8.9) has some interesting applications. It can be utilized, for example, to select the "pilot bus" for secondary voltage regulation and to decide the locations that are most advantageous for the installation of devices that regulate the voltage. The idea is to select the buses that are most sensitive to reactive power injections and thus require the least amount of reactive power to achieve the biggest variations of the voltage. Note, however, that this information can be also effectively obtained by computing the short-circuit power of network buses using the admittance matrix $\bar{\mathbf{Y}}_{\mathrm{bus}}$ of the system.

The metric to obtain such an information is to calculate the eigenvalues of $\mathbf{G}_{\mathrm{R}}$ and their participation factors to the bus voltages. With this aim, (8.9) can be written as:

$$\Delta \boldsymbol{v} = \mathbf{G}_{\mathrm{R}}^{-1} \, \Delta \boldsymbol{Q} \,. \tag{8.11}$$

Then, if $\mathbf{J}_b$, $\mathbf{Q}$ and $\mathbf{P}$ are the diagonal matrix of the eigenvalues of $\mathbf{G}_{\mathrm{R}}$ and the associated right and left eigenvector matrices, respectively, one has:

$$\Delta \boldsymbol{v} = \mathbf{Q} \mathbf{J}_b^{-1} \mathbf{P} \, \Delta \boldsymbol{Q} \,, \tag{8.12}$$

from where one can obtain the sensitivity of the voltage v_h with respect to the reactive power Q_k:

$$\frac{\partial v_h}{\partial Q_k} = \sum_i^b \frac{\mathrm{u}_{hi} \, \mathrm{w}_{ik}}{\lambda_i} \,, \tag{8.13}$$

where u_{hi} and w_{ik} are the elements of matrices $\mathbf{Q}$ and $\mathbf{P}$, respectively.

The expression (8.13) has been also utilized, in occasions incorrectly, to assess the "voltage stability" of the system. It is important to note that the Lyapunov stability criterion does not apply in this case, as the starting equations are algebraic and no dynamic is actually associated with (8.6). As a matter of fact, a mode of $\mathbf{G}_{\mathrm{R}}$ is "stable" if its eigenvalue is positive. This can be easily understood by looking at (8.13): positive eigenvalues mean that to an increase of the reactive power injected at a node will follow an increase of the voltage. This is what it is expected from voltage regulation.

On the other hand, if the eigenvalue is negative, increasing the voltage will further decrease the voltage and lead to an "unstable," possibly unrecoverable situation (voltage collapse). Special care is also to be paid to the case for which an eigenvalue of $\mathbf{G}_{\rm R}$ approaches zero. In this case the sensitivities (8.13) become large and tend to infinity if an eigenvalue tends to zero. This phenomenon, however, does not mean that it is possible to vary the voltages with arbitrarily small reactive power variations but, rather, that the system is close to a saddle-node bifurcation and, hence, to its maximum loading condition – see the "foldings" of the continuation curve of Figure 1.13.

Example 8.2 (Power flow modal analysis). The four Jacobian matrices of the three-bus system are:

$$\frac{\partial \boldsymbol{g}_P}{\partial \boldsymbol{\theta}} = \begin{bmatrix} 13.6120 & -4.6505 & -8.9615 \\ -4.5011 & 13.6547 & -9.1536 \\ -8.5594 & -8.9982 & 17.5576 \end{bmatrix},$$

$$\frac{\partial \boldsymbol{g}_P}{\partial \boldsymbol{v}} = \begin{bmatrix} 1.7048 & -0.2895 & -1.4152 \\ 1.2047 & -1.3349 & 0.1302 \\ 3.0080 & 1.6849 & -4.6930 \end{bmatrix},$$

$$\frac{\partial \boldsymbol{g}_Q}{\partial \boldsymbol{\theta}} = \begin{bmatrix} 4.2527 & 0.2839 & -9.0322 \\ -1.1928 & 1.4455 & -9.2258 \\ -2.9783 & -1.6519 & 17.2928 \end{bmatrix},$$

$$\frac{\partial \boldsymbol{g}_Q}{\partial \boldsymbol{v}} = \begin{bmatrix} 13.0491 & -4.5593 & 1.4264 \\ -4.4565 & 13.3704 & -0.1313 \\ -8.4747 & -8.8217 & -1.3173 \end{bmatrix}.$$

Hence:

$$\mathbf{G}_{\rm R} = \begin{bmatrix} 12.5063 & -4.6298 & -9.2236 \\ -4.8162 & 13.4624 & -9.3257 \\ -9.4789 & -9.2218 & 16.8863 \end{bmatrix}.$$

To these matrices, one has to remove the column and row corresponding to the slack bus (in this case, the second column and row) as its voltage magnitude and phase angle are, by definition, constant.[1] This leads to a 2×2 matrix with eigenvalues:

$$\lambda_1 = 5.092\,,$$
$$\lambda_2 = 24.300\,,$$

and participation factor matrix:

$$\mathbf{\Pi}_{\rm PF} = \begin{bmatrix} 0.614 & 0.386 \\ 0.386 & 0.614 \end{bmatrix},$$

[1] For the same reason, i.e. constant voltage, we should have removed also the line corresponding to bus 1 where the PV generator is connected, but we have kept bus 1 for the sake of example.

where the rows correspond to λ_1 and λ_2 and the columns to v_1 and v_3, respectively.

Since both eigenvalues are positive the system is "stable," i.e. an increase of the reactive power injections at the buses will increase the voltages. □

8.2 Generalized Participation Factors

We consider the following system:

$$\mathbf{E}\,\dot{\boldsymbol{x}}(t) = \mathbf{A}\,\boldsymbol{x}(t)\ , \tag{8.14}$$

where $\mathbf{E}, \mathbf{A} \in \mathbb{C}^{r\times m}$, $\boldsymbol{x} : \mathbb{R}^+ \mapsto \mathbb{C}^{m\times 1}$. The matrices $\mathbf{E}$ and $\mathbf{A}$ can be non-square ($r \neq m$) or square ($r = m$) with $\mathbf{E}$ regular, or singular, i.e. $\det(\mathbf{E}) = 0$. As proved in previous sections, regarding the existence, and uniqueness of solutions for a system of type (8.14), as well as its formula of solutions we provide as a reminder the following Theorem.

Theorem 8.1 (Existence of the solution of a regular pencil). We consider system (8.14) with a regular pencil, or a singular pencil with $r > m$ and that (6.3) holds. Let $\mathbf{J}_p$ be the Jordan matrix of the finite eigenvalues, and $\mathbf{Q}_p$ the matrix that contains all left linear independent eigenvectors related to the finite eigenvalues of $s\mathbf{E} - \mathbf{A}$. Then there exists a solution and is given by:

$$\boldsymbol{x}(t) = \mathbf{Q}_p\, \mathrm{e}^{\mathbf{J}_p t}\boldsymbol{z}_p(0)\ ,$$

where $\boldsymbol{z}_p(0) \in \mathbb{C}^{p\times 1}$ is a constant vector.

Let:

- $\lambda_i \in \mathbb{C}$, $i = 1, 2, \ldots, \nu$, be finite eigenvalue, and p_i be rank of the corresponding Jordan block, where $\sum_{i=1}^{\nu} p_i = p$.
- the infinite eigenvalue have algebraic multiplicity q.

We provide the following Theorem.

Theorem 8.2 (Solution of a pencil). We consider system (8.14) with a regular pencil, or a singular pencil with $r > m$ and for which (6.3) holds. Let λ_i, $i = 1, 2, \ldots, \nu$, be a finite eigenvalue of the pencil, p_i be the rank of the corresponding Jordan block, $\sum_{i=1}^{\nu} p_i = p$, and $\mathbf{u}_{i,j}$, $j = 1, 2, \ldots, p_i$ linear independent (generalized) eigenvectors. Then the general solution of (8.14) is given by:

$$\boldsymbol{x}(t) = \sum_{i=1}^{\nu} \mathrm{e}^{\lambda_i t} \sum_{j=1}^{p_i} \Big(\sum_{k=1}^{j} c_{i,j-(k-1)} t^{k-1} \Big) \mathbf{u}_{i,j}\ , \tag{8.15}$$

where $c_{i,j-(k-1)} \in \mathbb{C}$ are constants.

Proof. From Theorem 8.1 the solution of system (8.14) is given by:

$$\boldsymbol{x}(t) = \mathbf{Q}_p \, \mathrm{e}^{\mathbf{J}_p t} \boldsymbol{z}_p(0) \ .$$

The Jordan matrix has the form:

$$\mathbf{J}_p := \mathbf{J}_{p_1}(\lambda_1) \oplus \cdots \oplus \mathbf{J}_{p_\nu}(\lambda_\nu) \ ,$$

where

$$\mathbf{J}_{p_i}(\lambda_i) = \begin{bmatrix} \lambda_i & 1 & \dots & 0 & 0 \\ 0 & \lambda_i & \dots & 0 & 0 \\ \vdots & \vdots & \ddots & \vdots & \vdots \\ 0 & 0 & \dots & \lambda_i & 1 \\ 0 & 0 & \dots & 0 & \lambda_i \end{bmatrix} \in \mathbb{C}^{p_i \times p_i}, \quad i = 1, 2, \dots, \nu.$$

In addition:

$$\mathrm{e}^{\mathbf{J}_p t} := \mathrm{e}^{\mathbf{J}_{p_1}(\lambda_1)t} \oplus \cdots \oplus \mathrm{e}^{\mathbf{J}_{p_\nu}(\lambda_\nu)t} \ ,$$

and

$$e^{\boldsymbol{J}_{p_i}(\lambda_i)t} = \begin{bmatrix} e^{\lambda_i t} & e^{\lambda_i t} t & \dots & e^{\lambda_i t}\frac{t^{p_i-1}}{(p_i-1)!} & e^{\lambda_i t}\frac{t^{p_i}}{p_i!} \\ 0 & e^{\lambda_i t} & \dots & e^{\lambda_i t}\frac{t^{p_i-2}}{(p_i-2)!} & e^{\lambda_i t}\frac{t^{p_i-1}}{(p_i-1)!} \\ \vdots & \vdots & \ddots & \vdots & \vdots \\ 0 & 0 & \dots & e^{\lambda_i t} & e^{\lambda_i t} t \\ 0 & 0 & \dots & 0 & e^{\lambda_i t} \end{bmatrix} \in \mathbb{C}^{p_i \times p_i}, \ i = 1, 2, \dots, \nu.$$

The matrix $\mathbf{Q}_p$ has as columns the p linear independent (generalized) eigenvectors, and can be written in the form:

$$\mathbf{Q}_p = \left[\ \mathbf{u}_{1,p_1} \quad \dots \quad \mathbf{u}_{1,2} \quad \mathbf{u}_{1,1} \quad \dots \quad \mathbf{u}_{\nu,p_\nu} \quad \dots \quad \mathbf{u}_{\nu,2} \quad \mathbf{u}_{\nu,1} \ \right] \ ,$$

where $\mathbf{u}_{i,j}$, $j = 1, 2, \dots, p_i$ linear independent eigenvectors of λ_i, $i = 1, 2, \dots, \nu$. Finally, $\boldsymbol{z}_p(0)$ can be written as:

$$\boldsymbol{z}_p(0) = \left[\ c_{1,p_1} \quad \dots \quad c_{1,2} \quad c_{1,1} \quad \dots \quad c_{\nu,p_\nu} \quad \dots \quad c_{\nu,2} \quad c_{\nu,1} \ \right]^{\mathrm{T}} \ ,$$

where $c_{i,j} \in \mathbb{C}$, $\mathbf{u}_{i,j}$, $i = 1, 2, \dots, \nu$, $j = 1, 2, \dots, p_i$, constants. If we replace the above expressions in the general solution we arrive at (8.15). The proof is completed. ■

Corollary 8.1 (Solution of a pencil). We consider system (8.14) with a regular pencil, or a singular pencil with $r > m$ and for which (6.3) holds. Let the finite eigenvalues be either distinct, or with algebraic multiplicity equal to geometric, i.e. $p_i = 1$ is the rank of corresponding Jordan block. Then, in Theorem 8.2, $\nu = p$, $\mathbf{u}_i = \mathbf{u}_{i,j}$, and the general solution of (8.14) can be written as:

$$\boldsymbol{x}(t) = \sum_{i=1}^{p} \mathbf{u}_i \, \mathrm{e}^{\lambda_i t} c_i \ ,$$

where $c_i \in \mathbb{C}$, constants.

Based on the above results, we now provide a Theorem about the PFs of system (8.14).

Theorem 8.3 (Solution of a pencil). We consider system (8.14) with a regular pencil, or a singular pencil with $r > m$ and for which (6.3) holds. Let λ_i, $i = 1, 2, \ldots, \nu$, be a finite eigenvalue of the pencil, p_i be rank of the corresponding Jordan block, $\sum_{i=1}^{\nu} p_i = p$, and $\mathbf{w}_{i,j}$, $\mathbf{u}_{i,j}$, $j = 1, 2, \ldots, p_i$ left, right respectively linear independent (generalized) eigenvectors. Then:

- The solution of (8.14) with initial condition $\boldsymbol{x}(0)$ is given by:

$$\boldsymbol{x}(t) = \sum_{i=1}^{\nu} \mathrm{e}^{\lambda_i t} \sum_{j=1}^{p_i} \Big(\sum_{k=1}^{j} t^{k-1} \mathbf{w}_{i,j-(k-1)} \mathbf{E} \, \boldsymbol{x}(0) \Big) \mathbf{u}_{i,j} \ .$$

- Let $x_\mu(t)$ be the μ-th element of $\boldsymbol{x}(t)$. Then the participation of the h-th eigenvalue, $h = 1, 2, \ldots, \nu$ in $x_\mu(t)$, $\mu = 1, 2, \ldots, m$, is given by:

$$\frac{\partial x_\mu(t)}{\partial \, \mathrm{e}^{\lambda_h t}} = \sum_{j=1}^{p_h} \Big(\sum_{k=1}^{j} t^{k-1} \mathbf{w}_{h,j-(k-1)} \mathbf{E} \, \boldsymbol{x}(0) \Big) \mathrm{u}_{h,j}^{(\mu)} \ , \quad \text{(Participation Factors)}$$

  where $\mathrm{u}_{h,j}^{(\mu)}$ is the μ-th element of the eigenvector $\mathbf{u}_{h,j}$.

Proof. By using the transformation $\boldsymbol{x}(t) = \mathbf{Q}\boldsymbol{z}(t)$ from the proof in Theorem 8.1, we have $\boldsymbol{x}(t) = \mathbf{Q}_p \boldsymbol{z}_p(t)$ or, equivalently,

$$\boldsymbol{x} = \mathbf{Q}_p \boldsymbol{z}_p \ .$$

It is known from the previous sections that there exist $\mathbf{P}_1 \in \mathbb{C}^{p \times m}$, and $\mathbf{Q}_p \in \mathbb{C}^{m \times p}$, where $\mathbf{P}_1$ is a matrix with rows p linear independent (generalized) left eigenvectors of the p finite eigenvalues of $s\mathbf{E} - \mathbf{A}$; $\mathbf{Q}_p$ is a matrix with columns p linear independent (generalized) right eigenvectors of the p finite eigenvalues of $s\mathbf{E} - \mathbf{A}$; such that $\mathbf{P}_1 \mathbf{E} \mathbf{Q}_p = \mathbf{I}_p$. By multiplying the above expression by $\mathbf{P}_1 \mathbf{E}$ we have:

$$\mathbf{P}_1 \mathbf{E} \, \boldsymbol{x} = \mathbf{P}_1 \mathbf{E} \mathbf{Q}_p \, \boldsymbol{z}_p \ ,$$

or, equivalently,

$$\boldsymbol{z}_p = \mathbf{P}_1 \mathbf{E} \, \boldsymbol{x} \ .$$

Hence:

$$\boldsymbol{z}_p(0) = \mathbf{P}_1 \mathbf{E} \, \boldsymbol{x}(0) \ .$$

The matrix $\mathbf{P}_1$ has as rows the p linear independent (generalized) left eigenvectors, and can be written in the form:

$$\mathbf{P}_1 = \begin{bmatrix} \mathbf{w}_{1,p_1} \\ \vdots \\ \mathbf{w}_{1,2} \\ \mathbf{w}_{1,1} \\ \vdots \\ \mathbf{w}_{\nu,p_\nu} \\ \vdots \\ \mathbf{w}_{\nu,2} \\ \mathbf{w}_{\nu,1} \end{bmatrix},$$

where $\mathbf{w}_{i,j}$, $j = 1, 2, \ldots, p_i$ linear independent left eigenvectors of λ_i, $i = 1, 2, \ldots, \nu$. Let $x_\mu(t)$ be the μ-th element of $\boldsymbol{x}(t)$, then by replacing the above expressions into the general solution given in Theorem 8.2, we arrive at:

$$x_\mu(t) = \sum_{i=1}^{\nu} \mathrm{e}^{\lambda_i t} \sum_{j=1}^{p_i} \big(\sum_{k=1}^{j} t^{k-1} \mathbf{w}_{i,j-(k-1)} \mathbf{E}\, \boldsymbol{x}(0)\big) \mathrm{u}_{i,j}^{(\mu)} .$$

Furthermore:

$$\frac{\partial x_\mu(t)}{\partial\, \mathrm{e}^{\lambda_h t}} = \sum_{j=1}^{p_h} \big(\sum_{k=1}^{j} t^{k-1} \mathbf{w}_{h,j-(k-1)} \mathbf{E}\, \boldsymbol{x}(0)\big) \mathrm{u}_{h,j}^{(\mu)} .$$

which are the PFs, i.e. the participation of the h-th eigenvalue, $h = 1, 2, \ldots, \nu$, in $x_\mu(t)$, $\mu = 1, 2, \ldots, m$. The proof is completed. ■

Corollary 8.2 (Solution of a pencil)**.** We consider system (8.14) with a regular pencil, or a singular pencil with $r > m$ and that (6.3) holds. Let the finite eigenvalues be either distinct, or with algebraic multiplicity equal to geometric, i.e. $p_i = 1$ is the rank of corresponding Jordan block. Then in Theorem 8.3, we have $\nu = p$, $\mathbf{u}_{i,j} = \mathbf{u}_i$, and:

1. The solution of (8.14) with initial condition $\boldsymbol{x}(0)$ is given by:

$$\boldsymbol{x}(t) = \sum_{i=1}^{p} \mathbf{w}_i \mathbf{E}\, \boldsymbol{x}(0) \mathbf{u}_i\, \mathrm{e}^{\lambda_i t} .$$

2. Let $x_\mu(t)$ be the μ-th element of $\boldsymbol{x}(t)$. Then the participation of the h-th eigenvalue, $h = 1, 2, \ldots, p$ in $x_\mu(t)$, $\mu = 1, 2, \ldots, m$, is given by:

$$\frac{\partial x_\mu(t)}{\partial\, \mathrm{e}^{\lambda_h t}} = \mathbf{w}_h \mathbf{E}\, \boldsymbol{x}(0) \mathrm{u}_h^{(\mu)} , \quad \text{(Participation Factors)} \qquad (8.16)$$

where $\mathrm{u}_h^{(\mu)}$ is the μ-th element of the eigenvector u_h.

Remark 8.1. Applying appropriate initial conditions in (8.16), i.e. $x_\mu(0) = 1$, and $x_i(0) = 0$, $i \neq \mu$, allows obtaining the PFs in the classical sense. In this case, the (critical) participation matrix, i.e. the part of the participation matrix that is associated with the most critical eigenvalues of the system, can be expressed as:

$$\mathbf{\Pi}_{\mathrm{PF},\kappa} = \mathbf{P}_\kappa^{\mathrm{T}} \circ (\mathbf{E}\mathbf{Q}_\kappa) \ , \tag{8.17}$$

where κ is the number of critical eigenvalues and $\mathbf{Q}_\kappa$, $\mathbf{P}_\kappa$ are the corresponding right and left modal matrices.

Remark 8.2. By applying a simple Möbius transform into (8.14), we arrive at the system $\mathbf{A}\dot{\hat{\boldsymbol{x}}} = \mathbf{E}\,\hat{\boldsymbol{x}}$ which is the dual system of (8.14). Let $x_\mu(t)$ be the μ-th element of $\boldsymbol{x}(t)$, and $\hat{x}_\mu(t)$ be the μ-th element of $\hat{\boldsymbol{x}}(t)$. Then the participation of the infinite eigenvalue of $s\mathbf{E} - \mathbf{A}$ in $x_\mu(t)$, $\mu = 1, 2, \ldots, m$, is equal to the participation of the zero eigenvalue of $z\mathbf{A} - \mathbf{E}$ in $\hat{x}_\mu(t)$, $\mu = 1, 2, \ldots, m$. This is a direct result from the duality between (8.14) and its dual system, or, additionally, between their pencils $s\mathbf{E} - \mathbf{A}$, and $z\mathbf{A} - \mathbf{E}$ respectively, see [120]. As a consequence through transformation $s \longrightarrow \frac{1}{z}$:

- A zero eigenvalue of $s\mathbf{E} - \mathbf{A}$ is an infinite eigenvalue of $\hat{s}\mathbf{A} - \mathbf{E}$;
- A non-zero finite eigenvalue λ_i defines a non-zero finite eigenvalue $\frac{1}{\lambda_i}$ of $z\mathbf{A} - \mathbf{E}$;
- An infinite eigenvalue of $s\mathbf{E} - \mathbf{A}$ is a zero eigenvalue of $\hat{s}\mathbf{A} - \mathbf{E}$.

Note that an eigenvector (left, or right) of the infinite eigenvalue of $s\mathbf{E} - \mathbf{A}$ is also an eigenvector of the zero eigenvalue of $z\mathbf{A} - \mathbf{E}$. Some examples. We may use the result in Theorem 8.3, and Corollary 8.2 to define the PFs for a system of differential equations. Note that, in classical modal participation analysis, the participation factors, i.e. the participation of the h-th eigenvalue, $h = 1, 2, \ldots, \nu$, in $x_\mu(t)$, $\mu = 1, 2, \ldots, m$, are conventionally determined by specifying $x_\mu(0) = 1$, and $x_i(0) = 0$, $i \neq \mu$, see [140].

Example 8.3 (Generalized participation factors)**.** We consider system (8.14) with

$$\mathbf{E} = \begin{bmatrix} 12 & -3 & 0 & 0 & 0 \\ 4 & 1 & -1 & 3 & 0 \\ 0 & -4 & -5 & 1 & 0 \\ 8 & 2 & -5 & 9 & 0 \\ 0 & 0 & 0 & 0 & 0 \end{bmatrix}, \quad \mathbf{A} = \begin{bmatrix} -17 & 8 & -2 & 5 & 3 \\ -7 & -3 & 3 & -8 & 1 \\ 13 & 9 & 9 & 3 & 1 \\ -12 & -7 & 13 & -22 & 0 \\ 1 & 0 & 0 & 0 & 1 \end{bmatrix}.$$

The pencil $s\mathbf{E} - \mathbf{A}$ has $\nu = 2$ finite eigenvalues $\lambda_1 = -2$, $\lambda_2 = -3$, of algebraic multiplicity $p_1 = 2$, $p_2 = 1$ and infinite eigenvalues λ_3, λ_4. The geometric multiplicity κ_i of the finite eigenvalue λ_i is found as the dimension of the null space of $\lambda_i\mathbf{E} - \mathbf{A}$. In our case, $\kappa_1 = 1$, $\kappa_2 = 1$. The right and left eigenvectors of $s\mathbf{E} - \mathbf{A}$ associated with the finite eigenvalue $\lambda_1 = -2$ are:

$$\mathbf{u}_{1,1} = \begin{bmatrix} 0 \\ -1 \\ -1 \\ 0 \\ 0 \end{bmatrix}, \qquad \mathbf{u}_{1,2} = \begin{bmatrix} 0.0049 \\ -3.282 \cdot 10^7 \\ -3.282 \cdot 10^7 \\ 0 \\ 0.0049 \end{bmatrix},$$

$$\mathbf{w}_{1,1} = \begin{bmatrix} -0.2308 \\ -0.3846 \\ 0.0769 \\ 0 \\ 1 \end{bmatrix}^{\mathrm{T}}, \qquad \mathbf{w}_{1,2} = \begin{bmatrix} -0.1426 \\ -0.2376 \\ 0.0475 \\ 0 \\ 0.6178 \end{bmatrix}^{\mathrm{T}},$$

where $\mathbf{u}_{1,2}$, $\mathbf{w}_{1,2}$ are generalized eigenvectors determined from $(\mathbf{A}-\lambda_1\mathbf{E})\mathbf{u}_{12} = \mathbf{E}\mathbf{u}_{11}$ and $\mathbf{w}_{12}(\mathbf{A}-\lambda_1\mathbf{E}) = \mathbf{w}_{11}\mathbf{E}$ respectively. The right and left eigenvectors of $s\mathbf{E} - \mathbf{A}$ associated with the finite eigenvalue $\lambda_2 = -3$ are:

$$\mathbf{u}_{2,1} = \begin{bmatrix} 0 \\ 1 \\ -0.5 \\ 0 \\ 0 \end{bmatrix}, \qquad \mathbf{w}_{2,1} = \begin{bmatrix} -0.3333 \\ 1 \\ 0.1111 \\ 0 \\ -0.1111 \end{bmatrix}^{\mathrm{T}}.$$

The sensitivities $\frac{\partial x_\mu(t)}{\partial \mathrm{e}^{\lambda_h t}}$ are obtained from Theorem 8.3 as follows:

$$\frac{\partial x_\mu(t)}{\partial \mathrm{e}^{\lambda_h t}} = \sum_{j=1}^{p_h} \Big(\sum_{k=1}^{j} t^{k-1} \mathbf{w}_{h,j-(k-1)} \mathbf{E}\, \boldsymbol{x}(0) \Big) \mathrm{u}_{h,j}^{(\mu)} .$$

For λ_1 and λ_2 we have respectively:

$$\begin{aligned} \frac{\partial x_\mu(t)}{\partial \mathrm{e}^{\lambda_1 t}} &= \sum_{j=1}^{2} \Big(\sum_{k=1}^{j} t^{k-1} \mathbf{w}_{1,j-(k-1)} \mathbf{E}\, \boldsymbol{x}(0) \Big) \mathrm{u}_{1,j}^{(\mu)} \\ &= \mathbf{w}_{1,1}\mathbf{E}\, \boldsymbol{x}(0) \mathrm{u}_{1,1}^{(\mu)} + \Big(\sum_{k=1}^{2} t^{k-1} \mathbf{w}_{1,2-(k-1)} \mathbf{E}\, \boldsymbol{x}(0) \Big) \mathrm{u}_{1,2}^{(\mu)} \\ &= \mathbf{w}_{1,1}\mathbf{E}\, \boldsymbol{x}(0) \mathrm{u}_{1,1}^{(\mu)} + \mathbf{w}_{1,2}\mathbf{E}\, \boldsymbol{x}(0) \mathrm{u}_{1,2}^{(\mu)} + t\mathbf{w}_{1,1}\mathbf{E}\, \boldsymbol{x}(0) \mathrm{u}_{1,2}^{(\mu)} , \end{aligned}$$

$$\frac{\partial x_\mu(t)}{\partial \mathrm{e}^{\lambda_2 t}} = \mathbf{w}_{2,1}\mathbf{E}\, \boldsymbol{x}(0) \mathrm{u}_{2,1}^{(\mu)} .$$

Consider $x_\mu(0) = 1$, and $x_i(0) = 0$, $i \neq \mu$, which lead to the PFs related to the system finite modes. We have the following:

- For $\frac{\partial x_1(t)}{\partial \mathrm{e}^{\lambda_h t}}$, we have $\boldsymbol{x}(0) = \begin{bmatrix} 1 & 0 & 0 & 0 & 0 \end{bmatrix}^{\mathrm{T}}$. Hence,

$$\pi_{1,1} = 0.0130 + 0.0209t , \qquad \pi_{1,2} = 0 .$$

- For $\dfrac{\partial x_2(t)}{\partial \, e^{\lambda_h t}}$, we have $\boldsymbol{x}(0) = \begin{bmatrix} 0 & 1 & 0 & 0 & 0 \end{bmatrix}^{\mathrm{T}}$. Hence,

$$\pi_{2,1} = 0.3290 + 1.0839t \ , \qquad \pi_{2,2} = 0.6667 \ .$$

- For $\dfrac{\partial x_3(t)}{\partial \, e^{\lambda_h t}}$, we have $\boldsymbol{x}(0) = \begin{bmatrix} 0 & 0 & 1 & 0 & 0 \end{bmatrix}^{\mathrm{T}}$. Hence,

$$\pi_{3,1} = 0.6580 + 2.1678t \ , \qquad \pi_{3,2} = 0.3333 \ .$$

- For $\dfrac{\partial x_4(t)}{\partial \, e^{\lambda_h t}}$, we have $\boldsymbol{x}(0) = \begin{bmatrix} 0 & 0 & 0 & 1 & 0 \end{bmatrix}^{\mathrm{T}}$. Hence,

$$\pi_{4,1} = 0 \ , \qquad \pi_{4,2} = 0 \ .$$

- For $\dfrac{\partial x_5(t)}{\partial \, e^{\lambda_h t}}$, we have $\boldsymbol{x}(0) = \begin{bmatrix} 0 & 0 & 0 & 0 & 1 \end{bmatrix}^{\mathrm{T}}$. Hence,

$$\pi_{5,1} = 0 \ , \qquad \pi_{5,2} = 0 \ .$$

The results are summarized in Table 8.3, where we assumed $t \to 0$. The pencil $s\mathbf{E} - \mathbf{A}$ has dimensions 5×5 and its rank is equal to 3, thus there exist $5 - 3 = 2$ variables the participation of which to the system finite eigenvalues is zero. These variables are x_4 and x_5. In addition, Table 8.3 shows that x_3 is dominant in λ_1, while x_2 is dominant in λ_2. □

TABLE 8.3: PFs associated with finite modes.

	x_1	x_2	x_3	x_4	x_5
λ_1	0.0130	0.3290	0.6580	0	0
λ_2	0	0.6667	0.3333	0	0

8.3 Participation Factors of Algebraic Variables

For a system of DAEs, the form of (8.17) imposes that the PFs of the algebraic variables in the finite eigenvalues of the system are null. This is a consequence of the fact that the coefficients of the first derivatives of the algebraic variables are zero, which implies that the algebraic variables introduce only infinite modes to the system. Nevertheless, the algebraic variables constrain a power system and, in this sense, do participate in the system finite modes.

In this section, we describe an approach to measure the participation of algebraic variables in power system modes. These can be algebraic variables included in the DAE system model, or, in general, any algebraic outputs that is defined as a function of the states and algebraic variables of the DAE system.

Consider system (1.160), which for zero input matrix is:

$$\Delta\dot{\boldsymbol{x}} = \mathbf{A}_{\mathrm{S}}\,\Delta\boldsymbol{x}\,, \tag{8.18}$$

where the algebraic variables have been eliminated using the relationship:

$$\Delta\boldsymbol{y} = -\left(\frac{\partial \boldsymbol{g}}{\partial \boldsymbol{y}}\right)^{-1}\frac{\partial \boldsymbol{g}}{\partial \boldsymbol{x}}\Delta\boldsymbol{x}\,. \tag{8.19}$$

The Jacobian matrices in (8.19) and in the following equations are calculated at $(\boldsymbol{x}_o, \boldsymbol{y}_o)$.

Let us define the output vector $\boldsymbol{w}$, $\boldsymbol{w} \in \mathbb{R}^{q\times 1}$ such that:

$$\boldsymbol{w} = \boldsymbol{h}(\boldsymbol{x}, \boldsymbol{y})\,,$$

where $\boldsymbol{h}$ ($\boldsymbol{h} : \mathbb{R}^{(n+l)\times 1} \mapsto \mathbb{R}^{q\times 1}$) is a non-linear function of $\boldsymbol{x}$, $\boldsymbol{y}$. Then differentiation around $(\boldsymbol{x}_o, \boldsymbol{y}_o)$ yields:

$$\Delta\boldsymbol{w} = \frac{\partial \boldsymbol{h}}{\partial \boldsymbol{x}}\Delta\boldsymbol{x} + \frac{\partial \boldsymbol{h}}{\partial \boldsymbol{y}}\Delta\boldsymbol{y}\,. \tag{8.20}$$

Substitution of (8.19) to the last equation gives:

$$\Delta\boldsymbol{w} = \mathbf{C}\,\Delta\boldsymbol{x}\,, \tag{8.21}$$

where $\mathbf{C}$, $\mathbf{C} \in \mathbb{R}^{q\times n}$, is the output matrix, defined as:

$$\mathbf{C} = \frac{\partial \boldsymbol{h}}{\partial \boldsymbol{x}} - \frac{\partial \boldsymbol{h}}{\partial \boldsymbol{y}}\left(\frac{\partial \boldsymbol{g}}{\partial \boldsymbol{y}}\right)^{-1}\frac{\partial \boldsymbol{g}}{\partial \boldsymbol{x}}\,.$$

Let Δw_μ be the μ-th system output. The following expression provides the PF of Δw_μ in the mode λ_i:

$$\hat{\pi}_{\mu,i} = \frac{\partial \Delta w_\mu}{\partial\, \mathrm{e}^{\lambda_i t}}\,. \tag{8.22}$$

From the state-space viewpoint, $\hat{\pi}_{\mu,i}$ expresses the residue (or the joint observability/controllability) of the i-th mode, when the input is, exactly as it holds for $\pi_{k,i}$, a perturbation in the differential equation that defines $\dot{x}_k$. The output however is Δw_μ, which can be, in principle, any function of the system state variables. The fact that the perturbation that leads from (8.16) and (8.22) to the classical PFs is the same, is also the reason that $\hat{\pi}_{\mu,i}$ is called PF.

Proposition 8.1 (Participation factors of output variables)**.** Let the PF $\hat{\pi}_{\mu,i}$ be the μ-th row, i-th column element of the participation matrix $\hat{\mathbf{\Pi}}_{\mathrm{PF},(\boldsymbol{w})}$. Then:

$$\hat{\mathbf{\Pi}}_{\mathrm{PF},(\boldsymbol{w})} = \mathbf{C}\mathbf{\Pi}_{\mathrm{PF}} \ . \tag{8.23}$$

Proof. Let $\mathbf{c}_{\mu} = \begin{bmatrix} \mathrm{c}_{\mu,1} & \dots & \mathrm{c}_{\mu,n} \end{bmatrix}$ be the μ-th row of $\mathbf{C}$. Then, we have for Δw_{μ}:

$$\Delta w_{\mu} = \mathbf{c}_{\mu}\Delta \boldsymbol{x} = \mathrm{c}_{\mu,1}\Delta x_1 + \mathrm{c}_{\mu,2}\Delta x_2 + \cdots + \mathrm{c}_{\mu,n}\Delta x_n \ .$$

Partial differentiation over $\mathrm{e}^{\lambda_i t}$ leads to:

$$\begin{aligned} \frac{\partial \Delta w_{\mu}}{\partial\, \mathrm{e}^{\lambda_i t}} &= \mathrm{c}_{\mu,1}\frac{\partial \Delta x_1}{\partial\, \mathrm{e}^{\lambda_i t}} + \mathrm{c}_{\mu,2}\frac{\partial \Delta x_2}{\partial\, \mathrm{e}^{\lambda_i t}} + \cdots + \mathrm{c}_{\mu,n}\frac{\partial \Delta x_n}{\partial\, \mathrm{e}^{\lambda_i t}} + \\ &\quad + \frac{\partial \mathrm{c}_{\mu,1}}{\partial\, \mathrm{e}^{\lambda_i t}}\Delta x_1 + \frac{\partial \mathrm{c}_{\mu,2}}{\partial\, \mathrm{e}^{\lambda_i t}}\Delta x_2 + \cdots + \frac{\partial \mathrm{c}_{\mu,n}}{\partial\, \mathrm{e}^{\lambda_i t}}\Delta x_n \\ \Rightarrow \hat{\pi}_{\mu,i} &= \mathrm{c}_{\mu,1}\pi_{1,i} + \mathrm{c}_{\mu,2}\pi_{2,i} + \cdots + \mathrm{c}_{\mu,n}\pi_{n,i} \ , \end{aligned}$$

where $\dfrac{\partial \mathrm{c}_{\mu,1}}{\partial\, \mathrm{e}^{\lambda_i t}} = \dfrac{\partial \mathrm{c}_{\mu,2}}{\partial\, \mathrm{e}^{\lambda_i t}} = \cdots = \dfrac{\partial \mathrm{c}_{\mu,n}}{\partial\, \mathrm{e}^{\lambda_i t}} = 0$, since the elements of $\mathbf{C}$ do not depend on functions of t. By applying the same steps for all outputs and representing in matrix form, we arrive at (8.23). The proof is completed. ■

The main feature of (8.23) is that it allows defining the participation matrix not only of the algebraic variables of the DAE, but also of any defined output vector that is a function of the system state and algebraic variables. One has only to specify the gradients $\frac{\partial \boldsymbol{h}}{\partial \boldsymbol{x}}$ and $\frac{\partial \boldsymbol{h}}{\partial \boldsymbol{y}}$ at the operating point, and then calculate the output matrix $\mathbf{C}$. The proposed participation matrix $\hat{\mathbf{\Pi}}_{\mathrm{PF},(\boldsymbol{w})}$ provides meaningful information for the system coupling.

Remark 8.3. We enumerate the following important special cases for the participation matrix of (8.23):

- *State variables:* If $\boldsymbol{w} = \boldsymbol{x}$, the gradients in (8.20) become:

  $$\frac{\partial \boldsymbol{h}}{\partial \boldsymbol{x}} = \mathbf{I}_n \ , \qquad \frac{\partial \boldsymbol{h}}{\partial \boldsymbol{y}} = \mathbf{0}_{n,l} \ .$$

  The output matrix is $\mathbf{C} = \mathbf{I}_n$ and hence the participation matrix of the system states is, as to be expected:

  $$\hat{\mathbf{\Pi}}_{\mathrm{PF},(\boldsymbol{x})} = \mathbf{\Pi}_{\mathrm{PF}} \ . \tag{8.24}$$

- *Algebraic variables:* If $\boldsymbol{w} = \boldsymbol{y}$, the gradients in (8.20) become:

  $$\frac{\partial \boldsymbol{h}}{\partial \boldsymbol{x}} = \mathbf{0} \ , \qquad \frac{\partial \boldsymbol{h}}{\partial \boldsymbol{y}} = \mathbf{I}_l \ .$$

The output matrix is $\mathbf{C} = -(\frac{\partial \boldsymbol{g}}{\partial \boldsymbol{y}})^{-1}\frac{\partial \boldsymbol{g}}{\partial \boldsymbol{x}}$. Thus:

$$\hat{\mathbf{\Pi}}_{\mathrm{PF},(\boldsymbol{y})} = -\left(\frac{\partial \boldsymbol{g}}{\partial \boldsymbol{y}}\right)^{-1} \frac{\partial \boldsymbol{g}}{\partial \boldsymbol{x}} \mathbf{\Pi}_{\mathrm{PF}} \,, \tag{8.25}$$

which is the participation matrix of the algebraic variables in system modes included in the DAE model.

- *Rates of change of state variables:* If we have the output $\boldsymbol{w} = \dot{\boldsymbol{x}} = \boldsymbol{f}(\boldsymbol{x}, \boldsymbol{y})$, we obtain the participation matrix of the derivatives of the state variables in system modes. The RoCoF of the synchronous machines is a relevant case. The gradients in (8.20) become:

$$\frac{\partial \boldsymbol{h}}{\partial \boldsymbol{x}} = \frac{\partial \boldsymbol{f}}{\partial \boldsymbol{x}} \,, \qquad \frac{\partial \boldsymbol{h}}{\partial \boldsymbol{y}} = \frac{\partial \boldsymbol{f}}{\partial \boldsymbol{y}} \,.$$

The output matrix is $\mathbf{C} = \mathbf{A}_{\mathrm{S}}$. Thus:

$$\hat{\mathbf{\Pi}}_{\mathrm{PF},(\dot{\boldsymbol{x}})} = \mathbf{A}_{\mathrm{S}} \mathbf{\Pi}_{\mathrm{PF}} \,. \tag{8.26}$$

- *Parameters:* Finally, consider the scalar output $w = \varrho$, where ϱ is a defined parameter. If ϱ appears only in the j-th algebraic equation $0 = g_j(\boldsymbol{x}, \boldsymbol{y}, \varrho)$, then we can obtain the participation vector of ϱ in the system modes. Linearization of the j-th algebraic equation around the operating point yields:

$$0 = \frac{\partial g_j}{\partial \boldsymbol{x}} \Delta \boldsymbol{x} + \frac{\partial g_j}{\partial \boldsymbol{y}} \Delta \boldsymbol{y} + \frac{\partial g_j}{\partial \varrho} \Delta \varrho \,, \tag{8.27}$$

where $\frac{\partial g_j}{\partial \boldsymbol{x}} \in \mathbb{R}^{1 \times n}$, $\frac{\partial g_j}{\partial \boldsymbol{y}} \in \mathbb{R}^{1 \times l}$ and $\frac{\partial g_j}{\partial \varrho} \in \mathbb{R}_{\neq 0}$. Solving (8.27) for $\Delta \varrho$ and comparing with (8.20), we obtain that

$$\frac{\partial \boldsymbol{h}}{\partial \boldsymbol{x}} = -\frac{\partial g_j}{\partial \boldsymbol{x}} / \frac{\partial g_j}{\partial \varrho} \,,$$

and

$$\frac{\partial \boldsymbol{h}}{\partial \boldsymbol{y}} = -\frac{\partial g_j}{\partial \boldsymbol{y}} / \frac{\partial g_j}{\partial \varrho} \,.$$

The participation vector is obtained from (8.23) for

$$\mathbf{C} = \left[-\frac{\partial g_j}{\partial \boldsymbol{x}} + \frac{\partial g_j}{\partial \boldsymbol{y}} \left(\frac{\partial \boldsymbol{g}}{\partial \boldsymbol{y}}\right)^{-1} \frac{\partial \boldsymbol{g}}{\partial \boldsymbol{x}} \right] / \frac{\partial g_j}{\partial \varrho} \,.$$

Notice, finally, that once the eigenvalue analysis is completed and the modal matrices are known, calculating the proposed participation matrices involves few matrix multiplications. From the computational burden viewpoint, the cost of calculating the PFs is marginal compared to the eigenvalue analysis.

Example 8.4 (All-island Irish transmission system). In this section we consider a real-world model of the all-island Irish transmission system (AIITS). The topology and the steady-state operation data of the system have been provided by the Irish transmission system operator, EirGrid Group, whereas the dynamic data have been defined based on our knowledge about the technology of the generators and the controllers. The system consists of 1,479 buses, 796 lines, 1,055 transformers, 245 loads, 22 synchronous machines, with AVRs and TGs, 6 PSSs and 176 wind power plants. The dynamic order of the system is $n = 1,443$.

The eigenvalue analysis shows that the system is stable when subject to small disturbances. The system presents both local machine modes and inter-machine modes. Recall that, a local machine mode refers to a single machine oscillating against the rest of the system. On the other hand, an inter-machine mode refers to a group of machines of the same area oscillating against each other [87]. In the remainder of this section, we show two modes with different damping ratios and natural frequencies. The examined modes are summarized in Table 8.4.

TABLE 8.4: Relevant modes and dominant states for the AIITS.

Mode	Mode 1		Mode 2	
Eigenvalue	$-0.586 \pm \jmath 7.248$		$-0.722 \pm \jmath 4.618$	
f_n (Hz)	1.16		0.74	
ζ (%)	8.06		15.44	
Type	Local		Inter-machine	
Dominant states	State	$\lvert\pi\rvert_{\max}$	State	$\lvert\pi\rvert_{\max}$
1st	$\delta_{\mathrm{r},16}$	0.4456	$\omega_{\mathrm{r},2}$	0.2883
2nd	$\omega_{\mathrm{r},16}$	0.4456	$\delta_{\mathrm{r},2}$	0.2872

Mode 1 has eigenvalue $-0.586 \pm \jmath 7.248$, with natural frequency 1.16 Hz and damping ratio 8.06%. The dominant states in this mode are the rotor angle and speed of machine 16. The PFs of these states sum to 0.8912. The mode is local with machine 16 oscillating against the rest of the system. Mode 2 has eigenvalue $-0.722 \pm \jmath 4.618$, with frequency 0.74 Hz and damping ratio 15.44%. The mostly participating states are the rotor speed and angle of machine 2. The corresponding PFs sum to 0.5755. The natural frequency and the distribution of the PFs indicate that this is an inter-machine mode [87].

The Python module *graph-tool* [139] is utilized to generate a graph of the studied network. The resulting graph has 1,479 vertices, which correspond to the system buses and 1,851 edges, which correspond to lines and transformers. Note that the coordinates of the graph vertices and edges do not represent the actual geography of the system. For the considered modes, we calculate the participation matrices of the bus active power injections. Then, the sizes of the graph vertices are adjusted with respect to the magnitude of the calculated PFs.

The generated graph with the PFs of all bus active power injections in the local Mode 1 is illustrated in Figure 8.2. The mostly participating active power injection is the one of the bus 552, that is adjacent to machine 16, with $|\hat{\pi}|_{max} = 0.3218$.

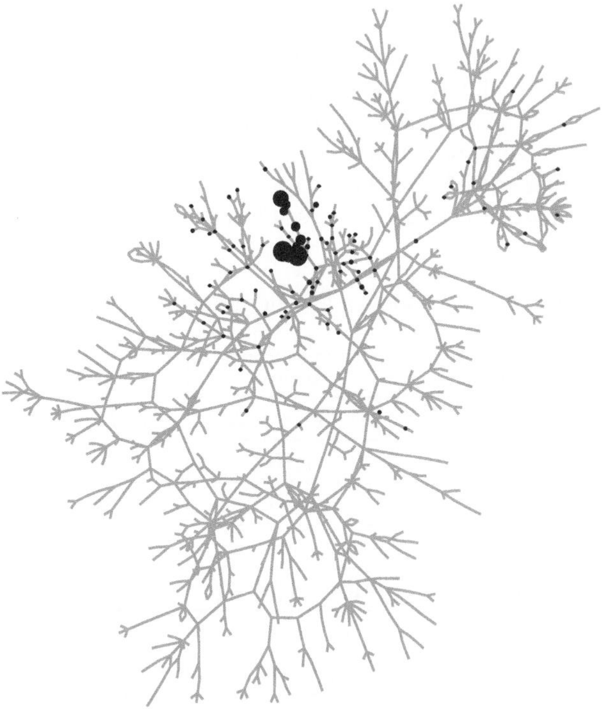

FIGURE 8.2: Participation of bus active power injections in Mode 1 for the AIITS.

The PFs of all bus active power injections in the inter-machine Mode 2 is illustrated in Figure 8.3. The mostly participating active power injection is the one of bus 1405, that is close to the synchronous machine 2, with $|\hat{\pi}|_{max} = 0.2508$. Figure 8.3 shows that the lower frequency oscillations spread over the power system. In fact, there are several buses in a large area that have a high participation in the inter-machine mode. □

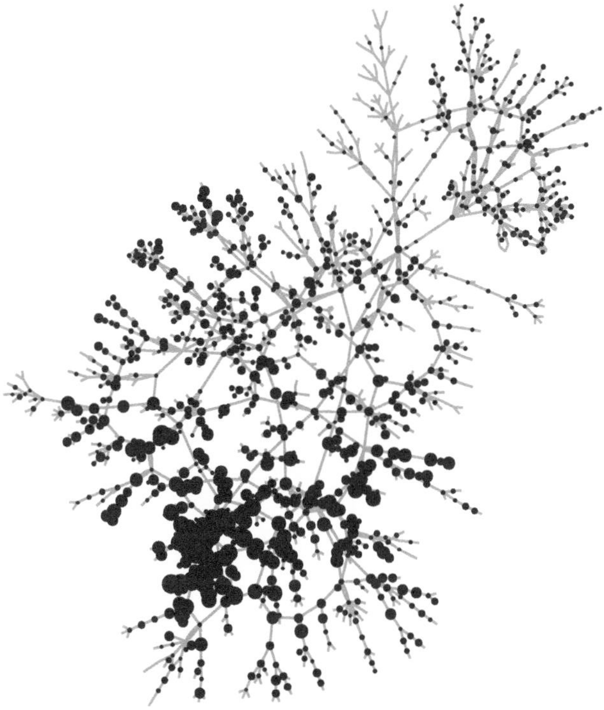

FIGURE 8.3: Participation of bus active power injections in Mode 2 for the AIITS.

Part III

Non-Linear Eigenvalue Problems

9

Polynomial Eigenvalue Problem

9.1 Formulation

In this chapter we consider a generalized system of differential equations of higher order:

$$\mathbf{A}_n \,\tilde{\boldsymbol{x}}^{[n]}(t) + \mathbf{A}_{n-1}\,\tilde{\boldsymbol{x}}^{[n-1]}(t) + \cdots + \mathbf{A}_1\,\dot{\tilde{\boldsymbol{x}}}(t) + \mathbf{A}_0\,\tilde{\boldsymbol{x}}(t) = \boldsymbol{U}(t)\,, \qquad (9.1)$$

where $\mathbf{A}_i \in \mathbb{C}^{r\times m}$ with $i = 0, 1, \ldots, n$, $\hat{\boldsymbol{x}} \in \mathbb{C}^{m\times 1}$, and $\boldsymbol{U} \in \mathbb{C}^{r\times 1}$.

Definition 9.1 (Regular and singular pencil). Given $\mathbf{A}_i, i = 0, 1, \ldots, n \in \mathbb{C}^{r\times m}$, and an arbitrary $s \in \mathbb{C}$, the matrix pencil $s^n\mathbf{A}_n + s^{n-1}\mathbf{A}_{n-1} + \cdots + s\mathbf{A}_1 + \mathbf{A}_0$ is called:

1. Regular when $r = m$ and $\det(s^n\mathbf{A}_n + s^{n-1}\mathbf{A}_{n-1} + \cdots + s\mathbf{A}_1 + \mathbf{A}_0) \not\equiv 0$;
2. Singular when $r \neq m$, or $r = m$ and $\det(s^n\mathbf{A}_n + s^{n-1}\mathbf{A}_{n-1} + \cdots + s\mathbf{A}_1 + \mathbf{A}_0) \equiv 0$.

In the next Lemma we will use the following notation. Let $[I_{ij}]_{i=1,2,\ldots,r}^{j=1,2,\ldots,m}$ be an element of the matrix $\mathbf{I}_{r,m}$ in the i-th row, j-th column, with $I_{ij} = 1$ for $i = j$, and $I_{ij} = 0$ for $i \neq j$.

Lemma 9.1 (Formulation of differential equations). System (9.1) can be reformulated in the following generalized system of differential equations of first order:

$$\mathbf{E}\,\dot{\boldsymbol{x}}(t) = \mathbf{A}\,\boldsymbol{x}(t) + \boldsymbol{V}(t)\,, \qquad (9.2)$$

where

$$\mathbf{E} = \begin{bmatrix} \mathbf{I}_{r,m} & \mathbf{0}_{r,m} & \cdots & \mathbf{0}_{r,m} & \mathbf{0}_{r,m} \\ \mathbf{0}_{r,m} & \mathbf{I}_{r,m} & \cdots & \mathbf{0}_{r,m} & \mathbf{0}_{r,m} \\ \vdots & \vdots & \ddots & \vdots & \vdots \\ \mathbf{0}_{r,m} & \mathbf{0}_{r,m} & \cdots & \mathbf{I}_{r,m} & \mathbf{0}_{r,m} \\ \mathbf{0}_{r,m} & \mathbf{0}_{r,m} & \cdots & \mathbf{0}_{r,m} & \mathbf{A}_n \end{bmatrix} \in \mathbb{C}^{nr\times nm}\,,$$

and

$$\mathbf{A} = \begin{bmatrix} \mathbf{0}_{r,m} & \mathbf{I}_{r,m} & \dots & \mathbf{0}_{r,m} & \mathbf{0}_{r,m} \\ \mathbf{0}_{r,m} & \mathbf{0}_{r,m} & \dots & \mathbf{0}_{r,m} & \mathbf{0}_{r,m} \\ \vdots & \vdots & \ddots & \vdots & \vdots \\ \mathbf{0}_{r,m} & \mathbf{0}_{r,m} & \dots & \mathbf{0}_{r,m} & \mathbf{I}_{r,m} \\ -\mathbf{A}_0 & -\mathbf{A}_1 & \dots & -\mathbf{A}_{n-2} & -\mathbf{A}_{n-1} \end{bmatrix} \in \mathbb{C}^{nr \times nm}.$$

Furthermore where

$$\boldsymbol{x}(t) = \begin{bmatrix} \boldsymbol{x}_1 \\ \boldsymbol{x}_2 \\ \vdots \\ \boldsymbol{x}_{n-1} \\ \boldsymbol{x}_n \end{bmatrix} \in \mathbb{C}^{nm \times 1}, \quad \boldsymbol{V}(t) = \begin{bmatrix} \mathbf{0}_{r,1} \\ \mathbf{0}_{r,1} \\ \vdots \\ \mathbf{0}_{r,1} \\ \boldsymbol{U}(t) \end{bmatrix} \in \mathbb{C}^{nr \times 1}, \quad \text{and} \quad \boldsymbol{x}_1(t) = \hat{\boldsymbol{x}}(t)\,.$$

Proof. Firstly we set:

$$\begin{aligned} \boldsymbol{x}_1 &= \tilde{\boldsymbol{x}}\,, \\ \boldsymbol{x}_2 &= \dot{\tilde{\boldsymbol{x}}}\,, \\ &\vdots \\ \boldsymbol{x}_{n-1} &= \tilde{\boldsymbol{x}}^{[n-2]}\,, \\ \boldsymbol{x}_n &= \tilde{\boldsymbol{x}}^{[n-1]}\,, \end{aligned}$$

whereby taking the derivatives we get

$$\begin{aligned} \dot{\boldsymbol{x}}_1 &= \dot{\tilde{\boldsymbol{x}}}\,, \\ \dot{\boldsymbol{x}}_2 &= \ddot{\tilde{\boldsymbol{x}}}\,, \\ &\vdots \qquad . \\ \dot{\boldsymbol{x}}_{n-1} &= \tilde{\boldsymbol{x}}^{[n-1]}\,, \\ \mathbf{A}_n\,\dot{\boldsymbol{x}}_n &= \mathbf{A}_n\,\tilde{\boldsymbol{x}}^{[n]}\,, \end{aligned}$$

or, equivalently,

$$\dot{\boldsymbol{x}}_1 = \boldsymbol{x}_2(t)\,,$$

$$\dot{\boldsymbol{x}}_2 = \boldsymbol{x}_3(t)\,,$$

$$\vdots \qquad .$$

$$\dot{\boldsymbol{x}}_{n-1} = \boldsymbol{x}_n(t)\,,$$

$$\mathbf{A}_n \dot{\boldsymbol{x}}_n = -\mathbf{A}_{n-1}\,\boldsymbol{x}_n - \cdots - \mathbf{A}_0\,\boldsymbol{x}_1 + \boldsymbol{U}\,.$$

The above equations can then be written in the matrix form (9.2). The proof is completed. ■

Theorem 9.1 (Equivalence between polynomial and linear pencils)**.** The pencils $s^n\mathbf{A}_n + s^{n-1}\mathbf{A}_{n-1} + \cdots + s\mathbf{A}_1 + \mathbf{A}_0$, $s\mathbf{E} - \mathbf{A}$ of systems (9.1), (9.2) respectively, have exactly the same finite eigenvalues.

Proof. We will prove this Theorem for a regular pencil. For a singular pencil the proof will be similar since the finite eigenvalues are obtained from similar sub-determinants to the regular pencil. Note that if $\mathbf{M}$ is a square matrix, and $\mathbf{M}_i$, $i = 1, 2, 3, 4$ are matrices (not necessary square) such that $\mathbf{M}$ can be written in the form

$$\mathbf{M} = \begin{bmatrix} \mathbf{M}_1 & \mathbf{M}_2 \\ \mathbf{M}_3 & \mathbf{M}_4 \end{bmatrix},$$

then, see [51], if $\mathbf{M}_1$ is square and invertible

$$\det(\mathbf{M}) = \det(\mathbf{M}_1)\det(\mathbf{M}_4 - \mathbf{M}_3\mathbf{M}_1^{-1}\mathbf{M}_2)\,.$$

For $r = m$ the pencil $s\mathbf{E} - \mathbf{A}$ is equal to

$$s\mathbf{E} - \mathbf{A} = \begin{bmatrix} s\mathbf{I}_m & -\mathbf{I}_m & \ldots & \mathbf{0}_{m,m} & \mathbf{0}_{m,m} \\ \mathbf{0}_{m,m} & s\mathbf{I}_m & \ldots & \mathbf{0}_{m,m} & \mathbf{0}_{m,m} \\ \vdots & \vdots & \ddots & \vdots & \vdots \\ \mathbf{0}_{m,m} & \mathbf{0}_{m,m} & \ldots & s\mathbf{I}_m & -\mathbf{I}_m \\ \mathbf{A}_0 & \mathbf{A}_1 & \ldots & \mathbf{A}_{n-2} & s\mathbf{A}_n + \mathbf{A}_{n-1} \end{bmatrix}.$$

Hence if we set $\mathbf{M}(s) := s\mathbf{E} - \mathbf{A}$, the pencil can be written in the form:

$$s\mathbf{E} - \mathbf{A} = \begin{bmatrix} \mathbf{M}_1(s) & \mathbf{M}_2 \\ \mathbf{M}_3 & \mathbf{M}_4(s) \end{bmatrix},$$

where

$$\mathbf{M}_1(s) = \begin{bmatrix} s\mathbf{I}_m & -\mathbf{I}_m & \ldots & \mathbf{0}_{m,m} \\ \mathbf{0}_{m,m} & s\mathbf{I}_m & \ldots & \mathbf{0}_{m,m} \\ \vdots & \vdots & \ddots & \vdots \\ \mathbf{0}_{m,m} & \mathbf{0}_{m,m} & \ldots & s\mathbf{I}_m \end{bmatrix} \in \mathbb{R}^{(n-1)m\times(n-1)m},$$

$$\mathbf{M}_2 = \begin{bmatrix} \mathbf{0}_{m,m} \\ \mathbf{0}_{m,m} \\ \vdots \\ -\mathbf{I}_m \end{bmatrix} \in \mathbb{R}^{(n-1)m \times m},$$

and

$$\mathbf{M}_3 = \begin{bmatrix} \mathbf{A}_0 & \mathbf{A}_1 & \dots & \mathbf{A}_{n-2} \end{bmatrix} \in \mathbb{C}^{m \times (n-1)m}, \quad \mathbf{M}_4(s) = s\mathbf{A}_n + \mathbf{A}_{n-1} \in \mathbb{C}^{m \times m}.$$

Then since

$$\mathbf{M}_1^{-1}(s) = \begin{bmatrix} s^{-1}\mathbf{I}_m & s^{-2}\mathbf{I}_m & \dots & s^{-(n-1)}\mathbf{I}_m \\ \mathbf{0}_{m,m} & s^{-1}\mathbf{I}_m & \dots & s^{-(n-2)}\mathbf{I}_m \\ \vdots & \vdots & \ddots & \vdots \\ \mathbf{0}_{m,m} & \mathbf{0}_{m,m} & \dots & s^{-1}\mathbf{I}_m \end{bmatrix} \in \mathbb{R}^{(n-1)m \times (n-1)m},$$

and

$$\mathbf{M}_1^{-1}(s)\mathbf{M}_2 = \begin{bmatrix} -s^{-(n-1)}\mathbf{I}_m \\ -s^{-(n-2)}\mathbf{I}_m \\ \vdots \\ -s^{-1}\mathbf{I}_m \end{bmatrix} \in \mathbb{R}^{(n-1)m \times m},$$

we have

$$\mathbf{M}_3\mathbf{M}_1^{-1}(s)\mathbf{M}_2 = \begin{bmatrix} \mathbf{A}_0 & \mathbf{A}_1 & \dots & \mathbf{A}_{n-2} \end{bmatrix} \begin{bmatrix} -s^{-(n-1)}\mathbf{I}_m \\ -s^{-(n-2)}\mathbf{I}_m \\ \vdots \\ -s^{-1}\mathbf{I}_m \end{bmatrix},$$

or, equivalently,

$$\mathbf{M}_3\mathbf{M}_1^{-1}(s)\mathbf{M}_2 = -s^{-(n-1)}\mathbf{A}_0 - s^{-(n-2)}\mathbf{A}_1 - \dots - s^{-1}\mathbf{A}_{n-2}.$$

In addition:

$$\mathbf{M}_4(s) - \mathbf{M}_3\mathbf{M}_1^{-1}(s)\mathbf{M}_2 = s\mathbf{A}_n + \mathbf{A}_{n-1} + s^{-1}\mathbf{A}_{n-2} + \dots + s^{-(n-2)}\mathbf{A}_1 + s^{-(n-1)}\mathbf{A}_0.$$

Hence,

$$\det(s\mathbf{E} - \mathbf{A}) = \det(\mathbf{M}_1(s))\det(\mathbf{M}_4(s) - \mathbf{M}_3\mathbf{M}_1^{-1}(s)\mathbf{M}_2),$$

or, equivalently,

$$\det(s\mathbf{E} - \mathbf{A}) = \det(s^{n-1}\mathbf{I}_m)\det(s\mathbf{A}_n + \mathbf{A}_{n-1} + s^{-1}\mathbf{A}_{n-2} + \dots + s^{-(n-2)}\mathbf{A}_1 + s^{-(n-1)}\mathbf{A}_0)$$

or, equivalently,

$$\det(s\mathbf{E} - \mathbf{A}) = \det(s^n\mathbf{A}_n + s^{n-1}\mathbf{A}_{n-1} + \dots + s\mathbf{A}_1 + \mathbf{A}_0).$$

The proof is completed. ■

Remark 9.1. In Lemma 9.1 we proved that system (9.1) can be reformulated into system (9.2), and in Theorem 9.1 we proved that the pencil of (9.1) has the same finite eigenvalues with the pencil of (9.2). Consequently there exist solutions for system (9.1) if and only if there exist solutions for (9.2). Hence there exist solutions for (9.1) if the pencil is regular, see Theorem 3.1; or, if the pencil is singular and for (9.2), (6.3) holds, see Theorem 6.1.

We consider now the matrices $\mathbf{P}$, $\mathbf{Q}$ as defined for the regular pencil $s\mathbf{E}-\mathbf{A}$ in (3.7), and for the singular pencil $s\mathbf{E}-\mathbf{A}$ with $r > m$ in (6.5). Then we can define the matrices $\mathbf{Q}^1$, $\mathbf{Q}_p^1$, and $\mathbf{Q}_\zeta^1$ as

$$\begin{aligned} \mathbf{Q} &= \begin{bmatrix} \mathbf{Q}^1 \\ \mathbf{Q}^2 \end{bmatrix} \in \mathbb{C}^{nm \times nm}, \\ \mathbf{Q}_p &= \begin{bmatrix} \mathbf{Q}_p^1 \\ \mathbf{Q}_p^2 \end{bmatrix} \in \mathbb{C}^{nm \times p}, \\ \mathbf{Q}_\zeta &= \begin{bmatrix} \mathbf{Q}_\zeta^1 \\ \mathbf{Q}_\zeta^2 \end{bmatrix} \in \mathbb{C}^{nm \times \zeta_2}, \end{aligned} \tag{9.3}$$

where $\mathbf{Q}^1 \in \mathbb{C}^{m \times nm}$, $\mathbf{Q}_p^1 \in \mathbb{C}^{m \times p}$, and $\mathbf{Q}_\zeta^1 \in \mathbb{C}^{m \times \zeta_2}$.

Theorem 9.2 (Existence of the solution)**.** Consider system (9.1). Then there always exists a solution for (9.1) if

1. the system has a regular pencil. In this case, the solution is given by
$$\tilde{\boldsymbol{x}}(t) = \mathbf{Q}_p^1 \mathrm{e}^{\mathbf{J}_p(t)} \boldsymbol{C} + \mathbf{Q}^1 \boldsymbol{K}(t); \tag{9.4}$$
2. $r > m$, and (6.3) holds for the pencil $s\mathbf{E}-\mathbf{A}$ of (9.2). In this case the solution is given by
$$\tilde{\boldsymbol{x}}(t) = \mathbf{Q}_p^1 \mathrm{e}^{\mathbf{J}_p(t)} \boldsymbol{C} + \mathbf{Q}^1 \boldsymbol{K}(t) + \mathbf{Q}_\zeta^1 \boldsymbol{z}_\zeta, \tag{9.5}$$

where $\boldsymbol{K}(t) = \begin{bmatrix} \int_0^t \mathrm{e}^{\mathbf{J}_p(t-s)} \mathbf{P}_1 \boldsymbol{U}(s) ds \\ -\sum_{i=0}^{q_*-1} \mathbf{H}_q^i \mathbf{P}_2 U^{[i]}(t) \end{bmatrix}$, $\boldsymbol{C} \in \mathbb{C}^{p \times 1}$ is constant vector, and $\mathbf{J}_p \in \mathbb{C}^{p \times p}$, $\mathbf{H}_q \in \mathbb{C}^{q \times q}$ are the Jordan matrices related to the finite, infinite eigenvalues respectively. The matrices $\mathbf{P}$, $\mathbf{Q}$ are defined for the regular pencil $s\mathbf{E}-\mathbf{A}$ in (3.7), and for the singular pencil $s\mathbf{E}-\mathbf{A}$ with $r > m$ in (6.5). The matrices $\mathbf{Q}^1$, $\mathbf{Q}_p^1$, and $\mathbf{Q}_\zeta^1$ are defined in (9.3).

Proof. From Theorem 3.1, there always exist a solution for system (9.2) if its pencil is regular. In this case its solution is given by (3.8):

$$\tilde{\boldsymbol{x}}(t) = \mathbf{Q}_p \mathrm{e}^{\mathbf{J}_p(t)} \boldsymbol{C} + \mathbf{Q} \boldsymbol{K}(t).$$

By using Lemma 9.1, if system (9.1) has a regular pencil the solution is given by:

$$\tilde{\boldsymbol{x}}(t) = \mathbf{Q}_p^1 \mathrm{e}^{\mathbf{J}_p(t)} \boldsymbol{C} + \mathbf{Q}^1 \boldsymbol{K}(t).$$

From Theorem 6.2, there always exist a solution for system (9.2) if $r > m$, and (6.3) holds. In this case its solution is given by (6.7):

$$\tilde{\boldsymbol{x}}(t) = \mathbf{Q}_p\, \mathrm{e}^{\mathbf{J}_p(t)}\boldsymbol{C} + \mathbf{Q}\boldsymbol{K}(t) + \mathbf{Q}_\zeta \boldsymbol{z}_\zeta\,.$$

By using Lemma 9.1, if for system (9.1) $r > m$, and (6.3) holds for the pencil $s\mathbf{E} - \mathbf{A}$ of (9.2), then the solution of (9.1) is given by:

$$\tilde{\boldsymbol{x}}(t) = \mathbf{Q}_p^1\, \mathrm{e}^{\mathbf{J}_p(t)}\boldsymbol{C} + \mathbf{Q}^1\boldsymbol{K}(t) + \mathbf{Q}_\zeta^1 \boldsymbol{z}_\zeta\,.$$

The proof is completed. ■

Remark 9.2. Let

$$\tilde{\boldsymbol{x}}(0),\ \dot{\tilde{\boldsymbol{x}}}(0), \ldots,\ \tilde{\boldsymbol{x}}^{[n-1]}(0) \tag{9.6}$$

be the initial conditions of (9.1). We set

$$\boldsymbol{x}(0) = \begin{bmatrix} \tilde{\boldsymbol{x}}(0) \\ \dot{\tilde{\boldsymbol{x}}}(0) \\ \vdots \\ \tilde{\boldsymbol{x}}^{[n-2]}(0) \\ \tilde{\boldsymbol{x}}^{[n-1]}(0) \end{bmatrix} \in \mathbb{C}^{nm\times 1}.$$

By using Corollary 3.2, the solution of (9.1) with a regular pencil is unique if and only if:

$$\boldsymbol{x}(0) \in \mathrm{colspan}\mathbf{Q}_p + \mathbf{Q}\boldsymbol{K}(0).$$

In case that the above relation holds, the unique solution is given by (9.4) and $\boldsymbol{C}$ is the unique solution of the linear system

$$\mathbf{Q}_p\boldsymbol{C} = [\boldsymbol{x}(0) - \mathbf{Q}\boldsymbol{K}(0)]\,.$$

If $r > m$, and (6.3) holds, by using Corollary 6.1 the solution of (9.1) is unique if and only if:

$$\boldsymbol{x}(0) \in \mathrm{colspan}\mathbf{Q}_p + \mathbf{Q}\boldsymbol{K}(0) + \mathbf{Q}_\zeta \boldsymbol{z}_\zeta\,.$$

Then in the general solution (9.5) of (9.1), $\boldsymbol{C}$ is the unique solution of the linear system

$$\mathbf{Q}_p\boldsymbol{C} = [\boldsymbol{x}(0) - \mathbf{Q}\boldsymbol{K}(0) - \mathbf{Q}_\zeta \boldsymbol{z}_\zeta]\,.$$

Remark 9.3. Let $\mathcal{H}(t)$ be the Heaviside function and

$$\kappa_1(t) = \mathcal{H}(t) - \mathcal{H}(-t) = \left\{ \begin{array}{ll} 1, & t > 0 \\ 0, & t = 0 \end{array} \right\},$$

$$\kappa_2(t) = \mathcal{H}(-t) = \left\{ \begin{array}{ll} 1, & t = 0 \\ 0, & t \neq 0 \end{array} \right\}.$$

Then, if the initial conditions (9.6) are non-consistent, i.e. the conditions for consistency in Remark 9.2 do not hold, then system (9.2) can be written as

$$\kappa_1(t)\,(\mathbf{E}\,\dot{\boldsymbol{x}}(t) - \boldsymbol{V}(t)) = \mathbf{A}\,\boldsymbol{x}(t) - \kappa_2(t)\,\mathbf{A}\,\boldsymbol{x}(0)\,, \quad t \geq 0\,.$$

This is a generalized linear matrix differential equation of first order, and although the initial conditions are given due to their inconsistency the solution of this system is not unique. Its solution for a regular pencil is given by:

$$\tilde{\boldsymbol{x}}(t) = \kappa_1(t)[\mathbf{Q}_p\,\mathrm{e}^{\mathbf{J}_p(t)}\boldsymbol{C} + \mathbf{Q}\boldsymbol{K}(t)] + \kappa_2(t)\boldsymbol{x}(0)\,, \quad t \geq 0\,.$$

If $r > m$, and (6.3) holds its solution is given by:

$$\tilde{\boldsymbol{x}}(t) = \kappa_1(t)\,[\mathbf{Q}_p\mathrm{e}^{\mathbf{J}_p(t)}\boldsymbol{C} + \mathbf{Q}\boldsymbol{K}(t) + \mathbf{Q}_\varsigma\boldsymbol{z}_\varsigma] + \kappa_2(t)\,\boldsymbol{x}(0)\,, \quad t \geq 0\,.$$

In both cases where $\boldsymbol{C} = \begin{bmatrix} c_1 & c_2 & \ldots & c_p \end{bmatrix}^{\mathrm{T}}$ is constant vector, it can not be defined, and hence the dimension of the solution vector space is p. In addition, we can rewrite system (9.1) in the following form:

$$\sum_{i=0}^{n} \mathbf{A}_i\,\tilde{\boldsymbol{x}}^{[i]}(t) = \boldsymbol{U}(t), \quad t > 0\,.$$

For $t \geq 0$ system (9.1) can take the following form:

$$\mathbf{A}_0\,\tilde{\boldsymbol{x}}(t) + \kappa_1(t)\left[\sum_{i=1}^{n} \mathbf{A}_i\tilde{\boldsymbol{x}}^{[i]}(t) - \boldsymbol{U}(t)\right] = \kappa_2(t)\,\mathbf{A}_0\sum_{i=0}^{n-1}\frac{t^i}{i!}\boldsymbol{x}^{[i]}(0)\,.$$

Combining the results of the above discussion, the solution of (9.1) with a regular pencil for non-consistent initial conditions (9.6) is given by:

$$\tilde{\boldsymbol{x}}(t) = \kappa_1(t)[\mathbf{Q}_p^1\,\mathrm{e}^{\mathbf{J}_p(t)}\boldsymbol{C} + Q^1\boldsymbol{K}(t)] + \kappa_2(t)\sum_{i=0}^{n-1}\frac{(t)^i}{i!}\boldsymbol{x}^{[i]}(0), \quad t \geq 0\,.$$

If $r > m$, and (6.3) holds its solution is given by:

$$\tilde{\boldsymbol{x}}(t) = \kappa_1(t)[\mathbf{Q}_p^1\,\mathrm{e}^{\mathbf{J}_p(t)}\boldsymbol{C} + Q^1\boldsymbol{K}(t) + \mathbf{Q}_\varsigma^1\boldsymbol{z}_\varsigma] + \kappa_2(t)\sum_{i=0}^{n-1}\frac{(t)^i}{i!}\boldsymbol{x}^{[i]}(0), \quad t \geq 0\,.$$

The dimension of the solution vector space is p.

Example 9.1 (Polynomial pencil). We assume the system:

$$\mathbf{A}_3\,\dddot{\tilde{\boldsymbol{x}}}(t) + \mathbf{A}_2\ddot{\tilde{\boldsymbol{x}}}(t) + \mathbf{A}_1\dot{\tilde{\boldsymbol{x}}}(t) + \mathbf{A}_0\,\tilde{\boldsymbol{x}}(t) = \mathbf{0}_{2,1}\,,$$

where

$$\mathbf{A}_3 = \begin{bmatrix} 1 & 1 \\ 0 & 0 \end{bmatrix}, \quad \mathbf{A}_2 = \begin{bmatrix} 2 & 1 \\ 0 & 0 \end{bmatrix}, \quad \mathbf{A}_1 = \begin{bmatrix} -2 & 3 \\ 1 & 1 \end{bmatrix}, \quad \mathbf{A}_0 = \begin{bmatrix} 4 & -2 \\ -1 & -1 \end{bmatrix}. \qquad (9.7)$$

We adopt the following notations

$$\begin{aligned} \boldsymbol{x}_1(t) &= \tilde{\boldsymbol{x}}(t)\,, \\ \boldsymbol{x}_2(t) &= \dot{\tilde{\boldsymbol{x}}}(t)\,, \\ \boldsymbol{x}_3(t) &= \ddot{\tilde{\boldsymbol{x}}}(t)\,. \end{aligned}$$

By taking the derivatives:

$$\begin{aligned} \dot{\boldsymbol{x}}_1(t) &= \dot{\tilde{\boldsymbol{x}}}(t) = \boldsymbol{x}_2(t)\,, \\ \dot{\boldsymbol{x}}_2(t) &= \ddot{\tilde{\boldsymbol{x}}}(t) = \boldsymbol{x}_3(t)\,, \\ \mathbf{A}_3\dot{\boldsymbol{x}}_3(t) &= \mathbf{A}_3\,\dddot{\tilde{\boldsymbol{x}}}(t) = -\mathbf{A}_2\boldsymbol{x}_3(t) - \mathbf{A}_1\boldsymbol{x}_2(t) - \mathbf{A}_0\boldsymbol{x}_1(t)\,. \end{aligned}$$

Equivalently, the above equations can be written in the following matrix form:

$$\mathbf{E}\,\dot{\boldsymbol{x}}(t) = \mathbf{A}\,\boldsymbol{x}(t)\,,$$

where $\boldsymbol{x}(t) = \begin{bmatrix} \boldsymbol{x}_1(t) & \boldsymbol{x}_2(t) & \boldsymbol{x}_3(t) \end{bmatrix}^{\mathrm{T}}$, and the coefficient matrices $\mathbf{E}, \mathbf{A}$ are given by

$$\mathbf{E} = \begin{bmatrix} \mathbf{I}_2 & \mathbf{0}_{2,2} & \mathbf{0}_{2,2} \\ \mathbf{0}_{2,2} & \mathbf{I}_2 & \mathbf{0}_{2,2} \\ \mathbf{0}_{2,2} & \mathbf{0}_{2,2} & \mathbf{A}_3 \end{bmatrix}, \qquad \mathbf{A} = \begin{bmatrix} \mathbf{0}_{2,2} & \mathbf{I}_2 & \mathbf{0}_{2,2} \\ \mathbf{0}_{2,2} & \mathbf{0}_{2,2} & \mathbf{I}_2 \\ -\mathbf{A}_0 & -\mathbf{A}_1 & -\mathbf{A}_2 \end{bmatrix},$$

and hence the pencil of the system is given by:

$$s\mathbf{E} - \mathbf{A} = s\begin{bmatrix} 1 & 0 & 0 & 0 & 0 & 0 \\ 0 & 1 & 0 & 0 & 0 & 0 \\ 0 & 0 & 1 & 0 & 0 & 0 \\ 0 & 0 & 0 & 1 & 0 & 0 \\ 0 & 0 & 0 & 0 & 1 & 1 \\ 0 & 0 & 0 & 0 & 0 & 0 \end{bmatrix} - \begin{bmatrix} 0 & 0 & 1 & 0 & 0 & 0 \\ 0 & 0 & 0 & 1 & 0 & 0 \\ 0 & 0 & 0 & 0 & 1 & 0 \\ 0 & 0 & 0 & 0 & 0 & 1 \\ -4 & 2 & 2 & -3 & -2 & -1 \\ 1 & 1 & -1 & -1 & 0 & 0 \end{bmatrix}.$$

The pencil $s\mathbf{E} - \mathbf{A}$ has three finite eigenvalues, $\lambda_1 = 1$, $\lambda_2 = 2$, $\lambda_3 = 3$ of algebraic multiplicity 1, and an infinite eigenvalue of algebraic multiplicity 3. The general solution of this system is given by:

$$\boldsymbol{x}(t) = \mathbf{Q}_p\,\mathrm{e}^{\mathbf{J}_p t}\boldsymbol{C}\,,$$

and for system (9.7):

$$\tilde{\boldsymbol{x}}(t) = \mathbf{Q}_p^1\,\mathrm{e}^{\mathbf{J}_p t}\boldsymbol{C}\,,$$

where

$$\mathbf{J}_p = \begin{bmatrix} 1 & 0 & 0 \\ 0 & 2 & 0 \\ 0 & 0 & 3 \end{bmatrix}, \quad \mathrm{e}^{\mathbf{J}_p t} = \begin{bmatrix} \mathrm{e}^{t} & 0 & 0 \\ 0 & \mathrm{e}^{2t} & 0 \\ 0 & 0 & \mathrm{e}^{3t} \end{bmatrix}, \quad \mathbf{Q}_p = \begin{bmatrix} 3 & 1 & 1 \\ -5 & -1 & -1 \\ 3 & 2 & 3 \\ -5 & -2 & -3 \\ 3 & 4 & 9 \\ -5 & -4 & -9 \end{bmatrix}.$$

Let the initial conditions of the system be

$$\tilde{\boldsymbol{x}}(0) = \begin{bmatrix} 1 \\ -3 \end{bmatrix}, \quad \dot{\tilde{\boldsymbol{x}}}(0) = \begin{bmatrix} -2 \\ 0 \end{bmatrix}, \quad \ddot{\tilde{\boldsymbol{x}}}(0) = \begin{bmatrix} -10 \\ 8 \end{bmatrix}.$$

Then

$$\boldsymbol{x}(0)^{\mathrm{T}} = \begin{bmatrix} 1 & -3 & -2 & 0 & -10 & 8 \end{bmatrix}^{\mathrm{T}}.$$

It is easy to observe that $\boldsymbol{x}(0) \in \operatorname{colspan}\mathbf{Q}_p$. Hence the initial conditions are consistent, the system has a unique solution and $\boldsymbol{C}$ can be obtained from $\boldsymbol{x}(0) = \mathbf{Q}_p\boldsymbol{C}$:

$$\boldsymbol{C} = \begin{bmatrix} 1 & -1 & -1 \end{bmatrix}^{\mathrm{T}}.$$

Then

$$\boldsymbol{x}(t) = \begin{bmatrix} 3\,\mathrm{e}^{t} - \mathrm{e}^{2t} - \mathrm{e}^{3t} \\ -5\,\mathrm{e}^{t} + \mathrm{e}^{2t} + \mathrm{e}^{3t} \\ 3\,\mathrm{e}^{t} - 2\,\mathrm{e}^{2t} - 3\,\mathrm{e}^{3t} \\ -5\,\mathrm{e}^{t} + 2\,\mathrm{e}^{2t} + 3\,\mathrm{e}^{3t} \\ 3\,\mathrm{e}^{t} - 4\,\mathrm{e}^{2t} - 9\,\mathrm{e}^{2t} \\ -5\,\mathrm{e}^{t} + 4\,\mathrm{e}^{2t} + 9\,\mathrm{e}^{3t} \end{bmatrix},$$

and hence the unique solution of system (9.7) is given by:

$$\tilde{\boldsymbol{x}}(t) = \begin{bmatrix} 3\,\mathrm{e}^{t} - \mathrm{e}^{2t} - \mathrm{e}^{3t} \\ -5\,\mathrm{e}^{t} + \mathrm{e}^{2t} + \mathrm{e}^{3t} \end{bmatrix}.$$

Next we assume the initial conditions

$$\tilde{\boldsymbol{x}}(0) = \begin{bmatrix} 0 \\ 0 \end{bmatrix}, \quad \dot{\tilde{\boldsymbol{x}}}(0) = \begin{bmatrix} 0 \\ 0 \end{bmatrix}, \quad \ddot{\tilde{\boldsymbol{x}}}(0) = \begin{bmatrix} 1 \\ 1 \end{bmatrix}.$$

Then

$$\boldsymbol{x}(0)^{\mathrm{T}} = \begin{bmatrix} 0 & 0 & 0 & 0 & 1 & 1 \end{bmatrix}^{\mathrm{T}}.$$

It is easy to observe that $\boldsymbol{x}(0) \notin \operatorname{colspan}\mathbf{Q}_p$. Hence the initial conditions are non-consistent and the solution of (9.7) is given by:

$$\tilde{\boldsymbol{x}}(t) = \kappa_1(t)\mathbf{Q}_p^1\,\mathrm{e}^{J_p t}\boldsymbol{C} + \kappa_2(t)\sum_{i=0}^{2}\frac{t_i}{i!}\dot{i}\tilde{\boldsymbol{x}}(0),$$

or, equivalently,

$$\tilde{\boldsymbol{x}}(t) = \kappa_1(t)\begin{bmatrix} 3\,\mathrm{e}^{t}c_1 + \mathrm{e}^{2t}c_2 + \mathrm{e}^{3t}c_3 \\ -5\,\mathrm{e}^{t}c_1 - \mathrm{e}^{2t}c_2 - \mathrm{e}^{3t}c_3 \end{bmatrix} + \kappa_2(t)\frac{t^2}{2}\begin{bmatrix} 1 \\ 1 \end{bmatrix}, \quad t \geq 0.$$

The dimension of the solution vector space of system (9.7) is 3. □

9.2 Quadratic Eigenvalue Problem

An important application of the theory presented in this chapter, are quadratic systems. These appear very often in physical systems: mechanical systems of the kind "mass-spring-damper," *RLC* circuits; and particularly relevant for this book, power systems where the dynamic elements are synchronous machines modeled using the classical model presented in Section 1.3.4.7. We illustrate the latter with the following example.

Example 9.2 (Two-machine system). We recall the two-machine system of Example 1.11, which is based on an approximation of the three-bus system described in Appendix A, where the synchronous machines are modeled using the classical second-order model:

$$\begin{aligned}
\omega_o^{-1}\,\dot{\delta}_{\mathrm{r},1} &= \omega_{\mathrm{r},1} - 1\,,\\
M_1\dot{\omega}_{\mathrm{r},1} &= P_{\mathrm{m},1} - D_1(\omega_{\mathrm{r},1}-1) - \frac{e'_{\mathrm{r,q},1}e'_{\mathrm{r,q},2}}{X_{12}}\sin(\delta_{\mathrm{r},1}-\delta_{\mathrm{r},2})\,,\\
\omega_o^{-1}\,\dot{\delta}_{\mathrm{r},2} &= \omega_{\mathrm{r},2} - 1\,,\\
M_2\dot{\omega}_{\mathrm{r},2} &= P_{\mathrm{m},2} - D_2(\omega_{\mathrm{r},2}-1) - \frac{e'_{\mathrm{r,q},1}e'_{\mathrm{r,q},2}}{X_{21}}\sin(\delta_{\mathrm{r},2}-\delta_{\mathrm{r},1})\,.
\end{aligned} \tag{9.8}$$

where $\boldsymbol{x} = [\delta_{\mathrm{r},1}\quad \omega_{\mathrm{r},1}\quad \delta_{\mathrm{r},2}\quad \omega_{\mathrm{r},2}]^{\mathrm{T}}$.

With the notation introduced in Section 9.1, (9.8) can be rewritten as:

$$\begin{aligned}
\omega_o^{-1}\,M_1\ddot{\delta}_{\mathrm{r},1} &= P_{\mathrm{m},1} - \omega_o^{-1}D_1\dot{\delta}_{\mathrm{r},1} - \frac{e'_{\mathrm{r,q},1}e'_{\mathrm{r,q},2}}{X_{12}}\sin(\delta_{\mathrm{r},1}-\delta_{\mathrm{r},2})\,,\\
\omega_o^{-1}M_2\ddot{\delta}_{\mathrm{r},2} &= P_{\mathrm{m},2} - \omega_o^{-1}D_2\dot{\delta}_{\mathrm{r},2} - \frac{e'_{\mathrm{r,q},1}e'_{\mathrm{r,q},2}}{X_{21}}\sin(\delta_{\mathrm{r},2}-\delta_{\mathrm{r},1})\,.
\end{aligned} \tag{9.9}$$

or, equivalently,

$$\begin{aligned}
\omega_o^{-1}\,M_1\ddot{\delta}_{\mathrm{r},1} + \omega_o^{-1}D_1\dot{\delta}_{\mathrm{r},1} + \frac{e'_{\mathrm{r,q},1}e'_{\mathrm{r,q},2}}{X_{12}}\sin(\delta_{\mathrm{r},1}-\delta_{\mathrm{r},2}) &= P_{\mathrm{m},1}\,,\\
\omega_o^{-1}M_2\ddot{\delta}_{\mathrm{r},2} + \omega_o^{-1}D_2\dot{\delta}_{\mathrm{r},2} + \frac{e'_{\mathrm{r,q},1}e'_{\mathrm{r,q},2}}{X_{21}}\sin(\delta_{\mathrm{r},2}-\delta_{\mathrm{r},1}) &= P_{\mathrm{m},2}\,.
\end{aligned} \tag{9.10}$$

we have used the fact that $\dot{\omega}_{\mathrm{r}} = \frac{d}{dt}(\omega_{\mathrm{r}} - 1)$.

The linearization of (9.10) at an equilibrium point $(\delta_{\mathrm{r},1,o}, \delta_{\mathrm{r},2,o})$, leads to a system in the form of (9.1):

$$\mathbf{A}_2\,\ddot{\tilde{\boldsymbol{x}}} + \mathbf{A}_1\,\dot{\tilde{\boldsymbol{x}}} + \mathbf{A}_0\,\tilde{\boldsymbol{x}} = \mathbf{0}_{2,1}\,, \tag{9.11}$$

where $\tilde{\boldsymbol{x}} = [\delta_{\mathrm{r},1} - \delta_{\mathrm{r},1,o}\quad \delta_{\mathrm{r},2} - \delta_{\mathrm{r},2,o}]^{\mathrm{T}}$ and

$$\mathbf{A}_2 = \begin{bmatrix} \omega_o^{-1}\,M_1 & 0 \\ 0 & \omega_o^{-1}\,M_2 \end{bmatrix}, \quad \mathbf{A}_1 = \begin{bmatrix} \omega_o^{-1}\,D_1 & 0 \\ 0 & \omega_o^{-1}\,D_2 \end{bmatrix},$$

$$\mathbf{A}_0 = \begin{bmatrix} K_1 & -K_1 \\ -K_2 & K_2 \end{bmatrix},$$

where

$$K_1 = \frac{e'_{\mathrm{r,q},1} e'_{\mathrm{r,q},2}}{X_{12}} \cos(\delta_{\mathrm{r},1,o} - \delta_{\mathrm{r},2,o}),$$
$$K_2 = \frac{e'_{\mathrm{r,q},1} e'_{\mathrm{r,q},2}}{X_{21}} \cos(\delta_{\mathrm{r},2,o} - \delta_{\mathrm{r},1,o}).$$

It is worth noticing that if load admittances are neglected, the problem (9.10) leads to a symmetric pencil as $X_{12} = X_{21}$ and, hence $K_1 = K_2$. This property can be exploited by specific solvers, e.g. [27].

In the past, it was not uncommon to study the transient response of power systems using exclusively the classical model of synchronous machines. This was due mainly to the limited computational capability of workstations [167, 133, 132]. Dedicated algorithms and industrial-grade software tools to study the small-signal stability of "large" power systems with synchronous machine dynamics, e.g. analysis of essentially spontaneous oscillations in power systems (AESOPS) [23] and program for eigenvalue analysis of large systems (PEALS) [178], were developed in the 1980s and extended in following years [152].

In more recent years, the developments of efficient algorithms for the eigenvalue analysis, the relatively inexpensive computing compatibility of modern workstations and clouds and the diversification of the dynamics of power systems which are not limited any more to the electromechanical oscillations of synchronous machines does not justify the simplifications adopted in (9.8).

For the reasons above, we do not consider further quadratic eigenvalue problems. We note, however, that in recent years, there has been some interest in the development of solvers for the solution of the general polynomial eigenvalue problem, e.g. [25, 26, 28]. □

9.3 Fractional Calculus

In this section we briefly outline the theory of fractional calculus, then discuss the stability of power systems with inclusion of fractional order controllers (FOCs), which lead to non-linear pencils with non-integer exponents, and finally present a case study that includes a fractional-order PSS.

Fractional calculus deals with the problem of extending the differentiation and integration operators d^n/dt^n, $\int_0^t d^n(\tau)$, $n \in \mathbb{N}$, for real (or complex) number powers. Consequently, the pencil of a system of fractional differential equations depends non-linearly on s. There exist several approaches that address this problem.

A precise formulation of fractional calculus is given by the Riemann-Liouville (R-L) definition. Consider a function $\phi : [0, \infty) \to \mathbb{R}$. The idea behind the R-L definition is to first consider the n-fold integration of $\phi(t)$ and then extend $n \in \mathbb{N}$ to any $\gamma \in \mathbb{R}^+$. In its derivative form, the R-L definition reads [129]:

$$\phi^{[\gamma]}(t) = \frac{1}{\Gamma(\mu-\gamma)} \frac{d^\mu}{dt^\mu} \Big(\int_0^t \frac{\phi(\tau)}{(t-\tau)^{\gamma-\mu+1}} d\tau \Big) , \tag{9.12}$$

where γ, $\mu - 1 < \gamma < \mu$, $\mu \in \mathbb{N}$, is the fractional order; and $\phi^{[\gamma]}(t) = d^\gamma \phi / dt^\gamma$. The Laplace transform of (9.12) is:

$$\mathcal{L}\{\phi^{[\gamma]}(t)\} = s^\gamma \Phi(s) - \sum_{j=0}^{\mu-1} s^j \phi^{[\gamma-j-1]}(0) , \tag{9.13}$$

where $s \in \mathbb{C}$. Equation (9.13) requires the knowledge of the fractional order initial conditions $\phi^{[\gamma-j-1]}(0)$, $j = 0, 1, \dots, \mu - 1$. This raises an issue for engineering systems since, currently, only integer order initial conditions are well understood and known for physical variables. Other properties of the R-L definition are also counter-intuitive in the sense of classical differentiation. For example, the R-L derivative of a constant function is typically unbounded at $t = 0$ [141].

With the aim of meeting the requirements of known physical variables and systems, (9.12) was revisited by Caputo [29]. Caputo definition of $\phi^{[\gamma]}(t)$ reads:

$$\phi^{[\gamma]}(t) = \frac{1}{\Gamma(\mu-\gamma)} \int_0^t \frac{\phi^{[\mu]}(\tau)}{(t-\tau)^{\gamma-\mu+1}} d\tau . \tag{9.14}$$

The Laplace transform of (9.14) is:

$$\mathcal{L}\{\phi^{[\gamma]}(t)\} = s^\gamma \Phi(s) - \sum_{j=0}^{\mu-1} s^{\gamma-j-1} \phi^{[j]}(0) . \tag{9.15}$$

Equation (9.15) requires the knowledge of the initial conditions $\phi^{[j]}(0)$, $j = 0, 1, \dots, \mu - 1$, which in this case are of integer order. This property is crucial for the solution of initial value problems. In fact, for the purpose of fractional control, that is of concern here, one needs to use a definition with integer order initial conditions. Caputo definition of fractional derivative given in (9.14), which is consistent for control applications and follows the properties of differentiation in the classical sense. For example, the Caputo fractional derivative of a constant function is zero.

Consider the linearized model described by (1.156). Let also the vector of the system output measurements $\boldsymbol{w}$, $\boldsymbol{w} \in \mathbb{R}^{q \times 1}$, be:

$$\boldsymbol{w} = \mathbf{C}\, \Delta\boldsymbol{\xi} + \mathbf{D}\, \Delta\boldsymbol{u} , \tag{9.16}$$

where $\mathbf{C} \in \mathbb{R}^{q\times m}$, $\mathbf{D} \in \mathbb{R}^{q\times p}$. Then, a multiple-input multiple-output FOC for the system that consists of (1.156) and (9.16), can be described by a set of fractional DAEs as follows:

$$\begin{aligned} \mathbf{E}_c\, \boldsymbol{x}_c{}^{[\gamma]} &= \mathbf{A}_c\, \boldsymbol{x}_c + \mathbf{B}_c\, \boldsymbol{w}\,, \\ \mathbf{0}_{p,1} &= \mathbf{C}_c\, \boldsymbol{x}_c + \mathbf{D}_c\, \boldsymbol{w} - \Delta \boldsymbol{u}\,, \end{aligned} \tag{9.17}$$

where γ is the controller's fractional order; $\boldsymbol{x}_c$, $\boldsymbol{x}_c \in \mathbb{R}^{\nu\times 1}$, is the vector of the controller states; $\mathbf{E}_c$, $\mathbf{A}_c \in \mathbb{R}^{\nu\times\nu}$, $\mathbf{B}_c \in \mathbb{R}^{\nu\times q}$, $\mathbf{C}_c \in \mathbb{R}^{p\times\nu}$, $\mathbf{D}_c \in \mathbb{R}^{p\times q}$. It is relevant to mention that there are FOCs that introduce multiple, distinct fractional orders. Without loss of generality, we have chosen here to keep the analysis the simplest possible. Combining (1.156), (9.16) and (9.17) yields the closed-loop system representation. In matrix form:

$$\begin{bmatrix} \mathbf{E} & \mathbf{0}_{m,\nu} & \mathbf{0}_{m,p} \\ \mathbf{0}_{\nu,m} & \mathbf{0}_{\nu,\nu} & \mathbf{0}_{\nu,p} \\ \mathbf{0}_{p,n} & \mathbf{0}_{p,\nu} & \mathbf{0}_{p,p} \end{bmatrix} \begin{bmatrix} \Delta\dot{\boldsymbol{\xi}} \\ \dot{\boldsymbol{x}}_c \\ \Delta\dot{\boldsymbol{u}} \end{bmatrix} + \begin{bmatrix} \mathbf{0}_{m,m} & \mathbf{0}_{m,\nu} & \mathbf{0}_{m,p} \\ \mathbf{0}_{\nu,m} & \mathbf{E}_c & \mathbf{0}_{\nu,p} \\ \mathbf{0}_{p,m} & \mathbf{0}_{p,\nu} & \mathbf{0}_{p,p} \end{bmatrix} \begin{bmatrix} \Delta\boldsymbol{\xi} \\ \boldsymbol{x}_c \\ \Delta\boldsymbol{u} \end{bmatrix}^{[\gamma]}$$
$$= \begin{bmatrix} \mathbf{A} & \mathbf{0}_{m,\nu} & \mathbf{B} \\ \mathbf{B}_c\mathbf{C} & \mathbf{A}_c & \mathbf{B}_c\mathbf{D} \\ \mathbf{D}_c\mathbf{C} & \mathbf{C}_c & \mathbf{D}_c\mathbf{D} - \mathbf{I}_p \end{bmatrix} \begin{bmatrix} \Delta\boldsymbol{\xi} \\ \boldsymbol{x}_c \\ \Delta\boldsymbol{u} \end{bmatrix},$$

or,

$$\mathbf{M}\,\dot{\hat{\boldsymbol{x}}} + \mathbf{M}_\gamma\, \hat{\boldsymbol{x}}^{[\gamma]} = \mathbf{A}_{cl}\, \hat{\boldsymbol{x}}\,, \tag{9.18}$$

where $\hat{\boldsymbol{x}} = [\Delta\boldsymbol{\xi}^{\mathrm{T}} \;\; \boldsymbol{x}_c{}^{\mathrm{T}} \;\; \Delta\boldsymbol{u}^{\mathrm{T}}]^{\mathrm{T}}$, and

$$\mathbf{M} = \begin{bmatrix} \mathbf{E} & \mathbf{0}_{m,\nu} & \mathbf{0}_{m,p} \\ \mathbf{0}_{\nu,m} & \mathbf{0}_{\nu,\nu} & \mathbf{0}_{\nu,p} \\ \mathbf{0}_{p,m} & \mathbf{0}_{p,\nu} & \mathbf{0}_{p,p} \end{bmatrix}, \qquad \mathbf{M}_\gamma = \begin{bmatrix} \mathbf{0}_{m,m} & \mathbf{0}_{m,\nu} & \mathbf{0}_{m,p} \\ \mathbf{0}_{\nu,m} & \mathbf{E}_c & \mathbf{0}_{\nu,p} \\ \mathbf{0}_{p,m} & \mathbf{0}_{p,\nu} & \mathbf{0}_{p,p} \end{bmatrix},$$

$$\mathbf{A}_{cl} = \begin{bmatrix} \mathbf{A} & \mathbf{0}_{m,\nu} & \mathbf{B} \\ \mathbf{B}_c\mathbf{C} & \mathbf{A}_c & \mathbf{B}_c\mathbf{D} \\ \mathbf{D}_c\mathbf{C} & \mathbf{C}_c & \mathbf{D}_c\mathbf{D} - \mathbf{I}_p \end{bmatrix}.$$

We study now the stability of (9.18). With this aim, we first provide the following property of Caputo fractional derivative [41, 166]:

Proposition 9.1 (Property of Caputo fractional derivative). Let $\boldsymbol{\phi}(t)$, $\boldsymbol{\phi}(t) \in \mathcal{C}^1[0,T]^m$ for some $T > 0$. Then:

$$[\boldsymbol{\phi}^{[a]}(t)]^{[b]} = [\boldsymbol{\phi}^{[b]}(t)]^{[a]} = \boldsymbol{\phi}^{[a+b]}(t)\,, \tag{9.19}$$

where $a, b \in \mathbb{R}^+$, and $a + b \leq 1$.

Note that (9.19) does not hold for the R-L derivative.
We rewrite (9.18) as:

$$\mathbf{M}\,\hat{\boldsymbol{x}}^{[\gamma+\beta]} + \mathbf{M}_\gamma\, \hat{\boldsymbol{x}}^{[\gamma]} = \mathbf{A}_{cl}\, \hat{\boldsymbol{x}}\,, \tag{9.20}$$

where $\gamma + \beta = 1$. Adopting the notation:

$$\boldsymbol{\psi}_1 = \hat{\boldsymbol{x}}, \qquad \boldsymbol{\psi}_2 = \hat{\boldsymbol{x}}^{[\gamma]},$$

we obtain $\boldsymbol{\psi}_1^{[\gamma]} = \hat{\boldsymbol{x}}^{[\gamma]} = \boldsymbol{\psi}_2$. Making use of (9.19), yields $\boldsymbol{\psi}_2^{[\beta]} = \hat{\boldsymbol{x}}^{[\gamma+\beta]}$. Substitution to (9.20) gives:

$$\begin{aligned} \mathbf{M}\boldsymbol{\psi}_2^{[\beta]} + \mathbf{M}_\gamma \boldsymbol{\psi}_2 &= \mathbf{A}_{cl}\boldsymbol{\psi}_1 \;\Rightarrow \\ \mathbf{M}\boldsymbol{\psi}_2^{[\beta]} &= \mathbf{A}_{cl}\boldsymbol{\psi}_1 - \mathbf{M}_\gamma \boldsymbol{\psi}_2 \,. \end{aligned} \tag{9.21}$$

Equation (9.21) can be rewritten as:

$$\begin{bmatrix} \mathbf{I}_\rho & \mathbf{0}_{\rho,\rho} \\ \mathbf{0}_{\rho,\rho} & \mathbf{M} \end{bmatrix} \begin{bmatrix} \boldsymbol{\psi}_1^{[\gamma]} \\ \boldsymbol{\psi}_2^{[\beta]} \end{bmatrix} = \begin{bmatrix} \mathbf{0}_{\rho,\rho} & \mathbf{I}_\rho \\ \mathbf{A}_{cl} & -\mathbf{M}_\gamma \end{bmatrix} \begin{bmatrix} \boldsymbol{\psi}_1 \\ \boldsymbol{\psi}_2 \end{bmatrix}, \tag{9.22}$$

or, equivalently,

$$\tilde{\mathbf{M}}\boldsymbol{\psi}^{\Delta} = \tilde{\mathbf{A}}\boldsymbol{\psi}\,, \tag{9.23}$$

where

$$\tilde{\mathbf{M}} = \begin{bmatrix} \mathbf{I}_\rho & \mathbf{0}_{\rho,\rho} \\ \mathbf{0}_{\rho,\rho} & \mathbf{M} \end{bmatrix}, \; \tilde{\mathbf{A}} = \begin{bmatrix} \mathbf{0}_{\rho,\rho} & \mathbf{I}_\rho \\ \mathbf{A}_{cl} & -\mathbf{M}_\gamma \end{bmatrix},$$

$\rho = n + \nu + p$; $\boldsymbol{\psi} = [\boldsymbol{\psi}_1 \;\; \boldsymbol{\psi}_2]^{\mathrm{T}}$; and $\boldsymbol{\psi}^{\Delta} = [\boldsymbol{\psi}_1^{[\gamma]} \;\; \boldsymbol{\psi}_2^{[\beta]}]^{\mathrm{T}}$.

Theorem 9.3 (Matrix pencil of a fractional order system)**.** Consider system (9.23). Then its matrix pencil is given by

$$\begin{bmatrix} s^\gamma\,\mathbf{I}_\rho & \mathbf{0}_{\rho,\rho} \\ \mathbf{0}_{\rho,\rho} & s^\beta\,\mathbf{I}_\rho \end{bmatrix} \tilde{\mathbf{M}} - \tilde{\mathbf{A}}\,. \tag{9.24}$$

Proof. Let $\mathcal{L}\{\boldsymbol{\psi}(t)\} = \boldsymbol{\Psi}(s)$. Using Caputo fractional derivative, by applying the Laplace transform $\mathcal{L}$ as defined in (9.15) for $\mu = 1$ into (9.23), we obtain

$$\tilde{\mathbf{M}}\,\mathcal{L}\{\boldsymbol{\psi}^{\Delta}(t)\} = \tilde{\mathbf{A}}\,\mathcal{L}\{\boldsymbol{\psi}(t)\}\,.$$

Furthermore:

$$\tilde{\mathbf{M}}\,\mathcal{L}\left\{ \begin{bmatrix} \boldsymbol{\psi}_1^{[\gamma]} \\ \boldsymbol{\psi}_2^{[\beta]} \end{bmatrix} \right\} = \tilde{\mathbf{A}}\,\mathcal{L}\{\boldsymbol{\psi}(t)\},$$

or, equivalently,

$$\tilde{\mathbf{M}} \left[\begin{array}{c} s^\gamma\,\mathcal{L}\{\boldsymbol{\psi}_1(t)\} - s^{\gamma-1}\,\boldsymbol{\psi}_1(0) \\ s^\beta\,\mathcal{L}\{\boldsymbol{\psi}_2(t)\} - s^{\beta-1}\,\boldsymbol{\psi}_2(0) \end{array} \right] = \tilde{\mathbf{A}}\,\mathcal{L}\{\boldsymbol{\psi}(t)\}\,,$$

or, equivalently,

$$\tilde{\mathbf{M}} \left[\begin{array}{c} s^\gamma\mathcal{L}\{\boldsymbol{\psi}_1(t)\} \\ s^\beta\mathcal{L}\{\boldsymbol{\psi}_2(t)\} \end{array} \right] - \tilde{\mathbf{M}} \left[\begin{array}{c} s^{\gamma-1}\boldsymbol{\psi}_1(0) \\ s^{\beta-1}\boldsymbol{\psi}_2(0) \end{array} \right] = \tilde{\mathbf{A}}\,\mathcal{L}\{\boldsymbol{\psi}(t)\}\,,$$

or, equivalently,

$$\tilde{\mathbf{M}} \begin{bmatrix} s^{\gamma}\,\mathbf{I}_{\rho} & \mathbf{0}_{\rho,\rho} \\ \mathbf{0}_{\rho,\rho} & s^{\beta}\,\mathbf{I}_{\rho} \end{bmatrix} \begin{bmatrix} \mathcal{L}\{\boldsymbol{\psi}_1(t)\} \\ \mathcal{L}\{\boldsymbol{\psi}_2(t)\} \end{bmatrix} -$$

$$\tilde{\mathbf{M}} \begin{bmatrix} s^{\gamma-1}\,\mathbf{I}_{\rho} & \mathbf{0}_{\rho,\rho} \\ \mathbf{0}_{\rho,\rho} & s^{\beta-1}\,\mathbf{I}_{\rho} \end{bmatrix} \begin{bmatrix} \boldsymbol{\psi}_1(0) \\ \boldsymbol{\psi}_2(0) \end{bmatrix} = \tilde{\mathbf{A}}\,\boldsymbol{\Psi}(s)\,,$$

or, equivalently,

$$\left(\tilde{\mathbf{M}} \begin{bmatrix} s^{\gamma}\,\mathbf{I}_{\rho} & \mathbf{0}_{\rho,\rho} \\ \mathbf{0}_{\rho,\rho} & s^{\beta}\,\mathbf{I}_{\rho} \end{bmatrix} - \tilde{\mathbf{A}}\right) \boldsymbol{\Psi}(s) = \begin{bmatrix} s^{\gamma-1}\,\mathbf{I}_{\rho} & \mathbf{0}_{\rho,\rho} \\ \mathbf{0}_{\rho,\rho} & s^{\beta-1}\,\mathbf{I}_{\rho} \end{bmatrix} \boldsymbol{\psi}(0)\,.$$

The proof is completed. ■

The eigenvalues of the matrix pencil (9.24) provide insight on the stability of system (9.23), or equivalently, of system (9.18). We finally provide the following Proposition, see [166]:

Proposition 9.2 (Asymptotic stability of fractional order systems)**.** Consider system (9.23). If $\tilde{\gamma} = \min\{\gamma, 1-\gamma\}$, and λ is an eigenvalue of the pencil (9.24), then system (9.23) is asymptotically stable if all eigenvalues λ satisfy:

$$|\text{Arg}(\lambda)| > \tilde{\gamma}\frac{\pi}{2}\ \text{(rad)}\,. \tag{9.25}$$

For linearized systems, as it is the case of power systems, the condition (9.25) guarantees stability in a neighborhood of the operating point utilized to calculate the pencil (9.24).

Example 9.3 (Fractional Order Control)**.** We provide an illustrative example on the small-signal stability analysis of power systems with inclusion of FOCs. The example is based on the well-known WSCC nine-bus system, the data of which are provided in [4]. The system consists of three synchronous machines, six transmission lines, three transformers and three loads, modeled as constant power consumption. Each machine provides primary voltage and frequency control through an AVR and a TG, respectively. The original system model does not include any fractional dynamics.

Suppose that a fractional-order (FO)-PSS is installed at the synchronous machine connected at bus 2. The FO-PSS employed has the following transfer function:

$$G_{\text{FOPSS}} = K\left(\frac{T_1 s^{\gamma}+1}{T_2 s^{\gamma}+1}\right)^2\,.$$

The controller input is the local rotor speed, while the output is an additional input to the algebraic equation of the local AVR reference. The FO-PSS can be written in the form of (9.17), where:

$$\mathbf{E}_c = \begin{bmatrix} T_2 & 0 & 0 & 0 \\ T_1 & 0 & 0 & 0 \\ 0 & 0 & T_2 & 0 \\ 0 & 0 & T_1 & 0 \end{bmatrix}, \qquad \mathbf{A}_c = \begin{bmatrix} -1 & 0 & 0 & 0 \\ -1 & 1 & 0 & 0 \\ 0 & 1 & -1 & 0 \\ 0 & 0 & -1 & 1 \end{bmatrix},$$

$$\mathbf{B}_c = \begin{bmatrix} K & 0 & 0 & 0 \end{bmatrix}^{\mathrm{T}}, \qquad \mathbf{C}_c = \begin{bmatrix} 0 & 0 & 0 & 1 \end{bmatrix}, \qquad \mathbf{D}_c = \begin{bmatrix} 0 \end{bmatrix}.$$

Suppose that $T_1 = 0.01$ s, $T_2 = 0.22$ s, $\gamma = 0.75$. Then, small-signal stability is assessed by calculating the eigenvalues of (9.24). If λ_i is a finite eigenvalue, then (9.25) suggests that the system is stable if:

$$|\mathrm{Arg}(\lambda_i)| > \tilde{\gamma}\frac{\pi}{2} = 0.393 \text{ rad} ,$$

where $\tilde{\gamma} = \min\{0.75, 0.25\} = 0.25$. The most critical eigenvalues of the closed-loop system are shown in Figure 9.1, where the shaded region is unstable. The equilibrium point of the system with the FO-PSS is stable. It is interesting to note that an equilibrium point of a fractional order system can be stable also if some of its eigenvalues have positive real part. □

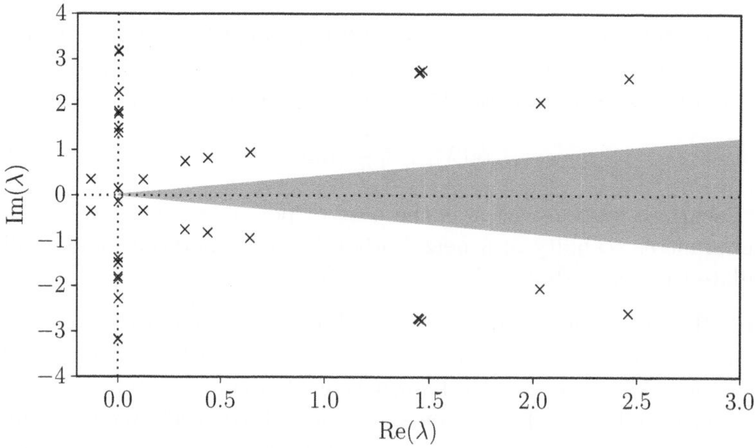

FIGURE 9.1: WSCC system with FO-PSS: most critical eigenvalues. The region of instability is shaded.

10

Delay Differential Equations

10.1 Formulation

This chapter discusses the non-linear eigenvalue problem (NEP) of systems modeled as a set of delay differential equations (DDEs). DDEs are a special case of functional differential equations (FDEs), which can be written in the form:

$$\mathbf{E}\,\dot{\boldsymbol{x}}(t) = \boldsymbol{\varphi}\left(t, \boldsymbol{x}(t), \boldsymbol{x}\big(\boldsymbol{\kappa}(t)\big), \dot{\boldsymbol{x}}\big(\boldsymbol{\kappa}(t)\big)\right) , \tag{10.1}$$

where $\boldsymbol{x} \in \mathbb{R}^{n\times 1}$ and $\boldsymbol{\kappa} = [\kappa_1, \kappa_2, \dots, \kappa_{k_{\mathrm{m}}}]^{\mathrm{T}}$. If (10.1) is a set of DDEs, then:

$$\kappa_i(t) = t - \tau_i \, , \quad i = 1, 2, \dots, k_{\mathrm{m}} \, ; \tag{10.2}$$

where τ_i represents a time delay.

This chapter focuses on the following autonomous DDEs with constant delays:

$$\begin{aligned} \mathbf{E}\,\dot{\boldsymbol{x}}(t) = \boldsymbol{f}\big(\boldsymbol{x}(t), &\boldsymbol{x}(t-\tau_1), \, \dot{\boldsymbol{x}}(t-\tau_1), \, \dots \\ &\boldsymbol{x}(t-\tau_i), \, \dot{\boldsymbol{x}}(t-\tau_i), \, \dots \\ &\boldsymbol{x}(t-\tau_{k_{\mathrm{m}}}), \, \dot{\boldsymbol{x}}(t-\tau_{k_{\mathrm{m}}})\big) \, . \end{aligned} \tag{10.3}$$

According to the dependency on delayed state variables $\dot{\boldsymbol{x}}(\boldsymbol{\kappa}(t))$, DDEs can be divided into two types: retarded delay differential equations (RDDEs) and neutral delay differential equations (NDDEs) [110]. The standard linearized formulation of RDDEs and NDDEs for eigenvalue analysis are presented and discussed in Sections 10.1.1 and 10.1.2.

10.1.1 Retarded Delay Differential Equations

In this section, we consider the linear autonomous RDDEs:

$$\mathbf{E}\,\dot{\boldsymbol{x}}(t) = \mathbf{A}_0\,\boldsymbol{x}(t) + \sum_{k=1}^{k_{\mathrm{m}}} \mathbf{A}_k\,\boldsymbol{x}(t-\tau_k) \, , \tag{10.4}$$

where $\mathbf{A}_k \in \mathbb{R}^{n\times n}$ for $k = 0, 1, 2, \dots k_{\mathrm{m}}$. The matrix $\mathbf{A}_0 = \mathbf{A}_{\mathrm{S}}$ is the state matrix as defined in Section 1.5, and matrices $\mathbf{A}_k$ are called *delay matrices.*

If the system of RDDEs (10.4) is stable for the no delay scenario, namely $\tau_k = 0 \; \forall k \le k_{\mathrm{m}}$, then there exists a constant value τ_c such that $\forall \tau \le \tau_c$, the system remains stable. The maximum value of τ_c is called the *delay margin* of the system. Note that the delay margin exists not only for RDDEs but also for all kinds of time-delay systems. If $\tau_c \to \infty$, then the time-delay system is *delay-independent stable.* The eigenvalue analysis of (10.4) allows assessing the stability of (10.4) with a specific τ_k and estimating the delay margin.

Let us assume that the general solution of (10.4) is:

$$\boldsymbol{x}(t) = \mathrm{e}^{\lambda t} \boldsymbol{\nu} \, , \tag{10.5}$$

so that

$$\boldsymbol{x}(t - \tau) = \mathrm{e}^{-\lambda \tau} \, \mathrm{e}^{\lambda t} \boldsymbol{\nu} \, . \tag{10.6}$$

Substituting (10.5) and (10.6) into the (10.4) and using the Laplace transform, we can deduce the matrix pencil:

$$\boldsymbol{\Delta}(s) = s\mathbf{E} - \mathbf{A}_0 - \sum_{k=1}^{k_{\mathrm{m}}} \mathbf{A}_k \, \mathrm{e}^{-s\tau_k} \, , \tag{10.7}$$

with characteristic equation:

$$\boldsymbol{\Delta}(s) \, \boldsymbol{\nu} = \mathbf{0} \, , \tag{10.8}$$

and, considering that $\boldsymbol{\nu}$ is non-trivial, one has:

$$\det(\boldsymbol{\Delta}(s)) = 0 \, , \tag{10.9}$$

which admits λ as a root.

Equation (10.9) is a transcendental problem that has infinite many eigenvalues due to the existence of the non-linear term $\mathrm{e}^{-\lambda \tau_k}$. Although it is impossible to obtain the whole spectrum of this transcendental problem, the following Lemma indicates that it is sufficient to find a finite number of the rightmost eigenvalues to define the stability of the time-delay system [110].

Lemma 10.1 (Spectrum of time-delay system)**.** The spectrum of a time-delay system has the following properties:

- The number of roots in any vertical strip of the complex plane is finite:

  $$\{\lambda \in \mathbb{C} : \alpha < \mathrm{Re}(\lambda) < \beta\} \, ,$$

  where $\alpha, \beta \in \mathbb{R}$.

- There exists $\gamma \in \mathbb{R}$ such that all the roots lay on the left-hand side of γ in the complex plane:

  $$\{\lambda \in \mathbb{C} : \mathrm{Re}(\lambda) < \gamma\} \, .$$

In practice, it is not possible to find an accurate value of γ. Nevertheless, based on Lemma 10.1, one can solve the stability analysis of system (10.4) by finding the rightmost roots of the characteristic equation (10.9). The delay margin of a time-delay system can then be estimated through tracking the trajectories of the rightmost eigenvalues with respect to the delays.

Lemma 10.2 (Rightmost eigenvalues of a time-delay system)**.** For a time-delay system at its delay margin, the rightmost eigenvalue λ_c satisfies the following condition:

$$\mathrm{Re}(\lambda_c) = 0\,, \qquad \frac{\partial \mathrm{Re}(\lambda_c)}{\partial \tau} > 0\,.$$

Remark 10.1. Lemma 10.1 implies that a time-delay system at its delay margin is weakly stable; and becomes unstable if the delay increases.

Remark 10.2. The eigenvalues of the pencil of a delay-independent stable system never move to the right-hand side of the imaginary axis on the complex plane regardless the magnitude of delays.

Finally, we recall Remark 1.1:

Remark 10.3. A regular pencil is called a stable pencil (or Hurwitz pencil) if every eigenvalue of the pencil has a strictly negative real part.

If the series in (10.4) is commensurate, i.e. $\tau_k = k\tau_1$, $k > 1$, the delay-independent stability of system (10.4) can be assessed through the following Proposition.

Proposition 10.1 (Delay-independent stability of RDDEs)**.** The system of RDDEs (10.4) with commensurate time delays is delay-independent stable if and only if the following conditions are satisfied:

1. The pencil $s\mathbf{E} - \mathbf{A}_0$ is stable;
2. The pencil $s\mathbf{E} - \mathbf{A}_0 - \sum_{k=1}^{k_\mathrm{m}} \mathbf{A}_k$ is stable;
3. The spectral radius of the matrix $\mathbf{F}^{-1}\mathbf{G}$ is less than 1, $\forall \omega > 0$, where

$$\mathbf{F}^{-1}\mathbf{G} = \begin{bmatrix} \boldsymbol{\Omega}\mathbf{A}_1 & \boldsymbol{\Omega}\mathbf{A}_2 & \dots & \boldsymbol{\Omega}\mathbf{A}_{k_\mathrm{m}-1} & \boldsymbol{\Omega}\mathbf{A}_{k_\mathrm{m}} \\ \mathbf{I}_n & \mathbf{0}_{n,n} & \dots & \mathbf{0}_{n,n} & \mathbf{0}_{n,n} \\ \mathbf{0}_{n,n} & \mathbf{I}_n & \dots & \mathbf{0}_{n,n} & \mathbf{0}_{n,n} \\ \vdots & \vdots & \ddots & \vdots & \vdots \\ \mathbf{0}_{n,n} & \mathbf{0}_{n,n} & \dots & \mathbf{I}_n & \mathbf{0}_{n,n} \end{bmatrix}$$

and $\boldsymbol{\Omega} = (\jmath\omega\mathbf{E} - \mathbf{A}_0)^{-1}$.

Proof. If $\tau \to \infty$ then (10.7) takes the form

$$\mathbf{\Delta}(s) = s\,\mathbf{E} - \mathbf{A}_0\,.$$

Hence, pencil $s\mathbf{E} - \mathbf{A}_0$ has to be stable in order to have stability for (10.4) at the equilibrium point. If $\tau = 0$ then (10.7) takes the form

$$\mathbf{\Delta}(s) = s\,\mathbf{E} - \mathbf{A}_0 - \mathbf{A}_1 - \cdots - \mathbf{A}_{k_\mathrm{m}}\,,$$

which means that pencil $s\mathbf{E} - \mathbf{A}_0 - \sum_{k=1}^{k_\mathrm{m}} \mathbf{A}_k$ has to be stable in order to have stability for (10.4) at the equilibrium point. By applying the Fourier transform $\mathcal{F}(\boldsymbol{X}) = \boldsymbol{X}(\omega)$ into (10.4), we get:

$$\jmath\omega\mathbf{E}\boldsymbol{X}(\omega) = \mathbf{A}_0\boldsymbol{X}(\omega) + \mathbf{A}_1\,\mathrm{e}^{-\jmath\omega\tau}\boldsymbol{X}(\omega) + \cdots + \mathbf{A}_{k_\mathrm{m}}\,\mathrm{e}^{-\jmath\omega k_\mathrm{m}\tau}\boldsymbol{X}(\omega)\,,$$

or, equivalently,

$$[\jmath\omega\mathbf{E} - \mathbf{A}_0 - \sum_{k=1}^{k_\mathrm{m}} \mathbf{A}_k\,\mathrm{e}^{-\jmath\omega k\tau}]\boldsymbol{X}(\omega) = \mathbf{0}_{n,1}\,.$$

Then $\mathbf{\Delta}(\jmath\omega) = \jmath\omega\mathbf{E} - \mathbf{A}_0 - \sum_{k=1}^{k_\mathrm{m}} \mathbf{A}_k\,\mathrm{e}^{-\jmath\omega k\tau}$ is the characteristic equation of (10.4). We adopt the following notation:

$$\begin{aligned}
\boldsymbol{x}_1(t) &= \boldsymbol{x}(t)\,,\\
\boldsymbol{x}_2(t) &= \boldsymbol{x}(t-\tau)\,,\\
\boldsymbol{x}_3(t) &= \boldsymbol{x}(t-2\tau)\,,\\
&\ldots,\\
\boldsymbol{x}_{k_\mathrm{m}-1}(t) &= \boldsymbol{x}(t-(k_\mathrm{m}-2)\tau)\,,\\
\boldsymbol{x}_{k_\mathrm{m}}(t) &= \boldsymbol{x}(t-(k_\mathrm{m}-1)\tau).
\end{aligned}$$

Furthermore

$$\begin{aligned}
\boldsymbol{x}_1(t-\tau) &= \boldsymbol{x}(t-\tau)\,,\\
\boldsymbol{x}_2(t-\tau) &= \boldsymbol{x}(t-2\tau)\,,\\
\boldsymbol{x}_3(t-\tau) &= \boldsymbol{x}(t-3\tau)\,,\\
&\vdots\\
\boldsymbol{x}_{k_\mathrm{m}}(t) &= \boldsymbol{x}(t-(k_\mathrm{m}-1)\tau)\,,\\
\mathbf{A}_{k_\mathrm{m}}\boldsymbol{x}_{k_\mathrm{m}}(t-\tau) &= \mathbf{A}_{k_\mathrm{m}}\boldsymbol{x}(t-k_\mathrm{m}\tau) = \mathbf{E}\dot{\boldsymbol{x}}(t) - \textstyle\sum_{k=0}^{k_\mathrm{m}-1} \mathbf{A}_k\boldsymbol{x}(t-k\tau)\,,
\end{aligned}$$

or, equivalently,

$$\begin{aligned}
\boldsymbol{x}_1(t-\tau) &= \boldsymbol{x}_2(t)\,,\\
\boldsymbol{x}_2(t-\tau) &= \boldsymbol{x}_3(t)\,,\\
\boldsymbol{x}_3(t-\tau) &= \boldsymbol{x}_4(t)\,,\\
&\vdots\\
\boldsymbol{x}_{k_\mathrm{m}-1}(t-\tau) &= \boldsymbol{x}_{k_\mathrm{m}}(t)\,,\\
\mathbf{A}_{k_\mathrm{m}}\boldsymbol{x}_{k_\mathrm{m}}(t-\tau) &= \mathbf{E}\dot{\boldsymbol{x}}_1(t) - \textstyle\sum_{k=0}^{k_\mathrm{m}-1} \mathbf{A}_k\boldsymbol{x}_{k+1}(t)\,,
\end{aligned}$$

or, equivalently, in matrix form

$$\mathbf{G}\tilde{\boldsymbol{X}}(t-\tau) = \mathbf{F}_1\dot{\tilde{\boldsymbol{X}}}(t) + \mathbf{F}_2\tilde{\boldsymbol{X}}(t),$$

where

$$\tilde{\boldsymbol{X}}(t) = \begin{bmatrix} \boldsymbol{x}_1(t) \\ \boldsymbol{x}_2(t) \\ \boldsymbol{x}_3(t) \\ \vdots \\ \boldsymbol{x}_{k_{\rm m}}(t) \end{bmatrix}, \mathbf{G} = \begin{bmatrix} \mathbf{I}_n & \mathbf{0}_{n,n} & \dots & \mathbf{0}_{n,n} & \mathbf{0}_{n,n} \\ \mathbf{0}_{n,n} & \mathbf{I}_n & \dots & \mathbf{0}_{n,n} & \mathbf{0}_{n,n} \\ \vdots & \vdots & \ddots & \vdots & \vdots \\ \mathbf{0}_{n,n} & \mathbf{0}_{n,n} & \dots & \mathbf{I}_n & \mathbf{0}_{n,n} \\ \mathbf{0}_{n,n} & \mathbf{0}_{n,n} & \dots & \mathbf{0}_{n,n} & \mathbf{A}_{k_{\rm m}} \end{bmatrix},$$

and $\mathbf{F}_1$, $\mathbf{F}_2$ are given by

$$\mathbf{F}_1 = \begin{bmatrix} \mathbf{0}_{n,n} & \mathbf{0}_{n,n} & \mathbf{0}_{n,n} & \dots & \mathbf{0}_{n,n} \\ \mathbf{0}_{n,n} & \mathbf{0}_{n,n} & \mathbf{0}_{n,n} & \dots & \mathbf{0}_{n,n} \\ \vdots & \vdots & \vdots & \ddots & \vdots \\ \mathbf{0}_{n,n} & \mathbf{0}_{n,n} & \mathbf{0}_{n,n} & \dots & \mathbf{0}_{n,n} \\ \mathbf{E} & \mathbf{0}_{n,n} & \mathbf{0}_{n,n} & \dots & \mathbf{0}_{n,n} \end{bmatrix},$$

and

$$\mathbf{F}_2 = \begin{bmatrix} \mathbf{0}_{n,n} & \mathbf{I}_n & \mathbf{0}_{n,n} & \dots & \mathbf{0}_{n,n} \\ \mathbf{0}_{n,n} & \mathbf{0}_{n,n} & \mathbf{I}_n & \dots & \mathbf{0}_{n,n} \\ \vdots & \vdots & \vdots & \ddots & \vdots \\ \mathbf{0}_{n,n} & \mathbf{0}_{n,n} & \mathbf{0}_{n,n} & \dots & \mathbf{I}_n \\ -\mathbf{A}_0 & -\mathbf{A}_1 & -\mathbf{A}_2 & \dots & -\mathbf{A}_{k_{\rm m}-1} \end{bmatrix}.$$

Hence, we proved that the characteristic equation of (10.4) is $\mathbf{\Delta}(\jmath\omega) = \jmath\omega\mathbf{F}_1 + \mathbf{F}_2 - e^{-\jmath\omega\tau}\mathbf{G}$. Then $\mathbf{\Delta}(\jmath\omega) = \jmath\omega\mathbf{E} - \mathbf{A}_0 - \sum_{k=1}^{k_{\rm m}} \mathbf{A}_k\, \mathrm{e}^{-\jmath\omega k\tau}$ is the characteristic equation of (10.4). We have that:

$$\mathbf{F} := \mathbf{F}(\jmath\omega) = \jmath\omega\mathbf{F}_1 + \mathbf{F}_2 = \begin{bmatrix} \mathbf{0}_{n,n} & \mathbf{I}_n & \mathbf{0}_{n,n} & \dots & \mathbf{0}_{n,n} \\ \mathbf{0}_{n,n} & \mathbf{0}_{n,n} & \mathbf{I}_n & \dots & \mathbf{0}_{n,n} \\ \vdots & \vdots & \vdots & \ddots & \vdots \\ \mathbf{0}_{n,n} & \mathbf{0}_{n,n} & \mathbf{0}_{n,n} & \dots & \mathbf{I}_n \\ \jmath\omega\mathbf{E} - \mathbf{A}_0 & -\mathbf{A}_1 & -\mathbf{A}_2 & \dots & -\mathbf{A}_{k_{\rm m}-1} \end{bmatrix}.$$

Then, $\mathbf{\Delta}(\jmath\omega) = \boldsymbol{F}(\jmath\omega) - e^{-\jmath\omega\tau}\boldsymbol{G}$. If $\|e^{-\jmath\omega\tau}\boldsymbol{F}^{-1}\boldsymbol{G}\| < 1$, $\forall\omega > 0$, (10.4) is delay-independent stable. Where

$$\mathbf{F}^{-1} = \begin{bmatrix} \mathbf{\Omega}\mathbf{A}_1 & \mathbf{\Omega}\mathbf{A}_2 & \dots & \mathbf{\Omega}\mathbf{A}_{k_{\rm m}-1} & \mathbf{\Omega} \\ \mathbf{I}_n & \mathbf{0}_{n,n} & \dots & \mathbf{0}_{n,n} & \mathbf{0}_{n,n} \\ \mathbf{0}_{n,n} & \mathbf{I}_n & \dots & \mathbf{0}_{n,n} & \mathbf{0}_{n,n} \\ \vdots & \vdots & \ddots & \vdots & \vdots \\ \mathbf{0}_{n,n} & \mathbf{0}_{n,n} & \dots & \mathbf{I}_n & \mathbf{0}_{n,n} \end{bmatrix},$$

and

$$\mathbf{F}^{-1}\mathbf{G} = \begin{bmatrix} \boldsymbol{\Omega}\mathbf{A}_1 & \boldsymbol{\Omega}\mathbf{A}_2 & \dots & \boldsymbol{\Omega}\mathbf{A}_{k_m-1} & \boldsymbol{\Omega}\mathbf{A}_n \\ \mathbf{I}_n & \mathbf{0}_{n,n} & \dots & \mathbf{0}_{n,n} & \mathbf{0}_{n,n} \\ \mathbf{0}_{n,n} & \mathbf{I}_n & \dots & \mathbf{0}_{n,n} & \mathbf{0}_{n,n} \\ \vdots & \vdots & \ddots & \vdots & \vdots \\ \mathbf{0}_{n,n} & \mathbf{0}_{n,n} & \dots & \mathbf{I}_n & \mathbf{0}_{n,n} \end{bmatrix} .$$

Furthermore, since $\forall \omega > 0$, $|e^{-\jmath\omega\tau}| = 1$ we have that

$$||e^{-\jmath\omega\tau}\mathbf{F}^{-1}\mathbf{G}|| = |e^{-\jmath\omega\tau}|||\mathbf{F}^{-1}\mathbf{G}|| = ||\mathbf{F}^{-1}\mathbf{G}||.$$

Hence (10.4) is stable independent of delay if $||\mathbf{F}^{-1}\mathbf{G}|| < 1$, $\forall\omega > 0$, or, additionally (10.4) is stable independent of delay if the spectral radius of $\mathbf{F}^{-1}\mathbf{G}$ is less than 1, $\forall\omega > 0$, since the spectral radius of a matrix is always less than any of its natural norms, see [51]. The proof is completed. ■

Remark 10.4. Note that the matrix $\mathbf{F}$ is invertible, and therefore, the eigenvalues of the matrix $\mathbf{F}^{-1}\mathbf{G}$ are the eigenvalues of the pencil $s\mathbf{F} - \mathbf{G}$. If the system is delay-dependent stable, it must have a delay margin. The spectral radius $\rho(\mathbf{F}^{-1}\mathbf{G})$ is numerically equal to $||\mathbf{F}^{-1}\mathbf{G}||_2$. In addition:

$$||\mathbf{F}^{-1}\mathbf{G}||_2 \le ||\mathbf{F}^{-1}||_2||\mathbf{G}||_2 . \tag{10.10}$$

Since $\mathbf{G}$ is independent of ω, it is easy to compute $||\mathbf{G}||_2$. Furthermore $\mathbf{F}$ is a block diagonal matrix, and can be deduced as:

$$||\mathbf{F}^{-1}||_{2,\max} = \max(1, ||\mathbf{F}_0^{-1}||_{2,\max}) ,$$

where

$$\mathbf{F}_0 = \jmath\omega\mathbf{E} - \mathbf{A}_0 .$$

For the special case that $\mathbf{E} = \mathbf{I}_n$, $\mathbf{A}_0$ is square and, if it represents the state matrix of a physical system, we can always diagonalize it, as follows:

$$\boldsymbol{\Sigma}_n = \mathbf{D}\mathbf{A}_0\mathbf{D}^{-1} ,$$

where $\boldsymbol{\Sigma}_n = \mathrm{diag}\{\sigma_1, \sigma_2, \dots, \sigma_n\}$, is a set the elements of which belong to the spectrum of $\mathbf{A}_0$. Hence, we have:

$$||\mathbf{F}_0^{-1}||_{2,\max} = ||\jmath\omega\mathbf{E} - \boldsymbol{\Sigma}_n||_{2,\max} = \max\left\{\left|\frac{1}{\jmath\omega - \sigma_k}\right|\right\}, \quad k \in [1, k_\mathrm{m}] .$$

Let $\sigma = \alpha + \jmath\beta$ such that $\alpha, \beta \in \mathbb{R}$. Then, we have:

$$\left|\left|\frac{1}{\jmath\omega - \sigma}\right|\right|_2 = \frac{\sqrt{\alpha^2 + (\beta + \omega)^2}}{\alpha^2 + (\beta + \omega)^2} .$$

It is straightforward that for $\omega = 0$, we can have:

$$||(\jmath\omega\mathbf{E} - \boldsymbol{\Sigma}_n)^{-1}||_{2,\max} = \frac{1}{|\sigma|_{\min}} ,$$

and therefore:

$$||\mathbf{F}^{-1}\mathbf{G}||_2 \leq \max\left\{1, \frac{1}{|\sigma|_{\min}}\right\} ||\mathbf{G}||_2 .$$

Meanwhile, this also implies that if the pencil $\mathbf{A}_0$ is regular then

$$\max\{\rho(\mathbf{F}^{-1}\mathbf{G})\} = \rho(\mathbf{F}^{-1})|_{\omega=0} .$$

Corollary 10.1 (Upper bound of the spectral radius)**.** The matrix $\mathbf{F}$ is non-singular, and the upper boundary of the spectral radius of $\mathbf{F}^{-1}\mathbf{G}$, or, equivalently of the pencil $s\mathbf{F} - \mathbf{G}$ can be obtained with $\omega = 0$.

10.1.2 Neutral Delay Differential Equations

This section considers linear autonomous NDDEs with the first derivative of the delayed variables:

$$\mathbf{E}_0\, \dot{\boldsymbol{x}}(t) + \sum_{i=1}^{i_{\rm m}} \mathbf{E}_i\, \dot{\boldsymbol{x}}(t - \tau_i) = \mathbf{A}_0\, \boldsymbol{x}(t) + \sum_{k=1}^{k_{\rm m}} \mathbf{A}_k\, \boldsymbol{x}(t - \tau_k) , \tag{10.11}$$

where the neutral delay matrices $\mathbf{E}_i, \mathbf{A}_k \in \mathbb{R}^{n \times n}$, for $i = 0, 1, \ldots, i_{\rm m}$ and $k = 0, 1, \ldots, k_{\rm m}$, respectively.

It is straightforward to deduce the matrix pencil of (10.11):

$$s\mathbf{H}(s) - \mathbf{A}_0 - \sum_{k=1}^{k_{\rm m}} \mathbf{A}_k\, \mathrm{e}^{s\tau_k} , \tag{10.12}$$

where:

$$\mathbf{H}(s) = \mathbf{E}_0 + \sum_{i=1}^{i_{\rm m}} \mathbf{E}_i\, \mathrm{e}^{-s\tau_i} . \tag{10.13}$$

As a time-delay system, the set of NDDEs (10.11) also follows Lemmas 10.1 and 10.2 provided in the previous section. Therefore, we can solve the stability analysis of the NDDE system by obtaining the rightmost eigenvalues through the corresponding characteristic equation of the matrix pencil (10.12).

Lemma 10.3 (Subset of the spectrum of the NDDEs)**.** The roots λ_n obtained from solving the following equation:

$$\det(\mathbf{H}(s)) = 0 , \tag{10.14}$$

is a subset of the spectrum of the NDDEs (10.11).

Remark 10.5. For each λ_n, we have the corresponding solution

$$\boldsymbol{x}(t) = \mathrm{e}^{\lambda_n t}\boldsymbol{\nu}_n \tag{10.15}$$

that leads to

$$\mathbf{E}_0\, \boldsymbol{x}(t) + \sum_{i=1}^{i_{\mathrm{m}}} \mathbf{E}_i\, \boldsymbol{x}(t - \tau_i) = \mathbf{0}_{n,1}\,, \tag{10.16}$$

and therefore

$$\mathbf{A}_0\, \boldsymbol{x}(t) + \sum_{k=1}^{k_{\mathrm{m}}} \mathbf{A}_k\, \boldsymbol{x}(t - \tau_k) = \mathbf{0}_{n,1}\,, \tag{10.17}$$

must hold. This implies that the NDDEs (10.11) stay at the equilibrium point (10.15). Therefore, the set of λ_n must be a subset of the spectrum of (10.11).

The NDDEs can be transformed into a set of comparison RDDEs that has the same stability properties. This requires theory on delay differential algebraic equations (DDAEs) which is described in Section 11.3 of the next chapter.

10.2 Padé Approximation

The Padé approximation is a widely used method to study time-delay systems. This section explains how it works for the eigenvalue analysis.

The Laplace transform of a delayed variable can be presented as:

$$x(t - \tau)\mathcal{H}(t - \tau) \quad \xrightarrow{\mathcal{L}} \quad \mathrm{e}^{-\tau s} X(s)\,, \tag{10.18}$$

where $\mathcal{H}(t)$ is the Heaviside unit step function. The Padé approximation is a numerical method to transform the term $\mathrm{e}^{-\tau s}$ and therefore to linearize the eigenvalue problem of time-delay systems.

The Padé approximation is deduced from the Taylor expansions in the frequency domain (Laplace transform):

$$\begin{aligned} \mathrm{e}^{-\tau s} &= 1 - \tau s + \frac{(\tau s)^2}{2!} - \frac{(\tau s)^3}{3!} + \cdots \\ &\approx \frac{b_0 + b_1 \tau s + \cdots + b_q (\tau s)^q}{a_0 + a_1 \tau s + \cdots + a_p (\tau s)^p} = \frac{N(s)}{D(s)}\,, \end{aligned} \tag{10.19}$$

where coefficients $a_1, \ldots, a_p$ and $b_1, \ldots, b_q$ can be obtained according to the rules of Taylor expansion [10].

For simplicity, let us consider a set of RDDEs with single delay:

$$\mathbf{E}\, \dot{\boldsymbol{x}}(t) = \mathbf{A}_0\, \boldsymbol{x}(t) + \mathbf{A}_d\, \boldsymbol{x}(t - \tau)\,, \tag{10.20}$$

where $\boldsymbol{x}(t) \in \mathbb{R}^{n \times 1}$.

According to the discussion in Section 10.1.1, we can deduce the characteristic equation of (10.20) with the following Padé approximation:

$$s\mathbf{E} - \mathbf{A}_0 - \mathbf{A}_d \frac{N(s)}{D(s)} = \mathbf{0}_{n,n} , \tag{10.21}$$

or, equivalently:

$$D(s)s\mathbf{E} - D(s)\mathbf{A}_0 - \mathbf{A}_d N(s) = \mathbf{0}_{n,n} . \tag{10.22}$$

Substituting (10.19) into (10.22), we obtain the characteristic equation:

$$\mathbf{K}_m s^m + \mathbf{K}_{m-1} s^{m-1} + \cdots + \mathbf{K}_1 s + \mathbf{K}_0 = \mathbf{0} , \tag{10.23}$$

where $m = \max(p, q)$ is the order of the equation and the parameter matrices $\mathbf{K}_i \in \mathbb{R}^{n\times n}$, $i = 1, 2, \dots m$, can be deduced through the coefficients a, b and $\mathbf{A}$, $\mathbf{A}_d$.

According to the matrix pencil theory introduced in Chapter 2, we can construct the following ODEs:

$$\dot{\boldsymbol{z}}(t) = \tilde{\mathbf{A}}\, \boldsymbol{z}(t) , \tag{10.24}$$

where $\boldsymbol{z} \in \mathbb{R}^{mn\times 1}$, $\tilde{\mathbf{A}} \in \mathbb{R}^{mn\times mn}$ and the corresponding polynomial matrix:

$$\tilde{\mathbf{A}}(s) = \mathbf{K}_m s^m + \mathbf{K}_{m-1} s^{m-1} + \cdots + \mathbf{K}_1 s + \mathbf{K}_0 .$$

The spectrum of the ODEs (10.24) are numerically equivalent to the roots of (10.21), which are the approximated eigenvalues of the RDDEs (10.21).

The discussion above proves that the Padé approximation can transform the NEP of the RDDEs (10.20) into a comparison LEP (10.24), at the cost of increasing the size of the problem by m times. It is easy to deduce that this conclusion also holds for NDDEs.

Increasing the order p, q (m) of the Padé approximation, the resulting phase error decreases. A high order, however, not only increases the computational burden but also leads to numerical issues that appear when solving the eigenvalue analysis of the time-delay system. In most cases, a good tradeoff is $p, q \le 10$.

Generally, $p \ge q$ is the common choice to approximate the time shifting in a time-delay system. In particular, $p = q$ allows obtaining the exact amplitude of the frequency response. In power system analysis, $p = q = 6$ is recommended to fulfill the tradeoff between computational burden and accuracy [116].

If $p = q$, the coefficients a_i and b_i can be obtained with the following iterative formula:

$$\begin{aligned} a_0 &= 1, \quad a_i = a_{i-1} \frac{p - i + 1}{(2p - i + 1)\, i} , \\ b_i &= (-1)^i\, a_i . \end{aligned} \tag{10.25}$$

10.3 Quasi-Polynomial Mapping-based Root-finder

The quasi-polynomial mapping-based root-finder (QPMR) is a technique developed by Vyhlídal and Zítek [173] and is a general tool for all kinds of NEPs. This method aims at computing all zeros for the following quasi-polynomial:

$$\mathfrak{h}(s) = \det(\boldsymbol{\Delta}(s)) . \tag{10.26}$$

If $\boldsymbol{\Delta}(s)$ in (10.26) is the matrix pencil of a set of DDEs, then the obtained zeros of (10.26) are its eigenvalues. The QPMR consists of the following steps:

- Step 1: Split $\mathfrak{h}(s)$ into its real and imaginary parts:

$$\mathcal{R}(\beta, \omega) = \mathrm{Re}(\mathfrak{h}(s)) , \tag{10.27}$$
$$\mathcal{I}(\beta, \omega) = \mathrm{Im}(\mathfrak{h}(s)) , \tag{10.28}$$

  with the assumption $s = \beta + \jmath\omega$.

- Step 2: Decide the mapping range $\boldsymbol{\mathcal{D}}$ for the search of zeros of $\mathcal{R}(\beta, \omega)$ and $\mathcal{I}(\beta, \omega)$. In reference [173], the bounded region is set to $\boldsymbol{\mathcal{D}} = [\beta_{\mathrm{min}}, \beta_{\mathrm{max}}] \times [0, \omega_{\mathrm{max}}]$. To reduce the computational burden, reference [172] provides an algorithm to exclude the zero-free ranges within $\boldsymbol{\mathcal{D}}$ for RDDEs.

- Step 3: Obtain the approximated zeros according to the intersection points $\lambda_a = \beta_a + \jmath\omega_a$ of the contours $\mathcal{R}(\beta, \omega) = 0$ and $\mathcal{I}(\beta, \omega) = 0$. This step is the most computationally demanding of the whole method. It requires to compute the values for $\mathcal{R}(\beta, \omega)$ and $\mathcal{I}(\beta, \omega)$ with a plenty of discrete β and ω within $\boldsymbol{\mathcal{D}}$. The contours can be obtained through the classical zero finders such as the bisection method. Alternatively, the Matlab function *contour* provides an easy way to map the contours.

- Step 4: The obtained roots can be used as the initial guess for the iterative solvers that find the zeros of the original characteristic equation $\mathfrak{h}(s) = 0$ with required accuracy.

The Matlab package developed by Applied Cybernetics Laboratory, CTU, implements the QPMR method above to search the eigenvalues for NEPs within a defined range [88, 111, 171, 172, 173]. This method, however, has two significant drawbacks. First, it is difficult to define the bounded region $\boldsymbol{\mathcal{D}}$. Such a region, in fact, has to be sufficiently big so that critical eigenvalues are not missed, but at the same time, sufficiently small to reduce the computational burden. The "inch-by-inch" search in Step 3, in fact, can quickly become extremely expensive due to the need to recompute the set of equations for each point within $\boldsymbol{\mathcal{D}}$. This issue, in turn, limits the ability to apply the QPMR to large-size systems, that is the case of real-world power systems.

Example 10.1 (Partial-element equivalent circuit). Reference [13] discusses a third-order partial element equivalent circuit (PEEC) that includes two mutually coupled inductances with retarded currents. The PEEC is shown in Figure 10.1.

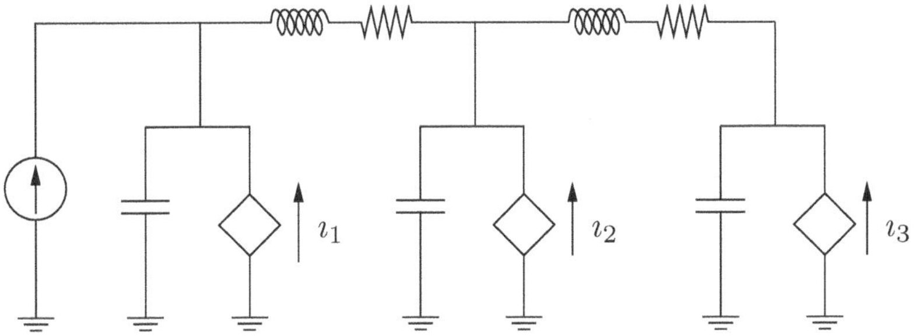

FIGURE 10.1: Partial-element equivalent circuit.

The currents of the PEEC undergo the following set of NDDEs:

$$\boldsymbol{i}(t) = \begin{bmatrix} i_1(t) \\ i_2(t) \\ i_3(t) \end{bmatrix} = \mathbf{L}\,\boldsymbol{\imath}(t) + \mathbf{M}\,\boldsymbol{\imath}(t-\tau) + \mathbf{N}\,\boldsymbol{i}(t-\tau)\,, \tag{10.29}$$

where parameter matrices $\mathbf{L}$, $\mathbf{M}$ and $\mathbf{N}$ are deduced according to the elements and topology of the circuit. A well-known numerical example assumes the following values for these matrices:

$$\frac{\mathbf{L}}{100} = \begin{bmatrix} -7 & 1 & 2 \\ 3 & -9 & 0 \\ 1 & 2 & -6 \end{bmatrix}, \quad \frac{\mathbf{M}}{100} = \begin{bmatrix} 1 & 0 & -3 \\ -0.5 & -0.5 & -1 \\ -0.5 & -1.5 & 0 \end{bmatrix}, \quad \mathbf{N} = \frac{1}{72}\begin{bmatrix} -1 & 5 & 2 \\ 4 & 0 & 3 \\ -2 & 4 & 1 \end{bmatrix}.$$

The NDDE system (10.11) is linear and has the following characteristic equation:

$$\det(s\mathbf{H}(s) - \mathbf{L} - \mathbf{M}\,\mathrm{e}^{-s\tau}) = 0\,, \tag{10.30}$$

where

$$\mathbf{H}(s) = (\mathbf{I}_3 - \mathbf{N}\,\mathrm{e}^{-s\tau})\,, \tag{10.31}$$

Consider the following equation:

$$\det(\mathbf{H}(s)) = 0\,. \tag{10.32}$$

The roots of (10.32) are a subset of the roots of (10.30) according to Lemma 10.3. Since the three eigenvalues of $\mathbf{N}$ are $-0.0762, 0.0029$ and 0.0733, to find the roots of (10.32), we rewrite $\det(\mathbf{H}(s))$ as:

$$(1 + 0.0762\,\mathrm{e}^{-s\tau})(1 - 0.0029\,\mathrm{e}^{-s\tau})(1 - 0.0733\,\mathrm{e}^{-s\tau}) = 0\,. \tag{10.33}$$

Let us consider the equation:

$$1 + 0.0762\,\mathrm{e}^{-s\tau} = 0\,. \tag{10.34}$$

The roots of (10.34) must be a subset of the solutions of (10.33). Assume $s = \beta + \jmath\omega$, where $\beta \in \mathbb{R}$ and $\omega \in \mathbb{R}^+ \cup \{0\}$, and substitute into (10.34). We obtain the following two equations for the real and imaginary parts:

$$\begin{aligned} 1 + 0.0762\,\mathrm{e}^{-\beta\tau}\cos(-\omega\tau) &= 0\,, \\ 0.0762\,\mathrm{e}^{-\beta\tau}\sin(-\omega\tau) &= 0\,. \end{aligned} \tag{10.35}$$

The general solution of (10.35) is:

$$\beta = -\frac{2.5744}{\tau}\,, \quad \omega = \frac{k\pi}{\tau}\,, \quad k = 1, 3, 5, \ldots \tag{10.36}$$

Similarly, the other two subsets of the roots of (10.33) are:

$$\beta = -\frac{5.843}{\tau}\,, \quad \omega = \frac{k\pi}{\tau}\,, \quad k = 0, 2, 4, \ldots \tag{10.37}$$

and

$$\beta = -\frac{2.6132}{\tau}\,, \quad \omega = \frac{k\pi}{\tau}\,, \quad k = 0, 2, 4, \ldots \tag{10.38}$$

According to (10.36)–(10.38), the roots of (10.33) have negative real parts for $\tau > 0$. Hence, the matrix $\mathbf{H}(s)$ is always stable for any positive delay, which is abbreviated as *delay-independent stable*, or *strong stable* in some references such as [111].

The roots above are obtained through an analytic solution of the characteristic equation. This only works for very simple systems. For relatively more complicated systems, QPMR and Padé approximation can help find the roots for a given τ. Figure 10.2 compares the roots of (10.33) with $\tau = 1$ s, distributed in the portion of the complex plane with ranges $\mathrm{Re} \in [-6.5, 0.5]$ and $\mathrm{Im} \in [-6.5\pi, 6.5\pi]$, obtained through an analytic solution, QPMR and tenth-order Padé approximations of $\mathbf{H}(s)$. The roots computed by QPMR are consistent with the analytic result, and therefore, accurate. The Padé approximation, however, has an unsatisfactory numerical accuracy, as it shows spurious rightmost roots.

The spectrum obtained with (10.32) is on the left-hand side of the complex plane for any delay according to the analytic results (10.36)–(10.38). It is thus relevant to discuss the delay margin of (10.29). Reference [13] proves that the model (10.29) is asymptotically stable for the trivial solution if the delay τ is *small* but leaves the exact delay margin of the system as an open question. The determination of such a margin has then been studied by means of several different Lyapunov-based approaches. References [78, 80, 102, 183] obtain 0.43 s, 1.1413 s, 1.5022 s and 1.6851 s, respectively, as delay margins of the PEEC (10.29).

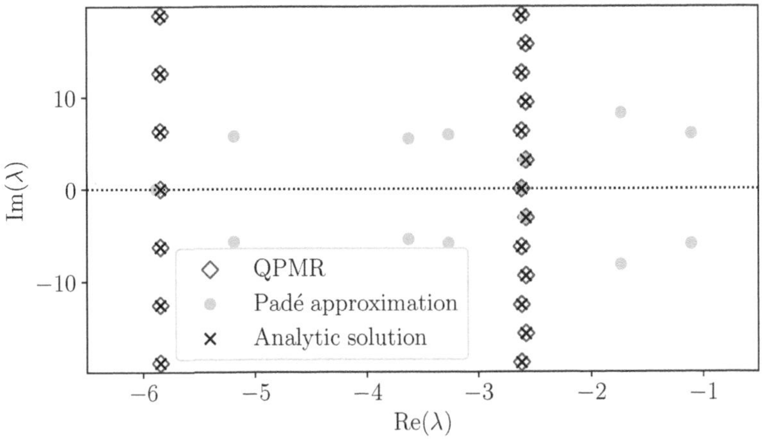

FIGURE 10.2: Roots of the PEEC with model (10.33) computed with different methods.

TABLE 10.1: Rightmost eigenvalues of (10.29) obtained by the QPMR within the range $\mathrm{Re}(\lambda) \in [-10, 10]$ and $\mathrm{Im}(\lambda) \in [-10\pi, 10\pi]$.

τ [s]	Rightmost Eigenvalue
0.43	$-1.7095 \pm \jmath 7.2591$
1.1413	$-0.6465 \pm \jmath 2.7460$
1.5022	$-0.4915 \pm \jmath 2.0875$
1.6851	$-0.4382 \pm \jmath 1.8613$

Table 10.1 shows the rightmost eigenvalues of (10.29) obtained with QPMR in the range $\mathrm{Re}(\lambda) \in [-10, 10]$ and $\mathrm{Im}(\lambda) \in [-10\pi, 10\pi]$. According to Table 10.1, the delay margins deduced in the references above are conservative. Example 11.3 will prove that this PEEC is, in fact, delay-independent stable. □

10.4 Spectrum Discretization

The spectrum discretization is another method to estimate the critical eigenvalues of time-delay systems. Unlike the Padé approximation and QPMR methods that can be applied for all kinds of time-delay systems, the spectrum discretization only applies to systems of RDDEs. The most relevant feature of the spectrum discretization method is its high accuracy and relatively light computational burden to solve large-size eigenvalue problems [116].

The method first transforms the non-linear finite-dimensional eigenvalue problem (10.4) into a linear infinite-dimensional eigenvalue problem, as discussed in Section 10.4.1. Then, it computes the rightmost eigenvalues of the infinite-dimensional problem through numerical discretization methods, as discussed in Section 10.4.2.

10.4.1 Infinite-Dimensional Eigenvalue Problem

Lemma 10.4 (Linear ODEs equivalent to RDDEs)**.** The set of RDDEs (10.4) can be reformulated as the following linear set of ODEs [110]:

$$\dot{\boldsymbol{x}} = \boldsymbol{\mathcal{A}} \boldsymbol{x} \, , \tag{10.39}$$

where $\boldsymbol{\mathcal{A}}$ is an infinitesimal operator. The corresponding characteristic equation is:

$$(s\mathbf{I}_\infty - \boldsymbol{\mathcal{A}})\, \mathbf{u} = \mathbf{0}_{\infty,1} \, . \tag{10.40}$$

The domain of $\boldsymbol{\mathcal{A}}$ is defined as:

$$\begin{aligned}
\mathcal{D}(\boldsymbol{\mathcal{A}}) =& \{ \upsilon \in \mathcal{C}([-\tau_m, 0], \mathbb{R}^{n \times 1}) : \\
& \frac{d\upsilon}{d\vartheta} \in \mathcal{C}([-\tau_m, 0], \mathbb{R}^{n \times 1}), \\
& \upsilon(0) = \mathbf{A}_0 \, \upsilon(0) + \sum_{i=1}^{i_{\mathrm{m}}} \mathbf{A}_i \, \upsilon(-\tau_i) \} \, , \\
\boldsymbol{\mathcal{A}} \upsilon =& \frac{d\upsilon}{d\vartheta} \; \vartheta \in [-\tau_m, 0] \, ,
\end{aligned}$$

where $\mathcal{C}$ is a Banach space for continuous functions; τ_m is the maximal delay among τ_i; υ is the initial function on the segment $[-\tau_m, 0]$; and ϑ is defined as:

$$\boldsymbol{x}(\upsilon)(\vartheta) = \upsilon(\vartheta), \qquad \forall \vartheta \in [-\tau_m, 0] \, .$$

Remark 10.6. The NEP of (10.4) is transformed into the LEP of (10.39). The spectrum of the operator $\boldsymbol{\mathcal{A}}$ contains all the roots of (10.9) and also follows Lemma 10.1.

For each eigenvalue λ of (10.39), it holds:

$$\boldsymbol{\mathcal{A}} \mathbf{z} = \lambda \, \mathbf{z} \, , \tag{10.41}$$

where:

$$\mathbf{z}(\vartheta) = \mathbf{u} \, e^{\lambda} \, . \tag{10.42}$$

One can always compute the finite rightmost eigenvalues that define the stability of the system through the numerical discretization of the infinitesimal operator $\boldsymbol{\mathcal{A}}$.

10.4.2 Approximated Eigenvalue Computation

This section is dedicated to the description of two approximated methods for the evaluation of the spectrum of $\boldsymbol{\mathcal{A}}$: the Chebyshev polynomial interpolation (Section 10.4.2.1) and the Krylov method (Section 10.4.2.2). Both are based on the Chebyshev discretization scheme, which has proved to be one of the most efficient numerical techniques for the determination of the approximated spectrum of infinitesimal operators [20]. Then, Section 10.4.3 describes a technique to improve the computational efficiency for RDDE systems with very sparse delay matrices $\mathbf{A}_i$.

10.4.2.1 Chebyshev Polynomial Interpolation

The polynomial interpolation method for finding the rightmost eigenvalues of the infinitesimal operator $\boldsymbol{\mathcal{A}}$ was firstly developed in [14]. This method consists in transforming the original problem into a comparison finite-dimensional LEP that theoretically has the same rightmost eigenvalues with the operator $\boldsymbol{\mathcal{A}}$.

Consider a discretized matrix $\boldsymbol{\mathcal{A}}_N$ such that:

$$(\lambda \mathbf{I}_{Nn} - \boldsymbol{\mathcal{A}}_N)\,\mathbf{u}_N = \mathbf{0}_{Nn,1}\,, \tag{10.43}$$

where N is the order of the Chebyshev discretization. The selection of $\boldsymbol{\mathcal{A}}_N$ ensures that the critical roots numerically approximate the ones of $\boldsymbol{\mathcal{A}}$.

Reference [14] provides a numerical method to compute the eigenvalues of $\boldsymbol{\mathcal{A}}_N$, which was later improved in [20, 21]. References [116, 118] further increased the efficiency of the Chebyshev polynomial interpolation method and applied the method on solving the small-signal stability analysis of power systems with inclusion of time delays.

The Chebyshev polynomial interpolation scheme provided in [116] is:

$$\boldsymbol{\mathcal{A}}_N = \left[\begin{array}{ccccc} \multicolumn{5}{c}{\hat{\boldsymbol{\Psi}} \otimes \mathbf{I}_n} \\ \hline \hat{\boldsymbol{A}}_N & \hat{\boldsymbol{A}}_{N-1} & \dots & \hat{\boldsymbol{A}}_1 & \hat{\boldsymbol{A}}_0 \end{array}\right], \tag{10.44}$$

where $\otimes$ indicates the Kronecker product. Assume a matrix $\mathbf{C}$ such that:

$$\mathbf{C} = \begin{bmatrix} c_{1,1} & c_{1,2} & \cdots & c_{1,n_c-1} & c_{1,n_c} \\ \vdots & \vdots & \ddots & \vdots & \vdots \\ c_{n_r,1} & c_{n_r,2} & \cdots & c_{n_r,n_c-1} & c_{n_r,n_c} \end{bmatrix}.$$

The Kronecker product of $\mathbf{C}$ with another matrix $\mathbf{D}$ is:

$$\mathbf{C} \otimes \mathbf{D} = \begin{bmatrix} c_{1,1}\mathbf{D} & c_{1,2}\mathbf{D} & \dots & c_{1,n_c-1}\mathbf{D} & c_{1,n_c}\mathbf{D} \\ \vdots & \vdots & \ddots & \vdots & \vdots \\ c_{n_r,1}\mathbf{D} & c_{n_r,2}\mathbf{D} & \dots & c_{n_r,n_c-1}\mathbf{D} & c_{n_r,n_c}\mathbf{D} \end{bmatrix}.$$

The matrix $\hat{\boldsymbol{\Psi}}$ is a matrix composed of the first $N-1$ rows of $\boldsymbol{\Psi}$ defined as:

$$\boldsymbol{\Psi} = -2\,\boldsymbol{\Xi}_N/\tau\,, \tag{10.45}$$

where Ξ_N is the Chebyshev differentiation matrix with dimensions $(N+1) \times (N+1)$. The i-th row, j-th column element of the differentiation matrix Ξ_N is defined as:

$$\Xi_{N_{(i,j)}} = \begin{cases} \dfrac{\xi_i(-1)^{i+j}}{\xi_j(x_i - x_j)}, \ i \neq j \\[2ex] -\dfrac{x_i}{2(1-x_i^2)}, \ i = j \neq 1, N-1, \\[2ex] \dfrac{2N^2+1}{6}, \ i = j = 0, \\[2ex] -\dfrac{2N^2+1}{6}, \ i = j = N, \end{cases} \tag{10.46}$$

where $\xi_0 = \xi_N = 2$ and $\xi_2 = \cdots = \xi_{N-1} = 1$, and:

$$x_k = \cos\frac{k\pi}{N}, \quad k = 0, \ldots, N\,.$$

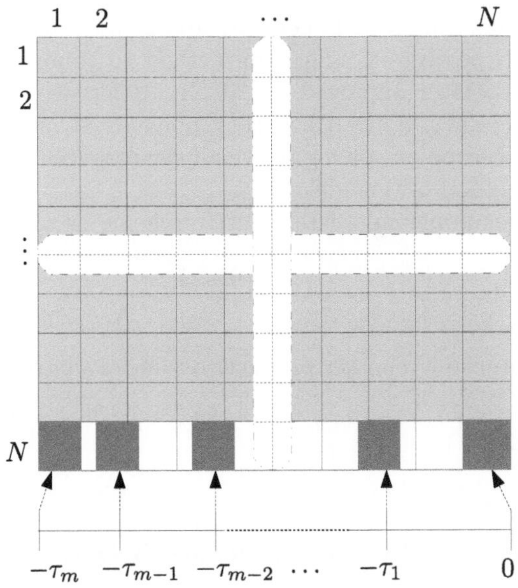

FIGURE 10.3: Representation of the Chebyshev discretization for a system with m delays $\tau_1 < \tau_2 < \cdots < \tau_{m-1} < \tau_m$. Each square slot of the grid is a $n \times n$ matrix. In the general case, the delays do not match exactly the slots and an interpolation between consecutive slots is required.

The matrices $\hat{\boldsymbol{A}}_0, \ldots, \hat{\boldsymbol{A}}_N$ are defined as follow. For the multiple delay case, assume that:

$$\tau_1 < \tau_2 < \cdots < \tau_{m-1} < \tau_m \, .$$

If the delays are commensurate, the state matrix $\mathbf{A}_0$ and delay matrices $\mathbf{A}_1, \ldots, \mathbf{A}_m$ are equally distributed on the last $N \times N$ sub-matrix of $\boldsymbol{\mathcal{A}}_{\boldsymbol{N}}$.

For example, consider a two-commensurate-delay system:

$$\dot{\boldsymbol{x}} = \mathbf{A}_0 \, \boldsymbol{x}(t) + \mathbf{A}_1 \, \boldsymbol{x}(t - \tau) + \mathbf{A}_2 \, \boldsymbol{x}(t - 2\tau) \, ,$$

the discretized matrix is:

$$\boldsymbol{\mathcal{A}}_{\boldsymbol{N}} = \left[\begin{array}{cccccccccc} \multicolumn{10}{c}{\hat{\Psi} \otimes \mathbf{I}_n} \\ \hline \mathbf{A}_2 & \mathbf{0}_n & \ldots & \mathbf{0}_n & \mathbf{A}_1 & \mathbf{0}_n & \ldots & \mathbf{0}_n & \mathbf{A}_0 \end{array} \right] .$$

If the delays are not commensurate, namely, there is no integral-multiple-time relationship between the delays, then one needs to consider the delay corresponding to the grid of the Chebyshev discretization:

$$\vartheta_j = (N - j)\Delta\tau \;\; j = 1, 2, \ldots, N \, , \tag{10.47}$$

where $\Delta\tau = \tau_m/(N-1)$. The delays that do not match the slots of the grid satisfy the condition:

$$\vartheta_j < \tau_k < \vartheta_{j+1} \, , \tag{10.48}$$

which leads to the matrices:

$$\hat{\boldsymbol{A}}_{j,k} = \frac{\tau_k - \vartheta_j}{\Delta\tau} \boldsymbol{A}_k \, , \qquad \hat{\boldsymbol{A}}_{j+1,k} = \frac{\vartheta_{j+1} - \tau_k}{\Delta\tau} \boldsymbol{A}_k \, . \tag{10.49}$$

Matrix $\hat{\boldsymbol{A}}_j$ is obtained as:

$$\hat{\boldsymbol{A}}_j = \sum_{k \in \Omega_k} \hat{\boldsymbol{A}}_{j,k} \, , \tag{10.50}$$

where Ω_k is the set of delays τ_k holding the condition (10.48). Figure 10.2 graphically illustrate the structure of matrix $\mathcal{A}_N$.

10.4.2.2 Krylov Method

Reference [79] develops an alternative Chebyshev discretization based eigenvalue computation approach. This method requires to solve a GEP that is obtained by projection on a Krylov subspace. This GEP-based numerical method is called *Krylov method.* Compared to the polynomial interpolation method discussed above, the Krylov method is supposed to achieve a better accuracy of the approximated eigenvalues. This method has also been applied to the small-signal stability analysis of power systems [182].

The Krylov method transforms the LEP (10.43) into the following GEP:

$$(\lambda \boldsymbol{\Pi}_N - \boldsymbol{\Sigma}_N)\boldsymbol{c} = \mathbf{0}_{(N+1)n,1} \, , \quad \boldsymbol{c} \in \mathbb{C}^{(N+1)n \times 1} , \quad \boldsymbol{c} \neq \mathbf{0}_{(N+1)n,1} \, , \tag{10.51}$$

where

$$
\mathbf{\Pi}_N = \frac{\tau_m}{4}\begin{bmatrix}
\frac{4}{\tau_m} & \frac{4}{\tau_m} & \frac{4}{\tau_m} & \cdots & & & \cdots & \frac{4}{\tau_m} \\
2 & 0 & -1 & & & & & \\
 & \frac{1}{2} & 0 & -\frac{1}{2} & & & & \\
 & & \frac{1}{3} & 0 & \ddots & & & \\
 & & & \frac{1}{4} & \ddots & & & \\
 & & & & \ddots & -\frac{1}{N-2} & & \\
 & & & & \ddots & 0 & -\frac{1}{N-1} & \\
 & & & & & & \frac{1}{N} & 0
\end{bmatrix} \otimes \mathbf{I}_n \, ; \tag{10.52}
$$

and

$$
\mathbf{\Sigma}_N = \begin{bmatrix}
\mathbf{R}_0 & \mathbf{R}_1 & \ldots & \mathbf{R}_N \\
 & \mathbf{I}_n & & \\
 & & \ddots & \\
 & & & \mathbf{I}_n
\end{bmatrix} , \tag{10.53}
$$

with

$$
\mathbf{R}_i = \mathbf{A}_0 + \sum_{k=1}^{m} \mathbf{A}_k T_i(-2\frac{\tau_k}{\tau_m} + 1) \, , \qquad i = 0, \ldots, N \, . \tag{10.54}
$$

The i-th order Chebyshev polynomial function T_i is defined as:

$$
\begin{aligned}
T_1(x) &= \frac{1}{2} U_1(x) \, , \\
T_i(x) &= \frac{1}{2}(U_i(x) - U_{i-2}(x)) \, , \qquad i > 2 \, ,
\end{aligned}
$$

where $U(x)$ is the Chebyshev polynomial of the second kind:

$$
\begin{aligned}
U_0(x) &= 1 \, , \\
U_1(x) &= 2x \, , \\
U_i(x) &= 2xU_{i-1} - U_{i-2} \, .
\end{aligned}
$$

Reference [79] provides an algorithm to find the optimal Chebyshev discretization order N. However, the resulting N increases rapidly with the growth of the size of the system. For example, for the IEEE 14-bus system whose state matrix is 52×52, the algorithm returns $N = 728$ as optimal order. However, with $N = 7$ the approximated critical eigenvalues already reach a satisfactory accuracy according to the Newton correction (see Section 10.5).

10.4.3 Methods for Sparse Delay Matrices

Based on the Krylov method, reference [94] proposes a spectral discretization method that largely improves the computational efficiency for RDDEs with very sparse delay matrices. We indicate such a technique as Sparse Discretization Method (SDM) in the remainder of the book. The SDM has relevant applications for the small-signal stability analysis of large power systems with inclusion of time delays.

Assume that the delay matrices $\mathbf{A}_i \in \mathbb{R}^{n\times n}$ in the set of linear RDDEs (10.4) are highly sparse and include only q ($q \ll n$) non-zero vectors. Each non-zero vector $\mathbf{a}_{i,k}$ equals to the k-th column of $\mathbf{A}_i$. We also define the set $\Omega_i = [k_{i,1}, k_{i,2}, \dots k_{i,q}]$. Consider the following characteristic equation:

$$s\,\mathbf{E}\,\boldsymbol{\nu} - \mathbf{A}_0\,\boldsymbol{\nu} - \sum_{i=1}^{i_{\rm m}} \sum_{k\in\Omega_i} \mathbf{a}_{i,k}\,e^{-s\tau_i}\,\nu_k\,, \tag{10.55}$$

where ν_k is the k-th element of $\boldsymbol{\nu}$.

The SDM applied to the N-th order Chebyshev scheme transforms (10.55) and leads to the following GEP, that has the same rightmost eigenvalues:

$$(s\tilde{\boldsymbol{\Pi}}_N - \tilde{\boldsymbol{\Sigma}}_N)\,\tilde{\boldsymbol{c}} = 0\,, \quad \tilde{\boldsymbol{c}} \in \mathbb{C}^{N_s\times 1}\,, \quad \tilde{\boldsymbol{c}} \neq 0\,, \tag{10.56}$$

where $N_s = n + (N+1)\sum_{i=1}^{i_{\rm m}} k_{m,i}$. The transformed matrix $\tilde{\boldsymbol{\Pi}}_N \in \mathbb{R}^{N_s\times N_s}$ is:

$$\tilde{\boldsymbol{\Pi}}_N = \begin{bmatrix} \mathbf{J}_{1,k_{1,1}} & & & & \\ & \ddots & & & \\ & & \mathbf{J}_{i,k} & & \\ & & & \ddots & \\ & & & \mathbf{J}_{i_{\rm m},k_{i_{\rm m},q}} & \\ & & & & \mathbf{E} \end{bmatrix}, \tag{10.57}$$

where $\mathbf{J}_{i,k} \in \mathbb{R}^{N+1\times N+1}$ is:

$$\mathbf{J}_{i,k} = \begin{bmatrix} \frac{\tau_i}{4} & 0 & -\frac{\tau_i}{4} & & & \\ & \ddots & \ddots & \ddots & & \\ & & \frac{\tau_i}{4} & 0 & -\frac{\tau_i}{4} & \\ & & & \ddots & \ddots & \\ & & & & & \frac{\tau_i}{4} \end{bmatrix}. \tag{10.58}$$

The matrix $\tilde{\boldsymbol{\Sigma}}_N \in \mathbb{R}^{N_s\times N_s}$ can be deduced as:

$$\tilde{\boldsymbol{\Sigma}}_N = \begin{bmatrix} \boldsymbol{\Gamma} & & & \\ -\boldsymbol{l}_1^{\rm T} & & & \mathbf{e}_{1,k_1} \\ & \ddots & & \vdots \\ & & \boldsymbol{\Gamma} & \\ & & -\boldsymbol{l}_1^{\rm T} & \mathbf{e}_{i_{\rm m},q} \\ \mathbf{a}_{1,k_{1,1}}\boldsymbol{l}_2^{\rm T} & \dots & \mathbf{a}_{i_{\rm m},k_{i_{\rm m},q}\}}\boldsymbol{l}_2^{\rm T} & \mathbf{A}_0 \end{bmatrix}, \tag{10.59}$$

where

$$\mathbf{e}_{i,k}\,\boldsymbol{\nu} = \nu_{i,k}\,,$$

and $\mathbf{\Gamma} = [\mathbf{0}_{N,1}, \mathbf{I}_N]$; the vectors $\boldsymbol{l}_1, \boldsymbol{l}_2 \in \mathbb{R}^{(N+1)\times 1}$:

$$\begin{aligned}
\boldsymbol{l}_1^{\mathrm{T}} &= \begin{bmatrix} 1 & 1 & 1 & \dots & 1 \end{bmatrix}, \\
\boldsymbol{l}_2^{\mathrm{T}} &= \begin{bmatrix} 1 & -1 & 1 & \dots & (-1)^N \end{bmatrix}.
\end{aligned}$$

The resulting eigenvalues of the GEP (10.56) are the approximated eigenvalues of the RDDEs and can achieve the same accuracy as the Krylov method by solving a much smaller comparison GEP. The SDM shows high efficiency compared to the other methods discussed in this section. All methods are compared in the examples presented in Chapter 11.

10.5 Newton Correction

The Newton correction is a tool to improve the accuracy of the eigenvalues obtained with any numerical method and has been successfully used in the eigenvalue analysis of DDEs [94, 110, 182].

The Newton correction takes the estimated eigenvalues and their corresponding eigenvectors as the initial guess for the roots of the characteristic equation, and computes accurately the roots of the characteristic equation through the Newton method (see Example 1.14). This section briefly introduces the working principle of the Newton correction for the following characteristic equation:

$$\mathbf{\Delta}(s)\boldsymbol{\nu} = \mathbf{0}_{m,1}\,. \tag{10.60}$$

If λ_o is the initial guess for a root of (10.60), the k-th iteration of the Newton correction applied to (10.60) is:

$$\begin{bmatrix} \delta\boldsymbol{\nu}_k \\ \delta\lambda_k \end{bmatrix} = \begin{bmatrix} \mathbf{\Delta}(\lambda_k) & \frac{d}{d\lambda}\mathbf{\Delta}(\lambda_k)\boldsymbol{\nu}_k \\ \boldsymbol{\nu}_0^H & 0 \end{bmatrix}^{-1} \begin{bmatrix} -\mathbf{\Delta}(\lambda_k)\boldsymbol{\nu}_k \\ 1 - \boldsymbol{\nu}_0^H\boldsymbol{\nu}_k \end{bmatrix},$$

and

$$\begin{aligned}
\lambda_k &= \lambda_{k-1} + \delta\lambda_{k-1} \\
\boldsymbol{\nu}_k &= \boldsymbol{\nu}_{k-1} + \delta\boldsymbol{\nu}_{k-1}\,.
\end{aligned}$$

Assuming that the tolerance for the corrected eigenvalue is ϵ, the stopping criterion of Newton correction is either:

$$\max(|\delta\lambda^k|, ||\delta\boldsymbol{\nu}^k||) \le \epsilon\,,$$

or

$$||\mathbf{\Delta}(\lambda^k)\boldsymbol{\nu}^k|| \le \epsilon\,.$$

The maximum number of iterations has to be chosen with care as it depends on both the accuracy of the initial guess and the mathematical properties of the characteristic equation.

The Newton correction is not necessary for eigenvalue problems that can be solved through direct methods, due to the limited improvement of the accuracy that and the relatively high computational burden. On the other hand, it can be utilized to evaluate the accuracy of approximated eigenvalue computation methods, e.g. the discretization methods for DDEs described in Section 10.4.2.

11

Systems with Constant Delays

11.1 Delay Differential-Algebraic Equations

This chapter focuses on delay differential algebraic equations (DDAEs) with constant delays. Similar to DDEs, DDAEs can also be separated to neutral and retarded types according to their dependence or not on the derivatives of the delayed state variables, respectively. There are also two kinds of retarded DDAEs. If delayed algebraic variables do not appear in algebraic equations, the set of retarded DDAEs is called index-1 Hessenberg form; otherwise, it is called non-index-1 Hessenberg form. The detailed mathematical formulations of these different kinds of DDAEs are presented in the remainder of this chapter.

As discussed in Chapter 10, SDMs are the most efficient and feasible methods to find the critical eigenvalues of large-size RDDEs. In this context, this chapter provides several techniques, that allow transforming different types of DDAEs and DDEs into equivalent RDDEs with same eigenvalues and, therefore, allow exploiting the SDMs – see Figure 11.1. Meanwhile, the Newton correction described in Section 10.5 is utilized to improve the accuracy of the estimated eigenvalues whenever necessary.

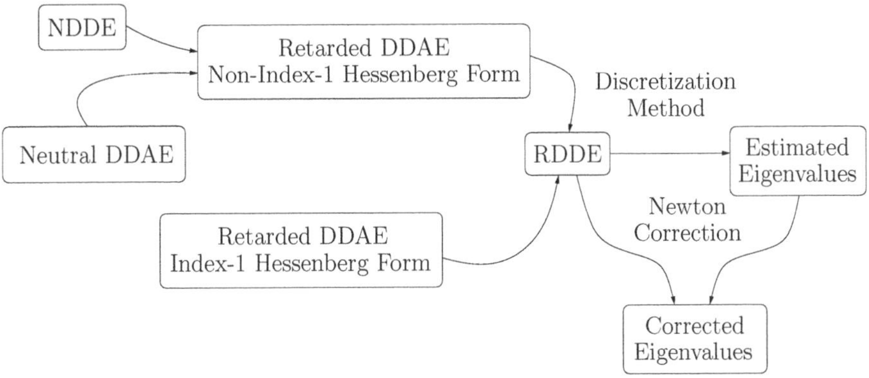

FIGURE 11.1: Sequence of transformations of DDEs and DDAEs into RDDEs that allows solving the eigenvalue analysis by means of an SDM and/or the Newton correction.

11.2 Retarded Index-1 Hessenberg Form DDAEs

The dynamic behavior of most power systems with inclusion of time delays can be modeled as a set of non-linear retarded DDAEs in index-1 Hessenberg form, as follows [118]:

$$\begin{aligned}\mathbf{E}\,\dot{\boldsymbol{x}}(t) &= \boldsymbol{f}\big(\boldsymbol{x}(t), \boldsymbol{y}(t), \boldsymbol{x}_d(t), \boldsymbol{y}_d(t)\big)\,, \\ \mathbf{0}_{l,1} &= \boldsymbol{g}\big(\boldsymbol{x}(t), \boldsymbol{y}(t), \boldsymbol{x}_d(t)\big)\,,\end{aligned} \tag{11.1}$$

where $\boldsymbol{x} \in \mathbb{R}^{n\times 1}$ are the state variables, $\boldsymbol{y} \in \mathbb{R}^{l\times 1}$ are the algebraic variables, t is the current simulation time, and where input variables $\boldsymbol{u}$ have been neglected for simplicity. The vectors $\boldsymbol{x}_d \in \mathbb{R}^{p\times 1}$ and $\boldsymbol{y}_d \in \mathbb{R}^{q\times 1}$, indicate delayed state and algebraic variables, respectively. Finally, $\boldsymbol{f} : \mathbb{R}^{(n+l+p+q)\times 1} \mapsto \mathbb{R}^{n\times 1}$ and $\boldsymbol{g} : \mathbb{R}^{(n+p+l)\times 1} \mapsto \mathbb{R}^{l\times 1}$.

For each element of the delayed variables shown in (11.1), we have:

$$\begin{aligned} x_{d,i} &= x_i(t-\tau_i)\,, \qquad & i &= 1, 2, \dots, p_i\,, \\ y_{d,j} &= y_j(t-\tau_j)\,, \qquad & j &= 1, 2, \dots, q_j\,, \end{aligned}$$

where $\tau_i, \tau_j \in \mathbb{R}^+$ represent constant delays. For simplicity, this section considers the single delay case. Hence:

$$\begin{aligned}\boldsymbol{x}_d(t) &= \boldsymbol{x}(t-\tau)\,, \\ \boldsymbol{y}_d(t) &= \boldsymbol{y}(t-\tau)\,,\end{aligned}$$

with $p = n$ and $q = l$.

If $(\boldsymbol{x}_o, \boldsymbol{y}_o, \boldsymbol{x}_o, \boldsymbol{y}_o)$ is an equilibrium point of (11.1), we have:

$$\begin{aligned}\mathbf{0}_{n,1} &= \boldsymbol{f}(\boldsymbol{x}_o, \boldsymbol{y}_o, \boldsymbol{x}_o, \boldsymbol{y}_o)\,, \\ \mathbf{0}_{l,1} &= \boldsymbol{g}(\boldsymbol{x}_o, \boldsymbol{y}_o, \boldsymbol{x}_o)\,.\end{aligned} \tag{11.2}$$

The linearization of (11.2) yields:

$$\mathbf{E}\,\Delta\dot{\boldsymbol{x}} = \frac{\partial \boldsymbol{f}}{\partial \boldsymbol{x}}\Delta\boldsymbol{x} + \frac{\partial \boldsymbol{f}}{\partial \boldsymbol{x}_d}\Delta\boldsymbol{x}_d + \frac{\partial \boldsymbol{f}}{\partial \boldsymbol{y}}\Delta\boldsymbol{y} + \frac{\partial \boldsymbol{f}}{\partial \boldsymbol{y}_d}\Delta\boldsymbol{y}_d\,, \tag{11.3}$$

$$\mathbf{0}_{l,1} = \frac{\partial \boldsymbol{g}}{\partial \boldsymbol{x}}\Delta\boldsymbol{x} + \frac{\partial \boldsymbol{g}}{\partial \boldsymbol{x}_d}\Delta\boldsymbol{x}_d + \frac{\partial \boldsymbol{g}}{\partial \boldsymbol{y}}\Delta\boldsymbol{y}\,, \tag{11.4}$$

where, as usual, the Jacobian matrices are computed at the equilibrium point and Δ indicates variable variations with respect to the equilibrium.

In the remainder of the chapter, we assume that the Jacobian matrix $\frac{\partial \boldsymbol{g}}{\partial \boldsymbol{y}}$ is full rank, i.e. $\det(\frac{\partial \boldsymbol{g}}{\partial \boldsymbol{y}}) \neq 0$.

Proposition 11.1 (RDDEs equivalent to index-1 Hessenberg form DDAEs). For the DDAEs in index-1 Hessenberg form (11.1), if the Jacobian matrix

$\frac{\partial \boldsymbol{g}}{\partial \boldsymbol{y}}$ is full rank, the linearized equations (11.3)-(11.4) can be simplified to the following RDDEs [118]:

$$\mathbf{E}\,\Delta\dot{\boldsymbol{x}} = \mathbf{A}_0\Delta\boldsymbol{x} + \mathbf{A}_1\Delta\boldsymbol{x}(t-\tau) + \mathbf{A}_2\Delta\boldsymbol{x}(t-2\tau)\,, \tag{11.5}$$

where:

$$\mathbf{A}_0 = \frac{\partial \boldsymbol{f}}{\partial \boldsymbol{x}} - \frac{\partial \boldsymbol{f}}{\partial \boldsymbol{y}}\left(\frac{\partial \boldsymbol{g}}{\partial \boldsymbol{y}}\right)^{-1}\frac{\partial \boldsymbol{g}}{\partial \boldsymbol{x}}\,, \tag{11.6}$$

$$\mathbf{A}_1 = \frac{\partial \boldsymbol{f}}{\partial \boldsymbol{x}_d} - \frac{\partial \boldsymbol{f}}{\partial \boldsymbol{y}}\left(\frac{\partial \boldsymbol{g}}{\partial \boldsymbol{y}}\right)^{-1}\frac{\partial \boldsymbol{g}}{\partial \boldsymbol{x}_d} - \frac{\partial \boldsymbol{f}}{\partial \boldsymbol{y}_d}\left(\frac{\partial \boldsymbol{g}}{\partial \boldsymbol{y}}\right)^{-1}\frac{\partial \boldsymbol{g}}{\partial \boldsymbol{x}}\,, \tag{11.7}$$

$$\mathbf{A}_2 = -\frac{\partial \boldsymbol{f}}{\partial \boldsymbol{y}_d}\left(\frac{\partial \boldsymbol{g}}{\partial \boldsymbol{y}}\right)^{-1}\frac{\partial \boldsymbol{g}}{\partial \boldsymbol{x}_d}\,. \tag{11.8}$$

Proof. From (11.4), one has:

$$\Delta\boldsymbol{y} = -\left(\frac{\partial \boldsymbol{g}}{\partial \boldsymbol{y}}\right)^{-1}\frac{\partial \boldsymbol{g}}{\partial \boldsymbol{x}}\Delta\boldsymbol{x} - \left(\frac{\partial \boldsymbol{g}}{\partial \boldsymbol{y}}\right)^{-1}\frac{\partial \boldsymbol{g}}{\partial \boldsymbol{x}_d}\Delta\boldsymbol{x}_d\,. \tag{11.9}$$

Substituting (11.9) into (11.3), one has:

$$\begin{aligned}\mathbf{E}\,\Delta\dot{\boldsymbol{x}} = &\left[\frac{\partial \boldsymbol{f}}{\partial \boldsymbol{x}} - \frac{\partial \boldsymbol{f}}{\partial \boldsymbol{y}}\left(\frac{\partial \boldsymbol{g}}{\partial \boldsymbol{y}}\right)^{-1}\frac{\partial \boldsymbol{g}}{\partial \boldsymbol{x}}\right]\Delta\boldsymbol{x} \\ &+\left[\frac{\partial \boldsymbol{f}}{\partial \boldsymbol{x}_d} - \frac{\partial \boldsymbol{f}}{\partial \boldsymbol{y}}\left(\frac{\partial \boldsymbol{g}}{\partial \boldsymbol{y}}\right)^{-1}\frac{\partial \boldsymbol{g}}{\partial \boldsymbol{x}_d}\right]\Delta\boldsymbol{x}_d + \frac{\partial \boldsymbol{f}}{\partial \boldsymbol{y}_d}\Delta\boldsymbol{y}_d\,.\end{aligned} \tag{11.10}$$

The expression (11.10) still depends explicitly on $\Delta\boldsymbol{y}_d$. To remove such a dependency, let us consider the algebraic equations $\boldsymbol{g}$ computed at $(t-\tau)$. Since algebraic constraints have to be always satisfied, the following steady-state condition must hold:

$$\mathbf{0}_{l,1} = \boldsymbol{g}\big(\boldsymbol{x}(t-\tau), \boldsymbol{x}_d(t-\tau), \boldsymbol{y}(t-\tau)\big)\,. \tag{11.11}$$

Then, observing that $\boldsymbol{x}_d = \boldsymbol{x}(t-\tau)$, $\boldsymbol{y}_d = \boldsymbol{y}(t-\tau)$, and $\boldsymbol{x}_{dd} = \boldsymbol{x}_d(t-\tau) = \boldsymbol{x}(t-2\tau)$, differentiating (11.11) leads to:

$$\mathbf{0}_{l,1} = \frac{\partial \boldsymbol{g}}{\partial \boldsymbol{x}}\Delta\boldsymbol{x}_d + \frac{\partial \boldsymbol{g}}{\partial \boldsymbol{x}_d}\Delta\boldsymbol{x}_{dd} + \frac{\partial \boldsymbol{g}}{\partial \boldsymbol{y}}\Delta\boldsymbol{y}_d\,. \tag{11.12}$$

In steady-state, for any instant t_o, $\boldsymbol{x}(t_o) = \boldsymbol{x}(t_o-\tau) = \boldsymbol{x}(t_o-2\tau) = \boldsymbol{x}_o$ and $\boldsymbol{y}(t_o) = \boldsymbol{y}_d(t_o) = \boldsymbol{y}_o$. Hence, the Jacobian matrices in (11.12) are the same as in (11.4). Equation (11.12) can be rewritten as:

$$\Delta\boldsymbol{y}_d = -\left(\frac{\partial \boldsymbol{g}}{\partial \boldsymbol{y}}\right)^{-1}\frac{\partial \boldsymbol{g}}{\partial \boldsymbol{x}}\Delta\boldsymbol{x}_d - \left(\frac{\partial \boldsymbol{g}}{\partial \boldsymbol{y}}\right)^{-1}\frac{\partial \boldsymbol{g}}{\partial \boldsymbol{x}_d}\Delta\boldsymbol{x}_{dd}\,, \tag{11.13}$$

and, substituting (11.13) into (11.10), one obtains:

$$
\begin{aligned}
\mathbf{E}\,\Delta\dot{x} = & \left[\frac{\partial \boldsymbol{f}}{\partial \boldsymbol{x}} - \frac{\partial \boldsymbol{f}}{\partial \boldsymbol{y}}\left(\frac{\partial \boldsymbol{g}}{\partial \boldsymbol{y}}\right)^{-1}\frac{\partial \boldsymbol{g}}{\partial \boldsymbol{x}}\right]\Delta\boldsymbol{x} \\
& + \left[\frac{\partial \boldsymbol{f}}{\partial \boldsymbol{x}_d} - \frac{\partial \boldsymbol{f}}{\partial \boldsymbol{y}}\left(\frac{\partial \boldsymbol{g}}{\partial \boldsymbol{y}}\right)^{-1}\frac{\partial \boldsymbol{g}}{\partial \boldsymbol{x}_d} - \frac{\partial \boldsymbol{f}}{\partial \boldsymbol{y}_d}\left(\frac{\partial \boldsymbol{g}}{\partial \boldsymbol{y}}\right)^{-1}\frac{\partial \boldsymbol{g}}{\partial \boldsymbol{x}}\right]\Delta\boldsymbol{x}_d \qquad (11.14) \\
& - \left[\frac{\partial \boldsymbol{f}}{\partial \boldsymbol{y}_d}\left(\frac{\partial \boldsymbol{g}}{\partial \boldsymbol{y}}\right)^{-1}\frac{\partial \boldsymbol{g}}{\partial \boldsymbol{x}_d}\right]\Delta\boldsymbol{x}_{dd}\,,
\end{aligned}
$$

which leads to the definitions of $\mathbf{A}_0$, $\mathbf{A}_1$ and $\mathbf{A}_2$ given in (11.6), (11.7) and (11.8), respectively. The proof is completed. ■

Remark 11.1. $\mathbf{A}_0$ is formally defined as the conventional state matrix of non-delayed systems.

Remark 11.2. The small-signal stability of (11.1) at the equilibrium point (11.2) can be solved by estimating the rightmost eigenvalues of (11.5).

Since (11.5) is a linear set of RDDEs of the type of (10.4) with 2 commensurate delays, its eigenvalues can be approximated with the SDM introduced in Section 10.4.

Example 11.1 (Three-bus system with delayed PSS)**.** This example discusses the small-signal stability analysis of the three-bus system modeled as a set of DDAEs in index-1 Hessenberg form with constant delays.

Example 4.3 in Chapter 4 discusses the three-bus system with inclusion of a well-designed PSS that improves the damping of the electromechanical modes. This example discusses the impact of a constant delay appearing in the input signal of the PSS. The control scheme of the delayed PSS is shown in Figure 11.2. The parameters of the three-bus system are given in Appendix A whereas the data of the PSS are the same as those utilized in Example 4.3.

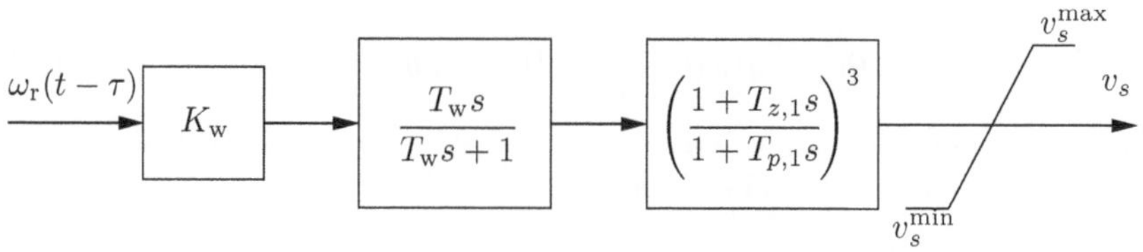

FIGURE 11.2: Control scheme of PSS with delayed input signal.

With the delayed PSS, the small-signal model of the system can be represented as:

$$
\mathbf{E}\,\dot{\boldsymbol{x}}(t) = \mathbf{A}_0\,\boldsymbol{x}(t) + \mathbf{A}_1\,\boldsymbol{x}(t-\tau) + \mathbf{A}_2\,\boldsymbol{x}(t-2\tau)\,, \qquad (11.15)
$$

where the delayed matrix $\mathbf{A}_1 \in \mathbb{R}^{23\times 23}$ is a very sparse matrix with 5 non-zero elements as shown in Table 11.1 and $\mathbf{A}_2 = \mathbf{0}_{23,23}$ as, for the non-linear

DDAE model of the power system $\frac{\partial \boldsymbol{f}}{\partial \boldsymbol{y}_d} = \mathbf{0}_{23,23}$, which means the (11.15) is a single-delay system.

TABLE 11.1: Three-bus system: elements of the sparse matrix $\boldsymbol{A}_1$.

Value	Row	Column
1003359.057	16	2
0.77857	20	2
108.242	21	2
543.469	22	2
2728.677	23	2

Compared to the system with non-delayed PSS, we have:

$$\mathbf{A} = \mathbf{A}_0 + \mathbf{A}_1 + \mathbf{A}_2 \,, \tag{11.16}$$

where $\mathbf{A}$ is the Jacobian matrix $\frac{\partial \boldsymbol{f}}{\partial \boldsymbol{x}}$ of the non-delayed system.

Using the SDM described in Chapter 10, namely the Chebyshev polynomial interpolation and the Krylov method, we can calculate the delay margin of the power system (11.16) according to Lemma 10.2. Figure 11.3 shows the variations of the real part of the rightmost eigenvalues of (11.16) as functions of τ in the range from 0 to 100 ms. The order of both discretization methods is $N = 5$.

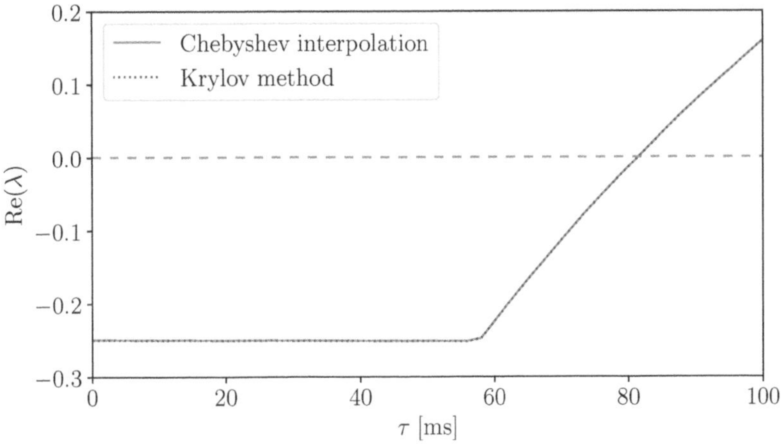

FIGURE 11.3: Three-bus system: real part of the rightmost eigenvalue of the power system (11.16) as function of τ.

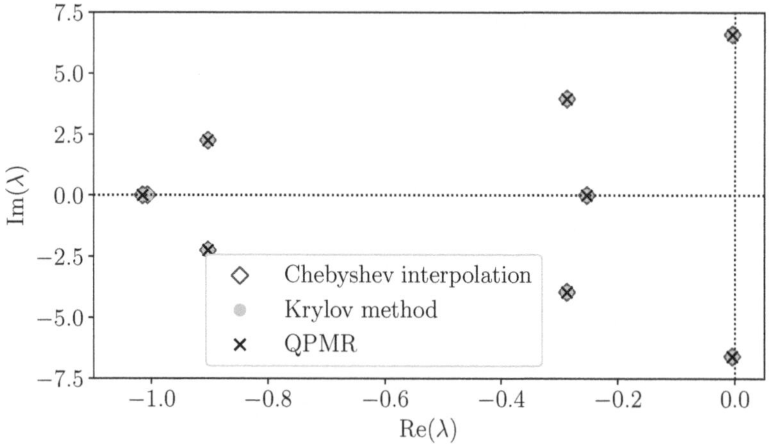

FIGURE 11.4: Three-bus system: rightmost eigenvalues of the power system (11.16) with $\tau = 81$ ms.

Figure 11.3 shows that the Chebyshev polynomial interpolation and the Krylov method obtain the same real parts of the rightmost eigenvalues for the RDDEs with single delay (11.16). According to several tests, we have found that, for $\tau \leq 58$ ms, the rightmost eigenvalues do not move; for $\tau > 58$ ms, the rightmost eigenvalues move towards the right-hand side on the complex plane and cross the imaginary axis for $\tau = 81$ ms. The delay margin of the power system, therefore, is approximately 81 ms, for which the rightmost eigenvalue is $-0.0045 \pm \jmath 6.5948$.

The identical trajectories of $\text{Re}(\lambda)$ against τ obtained through the different methods presented in Figure 11.3 imply that these two methods are accurate in solving the eigenvalue analysis of RDDEs with single constant delays. To further investigate the accuracy of these two methods, Figure 11.4 compares the critical eigenvalues with $\tau = 155$ ms obtained through Chebyshev polynomial interpolation, Krylov and QPMR methods. The QPMR range is $\text{Re}(\lambda) \times \text{Im}(\lambda) = [-1.05, 0.05] \times [-7, 7]$. Figure 11.4 confirms that the delay margin of the power system estimated by the various discretization methods is accurate. □

11.3 Retarded Non-Index-1 Hessenberg Form DDAEs

This section describes retarded DDAEs in non-index-1 Hessenberg form:

$$\begin{aligned} \mathbf{E}\,\dot{\boldsymbol{x}}(t) &= \boldsymbol{f}\big(\boldsymbol{x}(t), \boldsymbol{y}(t), \boldsymbol{x}_d(t), \boldsymbol{y}_d(t)\big) \\ \mathbf{0}_{l,1} &= \boldsymbol{g}\big(\boldsymbol{x}(t), \boldsymbol{y}(t), \boldsymbol{x}_d(t), \boldsymbol{y}_d(t)\big)\,, \end{aligned} \tag{11.17}$$

where $\boldsymbol{x} \in \mathbb{R}^{n\times 1}$ are the state variables, $\boldsymbol{y} \in \mathbb{R}^{l\times 1}$ are the algebraic variables, t is the current simulation time, and where input variables $\boldsymbol{u}$ have been neglected for simplicity. The vectors $\boldsymbol{x}_d \in \mathbb{R}^{p\times 1}$ and $\boldsymbol{y}_d \in \mathbb{R}^{q\times 1}$, indicate delayed state and algebraic variables, respectively. Finally, $\boldsymbol{f} : \mathbb{R}^{(n+l+p+q)\times 1} \mapsto \mathbb{R}^{n\times 1}$ and $\boldsymbol{g} : \mathbb{R}^{(n+p+l)\times 1} \mapsto \mathbb{R}^{l\times 1}$.

The major difference between the index-1 and non-index-1 Hessenberg form DDAEs is the dependency of the algebraic equations $\boldsymbol{g}$ on the delayed algebraic variables $\boldsymbol{y}_d$. In analogy with index-1 Hessenberg form DDAEs, also in this section we consider for simplicity but without loss of generality the single delay case. Hence:

$$\begin{aligned}\boldsymbol{x}_d(t) &= \boldsymbol{x}(t-\tau)\,,\\ \boldsymbol{y}_d(t) &= \boldsymbol{y}(t-\tau)\,,\end{aligned}$$

with $p = n$ and $q = l$.

Let us assume a stationary solution $(\boldsymbol{x}_o, \boldsymbol{y}_o, \boldsymbol{x}_o, \boldsymbol{y}_o)$ of (11.17) such that:

$$\begin{aligned}\mathbf{0}_{n,1} &= \boldsymbol{f}(\boldsymbol{x}_o, \boldsymbol{y}_o, \boldsymbol{x}_o, \boldsymbol{y}_o)\,,\\ \mathbf{0}_{l,1} &= \boldsymbol{g}(\boldsymbol{x}_o, \boldsymbol{y}_o, \boldsymbol{x}_o, \boldsymbol{y}_o)\,.\end{aligned} \tag{11.18}$$

The small-signal stability model of (11.17) near the equilibrium point is:

$$\mathbf{E}\,\Delta\dot{\boldsymbol{x}} = \frac{\partial \boldsymbol{f}}{\partial \boldsymbol{x}}\Delta\boldsymbol{x} + \frac{\partial \boldsymbol{f}}{\partial \boldsymbol{x}_d}\Delta\boldsymbol{x}_d + \frac{\partial \boldsymbol{f}}{\partial \boldsymbol{y}}\Delta\boldsymbol{y} + \frac{\partial \boldsymbol{f}}{\partial \boldsymbol{y}_d}\Delta\boldsymbol{y}_d\,, \tag{11.19}$$

$$\mathbf{0}_{l,1} = \frac{\partial \boldsymbol{g}}{\partial \boldsymbol{x}}\Delta\boldsymbol{x} + \frac{\partial \boldsymbol{g}}{\partial \boldsymbol{x}_d}\Delta\boldsymbol{x}_d + \frac{\partial \boldsymbol{g}}{\partial \boldsymbol{y}}\Delta\boldsymbol{y} + \frac{\partial \boldsymbol{g}}{\partial \boldsymbol{y}_d}\Delta\boldsymbol{y}_d\,. \tag{11.20}$$

Proposition 11.2 (RDDEs equivalent to non-index-1 Hessenberg form DDAEs)**.** If the Jacobian matrix $\frac{\partial \boldsymbol{g}}{\partial \boldsymbol{y}}$ is full rank, the linearized set of DDAEs (11.19)-(11.20) in non-index-1 Hessenberg form can be transformed into the following set of RDDEs [119]:

$$\mathbf{E}\,\Delta\dot{\boldsymbol{x}} = \mathbf{A}_0\,\Delta\boldsymbol{x} + \mathbf{A}_1\,\Delta\boldsymbol{x}(t-\tau) + \sum_{k=2}^{\infty}\mathbf{A}_k\,\Delta\boldsymbol{x}(t-k\tau)\,, \tag{11.21}$$

where:

$$\mathbf{A}_0 = \frac{\partial \boldsymbol{f}}{\partial \boldsymbol{x}} - \frac{\partial \boldsymbol{f}}{\partial \boldsymbol{y}}\left(\frac{\partial \boldsymbol{g}}{\partial \boldsymbol{y}}\right)^{-1}\frac{\partial \boldsymbol{g}}{\partial \boldsymbol{x}}\,, \tag{11.22}$$

$$\mathbf{A}_1 = \frac{\partial \boldsymbol{f}}{\partial \boldsymbol{x}_d} - \frac{\partial \boldsymbol{f}}{\partial \boldsymbol{y}}\left(\frac{\partial \boldsymbol{g}}{\partial \boldsymbol{y}}\right)^{-1}\frac{\partial \boldsymbol{g}}{\partial \boldsymbol{x}_d} - \frac{\partial \boldsymbol{f}}{\partial \boldsymbol{y}_d}\left(\frac{\partial \boldsymbol{g}}{\partial \boldsymbol{y}}\right)^{-1}\frac{\partial \boldsymbol{g}}{\partial \boldsymbol{x}}\,, \tag{11.23}$$

$$\mathbf{A}_k = -\mathbf{F}_1\mathbf{G}_3^{k-2}\mathbf{G}_4\,, \qquad k \geq 2\,, \tag{11.24}$$

where

$$\mathbf{G}_1 = -\left(\frac{\partial \boldsymbol{g}}{\partial \boldsymbol{y}}\right)^{-1}\frac{\partial \boldsymbol{g}}{\partial \boldsymbol{x}}\,, \qquad \mathbf{G}_2 = -\left(\frac{\partial \boldsymbol{g}}{\partial \boldsymbol{y}}\right)^{-1}\frac{\partial \boldsymbol{g}}{\partial \boldsymbol{x}_d}\,,$$

$$\mathbf{G}_3 = -\left(\frac{\partial \boldsymbol{g}}{\partial \boldsymbol{y}}\right)^{-1} \frac{\partial \boldsymbol{g}}{\partial \boldsymbol{y}_d}, \qquad \mathbf{G}_4 = \mathbf{G}_2 + \mathbf{G}_3\mathbf{G}_1,$$
$$\mathbf{F}_1 = \frac{\partial \boldsymbol{f}}{\partial \boldsymbol{y}}\mathbf{G}_3 + \frac{\partial \boldsymbol{f}}{\partial \boldsymbol{y}_d}.$$

Proof. From (11.20), we have

$$\Delta \boldsymbol{y} = -\left(\frac{\partial \boldsymbol{g}}{\partial \boldsymbol{y}}\right)^{-1} \frac{\partial \boldsymbol{g}}{\partial \boldsymbol{x}}\Delta \boldsymbol{x} - \left(\frac{\partial \boldsymbol{g}}{\partial \boldsymbol{y}}\right)^{-1} \frac{\partial \boldsymbol{g}}{\partial \boldsymbol{x}_d}\Delta \boldsymbol{x}_d - \left(\frac{\partial \boldsymbol{g}}{\partial \boldsymbol{y}}\right)^{-1} \frac{\partial \boldsymbol{g}}{\partial \boldsymbol{y}_d}\Delta \boldsymbol{y}_d, \tag{11.25}$$

or, equivalently,

$$\Delta \boldsymbol{y} = \mathbf{G}_1\Delta \boldsymbol{x} + \mathbf{G}_2\Delta \boldsymbol{x}_d + \mathbf{G}_3\Delta \boldsymbol{y}_d. \tag{11.26}$$

Note that $\Delta \boldsymbol{y}$ depends on $\Delta \boldsymbol{y}_d$, which, based on the same (11.26), can be written as

$$\Delta \boldsymbol{y}_d = \mathbf{G}_1\Delta \boldsymbol{x}_d + \boldsymbol{G}_2\Delta \boldsymbol{x}_{dd} + \mathbf{G}_3\Delta \boldsymbol{y}_{dd}, \tag{11.27}$$

where $\boldsymbol{x}_{dd} = \boldsymbol{x}(t-2\tau)$ and $\boldsymbol{y}_{dd} = \boldsymbol{y}(t-2\tau)$. In the same vein, $\Delta \boldsymbol{y}_d$ depends on $\Delta \boldsymbol{y}_{dd}$ and so on. Hence, (11.26) can be rewritten as follows:

$$\Delta \boldsymbol{y} = \mathbf{G}_1\Delta \boldsymbol{x} + \mathbf{G}_4\Delta \boldsymbol{x}_d + \mathbf{G}_3\mathbf{G}_2\Delta \boldsymbol{x}_{dd} + \mathbf{G}_3^2\Delta \boldsymbol{y}_{dd}, \tag{11.28}$$

or, equivalently,

$$\begin{aligned}\Delta \boldsymbol{y} = &\mathbf{G}_1\Delta \boldsymbol{x} + \mathbf{G}_4\Delta \boldsymbol{x}_d + \mathbf{G}_3\mathbf{G}_4\Delta \boldsymbol{x}_{dd} + \\ &\mathbf{G}_3^2\mathbf{G}_2\Delta \boldsymbol{x}_{ddd} + \mathbf{G}_3^3\Delta \boldsymbol{y}_{ddd},\end{aligned}$$

or, equivalently,

$$\begin{aligned}\Delta \boldsymbol{y} = \mathbf{G}_1\Delta \boldsymbol{x} + \sum_{k=1}^{\infty}[\mathbf{G}_3^{k-1}\mathbf{G}_4\Delta \boldsymbol{x}(t-k\tau)] + \\ \mathbf{G}_3^{n}\mathbf{G}_2\Delta \boldsymbol{x}(t-(n+1)\tau) + \\ \mathbf{G}_3^{n+1}\Delta \boldsymbol{y}(t-(n+1)\tau).\end{aligned}$$

If the condition $\rho(\mathbf{G}_3) < 1$ holds, then

$$\Delta \boldsymbol{y} = \mathbf{G}_1\Delta \boldsymbol{x} + \sum_{k=1}^{\infty}[\mathbf{G}_3^{k-1}\mathbf{G}_4\Delta \boldsymbol{x}(t-k\tau)].$$

Substituting the above expression into (11.19), we obtain

$$\begin{aligned}\mathbf{E}\,\Delta\dot{\boldsymbol{x}} = &\left(\frac{\partial \boldsymbol{f}}{\partial \boldsymbol{x}} + \frac{\partial \boldsymbol{f}}{\partial \boldsymbol{y}}\mathbf{G}_1\right)\Delta\boldsymbol{x} + \left(\frac{\partial \boldsymbol{f}}{\partial \boldsymbol{x}_d} + \frac{\partial \boldsymbol{f}}{\partial \boldsymbol{y}_d}\mathbf{G}_1\right)\Delta\boldsymbol{x}_d + \\ &\frac{\partial \boldsymbol{f}}{\partial \boldsymbol{y}}\sum_{k=1}^{\infty}[\mathbf{G}_3^{k-1}\mathbf{G}_4\Delta\boldsymbol{x}(t-k\tau)] + \\ &\frac{\partial \boldsymbol{f}}{\partial \boldsymbol{y}_d}\sum_{k=1}^{\infty}[\mathbf{G}_3^{k-1}\mathbf{G}_4\boldsymbol{\Delta x}(t-(k+1)\tau)]\,.\end{aligned} \tag{11.29}$$

By taking into account that

$$\frac{\partial \boldsymbol{f}}{\partial \boldsymbol{y}}\sum_{k=1}^{\infty}[\mathbf{G}_3^{k-1}\mathbf{G}_4\Delta\boldsymbol{x}(t-k\tau)] =$$

$$\frac{\partial \boldsymbol{f}}{\partial \boldsymbol{y}}\mathbf{G}_4\Delta\boldsymbol{x}_d + \frac{\partial \boldsymbol{f}}{\partial \boldsymbol{y}}\sum_{k=2}^{\infty}[\mathbf{G}_3^{k-1}\mathbf{G}_4\Delta\boldsymbol{x}(t-k\tau)]\,,$$

and

$$\frac{\partial \boldsymbol{f}}{\partial \boldsymbol{y}_d}\sum_{k=1}^{\infty}[\mathbf{G}_3^{k-1}\mathbf{G}_4\boldsymbol{\Delta x}(t-(k+1)\tau)] =$$

$$\frac{\partial \boldsymbol{f}}{\partial \boldsymbol{y}_d}\sum_{k=2}^{\infty}[\mathbf{G}_3^{k-2}\mathbf{G}_4\boldsymbol{\Delta x}(t-k\tau)]\,,$$

the equation (11.29) takes the form

$$\begin{aligned}\mathbf{E}\,\Delta\dot{\boldsymbol{x}} = &\left(\frac{\partial \boldsymbol{f}}{\partial \boldsymbol{x}} + \frac{\partial \boldsymbol{f}}{\partial \boldsymbol{y}}\mathbf{G}_1\right)\Delta\boldsymbol{x} + \\ &\left(\frac{\partial \boldsymbol{f}}{\partial \boldsymbol{x}_d} + \frac{\partial \boldsymbol{f}}{\partial \boldsymbol{y}_d}\mathbf{G}_1 + \frac{\partial \boldsymbol{f}}{\partial \boldsymbol{y}}\mathbf{G}_4\right)\Delta\boldsymbol{x}_d + \\ &\frac{\partial \boldsymbol{f}}{\partial \boldsymbol{y}}\sum_{k=2}^{\infty}[\mathbf{G}_3^{k-1}\mathbf{G}_4\Delta\boldsymbol{x}(t-k\tau)] + \\ &\frac{\partial \boldsymbol{f}}{\partial \boldsymbol{y}_d}\sum_{k=2}^{\infty}[\mathbf{G}_3^{k-2}\mathbf{G}_4\Delta\boldsymbol{x}(t-k\tau)]\,,\end{aligned}$$

or, equivalently,

$$\begin{aligned}\mathbf{E}\,\Delta\dot{\boldsymbol{x}} = &\left(\frac{\partial \boldsymbol{f}}{\partial \boldsymbol{x}} + \frac{\partial \boldsymbol{f}}{\partial \boldsymbol{y}}\mathbf{G}_1\right)\Delta\boldsymbol{x} + \\ &\left(\frac{\partial \boldsymbol{f}}{\partial \boldsymbol{x}_d} + \frac{\partial \boldsymbol{f}}{\partial \boldsymbol{y}_d}\mathbf{G}_1 + \frac{\partial \boldsymbol{f}}{\partial \boldsymbol{y}}\mathbf{G}_4\right)\Delta\boldsymbol{x}_d +\end{aligned}$$

$$\sum_{k=2}^{\infty} \left[\frac{\partial \boldsymbol{f}}{\partial \boldsymbol{y}} \mathbf{G}_3^{k-2} \mathbf{G}_4 \Delta \boldsymbol{x}(t - k\tau) \right] +$$
$$\sum_{k=2}^{\infty} \left[\frac{\partial \boldsymbol{f}}{\partial \boldsymbol{y}_d} \mathbf{G}_3^{k-2} \mathbf{G}_4 \Delta \boldsymbol{x}(t - k\tau) \right],$$

or, equivalently,

$$\mathbf{E}\, \Delta \dot{\boldsymbol{x}} = \left(\frac{\partial \boldsymbol{f}}{\partial \boldsymbol{x}} + \frac{\partial \boldsymbol{f}}{\partial \boldsymbol{y}} \mathbf{G}_1 \right) \Delta \boldsymbol{x} +$$
$$\left(\frac{\partial \boldsymbol{f}}{\partial \boldsymbol{x}_d} + \frac{\partial \boldsymbol{f}}{\partial \boldsymbol{y}_d} \mathbf{G}_1 + \frac{\partial \boldsymbol{f}}{\partial \boldsymbol{y}} \mathbf{G}_4 \right) \Delta \boldsymbol{x}_d +$$
$$\sum_{k=2}^{\infty} [\mathbf{F}_1 \mathbf{G}_3^{k-2} \mathbf{G}_4 \Delta \boldsymbol{x}(t - k\tau)] .$$

The proof is completed. ■

The comparison RDDEs (11.21) of the retarded DDAEs (11.17) introduces infinitely many delays. Thus, the rightmost eigenvalues of the (11.21) can be estimated if and only if the series (11.24) converges.

Proposition 11.3 (Convergence of the series of delay matrices)**.** The series (11.24) converges if and only if $||\mathbf{G}_3|| < 1$, or, equivalently, $\rho(\mathbf{G}_3) < 1$, where $\rho(\cdot)$ is the spectral radius of a matrix, i.e. the maximal absolute value of its spectrum. The convergence of the series is a necessary condition of the stability of system (11.21).

Proof. For a stable system, one has that $\mathrm{Re}(\lambda) < 0$, or, equivalently, since $\tau > 0$, $\tau\mathrm{Re}(\lambda) < 0$ and

$$\tau\mathrm{Re}(\lambda) + \jmath\tau\mathrm{Im}(\lambda) < \jmath\tau\mathrm{Im}(\lambda) ,$$

or, equivalently,

$$\mathrm{e}^{\tau\mathrm{Re}(\lambda) + \jmath\tau\mathrm{Im}(\lambda)} < \mathrm{e}^{\jmath\tau\mathrm{Im}(\lambda)} ,$$

or, equivalently,

$$|\,\mathrm{e}^{\tau\mathrm{Re}(\lambda) + \jmath\tau\mathrm{Im}(\lambda)}| < |\,\mathrm{e}^{\jmath\tau\mathrm{Im}(\lambda)}| ,$$

or, equivalently,

$$|\,\mathrm{e}^{\tau[\mathrm{Re}(\lambda) + \imath\mathrm{Im}(\lambda)]}| < |\,\mathrm{e}^{\jmath\tau\mathrm{Im}(\lambda)}| .$$

Considering the nature of the exponential function, one has:

$$|\,\mathrm{e}^{\jmath\tau\mathrm{Im}(\lambda)}| \leqslant 1 ,$$

thus,

$$|\,\mathrm{e}^{\tau\lambda}| < 1 .$$

The corresponding matrix pencil of (11.21) is:

$$\mathbf{\Delta}(s) = s\mathbf{E} - \mathbf{A}_0 - \mathbf{A}_1\,\mathrm{e}^{-s\tau} + \sum_{k=2}^{\infty} \mathbf{A}_k\,\mathrm{e}^{-sk\tau}\,. \tag{11.30}$$

If λ is a root of (11.30), the matrix series in (11.30) can be written as:

$$\sum_{k=2}^{\infty} \mathrm{e}^{-\lambda k\tau}\mathbf{G}_3^{k-1}\mathbf{G}_4 = \Big(\sum_{k=1}^{\infty}[\mathrm{e}^{-\lambda(k+1)\tau}\mathbf{G}_3^{k}]\Big)\mathbf{G}_4\,.$$

Hence the matrix series $\sum_{k=2}^{\infty} \mathrm{e}^{-\lambda k\tau}\mathbf{G}_3^{k-1}\boldsymbol{D}$ converges if and only if $\sum_{k=1}^{\infty} \mathrm{e}^{-\lambda(k+1)\tau}\mathbf{G}_3^{k}$ converges. By applying the d'Alembert criterion, $\sum_{k=1}^{\infty} \mathrm{e}^{-\lambda(k+1)\tau}\mathbf{G}_3^{k}$ converges if:

$$\lim_{k\to+\infty}\frac{\big\|\,\mathrm{e}^{-\lambda(k+2)\tau}\mathbf{G}_3^{(k+1)}\big\|}{\big\|\,\mathrm{e}^{-\lambda(k+1)\tau}\mathbf{G}_3^{k}\big\|} < 1\,,$$

or, equivalently,

$$|\,\mathrm{e}^{-\lambda\tau}|\lim_{k\to+\infty}\frac{\big\|\mathbf{G}_3^{(k+1)}\big\|}{\big\|\mathbf{G}_3^{k}\big\|} < 1\,,$$

by using $\|\mathbf{G}_3^{k+1}\| \le \|\mathbf{G}_3^{k}\|$ we get

$$|\,\mathrm{e}^{-\lambda\tau}|\lim_{k\to+\infty}\frac{\big\|\mathbf{G}_3^{(k+1)}\big\|}{\big\|\mathbf{G}_3^{k}\big\|} \le \mathrm{e}^{-\lambda\tau}\lim_{k\to+\infty}\frac{\big\|\mathbf{G}_3^{(k+1)}\big\|\big\|\mathbf{G}_3\big\|}{\big\|\mathbf{G}_3^{k}\big\|} < 1\,,$$

or, equivalently,

$$\lim_{k\to+\infty}\frac{\big\|\mathbf{G}_3^{k}\big\|\big\|\mathbf{G}_3\big\|}{\big\|\mathbf{G}_3^{k}\big\|} < |\,\mathrm{e}^{\lambda\tau}| < 1\,,$$

or, equivalently,

$$\big\|\mathbf{G}_3\big\| < 1\,.$$

Hence, the matrix series $\sum_{k=2}^{\infty} \mathrm{e}^{-\lambda k\tau}\mathbf{G}_3^{k-1}\mathbf{G}_4$ in (11.24) converges if

$$\rho(\mathbf{G}_3) = \rho\left(\frac{\partial \boldsymbol{g}}{\partial \boldsymbol{y}}\right)^{-1}\frac{\partial \boldsymbol{g}}{\partial \boldsymbol{y}_d} < 1$$

holds. The proof is completed. ∎

Corollary 11.1 (Truncation of the series of delay matrices)**.** The RDDEs (11.30) with a converging series can be truncated at a proper value k_{m} without loss of accuracy of the resulting roots. One can, thus, approximate (11.30) as:

$$\mathbf{E}\,\Delta\dot{\boldsymbol{x}} = \mathbf{A}_0\Delta\boldsymbol{x} + \mathbf{A}_1\Delta\boldsymbol{x}(t-\tau) + \sum_{k=2}^{k_{\mathrm{m}}} \mathbf{A}_k\Delta\boldsymbol{x}(t-k\tau)\,. \tag{11.31}$$

The selection of $k_{\rm m}$ requires:

$$\frac{|\rho(\mathbf{A}_0 + \sum_{k=1}^{k_{\rm m}} \mathbf{A}_k) - \rho(\mathbf{A}_0 + \sum_{k=1}^{k_{\rm m}+1} \mathbf{A}_k)|}{\rho(\mathbf{A}_0 + \sum_{k=1}^{k_{\rm m}} \mathbf{A}_k)} < \epsilon\,, \tag{11.32}$$

where ϵ is a given tolerance.

Remark 11.3. The selected ϵ needs to be small enough to avoid the approximation errors resulting from the truncation and also big enough to reduce the computational burden and rounding errors. According to a variety of numerical tests, $\epsilon = 10^{-6}$ appears as an adequate value.

Remark 11.4. The RDDEs (11.31) are in the form of standard RDDEs (10.4), which can be solved by the spectrum discretization methods discussed in Chapter 10.4. The critical eigenvalues obtained from (11.31) are expected to be equal to the eigenvalues of the corresponding retarded DDAEs in non-index-1 Hessenberg form (11.17).

Remark 11.5. The delays $(\tau, 2\tau, \ldots, k_{\rm m}\tau)$ included in (11.31) are commensurate. The delay-independent stability criterion concerning the system with commensurate delays (see Proposition 10.1), therefore, is suitable for the linear RDDEs (11.31).

Remark 11.6. A way to avoid the infinite delay series of the non-index-1 Hessenberg form system is to transform the original linearized model (11.19)-(11.20) into a neutral time-delay system. This point is discussed in Section 11.6.

11.4 Modeling Transmission Lines as a Continuum

This section discusses transmission lines, in particular long ones, modeled as a set of retarded DDAEs in non-index-1 Hessenberg form. The model of the transmission line with distributed parameters is presented first. Then, Example 11.2 solves the small-signal stability analysis of the three-bus system with inclusion of such a model of transmission lines.

A long transmission line can be modeled as a continuum, as follows:

$$\begin{aligned} \frac{\partial\, v(\ell,t)}{\partial \ell} &= R\,\imath(\ell,t) + L\frac{\partial\, \imath(\ell,t)}{\partial t}\,, \\ \frac{\partial\, \imath(\ell,t)}{\partial \ell} &= G\,v(\ell,t) + C\frac{\partial\, v(\ell,t)}{\partial t}\,, \end{aligned} \tag{11.33}$$

where R, L, C and G are the resistance, inductance, capacitance and conductance per unit length, respectively.

The boundary conditions of (11.33) are:

$$\begin{aligned} v(0,t) &= v_i(t)\,, & v(\ell_{ij},t) &= v_j(t)\,, \\ \imath(0,t) &= \imath_i(t)\,, & \imath(\ell_{ij},t) &= i_j(t) = -\imath_i(t)\,, \end{aligned} \tag{11.34}$$

where ℓ_{ij} is the total length of the line. The general solution of the boundary value problem for a power system with several lines is too complex to deduce. Some simplifications, therefore, are used.

A commonly accepted assumption is to use fast balanced time-varying phasors. The boundary value problem becomes:

$$\begin{aligned} \frac{\partial\,\bar{v}(\ell,t)}{\partial \ell} &= R\,\bar{\imath}(\ell,t) + L\frac{\partial\,\bar{\imath}(\ell,t)}{\partial t} + \jmath\,\omega_o L\,\bar{\imath}(\ell,t)\,, \\ \frac{\partial\,\bar{\imath}(\ell,t)}{\partial \ell} &= G\,\bar{v}(\ell,t) + C\frac{\partial\,\bar{v}(\ell,t)}{\partial t} + \jmath\,\omega_o C\,\bar{v}(\ell,t)\,, \\ \bar{v}(0,t) &= \bar{v}_i(t)\,, \qquad \bar{v}(\ell_{ij},t) = \bar{v}_j(t)\,, \\ \bar{\imath}(0,t) &= \bar{\imath}_i(t)\,, \qquad \bar{\imath}(\ell_{ij},t) = \bar{\imath}_j(t)\,, \end{aligned} \tag{11.35}$$

where ω_o is the system reference angular frequency.

By assuming $G \approx 0$, we obtain an explicit solution of (11.35) [77].

Let us define the following quantities:

- Characteristic admittance $Y_c = \sqrt{C/L}$.
- Time delay (or *traveling time*) $\tau_{ij} = \ell_{ij}\sqrt{LC}$, i.e. the time required by a wave to pass through the line at the wave speed $1/\sqrt{LC}$.
- Phase shift $\phi_{ij} = \omega_o \tau_{ij}$ and attenuation factor $\alpha_{ij} = \frac{R\ell_{ij}}{2} Y_c$.

Then, (11.35) has the solution:

$$\begin{aligned} 0 = &-\bar{\imath}_i(t) + \bar{\imath}_i(t - 2\tau_{ij})\,\mathrm{e}^{-2(\alpha_{ij}+\jmath\,\phi_{ij})} - Y_c\bar{w}_i(t) \\ &+ 2Y_c\bar{w}_j(t - \tau_{ij})\,\mathrm{e}^{-(\alpha_{ij}+\jmath\,\phi_{ij})} \\ &- Y_c\bar{w}_i(t - 2\tau_{ij})\,\mathrm{e}^{-(\alpha_{ij}+\jmath\,\phi_{ij})}\,, \\ 0 = &-\bar{\imath}_j(t) + \bar{\imath}_j(t - 2\tau_{ij})\,\mathrm{e}^{-2(\alpha_{ij}+\jmath\,\phi_{ij})} - Y_c\bar{w}_j(t) \\ &+ 2Y_c\bar{w}_i(t - \tau_{ij})\,\mathrm{e}^{-(\alpha_{ij}+\jmath\,\phi_{ij})} \\ &- Y_c\bar{w}_j(t - 2\tau_{ij})\,\mathrm{e}^{-2(\alpha_{ij}+\jmath\,\phi_{ij})}\,, \end{aligned} \tag{11.36}$$

where $\bar{w}_i$ and $\bar{w}_j$ satisfy the following complex differential equations:

$$\begin{aligned} \dot{\bar{w}}_i - \dot{\bar{v}}_i &= -\jmath\,\omega_o\bar{w}_i - \frac{R}{2L}\bar{w}_i + \jmath\,\omega_o\bar{v}_i\,, \\ \dot{\bar{w}}_j - \dot{\bar{v}}_j &= -\jmath\,\omega_o\bar{w}_j - \frac{R}{2L}\bar{w}_j + \jmath\,\omega_o\bar{v}_j\,. \end{aligned} \tag{11.37}$$

The set of FDEs (11.36)–(11.37) defines an approximated distributed-parameter transmission line model. If we further simplify this model with the assumption that there is no power loss on the line, namely $R \approx 0$, we deduce a lossless distributed-parameter transmission line model as:

$$\begin{aligned} 0 = & -\bar{\imath}_i(t) + \bar{\imath}_i(t - 2\tau_{ij})\,\mathrm{e}^{-\jmath 2\phi_{ij}} - Y_c\bar{v}_i(t) \\ & + 2Y_c\bar{v}_j(t - \tau_{ij})\,\mathrm{e}^{-\jmath \phi_{ij}} - Y_c\bar{v}_i(t - 2\tau_{ij})\,\mathrm{e}^{-\jmath 2\phi_{ij}} , \\ 0 = & -\bar{\imath}_j(t) + \bar{\imath}_j(t - 2\tau_{ij})\,\mathrm{e}^{-\jmath 2\phi_{ij}} - Y_c\bar{v}_j(t) \\ & + 2Y_c\bar{v}_i(t - \tau_{ij})\,\mathrm{e}^{-\jmath \phi_{ij}} - Y_c\bar{v}_j(t - 2\tau_{ij})\,\mathrm{e}^{-\jmath 2\phi_{ij}} . \end{aligned} \tag{11.38}$$

Model (11.36)–(11.37) or its lossless counterpart (11.37)–(11.38) lead to power system models in the form of (11.17).

The power system with the long transmission line modeled as (11.36)–(11.37) can be formulated as:

$$\begin{aligned} \dot{\boldsymbol{x}} &= \boldsymbol{f}(\boldsymbol{x}, \boldsymbol{y}) , \\ \mathbf{0}_{l,1} &= \boldsymbol{g}(\boldsymbol{x}, \boldsymbol{y}, \boldsymbol{x}_d, \boldsymbol{y}_d) . \end{aligned} \tag{11.39}$$

where delayed quantities are transmission line transient voltages $\boldsymbol{x}_d = [\boldsymbol{w}_{\mathrm{re},d}, \boldsymbol{w}_{\mathrm{im},d}]$, with $\bar{\boldsymbol{w}} = \boldsymbol{w}_{\mathrm{re}} + \jmath\,\boldsymbol{w}_{\mathrm{im}}$ and currents $\boldsymbol{y}_d = [\boldsymbol{\imath}_{\mathrm{re},d}, \boldsymbol{\imath}_{\mathrm{im},d}]$, with $\bar{\boldsymbol{\imath}} = \boldsymbol{\imath}_{\mathrm{re}} + \jmath\,\boldsymbol{\imath}_{\mathrm{im}}$.

The power system with the lossless long line modeled as (11.38) can be formulated as:

$$\begin{aligned} \dot{\boldsymbol{x}} &= \boldsymbol{f}(\boldsymbol{x}, \boldsymbol{y}) , \\ \mathbf{0}_{l,1} &= \boldsymbol{g}(\boldsymbol{x}, \boldsymbol{y}, \boldsymbol{y}_d) , \end{aligned} \tag{11.40}$$

where delayed quantities are only transmission line currents $\boldsymbol{y}_d = [\boldsymbol{\imath}_{\mathrm{re},d}, \boldsymbol{\imath}_{\mathrm{im},d}]$.

Note that, in (11.39) and (11.40) the complex expressions (11.36)–(11.37) and (11.38), respectively, are split into their real and imaginary parts to obtain a set of real DAEs.

From (11.39), we have, for each delay τ_{ij}:

$$\begin{aligned} &\frac{\partial \boldsymbol{f}}{\partial \boldsymbol{x}} \neq \mathbf{0}_{n,n} , \quad \frac{\partial \boldsymbol{f}}{\partial \boldsymbol{x}_{d,ij}} = \mathbf{0}_{n,p} , \quad \frac{\partial \boldsymbol{f}}{\partial \boldsymbol{y}} \neq \mathbf{0}_{n,l} , \quad \frac{\partial \boldsymbol{f}}{\partial \boldsymbol{y}_{d,ij}} = \mathbf{0}_{n,q} , \\ &\frac{\partial \boldsymbol{g}}{\partial \boldsymbol{x}} \neq \mathbf{0}_{l,n} , \quad \frac{\partial \boldsymbol{g}}{\partial \boldsymbol{x}_{d,ij}} \neq \mathbf{0}_{l,p} , \quad \frac{\partial \boldsymbol{g}}{\partial \boldsymbol{y}} \neq \mathbf{0}_{l,l} , \quad \frac{\partial \boldsymbol{g}}{\partial \boldsymbol{y}_{d,ij}} \neq \mathbf{0}_{l,q} , \end{aligned}$$

and, according to Proposition 11.2, we obtain:

$$\mathbf{A}_{k,ij} = \frac{\partial \boldsymbol{f}}{\partial \boldsymbol{y}} \mathbf{G}_{3,ij}^{k-1} \boldsymbol{G}_{4,ij} , \quad k \geq 1 , \quad ij \in \mathbb{L}_{ij} , \tag{11.41}$$

where $\mathbb{L}_{ij}$ is the set of transmission lines. Hence, this is a multi-delay system – delays τ_{ij} are as many as the transmission lines – and each delay generates an infinite series of non-null matrices $\mathbf{A}_{k,ij}$ associated to delays $k\tau_{ij}$, $k \geq 1$. Such matrices converge to $\mathbf{0}_{n,n}$ if and only if $\rho(\mathbf{G}_{3,ij}) < 1$ for $ij \in \mathbb{L}_{ij}$.

The lossless line model (11.40) shows same Jacobian matrices as (11.39) but for $\frac{\partial \boldsymbol{g}}{\partial \boldsymbol{x}_{d,ij}} = \mathbf{0}_{l,p}$, which leads to:

$$\mathbf{A}_{k,ij} = \frac{\partial \boldsymbol{f}}{\partial \boldsymbol{y}} \mathbf{G}_{3,ij}^{k} \mathbf{A}_{ij} , \quad k \geq 1 , \quad ij \in \mathbb{L}_{ij} . \tag{11.42}$$

Proposition 11.4 (Spectral radius of long transmission lines)**.** The spectral radius of transmission lines modeled with (11.36)–(11.37) satisfies the condition:

$$\rho(\mathbf{G}_{3,ij}) = \mathrm{e}^{-2\alpha_{ij}} \,. \tag{11.43}$$

Proof. Let us consider the structure of equations (11.36)–(11.37) and their Jacobian matrices $\frac{\partial \boldsymbol{g}}{\partial \boldsymbol{y}}$ and $\frac{\partial \boldsymbol{g}}{\partial \boldsymbol{y}_d}$. Current injections $\bar{\imath}_i$ and $\bar{\imath}_j$ are the only algebraic variables in (11.36) and appear as instantaneous and as delayed quantities, with delay $2\tau_{ij}$. While instantaneous currents also appear in other equations, at least in the current balances of network buses, delayed currents do not appear in any other equation of the system. Let us assume that (11.36) are split into their real and imaginary parts, and write Jacobian matrices $\frac{\partial \boldsymbol{g}}{\partial \boldsymbol{y}}$ and $\frac{\partial \boldsymbol{g}}{\partial \boldsymbol{y}_d}$ by separating the terms depending on the real and imaginary parts of $\bar{\imath}_i$ and $\bar{\imath}_j$, respectively. Then, we obtain:

$$\frac{\partial \boldsymbol{g}}{\partial \boldsymbol{y}} = \begin{bmatrix} \mathbf{J}_{q-4,q-4} & \mathbf{J}_{q-4,4} \\ \mathbf{0}_{4,q-4} & -\mathbf{I}_4 \end{bmatrix}, \tag{11.44}$$

$$\frac{\partial \boldsymbol{g}}{\partial \boldsymbol{y}_d} = \begin{bmatrix} \mathbf{0}_{q-4,q-4} & \mathbf{0}_{q-4,4} \\ \mathbf{0}_{4,q-4} & \mathbf{H}_4 \end{bmatrix}, \tag{11.45}$$

where $\mathbf{J}_{q-4,q-4}$ and $\mathbf{J}_{4,q-4}$ are sparse, non-null matrices and

$$\mathbf{H}_4 = \begin{bmatrix} \mathbf{H}_2 & \mathbf{0}_{2,2} \\ \mathbf{0}_{2,2} & \mathbf{H}_2 \end{bmatrix}, \tag{11.46}$$

where

$$\mathbf{H}_2 = \begin{bmatrix} \mathrm{e}^{-2\alpha_{ij}} \cos(2\phi_{ij}) & \mathrm{e}^{-2\alpha_{ij}} \sin(2\phi_{ij}) \\ -\,\mathrm{e}^{-2\alpha_{ij}} \sin(2\phi_{ij}) & \mathrm{e}^{-2\alpha_{ij}} \cos(2\phi_{ij}) \end{bmatrix}. \tag{11.47}$$

Given the structure of the Jacobian matrices above, we have:

$$\mathbf{G}_{3,ij} = -\left(\frac{\partial \boldsymbol{g}}{\partial \boldsymbol{y}}\right)^{-1} \frac{\partial \boldsymbol{g}}{\partial \boldsymbol{y}_d} = \begin{bmatrix} \mathbf{0}_{q-4,q-4} & \hat{\mathbf{J}}_{q-4,4} \\ \mathbf{0}_{4,q-4} & \mathbf{H}_4 \end{bmatrix}, \tag{11.48}$$

where $\hat{\mathbf{J}}_{q-4,4}$ is a sparse, non-null matrix. Hence $\mathbf{G}_{3,ij}$ has $q-4$ zero eigenvalues and two pairs of complex eigenvalues equal to $\mathrm{e}^{-2(\alpha_{ij} \pm \jmath\, \phi_{ij})}$. The proof is completed. ■

Corollary 11.2 (Spectral radius of long lossless transmission lines)**.** The spectral radius of transmission lines modeled as (11.38) satisfies the condition $\rho(\boldsymbol{G}_{3,ij}) = 1$.

Proof. Lossless transmission lines have $R = 0$, and, from the definition of the attenuation factor, $\alpha_{ij} = 0$. Hence, using Proposition 11.4, we obtain $\rho(\mathbf{G}_{3,ij}) = \mathrm{e}^0$. The proof is completed. ■

Example 11.2 (Three-bus system with a 10-km transmission line). This example considers the three-bus system discussed in Appendix A where the transmission line that connects buses 2 and 3 is assumed to have a length of 10 km and is modeled with distributed parameters as discussed above. We consider the following two scenarios:

- *A*. The line 2–3 is lossy and modeled as (11.36)–(11.37), with $R = 0.22$ Ω/km, $L = 0.25$ Ω/km and $C = 271$ μS/km. According to (11.48), we can deduce that $\rho(\mathbf{G}_{3,23}) = 0.4846$.
- *B*. The line 2–3 is lossless and modeled (11.38), with $R = 0$, $L = 0.25$ Ω/km and $C = 271$ μS/km. According to Corollary 11.2, in this scenario, $\rho(\mathbf{G}_{3,23}) = 1$.

As discussed in Section 11.4, the three-bus system is modeled as a set of retarded DDAEs in non-index-1 Hessenberg form. We can solve the small-signal stability analysis of the system by estimating the critical eigenvalues of its comparison RDDEs in the form of (11.21) (see Proposition 11.2). According to Corollary 11.1, the infinite constant delay series in (11.21) is truncated at a finite $k_{\rm m}$. Figures 11.5 and 11.6 show the real part of the rightmost eigenvalues of the two different scenarios as a function of $k_{\rm m}$. The eigenvalue analysis is solved with the Krylov method.

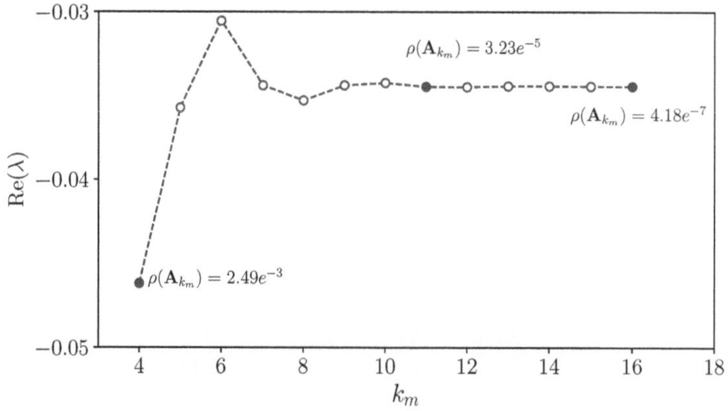

FIGURE 11.5: Real part of the rightmost eigenvalue as a function of $k_{\rm m}$ of the comparison RDDEs (11.31) for the three-bus system with a distributed-parameter 10-km transmission line.

Figure 11.5 shows $\rho(\mathbf{A}_{k_{\rm m}})$ for $k_{\rm m} = 4, 10, 16$. The real part of the rightmost eigenvalue converges at -0.0345 when $k_{\rm m} \geq 10$, which leads to small $\boldsymbol{A}_{k_{\rm m}}$ ($\rho(\mathbf{A}_{k_{\rm m}}) \leq 3.23\,\mathrm{e}^{-5}$). This means that with the transmission line 2–3 modeled as a continuum and considering losses (Scenario A), the three-bus system remains stable against small disturbances at the given operating point.

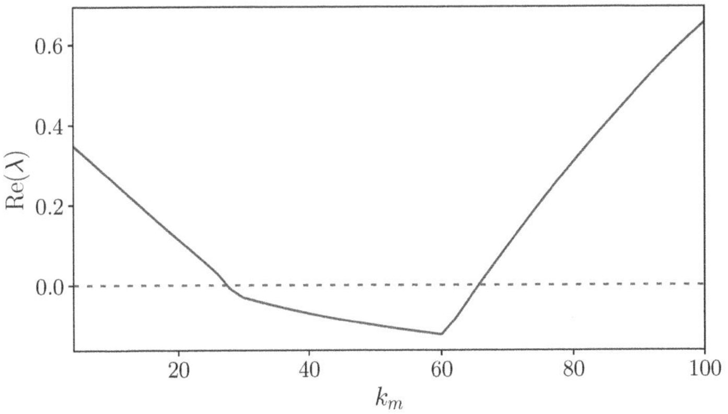

FIGURE 11.6: The real part of the rightmost eigenvalue as a function of k_{m} of the comparison RDDEs (11.31) for the three-bus system with a distributed-parameter lossy 10-km transmission line (Scenario A).

According to Proposition 11.2, the small-signal stability of the power system with lossless line modeled as a continuum (Scenario B), however, cannot be solved through the comparison RDDEs (11.21) for $\rho(\hat{\mathbf{G}}_{3,23}) = 1$ and correspondingly $\rho(\mathbf{A}_{k_{\mathrm{m}}}) = 0.01825$ holds for all $k_{\mathrm{m}} \in (2, \infty]$. Figure 11.6 shows the rightmost eigenvalue never converges, independently from k_{m}. Since Proposition 11.2 also points out that $\rho(\mathbf{G}_{3,23}) < 1$ is the necessary stability condition of the linear DDAEs, the three-bus system with lossless line modeled as a continuum is not stable following a small disturbance, which can be proved through time domain simulation.

Figure 11.7 shows the trajectories of the rotor speed of the synchronous machine following a small disturbance of the three-bus power system with different long transmission line models. The system remains stable with the lossy transmission line model (Scenario A), but collapses if the line is assumed to be lossless (Scenario B). □

11.5 Descriptor Transform for NDDEs

The descriptor transform developed by Fridman [49] depicts that the NDDEs can be transformed into retarded DDAEs under specific conditions. This section explains such a technique.

Let us consider the general autonomous NDDE model:

$$\dot{\boldsymbol{x}}(t) = \boldsymbol{f}\left(\boldsymbol{x}(t), \boldsymbol{x}_d(t), \dot{\boldsymbol{x}}_d(t)\right) , \tag{11.49}$$

where $\boldsymbol{x} \in \mathbb{R}^{n\times 1}$ and $\boldsymbol{f} : \mathbb{R}^{(n+2p)\times 1} \mapsto \mathbb{R}^{n\times 1}$.

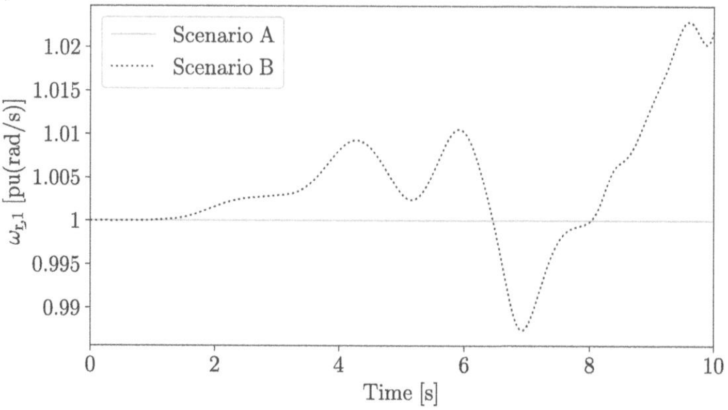

FIGURE 11.7: Trajectories of the three-bus system with different models of distributed-parameter transmission lines following a small disturbance.

For simplicity, we consider again the single delay case:

$$\boldsymbol{x}_d(t) = \boldsymbol{x}(t - \tau),$$

with $p = n$. Linearization of (11.49) at the stationary solution $\boldsymbol{x}_o$ leads to the following small-signal model:

$$\Delta\dot{\boldsymbol{x}} = \frac{\partial \boldsymbol{f}}{\partial \boldsymbol{x}}\Delta\boldsymbol{x} + \frac{\partial \boldsymbol{f}}{\partial \boldsymbol{x}_d}\Delta\boldsymbol{x}_d + \frac{\partial \boldsymbol{f}}{\partial \dot{\boldsymbol{x}}_d}\Delta\dot{\boldsymbol{x}}_d. \tag{11.50}$$

Equation (11.50) is in the form of the standard linear NDDEs (10.11) discussed in Section 10.1.2, with:

$$\mathbf{A}_0 = \frac{\partial \boldsymbol{f}}{\partial \boldsymbol{x}}, \qquad \mathbf{A}_{k_m} = \mathbf{A}_1 = \frac{\partial \boldsymbol{f}}{\partial \boldsymbol{x}_d}, \qquad \mathbf{E}_{i_m} = \mathbf{E}_1 = -\frac{\partial \boldsymbol{f}}{\partial \dot{\boldsymbol{x}}_d}.$$

Proposition 11.5 (Transformation of NDDEs into DDAEs). If $\frac{\partial \boldsymbol{f}}{\partial \boldsymbol{x}}$ is full rank, the NDDEs (11.50) have the same spectrum with the following retarded DDAEs in non-index-1 Hessenberg form at the same equilibrium point:

$$\begin{aligned} \dot{\boldsymbol{x}}(t) &= \boldsymbol{y}(t), \\ \mathbf{0}_{n,1} &= \boldsymbol{f}\big(\boldsymbol{x}(t), \boldsymbol{x}(t-\tau), \boldsymbol{y}(t-\tau)\big) - \boldsymbol{y}(t), \end{aligned} \tag{11.51}$$

where $\boldsymbol{y} \in \mathbb{R}^{n\times 1}$.

Proof. The DDAEs (11.51) can be written as

$$\tilde{\mathbf{E}}\begin{bmatrix} \dot{\boldsymbol{x}} \\ \mathbf{0}_{n,1} \end{bmatrix} = \begin{bmatrix} \boldsymbol{y} \\ \boldsymbol{f}(\boldsymbol{x}, \boldsymbol{x}_d, \boldsymbol{y}, \boldsymbol{y}_d) \end{bmatrix}, \tag{11.52}$$

where

$$\tilde{\mathbf{E}} = \begin{bmatrix} \mathbf{I}_n & \mathbf{0}_{n,n} \\ \mathbf{0}_{n,n} & \mathbf{0}_{n,n} \end{bmatrix} .$$

Then, differentiating (11.52) at the stationary solution yields

$$\tilde{\mathbf{E}} \begin{bmatrix} \dot{\boldsymbol{x}} \\ \mathbf{0}_{n,1} \end{bmatrix} = \begin{bmatrix} \Delta \boldsymbol{y} \\ \dfrac{\partial \boldsymbol{f}}{\partial \boldsymbol{x}} \Delta \boldsymbol{x} + \dfrac{\partial \boldsymbol{f}}{\partial \boldsymbol{x}_d} \Delta \boldsymbol{x}_d + \dfrac{\partial \boldsymbol{f}}{\partial \boldsymbol{y}} \Delta \boldsymbol{y} + \dfrac{\partial \boldsymbol{f}}{\partial \boldsymbol{y}_d} \Delta \boldsymbol{y}_d \end{bmatrix} ,$$

or, equivalently,

$$\tilde{\mathbf{E}} \begin{bmatrix} \dot{\boldsymbol{x}} \\ \mathbf{0}_{n,1} \end{bmatrix} = \tilde{\mathbf{A}}_0 \begin{bmatrix} \Delta \boldsymbol{x} \\ \Delta \boldsymbol{y} \end{bmatrix} + \tilde{\mathbf{A}}_1 \begin{bmatrix} \Delta \boldsymbol{x}_d \\ \Delta \boldsymbol{y}_d \end{bmatrix} , \tag{11.53}$$

where

$$\tilde{\mathbf{A}}_0 = \begin{bmatrix} \mathbf{0}_{n,n} & \mathbf{I}_n \\ \dfrac{\partial \boldsymbol{f}}{\partial \boldsymbol{x}} & \dfrac{\partial \boldsymbol{f}}{\partial \boldsymbol{y}} \end{bmatrix} , \qquad \tilde{\mathbf{A}}_1 = \begin{bmatrix} \mathbf{0}_{n,n} & \mathbf{0}_{n,n} \\ \dfrac{\partial \boldsymbol{f}}{\partial \boldsymbol{x}_d} & \dfrac{\partial \boldsymbol{f}}{\partial \boldsymbol{y}_d} \end{bmatrix} .$$

Then the characteristic equation of (11.53) is given by

$$\det\big(\lambda \tilde{\mathbf{E}} - \tilde{\mathbf{A}}_0 - \mathrm{e}^{-\lambda\tau} \tilde{\mathbf{A}}_1\big) = 0 ,$$

$$\det \left(\begin{bmatrix} \lambda \mathbf{I}_n & -\mathbf{I}_n \\ -\dfrac{\partial \boldsymbol{f}}{\partial \boldsymbol{x}} - \mathrm{e}^{-\lambda\tau} \dfrac{\partial \boldsymbol{f}}{\partial \boldsymbol{x}_d} & -\dfrac{\partial \boldsymbol{f}}{\partial \boldsymbol{y}} - \mathrm{e}^{-\lambda\tau} \dfrac{\partial \boldsymbol{f}}{\partial \boldsymbol{y}_d} \end{bmatrix} \right) = 0 .$$

Since $\boldsymbol{y} = \dot{\boldsymbol{x}}$ we have

$$\begin{aligned} \frac{\partial \boldsymbol{f}}{\partial \dot{\boldsymbol{x}}} &= \frac{\partial \boldsymbol{f}}{\partial \boldsymbol{x}} \frac{\partial \boldsymbol{x}}{\partial \dot{\boldsymbol{x}}} + \frac{\partial \boldsymbol{f}}{\partial \boldsymbol{x}_d} \frac{\partial \boldsymbol{x}_d}{\partial \dot{\boldsymbol{x}}} + \frac{\partial \boldsymbol{f}}{\partial \boldsymbol{y}} \frac{\partial \boldsymbol{y}}{\partial \dot{\boldsymbol{x}}} + \frac{\partial \boldsymbol{f}}{\partial \boldsymbol{y}_d} \frac{\partial \boldsymbol{y}_d}{\partial \dot{\boldsymbol{x}}} \\ &= \mathbf{0}_{n,n} + \mathbf{0}_{n,n} + \frac{\partial \boldsymbol{f}}{\partial \boldsymbol{y}} \mathbf{I}_n + \mathbf{0}_{n,n} \\ &= \frac{\partial \boldsymbol{f}}{\partial \boldsymbol{y}} . \end{aligned}$$

Similarly:

$$\frac{\partial \boldsymbol{f}}{\partial \dot{\boldsymbol{x}}_d} = \frac{\partial \boldsymbol{f}}{\partial \boldsymbol{y}_d} ,$$

and, thus, we have

$$\det \left(\begin{bmatrix} \lambda \mathbf{I}_n & -\mathbf{I}_n \\ -\dfrac{\partial \boldsymbol{f}}{\partial \boldsymbol{x}} - \mathrm{e}^{-\lambda\tau} \dfrac{\partial \boldsymbol{f}}{\partial \boldsymbol{x}_d} & -\dfrac{\partial \boldsymbol{f}}{\partial \dot{\boldsymbol{x}}} - \mathrm{e}^{-\lambda\tau} \dfrac{\partial \boldsymbol{f}}{\partial \dot{\boldsymbol{x}}_d} \end{bmatrix} \right) = 0 . \tag{11.54}$$

Let

$$\mathbf{F}_1 = -\frac{\partial \boldsymbol{f}}{\partial \boldsymbol{x}} - \mathrm{e}^{-\lambda\tau}\frac{\partial \boldsymbol{f}}{\partial \boldsymbol{x}_d}\,, \qquad \mathbf{F}_2 = -\frac{\partial \boldsymbol{f}}{\partial \dot{\boldsymbol{x}}} - \mathrm{e}^{-\lambda\tau}\frac{\partial \boldsymbol{f}}{\partial \dot{\boldsymbol{x}}_d}\,. \tag{11.55}$$

Then (11.54) can be written as

$$\det\left(\begin{bmatrix}\lambda\mathbf{I}_n & -\mathbf{I}_n \\ \mathbf{F}_1 & \mathbf{F}_2\end{bmatrix}\right) = 0\,,$$

or, equivalently,

$$\det\left(\begin{bmatrix}\lambda\mathbf{I}_n & \mathbf{0}_{n,n} \\ \mathbf{F}_1 & \mathbf{I}_n\end{bmatrix}\begin{bmatrix}\mathbf{I}_n & -\frac{1}{\lambda}\mathbf{I}_n \\ \mathbf{0}_{n,n} & \mathbf{F}_2 + \frac{1}{\lambda}\mathbf{F}_1\end{bmatrix}\right) = 0\,,$$

or, equivalently,

$$\det\left(\begin{bmatrix}\lambda\mathbf{I}_n & \mathbf{0}_{n,n} \\ \mathbf{F}_1 & \mathbf{I}_n\end{bmatrix}\right)\det\left(\begin{bmatrix}\mathbf{I}_n & -\frac{1}{\lambda}\mathbf{I}_n \\ \mathbf{0}_{n,n} & \mathbf{F}_2 + \frac{1}{\lambda}\mathbf{F}_1\end{bmatrix}\right) = 0\,.$$

Hence

$$\det\big(\lambda\mathbf{I}_n\big)\det\big(\mathbf{F}_2 + \frac{1}{\lambda}\mathbf{F}_1\big) = 0\,,$$

and

$$\det\big(\lambda\mathbf{F}_2 + \mathbf{F}_1\big) = 0\,.$$

Replacing (11.55) into the expression above, we obtain the following characteristic equation:

$$\det\left[\lambda\left(\frac{\partial \boldsymbol{f}}{\partial \boldsymbol{y}} + \mathrm{e}^{-\lambda\tau}\frac{\partial \boldsymbol{f}}{\partial \boldsymbol{y}_d}\right) + \frac{\partial \boldsymbol{f}}{\partial \boldsymbol{x}} + \mathrm{e}^{-\lambda\tau}\frac{\partial \boldsymbol{f}}{\partial \boldsymbol{x}_d}\right] = 0\,,$$

which is equal to the characteristic equation of (11.50). This proves that (11.50) and (11.51) have same solution and same spectrum since their stability is defined by the same characteristic equation. The proof is completed. ■

Remark 11.7. A descriptor transformation is reversible, that is, a set of retarded DDAEs in non-index-1 Hessenberg form can be always be transformed into a comparison NDDEs system [49].

Remark 11.8. According to the discussion in Section 11.3, the linearized form of the retarded DDAEs (11.51) can be further transformed into a set of RDDEs with infinitely many constant delays:

$$\Delta\dot{\boldsymbol{x}} = \mathbf{A}_0\Delta\boldsymbol{x} + \lim_{k_{\mathrm{m}}\to\infty}\sum_{k=1}^{k_{\mathrm{m}}}\mathbf{A}_k\Delta\boldsymbol{x}(t - k\tau)\,, \tag{11.56}$$

with:

$$\mathbf{A}_0 = \mathbf{G}_1, \quad \mathbf{A}_k = \mathbf{G}_3^{k-1}\mathbf{G}_4, \quad k \geq 2, \tag{11.57}$$

where

$$\mathbf{G}_1 = \frac{\partial \boldsymbol{f}}{\partial \boldsymbol{x}}, \quad \mathbf{G}_2 = \frac{\partial \boldsymbol{f}}{\partial \boldsymbol{x}_d}, \quad \mathbf{G}_3 = \frac{\partial \boldsymbol{f}}{\partial \boldsymbol{y}_d}, \quad \mathbf{G}_4 = \mathbf{G}_2 + \mathbf{G}_3\mathbf{G}_1 .$$

As discussed in Proposition 11.3 the series in (11.56) converges if and only if $\rho(\mathbf{G}_4) < 1$. The proper $k_{\rm m}$ to truncate the series in (11.56) can be found based on the fact that (10.11) holds the same stability properties as the transformed retarded non-index-1 Hessenberg form DDAEs (11.51).

Corollary 11.3 (Approximation of the eigenvalues of the NDDEs)**.** Consider the set of NDDEs (11.50) and that the transformation in Proposition 11.5 is feasible. If the infinite series in (11.56) converges, and there is a finite $k_{\rm m}$ such that $\mathbf{H}_m$ converges to $\hat{\mathbf{H}}$, where

$$\mathbf{H}_m = \mathbf{A}_0 + \sum_{k=1}^{k_{\rm m}} \mathbf{A}_k ,$$

and

$$\hat{\mathbf{H}} = -\left(\frac{\partial \boldsymbol{f}}{\partial \dot{\boldsymbol{x}}} + \frac{\partial \boldsymbol{f}}{\partial \dot{\boldsymbol{x}}_d}\right)^{-1} \left(\frac{\partial \boldsymbol{f}}{\partial \boldsymbol{x}} + \frac{\partial \boldsymbol{f}}{\partial \boldsymbol{x}_d}\right),$$

then the eigenvalues of (11.56) with the finite $k_{\rm m}$ approximate accurately the eigenvalues of the NDDEs.

Remark 11.9. Since a proper $k_{\rm m}$ ensures that the spectrum of (11.56) converges to (11.50), Corollary 11.3 considers the specific scenario $\tau = 0$, where the state matrix of the (11.50) is $\hat{\mathbf{H}}$, and the state matrix of the transformed RDDEs (11.56) is $\mathbf{H}_m$.

Remark 11.10. If $\mathbf{H}_m$ converges to $\hat{\mathbf{H}}$, we must have:

$$\rho(\mathbf{H}_m - \hat{\mathbf{H}}) \leq \epsilon , \tag{11.58}$$

where ϵ is a small tolerance close to zero.

According to the discussion above, the delay-independent stability criterion of the RDDEs with commensurate delay, i.e. Proposition 10.1, can be also extended to the NDDEs (10.11):

Proposition 11.6 (Delay-independent stability of NDDEs)**.** The linear NDDEs (11.50) are delay-independent stable if and only if the following conditions are satisfied:

- The matrix $\mathbf{A}_0 = \frac{\partial \boldsymbol{f}}{\partial \boldsymbol{x}}$ is stable;

- The matrix $\hat{\mathbf{H}}$ is stable;
- The spectral radius $\rho(\frac{\partial \boldsymbol{f}}{\partial \boldsymbol{y}_d}) < 1$.
- The spectral radius of the pencil $s\mathbf{F} - \mathbf{G}$ is less than 1, where

$$\mathbf{F} = \mathrm{diag}(\mathbf{I}_n, \ \ldots, \ \mathbf{I}_n, \ \jmath\omega\mathbf{I}_n - \mathbf{A}_0), \quad \forall\omega \in (0, \infty] ,$$

and

$$\mathbf{G} = \begin{bmatrix} \mathbf{0}_{n,n} & \mathbf{I}_n & \ldots & \mathbf{0}_{n,n} \\ \mathbf{0}_{n,n} & \mathbf{0}_{n,n} & \ldots & \mathbf{0}_{nn} \\ \vdots & \vdots & \ddots & \vdots \\ \mathbf{0}_{n,n} & \mathbf{0}_{n,n} & \ldots & \mathbf{I}_n \\ \mathbf{A}_{k_\mathrm{m}} & \mathbf{A}_{k_\mathrm{m}-1} & \ldots & \mathbf{A}_1 \end{bmatrix} ,$$

k_m is a finite number that ensures convergence from $\mathbf{H}_m$ to $\hat{\mathbf{H}}$, and $\mathbf{A}_1, \mathbf{A}_2 \ldots \mathbf{A}_{k_\mathrm{m}}$ are defined in (11.57).

Remark 11.11. According to Corollary 10.1, we can conclude that, if $\mathbf{A}_0$ is non-singular, we can always deduce the upper bound of the spectral radius of the pencil $s\mathbf{F} - \mathbf{G}$ as $\rho(\mathbf{F}^{-1}\mathbf{G})|_{\omega=0}$.

Example 11.3 (Partial-element equivalent circuit)**.** Example 10.1 illustrates a PEEC modeled as a set of NDDEs. This example revisits Example 10.1 by applying the descriptor transform to the PEEC model.

According to Proposition 11.5, we can deduce the comparison DDAEs of the original model (10.29):

$$\begin{aligned} \dot{\boldsymbol{\imath}}(t) &= \boldsymbol{y}(t) \\ \mathbf{0}_{3,1} &= \mathbf{L}\,\boldsymbol{\imath}(t) + \mathbf{M}\,\boldsymbol{\imath}(t-\tau) + \mathbf{N}\,\boldsymbol{y}(t-\tau) - \boldsymbol{y}(t) , \end{aligned} \tag{11.59}$$

where $\boldsymbol{y} \in \mathbb{R}^{3\times 1}$ is a vector of auxiliary algebraic variables.

Since (11.59) is a retarded DDAE system in non-index-1 Hessenberg form, according to Remark 11.8, it can be transformed into the following RDDEs:

$$\dot{\boldsymbol{\imath}}(t) = \mathbf{A}_0\,\boldsymbol{\imath}(t) + \lim_{k_\mathrm{m}\to\infty} \sum_{k=1}^{k_\mathrm{m}} \mathbf{A}_k\,\boldsymbol{\imath}(t - k\tau) , \tag{11.60}$$

where

$$\mathbf{A}_0 = \mathbf{L}, \quad \text{and} \quad \mathbf{A}_k = \mathbf{N}^{k-1}(\mathbf{M} + \mathbf{N}\mathbf{L}) .$$

Since $\rho(\mathbf{N}) = 0.0762 < 1$, the series in (11.60) can be truncated at a finite k_m without loss of the accuracy in the calculation of the eigenvalues. Table 11.2 shows the rightmost eigenvalues of (11.60) with different k_m for the scenario $\tau = 1$ s. According to the QPMR method, the rightmost eigenvalues of this scenario are $-0.7376 \pm \jmath 3.1329$.

TABLE 11.2: Approximated rightmost eigenvalues of the PEEC (10.29) obtained by solving the comparison RDDEs with different k_{m} with the Krylov method.

k_{m}	$\rho(\mathbf{H}_m - \hat{\mathbf{H}})$	Rightmost Eigenvalue	ϵ^* [%]
3	0.0951	$-0.7663 \pm \jmath 3.1323$	3.891
4	0.0069	$-0.7661 \pm \jmath 3.1323$	3.863
5	$5.11 \cdot 10^{-4}$	$-0.7338 \pm \jmath 3.1330$	0.515
6	$3.73 \cdot 10^{-5}$	$-0.7383 \pm \jmath 3.1329$	0.094
7	$2.75 \cdot 10^{-6}$	$-0.7375 \pm \jmath 3.1329$	0.0135
8	$2.01 \cdot 10^{-7}$	$-0.7377 \pm \jmath 3.1329$	0.0135
9	$1.48 \cdot 10^{-8}$	$-0.7376 \pm \jmath 3.1329$	0
10	$1.08 \cdot 10^{-9}$	$-0.7376 \pm \jmath 3.1329$	0

ϵ^*: Relative error of the real part.

Following Corollary 11.3, for the PEEC, we have:

$$\mathbf{H}_m = \mathbf{L} + \sum_{k=1}^{k_{\mathrm{m}}} (-\mathbf{N})^{k-2}(M + \mathbf{NL}),$$

$$\hat{\mathbf{H}} = (\mathbf{I}_3 - \mathbf{N})^{-1}(\mathbf{L} + \mathbf{M}).$$

The values of $\rho(\mathbf{H}_m - \hat{\mathbf{H}})$ for different k_{m} are also shown in Table 11.2. For $k_{\mathrm{m}} \geq 9$, the rightmost eigenvalue of (11.60) converges to that of the original PEEC model. $k_{\mathrm{m}} = 9$ also leads to a very small $\rho(\mathbf{H}_m - \hat{\mathbf{H}})$, which implies convergence from $\mathbf{H}_m$ to $\hat{\mathbf{H}}$.

Let us discuss the delay-independent stability of the PEEC for $k_{\mathrm{m}} = 9$ through the conditions given in Proposition 11.6:

- The matrix $\mathbf{A}_0 = \mathbf{L}$ is stable with rightmost eigenvalue -407.92.
- The matrix $\hat{\mathbf{H}}$ is stable with rightmost eigenvalue $-583.14 \pm \jmath 95.97$.
- The spectral radius is $\rho(\mathbf{N}) = 0.0762 < 1$.
- With $\omega = 0$, the upper bound of the spectral radius of the pencil $s\mathbf{F} - \mathbf{G}$ is 0.9424, which is less than 1.

All conditions are thus satisfied. These results are consistent with the discussions in Example 10.1. □

11.6 Neutral DDAEs

This section discusses a transform technique for DDAEs of neutral type. This is based on the descriptor transform introduced in the previous section.

Consider the following neutral DDAEs with a single delay:

$$\begin{aligned} \dot{\boldsymbol{x}}(t) &= \boldsymbol{f}\big(\boldsymbol{x}(t), \boldsymbol{y}(t), \boldsymbol{x}_d(t), \boldsymbol{y}_d(t), \dot{\boldsymbol{x}}_d(t)\big) \\ \mathbf{0}_{l,1} &= \boldsymbol{g}\big(\boldsymbol{x}(t), \boldsymbol{y}(t), \boldsymbol{x}_d(t), \boldsymbol{y}_d(t), \dot{\boldsymbol{x}}_d(t)\big), \end{aligned} \tag{11.61}$$

where $\boldsymbol{f} : \mathbb{R}^{(3n+2l)\times 1} \mapsto \mathbb{R}^{n\times 1}$ are the differential equations; $\boldsymbol{g} : \mathbb{R}^{(3n+2l)\times 1} \mapsto \mathbb{R}^{l\times 1}$ are the algebraic equations; and

$$\boldsymbol{x}_d(t) = \boldsymbol{x}(t-\tau), \quad \text{and} \quad \boldsymbol{y}_d(t) = \boldsymbol{y}(t-\tau).$$

Consistently with the descriptor transform technique given in Section 11.5, we introduce the following algebraic variables:

$$\dot{\boldsymbol{x}}(t) = \tilde{\boldsymbol{y}}(t). \tag{11.62}$$

With the transform (11.62), the neutral DDAEs (11.61) become:

$$\begin{aligned} \dot{\boldsymbol{x}}(t) &= \boldsymbol{f}\big(\boldsymbol{x}(t), \boldsymbol{y}(t), \boldsymbol{x}_d(t), \boldsymbol{y}_d(t), \tilde{\boldsymbol{y}}_d(t)\big) \\ \mathbf{0}_{l,1} &= \boldsymbol{g}\big(\boldsymbol{x}(t), \boldsymbol{y}(t), \boldsymbol{x}_d(t), \boldsymbol{y}_d(t), \tilde{\boldsymbol{y}}_d(t)\big). \end{aligned} \tag{11.63}$$

In order to transform (11.63) into a standard set of retarded DDAEs in non-index-1 Hessenberg form, we consider:

$$\hat{\boldsymbol{y}} = [\boldsymbol{y}, \ \tilde{\boldsymbol{y}}]^{\mathrm{T}}, \tag{11.64}$$

where $\hat{\boldsymbol{y}} \in \mathbb{R}^{(l+n)\times 1}$. Then, we can reformulate (11.63) as follows:

$$\begin{aligned} \dot{\boldsymbol{x}} &= \boldsymbol{\mathcal{F}}(\boldsymbol{x}, \boldsymbol{x}_d, \hat{\boldsymbol{y}}, \hat{\boldsymbol{y}}_d) \\ \mathbf{0}_{l,1} &= \boldsymbol{\mathcal{G}}(\boldsymbol{x}, \boldsymbol{x}_d, \hat{\boldsymbol{y}}, \hat{\boldsymbol{y}}_d), \end{aligned} \tag{11.65}$$

where $\boldsymbol{\mathcal{F}} : \mathbb{R}^{(4n+2l)\times 1} \mapsto \mathbb{R}^{n\times 1}$; $\boldsymbol{\mathcal{G}} : \mathbb{R}^{(4n+2l)\times 1} \mapsto \mathbb{R}^{(l+n)\times 1}$; $\boldsymbol{x} \in \mathbb{R}^{n\times 1}$, $\hat{\boldsymbol{y}} \in \mathbb{R}^{(l+n)\times 1}$. The transformed system (11.65) is in the form of retarded DDAEs (11.17) (see Section 11.3). The corresponding linearized equations of (11.65) at a given equilibrium point are:

$$\boldsymbol{x}(0) = \boldsymbol{x}_d(0) = \boldsymbol{x}_o, \qquad \hat{\boldsymbol{y}}(0) = \hat{\boldsymbol{y}}_d(0) = [\boldsymbol{y}_o, \boldsymbol{x}_o]^{\mathrm{T}},$$

can be deduced as:

$$\begin{aligned} \Delta\dot{\boldsymbol{x}} &= \frac{\partial \boldsymbol{\mathcal{F}}}{\partial \boldsymbol{x}}\Delta\boldsymbol{x} + \frac{\partial \boldsymbol{\mathcal{F}}}{\partial \boldsymbol{x}_d}\Delta\boldsymbol{x}_d + \frac{\partial \boldsymbol{\mathcal{F}}}{\partial \hat{\boldsymbol{y}}}\Delta\hat{\boldsymbol{y}} + \frac{\partial \boldsymbol{\mathcal{F}}}{\partial \hat{\boldsymbol{y}}_d}\Delta\hat{\boldsymbol{y}}_d \\ \mathbf{0}_{l+n,1} &= \frac{\partial \boldsymbol{\mathcal{G}}}{\partial \boldsymbol{x}}\Delta\boldsymbol{x} + \frac{\partial \boldsymbol{\mathcal{G}}}{\partial \boldsymbol{x}_d}\Delta\boldsymbol{x}_d + \frac{\partial \boldsymbol{\mathcal{G}}}{\partial \hat{\boldsymbol{y}}}\Delta\hat{\boldsymbol{y}} + \frac{\partial \boldsymbol{\mathcal{G}}}{\partial \hat{\boldsymbol{y}}_d}\Delta\hat{\boldsymbol{y}}_d. \end{aligned} \tag{11.66}$$

According to Proposition 11.5, we can further transform (11.66) into a set of RDDEs with infinitely many constant delays. Since we limit our discussion to the case of a single delay τ, we can deduce the following Corollary to summarize this transform:

Corollary 11.4 (Approximation of the eigenvalues of the neutral DDAEs). The eigenvalues of the neutral DDAEs (11.61) at the given point $(\boldsymbol{x}_o, \boldsymbol{y}_o)$ can be estimated through the comparison RDDEs:

$$\Delta\dot{\boldsymbol{x}} = \mathbf{A}_0 \Delta\boldsymbol{x} + \mathbf{A}_1 \Delta\boldsymbol{x}(t-\tau) + \sum_{k=2}^{\infty} \mathbf{A}_k \Delta\boldsymbol{x}(t-k\tau)\,, \tag{11.67}$$

where

$$\begin{aligned}
\mathbf{A}_0 &= \frac{\partial \boldsymbol{\mathcal{F}}}{\partial \boldsymbol{x}} + \frac{\partial \boldsymbol{\mathcal{F}}}{\partial \boldsymbol{y}} \mathbf{G}_1\,, \\
\mathbf{A}_1 &= \frac{\partial \boldsymbol{\mathcal{F}}}{\partial \boldsymbol{x}_d} + \frac{\partial \boldsymbol{\mathcal{F}}}{\partial \boldsymbol{y}_d} \mathbf{A} + \frac{\partial \boldsymbol{\mathcal{F}}}{\partial \boldsymbol{y}} \mathbf{G}_4\,, \\
\mathbf{A}_k &= \mathbf{F}_1 \mathbf{G}_3^{k-2} \mathbf{G}_4\,, \qquad k \geq 2\,,
\end{aligned}$$

with

$$\begin{aligned}
\mathbf{G}_1 &= -\left(\frac{\partial \boldsymbol{\mathcal{G}}}{\partial \boldsymbol{y}}\right)^{-1} \frac{\partial \boldsymbol{\mathcal{G}}}{\partial \boldsymbol{x}}\,, & \mathbf{G}_2 &= -\left(\frac{\partial \boldsymbol{\mathcal{G}}}{\partial \boldsymbol{y}}\right)^{-1} \frac{\partial \boldsymbol{\mathcal{G}}}{\partial \boldsymbol{x}_d}\,, \\
\mathbf{G}_3 &= -\left(\frac{\partial \boldsymbol{\mathcal{G}}}{\partial \boldsymbol{y}}\right)^{-1} \frac{\partial \boldsymbol{\mathcal{G}}}{\partial \boldsymbol{y}_d}\,, & \mathbf{G}_4 &= \mathbf{G}_2 + \mathbf{G}_3 \mathbf{G}_1\,, \\
\mathbf{F}_1 &= \frac{\partial \boldsymbol{\mathcal{F}}}{\partial \boldsymbol{y}} \mathbf{G}_3 + \frac{\partial \boldsymbol{\mathcal{F}}}{\partial \boldsymbol{y}_d}\,.
\end{aligned}$$

Remark 11.12. The series in (11.67) converges if $\rho(\mathbf{G}_3) < 1$. If the series converges, then it can be truncated at a finite k_{m}, and the rightmost eigenvalues can be found through the SDM discussed in Section 11.3.

11.7 Delay Compensation in Power Systems

In previous examples, we have shown how delays reduce, in general, the stability margin of the controllers and can lead a system to instability. A well-known technique to reduce the impact of delays is *delay compensation*.

Among the various solutions based on delay compensation, we consider in this section techniques that are based on the derivative of the measured signal. Such techniques are particularly relevant for the matter discussed in this chapter as they require, as we will show, information on the first derivative of delayed variables. The transient behavior of dynamic systems with derivative-based delay compensation, therefore, can be modeled through neutral DDAEs.

In this section, we describe two well-known model-independent derivative-based delay compensation methods that can be implemented in physical systems and, in particular, in power systems. These are the proportional-derivative (PD) delay compensation method [162] and the predictor delay compensation method [184].

PD Delay Compensation

Consider a signal $x(t)$, $x \in \mathbb{R}$. If the delayed signal is $x_d(t) = x(t-\tau)$, the signal compensated through a PD controller is:

$$x_{\mathrm{com_A}}(t) = x_d(t) + K_\tau \, \dot{x}_d(t) \,, \tag{11.68}$$

where K_τ is the compensation gain. The control diagram of this compensation is shown in Figure 11.8.

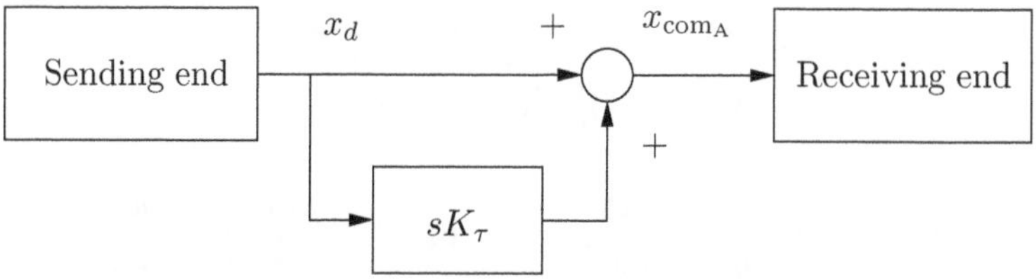

FIGURE 11.8: PD delay compensation diagram.

The best dynamic performance of the compensation is achieved for a value of K_τ that is around the magnitude of the delay τ. The rationale of this rule is given below. We have:

$$\begin{aligned} x(t) &= x(t - \tau + \tau) \\ &\approx x_d(t) + \tau \, \dot{x}(t) + \mathcal{O}(\tau^2) \,, \end{aligned} \tag{11.69}$$

where $\mathcal{O}(\tau^2) = \frac{\tau^2}{2}\ddot{x}(t)$. If x is a state variable and the constant delay τ is small, namely $\frac{\tau^2}{2}\ddot{x}(t) \to 0$, the following approximation holds:

$$x(t) \approx x_d(t) + \tau \, \dot{x}(t) \,. \tag{11.70}$$

At the receiving end, the time derivative of the non-delayed signal is unknown, so we assume the following approximation:

$$\dot{x}(t) \approx \dot{x}(t - \tau) \,, \tag{11.71}$$

which leads to rewrite (11.70) as:

$$x(t) \approx x_d(t) + \tau \, \dot{x}_d(t) \,. \tag{11.72}$$

Comparing (11.72) with (11.68), it appears that the optimal choice for the gain K_τ should be around τ. To evaluate the performance of the PD compensation method, we introduce $K_s = \tau K_\tau$.

Assuming a set of linear RDDEs (11.15) with a single constant delay and $\mathbf{A}_2 = \mathbf{0}_{n,n}$ and by applying the compensation on the delayed variable, we obtain the following set of NDDEs:

$$\dot{\boldsymbol{x}}(t) = \mathbf{A}_0\, \boldsymbol{x}(t) + \mathbf{A}_1\, \boldsymbol{x}(t-\tau) + K_\tau\, \mathbf{A}_1\, \dot{\boldsymbol{x}}(t-\tau)\,. \tag{11.73}$$

Since (11.73) is a typical set of NDDEs, according to Proposition 11.5, we can define the following comparison RDDEs of (11.73):

$$\dot{\boldsymbol{x}}(t) = \mathbf{A}_0\, \boldsymbol{x}(t) + \lim_{k_{\rm m}\to\infty} \sum_{k=1}^{k_{\rm m}} \hat{\mathbf{A}}_k\, \boldsymbol{x}(t-k\tau)\,, \tag{11.74}$$

where

$$\hat{\mathbf{A}}_k = (K_\tau \mathbf{A}_1)^{k-1}(\mathbf{A}_0 + K_\tau \mathbf{A}_1^2)\,. \tag{11.75}$$

If and only if the series of $\hat{\mathbf{A}}_k$ converges, i.e. $\rho(K_\tau \mathbf{A}_1) < 1$, we can truncate the series at a finite $k_{\rm m}$ in eigenvalue analysis.

Predictor delay compensation

Reference [184] proposes the following predictor delay compensation:

$$\dot{x}_{\rm com_B}(t) = \dot{x}_d(t) + K_a\big[x(t-\tau) - x_{\rm com_B}(t-\tau)\big]\,. \tag{11.76}$$

The corresponding control scheme is Figure 11.9.

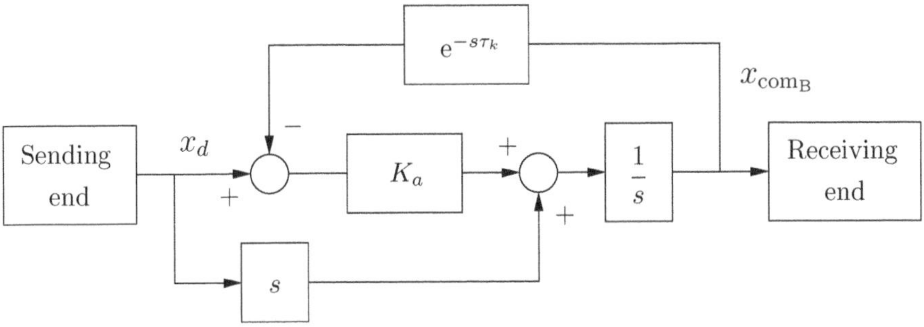

FIGURE 11.9: Predictor-based delay compensation diagram.

Implementing the predictor into a control system introduces an extra NDDE (11.76). With the descriptor transform, we reformulate the NDDE (11.76) as:

$$\begin{aligned} \dot{x} &= y\,, \\ \dot{x}_{\rm com_B}(t) &= y_d(t) + K_a\big[x(t-\tau) - x_{\rm com_B}(t-\tau)\big]\,. \end{aligned} \tag{11.77}$$

The comparison system of (11.77) is thus a set of DDAEs in non-index-1 Hessenberg form.

The estimation error of the predictor compensation is:

$$\varepsilon(t) = x(t) - x_{\mathrm{com}_\mathrm{B}}(t)\,. \tag{11.78}$$

According to (11.76) and (11.78), we can deduce:

$$\dot{\varepsilon}(t) = -K_a(x_d(t) - x_{\mathrm{com}_\mathrm{B}}(t-\tau)) + d(t)\,, \tag{11.79}$$

or, equivalently,

$$\dot{\varepsilon}(t) = -K_a \varepsilon(t-\tau) + d(t)\,, \tag{11.80}$$

where

$$d(t) = \dot{x}(t) - \dot{x}(t-\tau)\,.$$

The estimation error converges if and only if $\dot{\varepsilon}(\infty) \to 0$. Applying the Laplace transform to (11.80) leads to

$$s\varepsilon(s) + K_a \varepsilon(s)\,\mathrm{e}^{-s\tau} - d(s) = 0\,. \tag{11.81}$$

Clearly, the estimation error $\varepsilon(t)$ converges to zero if and only if the roots of (11.81) have negative real parts. For small τ that leads to $d(t) \approx 0$ and correspondingly $d(s) \to 0$, the pencil (11.81) becomes:

$$s + K_a\,\mathrm{e}^{-s\tau} = 0\,. \tag{11.82}$$

According to [63], with $K_a \in (0, \pi/2\tau)$, there always exist eigenvalues of (11.82) with negative real part. Therefore, the following Lemma holds:

Lemma 11.1 (Upper bound of the estimation error)**.** There must exist an upper bound $K_{s,m}$ for $K_s = \tau K_a$, that ensures the convergence of the estimation error of the predictor (11.76). If τ is *small* enough, we have $K_{s,m} = \pi/2$.

Remark 11.13. Although the upper bound for the small-delay scenario is deduced by Lemma 11.1, a general method to define *small* delays is still missing. Moreover, according to the numerical tests, such an upper bound may change and become larger as $d(t)$ increases. The upper bound $K_{s,m} = \frac{\pi}{2}$ is a conservative choice to ensure the convergence of the predictor error [184]. The exact boundary for various scenarios should be obtained through numerical tests.

Example 11.4 (Three-bus system with delay compensation)**.** In this example, we solve the small-signal stability analysis for three-bus system with the two derivative-based delay compensation methods discussed above. The small-signal stability analysis consists in computing the rightmost eigenvalues of their comparison DDAEs through the Krylov method.

In the following, we assume:

- For PD compensation, $K_s = \frac{K_\tau}{\tau}$;
- For predictor-based compensation, $Ks = \tau K_a$.

Figure 11.10 shows the stability maps for the three-bus system with PD and predictor-based delay compensation. Both compensation methods improve the delay margin of the power system. For $K_s \in [0.72, 0.96]$, the PD compensation method increases the delay margin from 81 ms to 144 ms. The predictor-based method performs less effectively, as it increases the delay margin to 127 ms with more strict requirements for the compensation gain. For the three-bus system, the choice of the compensation gain for predictor-based delay compensation is more difficult than that of the PD compensation. The scenarios that are stable without delay compensation can collapse by applying a predictor-based compensation without optimal gain, while the PD compensation avoids this issue at least for $K_s \in [0, 2]$.

To further discuss the different performance of these two compensation methods, Figure 11.11 shows the real part of the rightmost eigenvalue of the three-bus system with various compensation gains in the scenario that $\tau = 127$ ms. The PD compensation stabilizes the system for $K_s \in [0.5, 1.5]$, and provides the most negative rightmost eigenvalue for $K_s \in [0.9, 1.1]$. The predictor-based delay compensation can stabilize the system only for $K_s \in (2.0, 2.1]$.

Figure 11.12 shows the rightmost eigenvalues in the complex region $\mathrm{Re}(\lambda) \leq 1.5$, $|\mathrm{Im}(\lambda)| \leq 12$ for the three-bus system with $\tau = 127$ ms. The compensation gain for the two methods is the optimal choice obtained through a series of tests (see Figure 11.11), namely $K_s = 1$ for the PD and $K_s = 2.05$. With such optimal compensation gains, the two derivative-based delay compensation methods stabilize the system by eliminating the pair of positive eigenvalues but introduce poorly damped extra eigenvalues. □

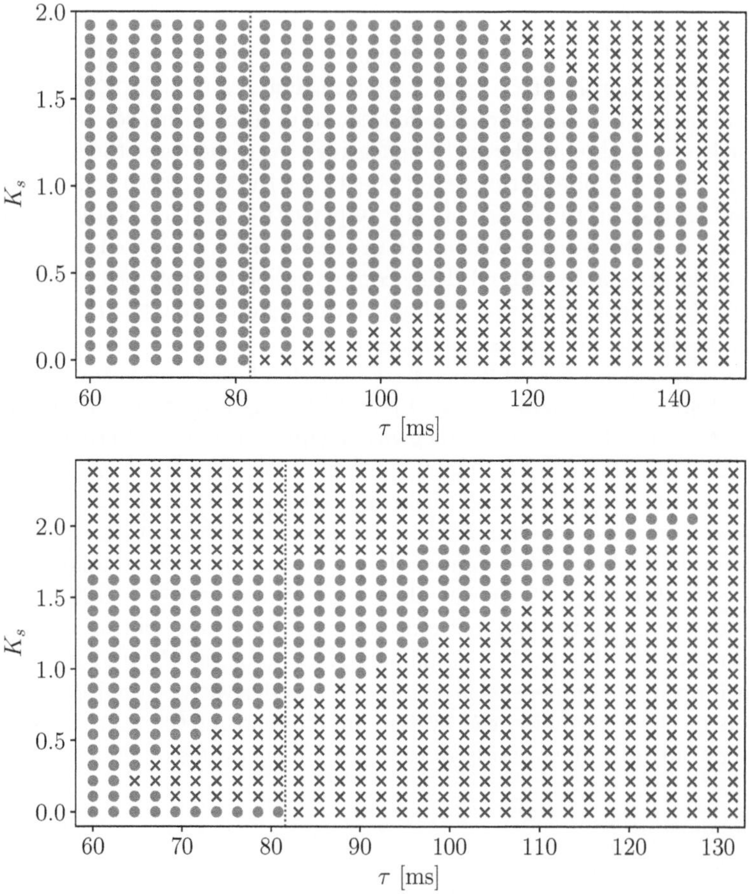

FIGURE 11.10: Stability maps for the three-bus system with constant delay τ compensated by the PD (upper panel) and the predictor-based compensations (lower panel). The crosses represent unstable scenarios, the circles represent stable scenarios. The dashed line is the delay margin for the three-bus system without compensation.

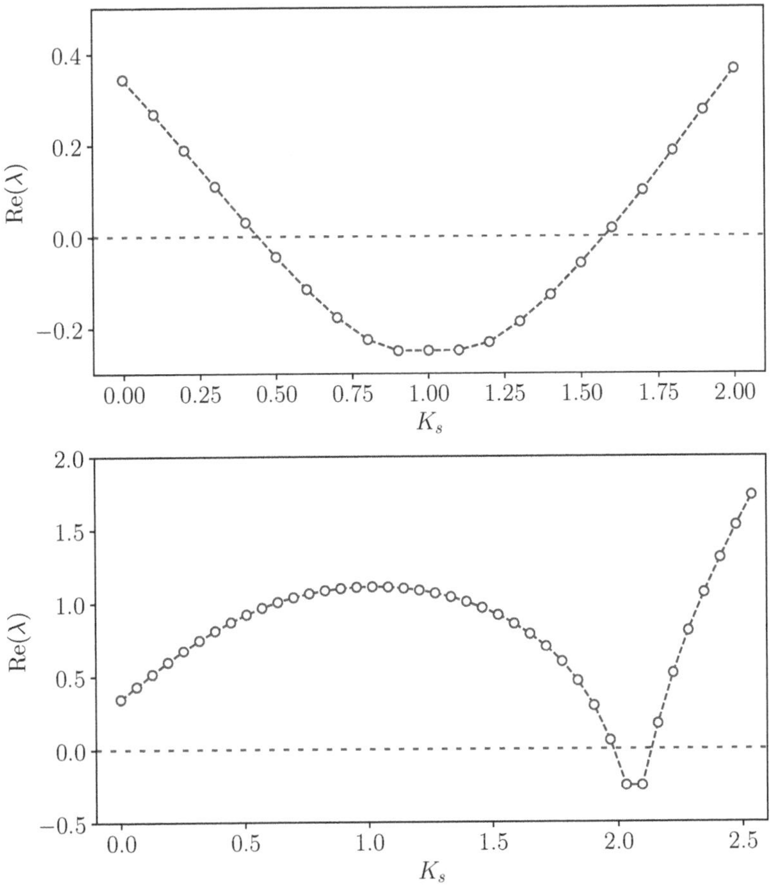

FIGURE 11.11: Real part of the rightmost eigenvalue as a function of compensation gain for the three-bus system with $\tau = 127$ ms and compensated by the PD (upper panel) and the predictor-based compensation (lower panel).

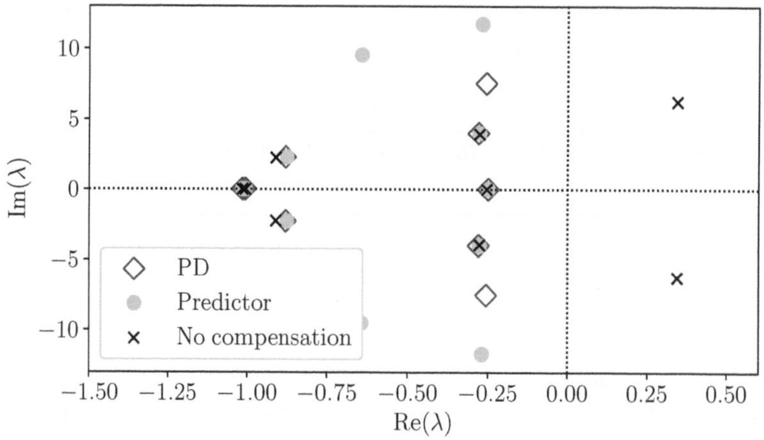

FIGURE 11.12: Rightmost eigenvalues for the three-bus system with $\tau = 127$ ms.

12

Systems with Time-Varying Delays

12.1 Analysis of Time-Varying Delay Systems

Chapters 10 and 11 discuss a variety of methods for the eigenvalue analysis of constant time-delay systems. Most delays that appear in physical systems, however, vary with time. The utilization of a constant delay model for the analysis of physical systems certainly helps understand their basic features and the potential impact of the delays. For the stability assessment of specific scenarios, however, the simplification of the delay model may lead to erroneous results due to the so-called quenching phenomenon (QP). The QP on time-delay systems was firstly discussed in [99], and a clear definition of QP is provided in [136], as follows:

Definition 12.1 (Quenching phenomenon). The quenching phenomenon (QP) of a delay system means that if the system is unstable with inclusion of constant delays $\tau_k \in [\tau_{\min}, \tau_{\max}]$, $k \in \mathbb{R}^+$, it may become stable if some of the constant delays τ_k are replaced by time-varying delays $\hat{\tau}_k(t) \in [\tau_{\min}, \tau_{\max}]$; and, on the other hand, if the system is stable with inclusion of τ_k, it may become unstable if some of τ_k are replaced by $\hat{\tau}_k(t)$.

In other words, assuming that constant and time-varying delays are within the same range and have the same mean values, they may have a different effect on the stability of the system.

A possible strategy to solve the eigenvalue analysis of a non-constant delay system is by transforming the non-constant delay system into a comparison constant delay system. In this vein, this chapter describes a technique to transform a time-varying delay system (TVDS) into a multiple constant delay system (MCDS) with the same spectrum. The steps of such a technique are shown in Figure 12.1. The TVDS is first transformed into an equivalent distributed delay system (DDS) and the latter into a comparison MCDS. Finally, the critical eigenvalues of the MCDS are estimated through any of the techniques described in Chapters 10 and 11. With this aim, in this chapter, we utilize the spectral discretization of RDDEs to solve the eigenvalue analysis of the comparison MCDS.

The following Corollary holds for TVDSs that can be transformed into MCDSs.

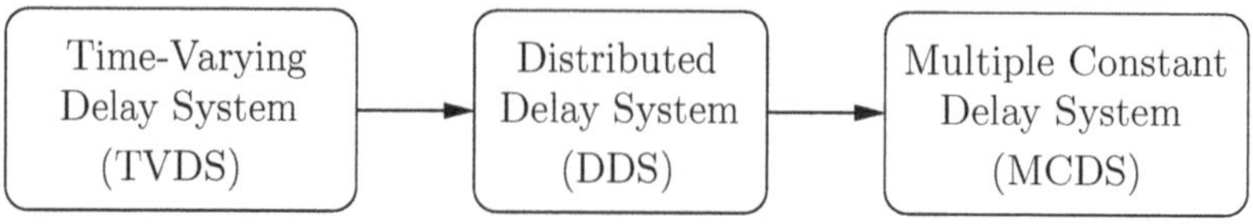

FIGURE 12.1: Steps to transform a TVDS into an MCDS.

Corollary 12.1 (Delay-independent stability of TVDSs)**.** If a system is delay-independent stable with the assumption that *all* the delays introduced in the system are constant, then the system that arises if the constant delays are replaced by time-varying delays of any type, is also delay-independent stable.

Remark 12.1. Corollary 12.1 implies that Proposition 10.1 also applies to TVDSs.

12.2 RDDEs with Distributed Delays

A distributed delay is a theoretical model that describes the probabilistic distribution of an ever-changing delay [104]. This model finds applications in control theory [125], economics [12], sociology [105] and pharmacodynamics [73]. It is usually presented as a continuous integral function with a specific probability distribution.

Consider a set of linear RDDEs with distributed delays in the following form:

$$\dot{\boldsymbol{x}}(t) = \mathbf{A}_0\, \boldsymbol{x}(t) + \sum_{k=1}^{k_\mathrm{m}} \mathbf{A}_k \int_{\tau_{\mathrm{min},k}}^{\tau_{\mathrm{max},k}} \mathrm{pdf}_i(\vartheta)\, \boldsymbol{x}(t-\vartheta)\, d\vartheta\,, \tag{12.1}$$

where $x \in \mathbb{R}^n$ and $\mathrm{pdf}_k(\vartheta)$ is the probability density function (PDF) of ϑ, such that:

$$\frac{1}{\tau_\mathrm{max} - \tau_\mathrm{min}} \int_{\tau_\mathrm{min}}^{\tau_\mathrm{max}} \mathrm{pdf}(\vartheta)\, d\vartheta = 1\,. \tag{12.2}$$

For simplicity, the following discussions are based on the single delay case of (12.1), namely $k_\mathrm{m} = 1$.

Based on the definition of integral function, the following corollary holds:

Corollary 12.2 (Comparison system of RDDEs with distributed delays)**.** The RDDEs with a distributed delay:

$$\dot{\boldsymbol{x}}(t) = \mathbf{A}_0\, \boldsymbol{x}(t) + \mathbf{A}_1 \int_{\tau_\mathrm{min}}^{\tau_\mathrm{max}} \mathrm{pdf}(\vartheta)\, \boldsymbol{x}(t-\vartheta)\, d\vartheta\,, \tag{12.3}$$

have the same spectrum as the comparison system:

$$\dot{\boldsymbol{x}}(t) = \mathbf{A}_0\, \boldsymbol{x}(t) + \mathbf{A}_1 \lim_{z_\mathrm{m}\to\infty} \mathfrak{h} \sum_{z=0}^{z_\mathrm{m}} \mathrm{pdf}(\Theta_z)\, \boldsymbol{x}(t-\Theta_z)\,, \tag{12.4}$$

where

$$\mathfrak{h} = \frac{\tau_{\max} - \tau_{\min}}{z_{\mathrm{m}}}, \qquad \Theta_z = \tau_{\min} + z\,\mathfrak{h}\,.$$

Remark 12.2. Corollary 12.2 proves that a DDS can be transformed into a comparison MCDS.

The comparison MCDS (12.4) is deduced based on a linear interpolation of the integral term of (12.3). If the number of the interpolation points (z_{m}) is large enough, the spectrum of the MCDS will converge to the original system. Numerically, there exists a finite number, say $\tilde{z}$, such that is $z_{\mathrm{m}} > \tilde{z}$, the spectrum of (12.4) converges to that of (12.3). Under this hypothesis, since (12.4) is a set of linear RDDEs with multiple constant delays, it can be studied through the eigenvalue analysis methods discussed in Chapter 10.

Higher order interpolation methods can help reduce the value of z_{m} that makes (12.4) converge to (12.3). Considering a generic interpolation method and a finite z_{m}, (12.4) can be rewritten as

$$\dot{\boldsymbol{x}}(t) = \mathbf{A}_0\,\boldsymbol{x}(t) + \mathbf{A}_1\,\mathfrak{h}\sum_{z=1}^{z_{\mathrm{m}}} \mathfrak{p}_z\,\mathrm{pdf}(\Theta_z)\,\boldsymbol{x}(t-\Theta_z)\,, \tag{12.5}$$

with $\mathfrak{p}_z$ are the coefficients of the interpolation. The quadratic and cubic interpolations are very common [159] and are outlined next.

The quadratic interpolation, which is also known as Lagrange polynomial interpolation or Simpson rule, requires that z_{m} is even. Assuming $z_{\mathrm{m}} = 2k$, $k \in \mathbb{Z}^+$ and $\kappa = \frac{1}{3}$, $\mathfrak{p}_z$ is defined as follows:

$$\mathfrak{p}_z = \begin{cases} \kappa, & \text{if} \quad z = 0 \quad \text{or} \quad z = z_{\mathrm{m}}\,, \\ 4\kappa, & \text{if} \quad z = 1+\gamma \quad \text{or} \quad z = z_{\mathrm{m}} - 1\,, \\ 2\kappa, & \text{if} \quad z = 2+\gamma\,, \end{cases} \tag{12.6}$$

with $\gamma = 0, 1, 2, \ldots, k-1$.

The cubic interpolation is also called Simpson second method or Simpson 3/8 Rule. Assuming $z_{\mathrm{m}} = 3\,k$, $k \in \mathbb{Z}^+$ and $\kappa = \frac{3}{8}$, $\mathfrak{p}_z$ is defined as follows:

$$\mathfrak{p}_z = \begin{cases} \kappa, & \text{if} \quad z = 0 \quad \text{or} \quad z = z_{\mathrm{m}}\,, \\ 3\kappa, & \text{if} \quad z = 1+\gamma \quad \text{or} \quad z = z_{\mathrm{m}} - 2\,, \\ 3\kappa, & \text{if} \quad z = 2+\gamma \quad \text{or} \quad z = z_{\mathrm{m}} - 1\,, \\ 2\kappa, & \text{if} \quad z = 3+\gamma\,, \end{cases} \tag{12.7}$$

with $\gamma = 0, 1, 2, \ldots, k-1$.

The value of z_{m} should be large enough to ensure the accuracy of the approximation, but also small enough to reduce the computational burden. The optimal interpolation of a system fulfills the required accuracy with the smallest z_{m}. The optimal value of z_{m} is expected to different for different systems. Reference [116] discusses the optimal z_{m} for each interpolation through numerical examples. Based on our experience, the accuracy and efficiency of the quadratic and cubic interpolation for what concerns eigenvalue analysis are similar.

Instead of considering the comparison system, one can also consider the following characteristic equation of (12.3):

$$\det\big(\mathbf{\Delta}(s)\big) = 0\,, \tag{12.8}$$

where

$$\mathbf{\Delta}(s) = s\mathbf{E} - \mathbf{A}_0 - \mathbf{A}_1 \int_{\tau_{\mathrm{min}}}^{\tau_{\mathrm{max}}} \mathrm{pdf}(\vartheta)\, \mathrm{e}^{-s\vartheta}\, d\vartheta\,. \tag{12.9}$$

If it is possible to find a function $\mathfrak{C}(\cdot)$ such that

$$\mathfrak{C}(\vartheta) = \mathrm{pdf}(\vartheta)\, \mathrm{e}^{-s\vartheta}\,, \tag{12.10}$$

then the pencil (12.9) can be transformed as:

$$\mathbf{\Delta}(s) = s\mathbf{I}_n - \mathbf{A}_0 - \mathbf{A}_1\big(\mathfrak{C}(\tau_{\mathrm{max}}) - \mathfrak{C}(\tau_{\mathrm{min}})\big)\,. \tag{12.11}$$

Unlike the approximated spectrum solved with the comparison system (12.4), (12.11) allows obtaining the accurate spectrum of the distributed delay system (12.3). Then, the characteristic equation (12.8) can be solved with the QPMR method discussed in Section 10.3.

12.3 RDDEs with Time-Varying Delays

A time-varying delay is a delay whose magnitude is a function of time. Time-varying delays can be straightforwardly implemented in time-domain simulations; while in steady-state analyses, including eigenvalue analysis, this time-varying function has to be transformed into a tractable form.

Reference [109] provides a theorem to transform time-varying periodic delays into distributed delays. By combining the mathematical proof of [109] with the nature of distributed delays, we deduce the following Proposition:

Proposition 12.1 (Transformation of time-varying delays into distributed delays)**.** Consider the following linear system with time-varying delays:

$$\dot{\boldsymbol{x}}(t) = \mathbf{A}_0\, \boldsymbol{x}(t) + \sum_{i=1}^{i_{\mathrm{max}}} \mathbf{A}_i\, \boldsymbol{x}\big(t - \tau_i(t)\big)\,, \tag{12.12}$$

where $\tau_i(t) : \mathbb{R}^+ \rightarrow [\tau_{\mathrm{min}}, \tau_{\mathrm{max}}], 0 \leq \tau_{\mathrm{min}} < \tau_{\mathrm{max}}$. If the delay $\tau(t)$ changes *fast* enough, the stability of (12.12) is the same as the following *comparison system*:

$$\dot{\boldsymbol{x}}(t) = \mathbf{A}_0\,\boldsymbol{x}(t) + \sum_{i=1}^{i_m} \mathbf{A}_i \int_{\tau_{\min}}^{\tau_{\max}} \mathrm{pdf}_i(\vartheta)\,\boldsymbol{x}(t-\vartheta)\,d\vartheta\,. \tag{12.13}$$

The matrix pencil of the comparison system is:

$$\boldsymbol{\Delta}(s) = s\mathbf{I}_n - \mathbf{A}_0 - \sum_{i=1}^{i_m} \mathbf{A}_i h(s)\,, \tag{12.14}$$

where

$$h(s) = \int_{\tau_{\min}}^{\tau_{\max}} \mathrm{e}^{-s\vartheta}\,\mathrm{pdf}(\vartheta)\,d\vartheta\,. \tag{12.15}$$

Proof. Let $\mathcal{L}\{\boldsymbol{x}(t)\} = \boldsymbol{X}(s)$ be the Laplace transform of $\boldsymbol{x}(t)$. Then, if the delays $\tau_i(t)$, $i = 1, \ldots, i_m$ change fast enough, such that we can assume a specific delay $\tau_i(t) = \vartheta$, by applying the Laplace transform into (12.12), we have

$$\mathcal{L}\{\dot{\boldsymbol{x}}(t)\} = \mathbf{A}_0\mathcal{L}\{\boldsymbol{x}(t)\} + \sum_{i=1}^{v} \mathbf{A}_i\mathcal{L}\{\boldsymbol{x}(t-\vartheta)\}\,.$$

By using $\mathcal{L}\{\dot{\boldsymbol{x}}(t)\} = s\boldsymbol{X}(s) - \boldsymbol{x}_o$, where $\boldsymbol{x}_o$ is the initial condition of (12.12); and $\mathcal{L}\{\boldsymbol{x}(t-\vartheta)\} = \mathrm{e}^{-s\vartheta}\boldsymbol{X}(s)$ for $t \geq \vartheta$ and 0 for $t < \vartheta$; the above expression takes the form

$$s\,\boldsymbol{X}(s) - \boldsymbol{x}_o = \mathbf{A}_0\,\boldsymbol{X}(s) + \sum_{i=1}^{i_m} \mathbf{A}_i\,\mathrm{e}^{-s\vartheta}\,\boldsymbol{X}(s)\,,$$

or, equivalently,

$$(s\,\boldsymbol{X}(s) - \mathbf{A}_0 - \sum_{i=1}^{i_m} \mathbf{A}_i\,\mathrm{e}^{-s\vartheta})\,\boldsymbol{X}(s) = \boldsymbol{x}_o\,, \tag{12.16}$$

and consequently

$$\lambda\boldsymbol{X}(s) - \mathbf{A}_0 - \sum_{i=1}^{i_m} \mathbf{A}_i\,\mathrm{e}^{-\lambda\vartheta} \tag{12.17}$$

is the characteristic polynomial. If we apply the Laplace transform into (12.13), we get

$$\mathcal{L}\{\dot{\boldsymbol{x}}(t)\} = \mathbf{A}_0\mathcal{L}\{\boldsymbol{x}(t)\} + \sum_{i=1}^{v} \mathbf{A}_i\mathcal{L}\left\{\int_{\tau_{min}}^{\tau_{max}} w_i(\vartheta)\boldsymbol{x}(t-\vartheta)d\vartheta\right\}\,,$$

whereby using into the above expression

$$\mathcal{L}\Big\{\int_{\tau_{min}}^{\tau_{max}} w_i(\vartheta)\boldsymbol{x}(t-\vartheta)d\vartheta\Big\} = \mathcal{L}\{w_i(t)\}\boldsymbol{X}(s),$$

and the definition of the Laplace transform of the PDF $w_i(t)$ of the specific delay ϑ:

$$\mathcal{L}\{w_i(t)\} = E[\mathrm{e}^{-s\vartheta}] = \mathrm{e}^{-s\vartheta},$$

we obtain (12.16), and consequently (12.17). The proof is completed. ■

The spectrum of the comparison distributed delay system is the expected spectrum of the original system according to Probability Theory [61]. In other words, the stability assertion obtained from the eigenvalue analysis of the time-varying system is a relevant but not deterministic assertion for the stability of the original system, unless the variation frequency of $\tau(t)$ is large enough to ensure the convergence from relevance to determination.

The boundary for such a convergence depends on the system and the time-varying delays and can be theoretically estimated in some scenarios. For example, [130] deduces a theoretical bound for Ω to ensure that the stability of a time-varying delay system with $\tau(t) = \tau_0 + \delta f_p(\Omega t)$ converges to its comparison distributed delay system. The determination of the theoretical bound, however, is a very complex problem to solve. Moreover, reference [109] states that the theoretical bound of Ω is conservative and thus suggests to determine the threshold through numerical tests. Based on a large set of simulations, we have concluded that the threshold is very sensitive to the different $\tau(t)$ as well as to the conditions of the system.

Remark 12.3. Proposition 12.1 implies that the spectrum of a time-varying delay system can be estimated through the following steps:

Step 1 Find the PDF pdf(τ) of the time-varying delay $\tau(t)$.

Step 2 Define the comparison distributed delay system according to Proposition 12.1.

Step 3 Solve the spectrum of the comparison distributed delay system according to the discussions in Section 12.2.

The first step above implies that knowing the PDF of the time-varying delay is necessary to determine the spectrum. The function pdf(.), however, is not always available for the real-world time-varying delays. In this circumstance, we can use an approximated PDF based on measurement data of the delay. Example 12.6 illustrates an application of such a technique.

Figure 12.2 shows the analytic PDF of a Gamma-distributed delay and its corresponding histogram of the measurement data that can be utilized to approximate the PDF. The points c_z are the mid-points of each bin of the histogram. The RDDE with the Gamma-distributed non-constant delay

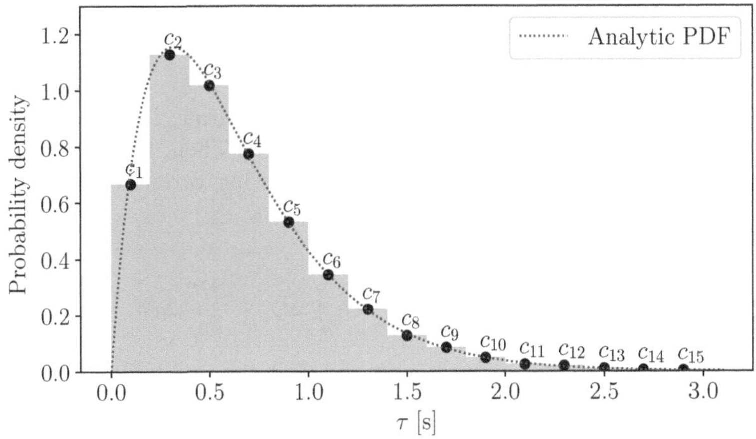

FIGURE 12.2: Histogram of a Gamma-distributed delay with analytic PDF.

system in the form of (12.12) can be firstly transformed into DDS (12.13), and then transformed into MCDS (12.4) with $(\Theta_z, \mathrm{pdf}(\Theta_z)) = c_z$. The critical eigenvalues of the MCDS can be calculated with the SDMs.

12.4 Effect of Time-Varying Delays on Power Systems

This section illustrates the effect of time delays on the OMIB model with inclusion of a delayed controller. The model of the system is presented in Section 12.4.1. Then, Section 12.4.2 discusses the existence of the QP in the power system by determining the analytical expression of the delay margin for constant and non-constant delays. A simplified power system with Gamma-distributed delay serves to prove the accuracy of numerical techniques proposed in Sections 12.2 and 12.3.

12.4.1 Simplified Power System Model

Let us consider the OMIB system with classical model of the synchronous machine described in Section 1.3.4.7. For clarity, the equations of the machine are recalled below:

$$
\begin{aligned}
\omega_o^{-1}\dot{\delta}_{\mathrm{r}} &= \omega_{\mathrm{r}} - 1\,, \\
M_1\dot{\omega}_{\mathrm{r}} &= P_{\mathrm{m}} - P_{\mathrm{e}} - D_1(\omega_{\mathrm{r}} - 1)\,,
\end{aligned}
\tag{12.18}
$$

where:

$$P_{\rm e} = \frac{e'_{\rm r,q} v_o}{X'_{\rm d}} \sin(\delta_{\rm r}) \, .$$

In the classical machine model $e'_{\rm r,q}$ is constant. However, in this section a simple PSS is assumed to make $e'_{\rm r,q} \propto \omega_{\rm r}$ in (12.18). Then, the linearization of the system at the equilibrium leads to the following variational equations [18]:

$$\begin{aligned} \omega_o^{-1} \Delta\dot{\delta}_{\rm r} &= \Delta\omega_{\rm r} \, , \\ M_1 \Delta\dot{\omega}_{\rm r} &= -\left(\frac{\partial P_{\rm e}}{\partial e'_{\rm r,q}} \frac{\partial e'_{\rm r,q}}{\partial \delta_{\rm r}} - D_1 \right) \Delta\omega_{\rm r} + \frac{\partial P_{\rm e}}{\partial \delta_{\rm r}} \Delta\delta_{\rm r} \, . \end{aligned} \tag{12.19}$$

For simplicity, let us assume that $D_1 = 0$ and that the input signal $\omega_{\rm r}$ to the PSS is delayed with delay τ, then (12.19) becomes:

$$\omega_o^{-1} \Delta\dot{\delta}_{\rm r}(t) = \Delta\omega_{\rm r} \, , \tag{12.20}$$

$$M_1 \Delta\dot{\omega}_{\rm r}(t) = -\frac{\partial P_{\rm e}}{\partial e'_{\rm r,q}} \frac{\partial e'_{\rm r,q}}{\partial \delta_{\rm r}} \Delta\omega_{\rm r}(t - \tau) + \frac{\partial P_{\rm e}}{\partial \omega_{\rm r}} \Delta\delta_{\rm r}(t) \, . \tag{12.21}$$

Substituting (12.20) into (12.21) and using the Laplace transform, we obtain:

$$s^2 + A \, s \, {\rm e}^{-\tau s} + K = 0 \, , \tag{12.22}$$

where

$$A = -\frac{\omega_o}{M_1} \frac{\partial P_{\rm e}}{\partial e'_{\rm r,q}} \frac{\partial e'_{\rm r,q}}{\partial \omega_{\rm r}} \, , \qquad K = \frac{\omega_o}{M_1} \frac{\partial P_{\rm e}}{\partial \delta_{\rm r}} \, .$$

where $\frac{\partial e'_{\rm r,q}}{\partial \omega_{\rm r}}$ and, hence, A is a function of the PSS control gain. Let us also assume that the OMIB is stable for $\tau = 0$, which is a condition that has to be expected in real-world systems.

The roots of (12.22) are:

$$\lambda = \frac{-\, A \pm \sqrt{A^2 - 4K}}{2} \, .$$

The system is asymptotically stable if and only if $\mathrm{Re}(\lambda) < 0$. There are two scenarios:

- If $A > 0$, then we need $\sqrt{A^2 - 4K} < A$; and thus, $K > 0$.
- If $A < 0$, then we need both $\sqrt{A^2 - 4K} > A$ and $\sqrt{A^2 - 4K} < A$, which can never be satisfied.

Hence the conditions $A, K > 0$ must hold for the OMIB to be stable for $\tau = 0$.

12.4.2 Stability Margin of Delay Models

This section discusses the stability margin of the OMIB system with delayed PSS and three delay models, namely, constant, square wave, and Gamma distributed.

Example 12.1 (Constant delay). If the delay τ_c is constant, at the delay margin τ_c^{m}, the characteristic equation (12.22) has a pair of pure imaginary roots $\lambda_{\mathrm{m}} = \pm \jmath \beta$, $\beta \in \mathbb{R}^+$. Substituting $\lambda = \jmath \beta$ into (12.22), and considering that $e^{-\jmath \tau \beta} = \cos(-\tau\beta) + \jmath \sin(-\tau\beta)$, we obtain:

$$0 = -\beta^2 + A\beta \sin(\tau_c^{\mathrm{m}}\beta) + K \,, \tag{12.23}$$

$$0 = \jmath A\beta \cos(\tau_c^{\mathrm{m}}\beta) \,. \tag{12.24}$$

The general solution of (12.24) is $\tau_c^{\mathrm{m}} = \frac{r\pi}{2\beta}$, where $r = 1, 3, 5 \ldots$ Since the delay margin focuses on the smallest delay that leads to pure imaginary rightmost eigenvalues, we limit the analysis to the case $r = 1$.

Then, substituting $\beta = \frac{\pi}{2\tau_c^{\mathrm{m}}}$ into (12.23), we have:

$$-\left(\frac{\pi}{2\tau_c^{\mathrm{m}}}\right)^2 + A\frac{\pi}{2\tau_c^{\mathrm{m}}} + K = 0 \,, \tag{12.25}$$

from where we obtain the delay margin:

$$\tau_c^{\mathrm{m}} = \frac{\pi}{A + \sqrt{A^2 + 4K}} \,. \tag{12.26}$$

□

Example 12.2 (Square-wave delay). Let us assume now that (12.22) includes a fast time-varying delay $\tau_v(t)$ and that the delay margin is $\bar{\tau}_v(t) = \tau_v^{\mathrm{m}}$. Since a fast time-varying delay can be transformed into multiple constant delays (see Section 12.2), the following set of equations hold:

$$\begin{aligned} -\beta^2 + A\beta \sum_{i=1}^{i_m} \mathfrak{p}_i \sin(\tau_i\beta) + K = 0 \,, \\ jA\beta \sum_{i=1}^{i_m} \mathfrak{p}_i \cos(\tau_i\beta) = 0 \,, \end{aligned} \tag{12.27}$$

where $\sum_{i=1}^{i_m} \mathfrak{p}_i = 1$ holds and the delays $\tau_1 < \tau_2 < \cdots < \tau_{i_m}$ form an arithmetic series and satisfy the condition $\frac{1}{i_m}\sum_{i=1}^{i_m} \tau_i = \tau_v^{\mathrm{m}}$.

Let us consider the simplest case of (12.27), namely:

$$i_m = 2 \,, \quad \text{and } \mathfrak{p}_1 = \mathfrak{p}_2 = \frac{1}{2} \,,$$

which implies a periodic square-wave delay (see [98] for details). In this scenario, assume that the two delays are $\tau_1 = \tau_v^{\mathrm{m}} - \kappa$ and $\tau_2 = \tau_v^{\mathrm{m}} + \kappa$. Since the delays are positive, $\tau_v^{\mathrm{m}} > \kappa > 0$ must hold. According to (12.27), we have:

$$\begin{aligned} -\beta^2 + A\beta\frac{1}{2}\big(\sin(\beta(\tau_v^{\mathrm{m}} - \kappa)) + \sin(\beta(\tau_v^{\mathrm{m}} + \kappa))\big) + K = 0\,, \\ \jmath A\beta\frac{1}{2}\big(\cos(\beta(\tau_v^{\mathrm{m}} - \kappa)) + \cos(\beta(\tau_v^{\mathrm{m}} + \kappa))\big) = 0\,. \end{aligned} \tag{12.28}$$

Similar to the constant delay case, from (12.28), at the delay margin, $\beta = \frac{\pi}{2\tau_m}$ and, thus, equation (12.28) can be rewritten as:

$$-\left(\frac{\pi}{2\tau_v^{\mathrm{m}}}\right)^2 + A\frac{\pi}{2\tau_v^{\mathrm{m}}}\cos\left(\frac{\pi\,\kappa}{2\,\tau_v^{\mathrm{m}}}\right) + K = 0\,. \tag{12.29}$$

Although the analytic solution of (12.29) cannot be obtained, by comparing (12.29) and (12.25), with $K, A, \tau_v^{\mathrm{m}}, \tau_c^{\mathrm{m}} > 0$ and $\cos(\frac{\pi\,\kappa}{2\,\tau_v^{\mathrm{m}}}) \in (0, 1]$, one can deduce that $\tau_v^{\mathrm{m}} > \tau_c^{\mathrm{m}}$ must hold. □

Example 12.3 (Gamma-distributed delay)**.** Finally, let us consider a Gamma-distributed delay system:

$$s^2 + A\,s\int_0^{\infty}\frac{\vartheta^{b-1}\,\mathrm{e}^{-\vartheta/a}}{a^b\,\Gamma(b)}\,\mathrm{e}^{-s\vartheta}\,d\vartheta + K = 0\,, \tag{12.30}$$

where a is the scale factor and b is the shape factor.

According to Corollary 12.2, we can deduce the characteristic equation of (12.30) as:

$$s^2 + A\,s\,h(s) + K = 0\,, \tag{12.31}$$

where $h(s) = (1 + a\,s)^{-b}$ (the details of the deduction of the expression of $h(s)$ for Gamma-distributed delay can be found in [97]).

The mean value of the Gamma distributed delay is:

$$\bar{\tau}_g = a\,b\,. \tag{12.32}$$

Considering the specific case for $b = 2$:

$$a = \frac{\bar{\tau}_g}{2}\,. \tag{12.33}$$

At the delay margin $\bar{\tau}_g = \tau_g^{\mathrm{m}}$, the system has an eigenvalue $\lambda = \jmath\,\beta$. With this condition, the following equations hold:

$$0 = -\beta^2 - \frac{2A\beta^2 a}{(1 - \beta^2 a^2)^2 + 4\beta^2 a^2} + K\,, \tag{12.34}$$

$$0 = 1 - \beta^2 a^2\,. \tag{12.35}$$

According to (12.33)–(12.35), with $A, K, \tau_g^{\rm m} > 0$, we have:

$$\tau_g^{\rm m} = \frac{8}{A + \sqrt{A^2 + 16K}} \, . \tag{12.36}$$

Comparing (12.36) with (12.26), assume a function:

$$\phi(A, K) = \frac{A + \sqrt{A^2 + 16K}}{A + \sqrt{A^2 + 4K}} \, . \tag{12.37}$$

There is $\tau_g^{\rm m} \geq \tau_c^{\rm m}$, if and only if:

$$\phi(A, K) \leq \frac{8}{\pi} \, . \tag{12.38}$$

Since $A, K > 0$, the maximal value of $\phi(A, K)$ is obtained for $A = 0$, i.e.:

$$\phi(A, K)_{\max} = 2 < \frac{8}{\pi} . \tag{12.39}$$

Thus, the condition $\tau_g^{\rm m} > \tau_c^{\rm m}$ must hold. □

The three examples above illustrate the nature of the QP: for a time-delay system, the stability margin for the mean value of a delay depends on how the delay varies with time.

The examples above also show the existence of cases for which, if the system is stable with a constant delay, it is also stable for the time-varying delays that have the same average as the constant delay. For these cases, solving the stability analysis with the constant time delay provides a conservative delay margin. However, a general method to define these cases without actually solving the stability analysis of the system with non-constant delay is still missing.

12.4.3 Power System Model with a Gamma-Distributed Delay

This section focuses on the numerical eigenvalue analysis of the characteristic equation (12.30), which represents the OMIB system with delayed PSS and a Gamma-distributed delay.

Example 12.4 (Eigenvalues for different interpolations)**.** Let us assume the OMIB model (12.30) with $A = 5$ and $K = 2$. According to (12.36), the delay margin of the system is $\tau_g^{\rm m} = 0.6375$ s. For this delay margin, one has $a = 0.3187$ s, $b = 2$ and $\lambda_{\rm m} = \pm \jmath 3.1375$.

Figure 12.3 shows the variations of the real part and the absolute value of the imaginary part of the rightmost eigenvalues of (12.30) as functions of $z_{\rm m}$. In Figure 12.3, "Analytic solution" indicates that the eigenvalue is deduced by (12.36); "Analytic distribution interpolation" indicates that the eigenvalue is solved by a comparison MCDS that consists in the interpolation of the analytic PDF; and "Numerical distribution interpolation" considers the interpolation points c_z, which are obtained from 2000 s time series with 100 Hz sampling rate.

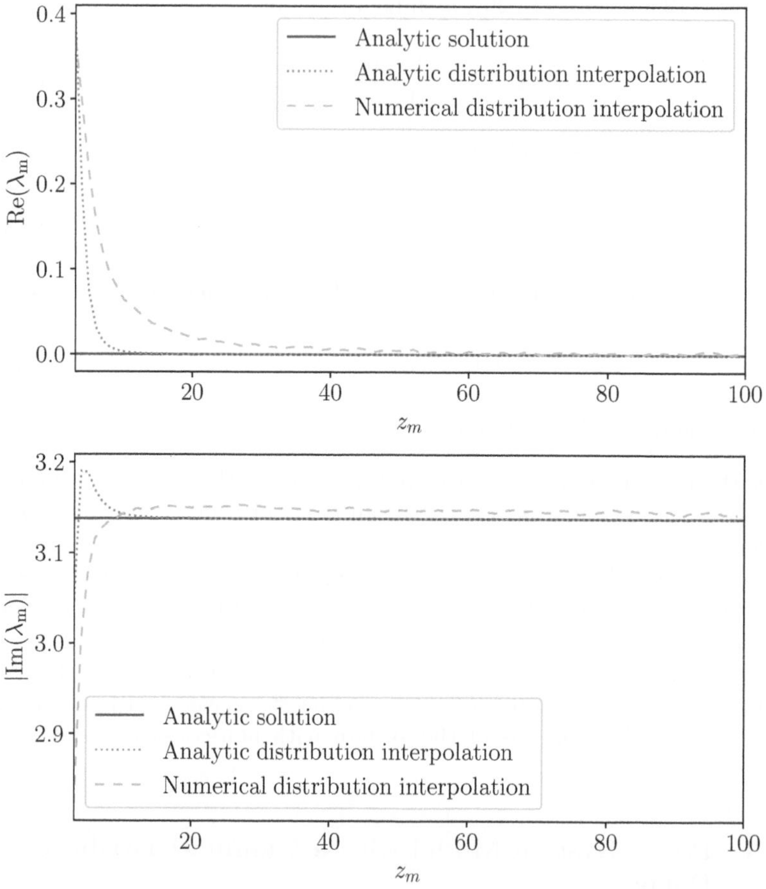

FIGURE 12.3: Real part of the rightmost eigenvalue of (12.30) with $A = 5$, $K = 2$, $a = 0.3187$ and $b = 2$ as function of z_m.

Figure 12.3 shows that, as the number of interpolation points z_m increases, all interpolation methods converge to the exact eigenvalue. The data-driven interpolation method, however, has a slower convergence rate and introduces a small offset in the imaginary part. □

Example 12.5 (Eigenvalues for different control gains)**.** To further investigate the accuracy of the proposed method on defining the stability margin, we consider another scenario with $K = 2$, $a = 0.15$ and $z_m = 42$. The real part of the rightmost eigenvalue as a function of A is shown in Figure 12.4. Note that the "Analytic solution" is obtained from applying the Newton method to (12.30) using as initial guess the values of the eigenvalues obtained with the "Analytic distribution interpolation" method.

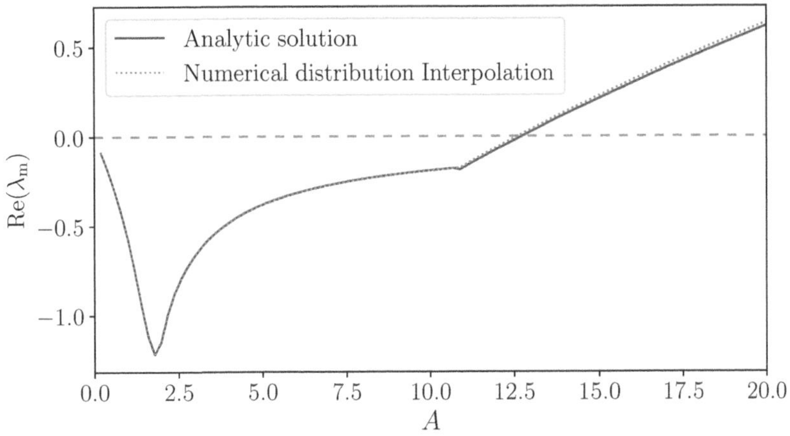

FIGURE 12.4: Real part of the rightmost eigenvalues of (12.30) with $K = 2$, $a = 0.15$, $b = 2$, and $z_m = 42$ as a function of A.

According to the results of Figure 12.4, the two interpolation methods find that the system becomes unstable if $A > 12.5$, which is consistent with the analytic solution. These results indicate that the interpolation methods yield an accurate estimation of the stability margin for systems with time-varying delays. □

12.5 Realistic Delay Models for Power System Analysis

Previous discussions in this chapter indicate that a precise model for time-varying delays is necessary to understand their impact on the stability of a system. In Chapter 11, the wide-area measurement system (WAMS) delay is simplified as constant. However, WAMS delays are the results of a series of processes along the data communication from the measurement device to the grid, including long-distance data delivery, data packet dropout, noise, communication network congestion, etc. [176]. Due to stochastic effects and the communication mechanism, WAMS delays are necessarily non-constant. As discussed above, due to the QP, the more realistic the delay model, the more accurate the stability assessment by means of the eigenvalue analysis.

With this aim, this section proposes a detailed model of realistic WAMS delays with full consideration of their physical nature.

12.5.1 Physical Structure of a WAMS

Let us assume that the WAMS measures a given quantity $x(t)$ of the power system. The signal is first measured by an appropriate device and digitalized. Then the signal is processed through a data package generator, transmitted and finally processed through a zero-order holder (ZOH). Finally, the resulting signal, say $x(t - \tau(t))$ is passed to due device/controller of the power system [176]. The steps of the measurement and communication process are illustrated in Figure 12.5.

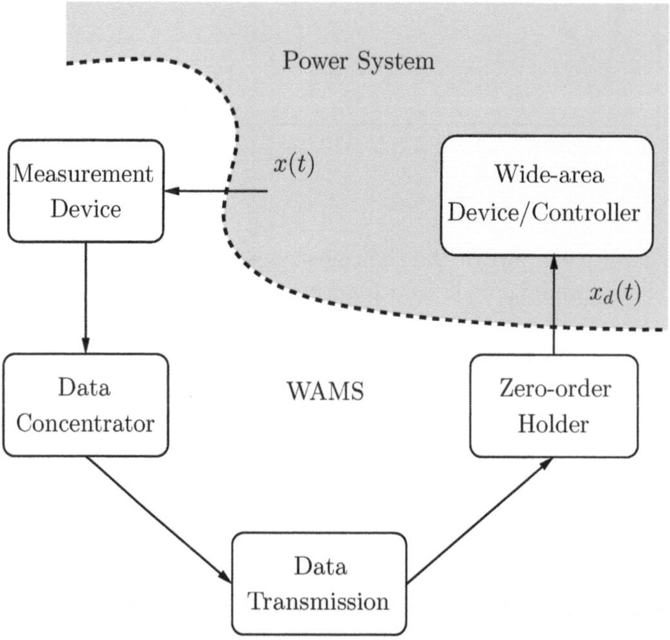

FIGURE 12.5: WAMS elements and their interaction with the power system.

The ZOH serves to hold the output of the controllers to avoid mistakes due to failures of the communication system. In this model, we actually assume that the ZOH holds the input signal of the WAMS controller. This assumption simplifies the analysis but provides identical effects as the real-world scenario.

In the model shown in Figure 12.5, all delays introduced in the WAMS-based control loop are treated as a part of the continuous model of τ, i.e. the digitalization of the signal is assumed to be fine enough to be treated as a continuous signal. This assumption is consistent with the typical time steps

(0.01 s or above) utilized to integrate power system models for transient stability analysis.

During the communication process, the data collected by the wide-area measurement devices are sent as discrete data packets. The data collecting and packing of the measurement devices, e.g. phasor measurement units (PMUs), lead to a delayed sending of each data packet, which is almost the same for each data packet. The magnitude of such a delay depends on the sampling rate, the data processing ability and the communication protocol of the measurement device [5]. The delivery of the data packets leads to time-varying latencies for the receiving signal, which are thoroughly discussed in the following section.

12.5.2 WAMS Delay Model

Let us consider the process with which the data packets are transmitted through the communication system and received by the WAMS controller. The variations in time of the WAMS delay during the period of the delivery of three successive data packets are qualitatively illustrated in Figure 12.6. The delay consists of two components, namely, τ_k and τ_p.

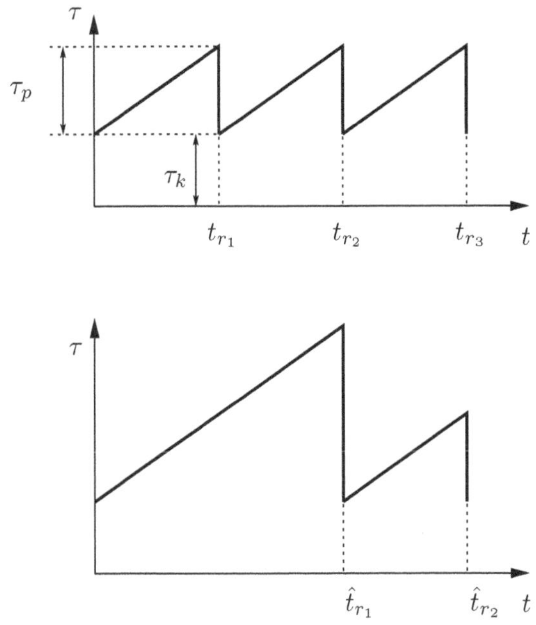

FIGURE 12.6: WAMS delay variations for the delivery of three successive data packets. Upper panel: all the packets arrive at the receiving end; lower panel: one packet loss.

Denoting with $t_{s,n}$ and $t_{r,n}$ the times at which the n-th data packet is sent and received, respectively, the communication delay is, by definition:

$$\tau_{k,n} = t_{s,n} - t_{r,n} \,. \tag{12.40}$$

The ZOH holds the last received data packet until the next one arrives, hence:

$$\tau_{p,n} = t_{r,n} - t_{r,n-1} \,. \tag{12.41}$$

The total delay τ of the n-th data packet is:

$$\tau(t) = t - t_{p,n-1} + \tau_{k,n-1} \,, \tag{12.42}$$

where $t \in [t_{r,n-1}, t_{r,n})$. If $t \to t_{r,n}$, we obtain the local maximum value of the delay as $\tau(t) = \tau_{p,n} + \tau_{k,n-1}$. Finally, at the time $t = t_{r,n}$, i.e. at the time at which the ZOH updates its signal, $\tau(t) = \tau_{p,n}$. Both τ_k and τ_p can be different for each data packet, due to the uncertainty of the communication system [97].

Moreover, a data packet may be lost during the data delivery, as shown in the lower panel of Figure 12.6. The time-varying delay model (12.42) is general enough to include this scenario if the sending time $t_{r,n}$ in (12.40) is obtained from the time stamp of the received data packet. To quantify the possibility of data packet loss, we introduce the variable $p \in [0, 1]$, abbreviated as *dropout rate* [97]. If $p = 0$, the data packet loss never occurs; and if $p = 1$, all the data packets are lost and never arrive at the receiving end.

According to Section 12.5.1, we also have to take into account the delay that is due by the data collection and processing of the measurement device, which can be regarded as a constant delay. Thus, the resulting expression of the WAMS delay is:

$$\tau_{\text{wams}} = \tau(t) + \tau_c \,, \tag{12.43}$$

where τ_c represents a constant delay that is assumed to be the same for each data packet.

In real-world power systems, the time-varying delay (12.43) can be defined through measurement data. In particular, measurement data allows defining the statistical properties of the communication system and, in turn, the PDF of the time-varying delay [100]. With the approximated PDF of τ_{wams}, we can solve the small-signal stability analysis according to the techniques described in Sections 12.3 and 12.4.

Example 12.6 (Three-bus system with WAMS delay model). This example revisits the small-signal stability of the three-bus system with inclusion of a delayed PSS that has been discussed in Example 11.1. In that example the WAMS delay is modeled as a constant. In the following, we consider the more realistic time-varying WAMS delay model introduced in Section 12.5.

To set up a numerical example of the time-varying WAMS delay model (12.43), we consider the following assumptions:

- The constant delay τ_c for each data packet is 25 ms.
- A data packet is sent every 20 ms.
- The time of each data packet delivery if not lost during the communication, namely the τ_p shown in the upper panel of Figure 12.6, is normally-distributed with mean 50 ms and standard deviation 6 ms.
- A data packet cannot arrive at the receiving end earlier then another packet that is sent in ahead, according to the rules of the communication protocol [176]. This condition is needed to amend the magnitude of the normally-distributed τ_p.
- The dropout rate for each packet is 10%.

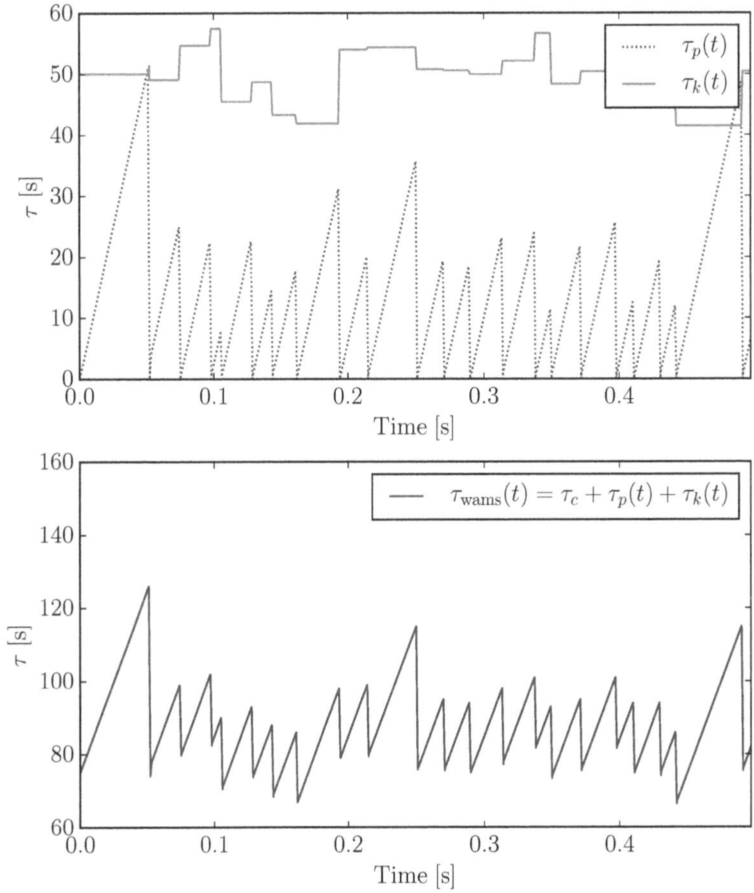

FIGURE 12.7: Trajectories of the time-varying WAMS delay.

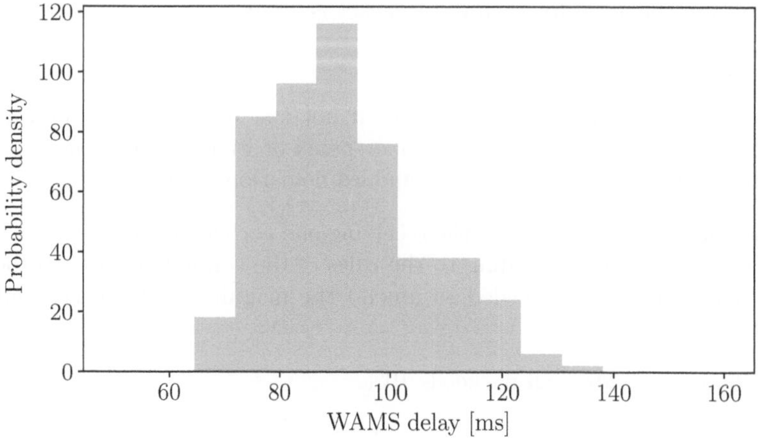

FIGURE 12.8: Histogram of the WAMS delay.

Figure 12.7 shows the trajectories of the time-varying WAMS delay obtained with the parameters and assumptions above. The stochastic network-induced issues and the randomly data packet dropout are all included in the trajectory shown in this figure. Figure 12.8 is the corresponding histogram of the time-varying delay obtained with a 2,000 s time series with sampling rate 1,000 Hz. According to the histogram, the mean value of this delay is 88.1 ms. The eigenvalue analysis discussed in this example are based on the histogram of Figure 12.8.

Next, we compare the time-varying model that emulates the physical behavior of the WAMS delay and its simplified constant delay model in the small-signal stability analysis of the three-bus system.

We firstly solve the eigenvalue analysis of the power system with the Krylov method. For the scenario that considers the time-varying delay, the system is transformed into a set of RDDEs with 15 commensurate constant delays in the range $[58, 165]$ ms, through the interpolation method presented in Remark 12.3 and the histogram shown in Figure 12.8. The comparison scenario considers a single constant delay of 88.1 ms.

Figure 12.9 shows the real part of the rightmost eigenvalues for various PSS gains $K_{\rm w} \in [7, 11]$. The two delay models show the same trend: the system with this WAMS delay becomes unstable if $K_{\rm w}$ is larger than a specific threshold. However, this threshold is different for different delay models. With the time-varying delay model, the threshold is approximately 10, while with the constant delay model is 10.2.

To evaluate which threshold is more accurate, we have solved a set of Monte-Carlo time domain simulations following a small angle disturbance on synchronous machine 1 with the time-varying WAMS delay shown in the lower panel of Figure 12.7.

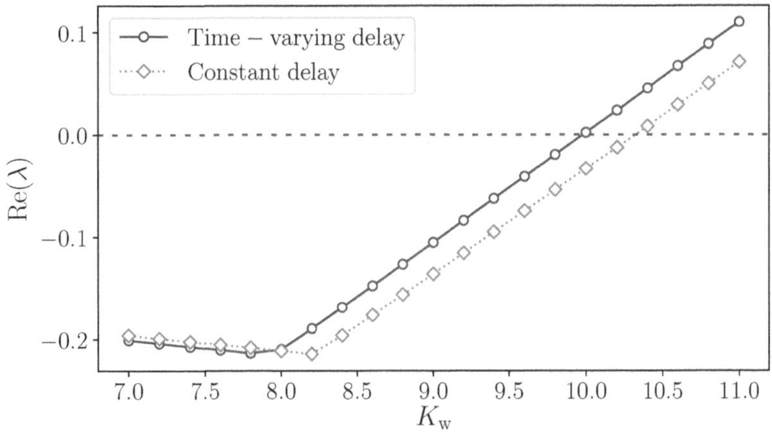

FIGURE 12.9: The real part of the rightmost eigenvalue of the three-bus system with different types of time delays as function of the PSS gain K_w.

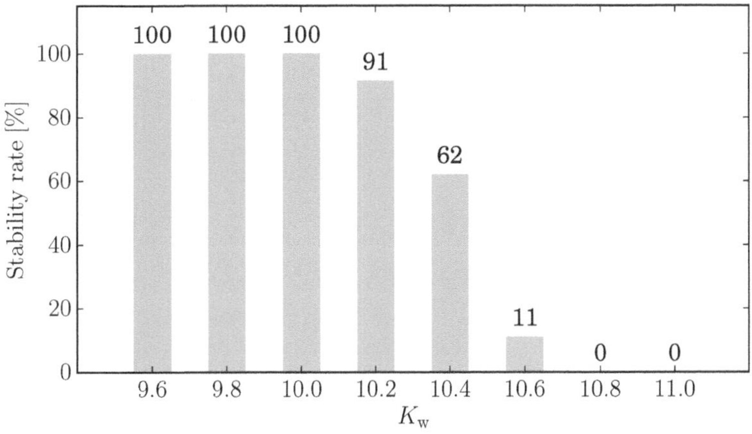

FIGURE 12.10: Stability rate of 200 Monte-Carlo tests for the three-bus system with the time-varying WAMS delay.

Figure 12.10 shows the stability rate for the $K_w \in [9.6, 11.0]$, where:

$$\text{Stability Rate} = \frac{\text{Number Stable Cases}}{\text{Number of Total Tests}}.$$

The number of total tests for each K_w is 200. With $K_w \leq 10$, the system achieves 100% stability rate, which is consistent with the eigenvalue analysis based on the transformed MCDS. On the other hand, as it can be seen, the threshold obtained with the simplified single constant delay model is slightly optimistic, i.e. using a gain value equal to this threshold may render the system unstable.

The discussion above indicates that the constant delay model is useful to understand the impact of a time delay on the stability of a power system. Moreover, thanks to its simplicity, a constant delay model is able to keep low the computational burden. To assess the stability of the power system in the critical scenario, however, a more precise delay model is necessary to improve the accuracy of the results. □

Part IV

Numerical Methods

13

Numerical Methods for Eigenvalue Problems

13.1 Introduction

We have seen that the eigenvalues of a matrix pencil $s\mathbf{E} - \mathbf{A}$ are the solutions of its characteristic equation:

$$\varpi(s) = \det(s\mathbf{E} - \mathbf{A}) = 0 \ , \tag{13.1}$$

where $s \in \mathbb{C}$. In the common scenario that $\varpi(s)$ is a polynomial of degree m, analytical solution of (13.1) is possible only if $m \leq 4$. For higher degrees, general formulas do not exist and only the application of a numerical method is possible. In addition, algorithms that, given a matrix pencil $s\mathbf{E} - \mathbf{A}$, explicitly determine the corresponding characteristic function $\varpi(s)$ and then numerically calculate its roots, may be extremely slow even for small problems.

Alternatively, we have also seen that the eigenvalues and the corresponding right eigenvectors of the pencil $s\mathbf{E} - \mathbf{A}$ are the solutions $(\lambda, \mathbf{u})$ of the problem:

$$(\lambda\mathbf{E} - \mathbf{A})\,\mathbf{u} = \mathbf{0}_{m,1} \ . \tag{13.2}$$

There is a rich literature on numerical algorithms that calculate eigenpairs of a given matrix pencil. Relevant monographs on the topic are, for example, [147] and [84]. However, not all available methods are suitable for small-signal stability analysis of power systems. In particular, many algorithms, for example the Davidson method [43], the implicitly restarted Lanczos method [157] and the locally optimal block preconditioned conjugate gradient method [81], focus only on symmetric (Hermitian) eigenvalue problems, whereas, as we have discussed in Parts II and III, the matrices that describe a linearized power system model are typically non-symmetric (non-Hermitian).

Methods for symmetric matrices are suitable for several relevant applications as symmetric eigenvalue problems arise in many engineering fields, e.g. the problem of mechanical vibrations. Algorithms for such problems are more abundant and more developed than algorithms for non-symmetric problems. This is also due to the fact that symmetric eigenvalue problems are generally well-conditioned, while non-symmetric eigenvalue problems are not.

Power systems yield non-symmetric pencils and, thus, the several algorithms available for symmetric problems have to be discarded *a priori*. In the remainder of this chapter, we provide a brief overview of numerical methods suitable for solving non-symmetric eigenvalue problems. The treatise is not intended to be complete but, rather, to describe the algorithms for which open source libraries are available and that worked for us for power system analysis.

13.2 Algorithms

A coarse taxonomy of algorithms for the solution of non-symmetric eigenvalue problems is as follows:

- *Vector iteration methods*, which, in turn, are separated to single and simultaneous vector iteration methods. Single vector iteration methods include the power method and its variants, such as the Rayleigh quotient iteration. Simultaneous vector iteration methods include the subspace iteration method and its variants, such as the inverse subspace method.
- *Schur decomposition methods*, which mainly include the QR algorithm, the QZ algorithm and their variants, such as the QR algorithm with shifts.
- *Krylov subspace methods*, which basically include the Arnoldi iteration and its variants, such as the implicitly restarted Arnoldi iteration and the Krylov-Schur method. In this category belong also preconditioned extensions of the Lanczos algorithm, such as the non-symmetric versions of the Generalized Davidson and Jacobi-Davidson methods.
- *Contour integration methods*, which basically include a moment-based Hankel method and a Rayleigh-Ritz-based projection method proposed by Sakurai and Sugiura; and the FEAST algorithm.

13.2.1 Vector Iteration Methods

Single Vector Iteration

The vector iteration or power method [169] is the oldest and probably the simplest and most intuitive numerical method for solving an eigenvalue problem.

Consider the LEP with matrix pencil $s\mathbf{I}_n - \mathbf{A}$, $\mathbf{A} \in \mathbb{R}^{n \times n}$. The main idea of the power method is that, if one starts with a vector $\mathbf{b}$ and repeatedly multiplies by $\mathbf{A}$, then the subsequence

$$\mathbf{Ab},\, \mathbf{A}^2\mathbf{b},\, \mathbf{A}^3\mathbf{b},\, \ldots, \tag{13.3}$$

converges to a multiple of the right eigenvector that corresponds to the eigenvalue of $\mathbf{A}$ with largest magnitude. To clearly see why and under which assumptions this is true, suppose that $\mathbf{b}$ is written as:

$$\mathbf{b} = \sum_{i=1}^{n} c_i \mathbf{u}_i \ , \tag{13.4}$$

where c_i, $i = 1, 2, \ldots, n$, is a constant; and $\mathbf{u}_i$ is a right eigenvector that corresponds to the eigenvalue λ_i. The eigenvalues are assumed to be sorted in descending order, i.e. $|\lambda_1| > |\lambda_2| > \ldots > |\lambda_n|$; and $\mathbf{A}^k \mathbf{u}_i = \lambda_i^k \mathbf{u}_i$. Then, the k-th term of (13.3) can be written as:

$$\begin{aligned}\mathbf{A}^k \mathbf{b} &= \sum_{i=1}^{n} c_i \mathbf{A}^k \mathbf{u}_i = \sum_{i=1}^{n} c_i \lambda_i^k \mathbf{u}_i \\ &= c_1 \lambda_1^k \left(\mathbf{u}_1 + \sum_{i=2}^{n} \frac{c_i \lambda_i^k \mathbf{u}_i}{c_1 \lambda_1^k} \right) \ . \end{aligned} \tag{13.5}$$

Under the assumption that the magnitude of λ_1 is strictly larger than that of all other eigenvalues, $\mathbf{A}^k \mathbf{b}$ in (13.5) linearly converges to $\mathbf{A}^k \mathbf{b} \rightarrow c_1 \lambda_1^k \mathbf{u}_1$, i.e. a multiple of the eigenvector $\mathbf{u}_1$. Note that, to guarantee the convergence, it is also required that $c_1 \neq 0$, which must be taken into account during the selection of the initial vector $\mathbf{b}$.

Applying the above idea, the method starts with an initial vector $\mathbf{b}^{(0)}$ with a non-zero component at the direction of the dominant eigenvalue. The vector is updated at each step of the iteration through multiplication with $\mathbf{A}$ and normalization. The k-th step of the iteration is as follows:

Power Iteration

$$\mathbf{b}^{(k)} = \frac{\mathbf{A}\mathbf{b}^{(k-1)}}{||\mathbf{A}\mathbf{b}^{(k-1)}||} \ , \quad k = 1, 2, 3, \ldots \tag{13.6}$$

Finally, under the above assumptions, $\mathbf{b}^{(k)}$ converges to $\mathbf{u}_1$, which corresponds to the eigenvalue of largest magnitude λ_1. Given the vector $\mathbf{u}_1$, an estimation of the eigenvalue λ_1 is given by the Rayleigh quotient.

In general, given a right eigenvector $\mathbf{u}_i$, the Rayleigh quotient provides an estimate of the corresponding eigenvalue λ_i as follows:

$$R(\mathbf{A}, \mathbf{u}_i) = \lambda_i = \frac{\mathbf{u}_i^{\mathrm{H}} \mathbf{A} \mathbf{u}_i}{\mathbf{u}_i^* \mathbf{u}_i} \ , \tag{13.7}$$

where $\mathbf{u}_i^{\mathrm{H}}$ is the conjugate transpose of $\mathbf{u}_i$.

The power method finds a right eigenvector that corresponds to the eigenvalue of largest magnitude. However, the eigenvalue of largest magnitude typically is not of interest for small-signal stability analysis. The inverse power method addresses this issue by using $(\mathbf{A} - \sigma \mathbf{I}_n)^{-1}$ instead of $\mathbf{A}$ in (13.6), where the spectral shift σ represents an initial guess of an eigenvalue λ_i. The k-th step of the inverse power method is as follows:

$$\textit{Inverse Power Iteration} \quad \mathbf{b}^{(k)} = \frac{(\mathbf{A} - \sigma\mathbf{I}_n)^{-1}\mathbf{b}^{(k-1)}}{||(\mathbf{A} - \sigma\mathbf{I}_n)^{-1}\mathbf{b}^{(k-1)}||}, \quad k = 1, 2, 3, \ldots \tag{13.8}$$

Provided that σ is a good initial guess of λ_i, $\mathbf{b}^{(k)}$ converges to the corresponding right eigenvector $\mathbf{u}_i$. Hence, the inverse power method is able to calculate the eigenvector corresponding to any single eigenvalue, by computing the dominant eigenvector of $(\mathbf{A} - \sigma\mathbf{I}_n)^{-1}$. Then, the eigenvalue λ_i is approximated by the Rayleigh quotient given by (13.7).

Despite the ability of the inverse power method to compute any eigenpair $(\lambda_i, \mathbf{u}_i)$, a bad selection of the spectral shift σ may lead to very slow convergence, or convergence to a different eigenvector than the one desired. The Rayleigh quotient iteration method is an extension of the inverse power method, that updates the applied spectral shift at each step of the iteration.

Starting from an initial eigenpair guess $(\sigma^{(0)}, \mathbf{b}^{(0)})$, the k-th step, $k = 1, 2, 3, \ldots$, of the Rayleigh quotient iteration is:

$$\textit{Rayleigh Quotient Iteration} \quad \begin{aligned} \mathbf{b}^{(k)} &= \frac{(\mathbf{A} - \sigma^{(k-1)}\mathbf{I}_n)^{-1}\mathbf{b}^{(k-1)}}{||(\mathbf{A} - \sigma^{(k-1)}\mathbf{I}_n)^{-1}\mathbf{b}^{(k-1)}||}, \\ \sigma^{(k)} &= \frac{(\mathbf{b}^{(k)})^{\mathrm{H}}\mathbf{A}\mathbf{b}^{(k)}}{(\mathbf{b}^{(k)})^{\mathrm{H}}\mathbf{b}^{(k)}}. \end{aligned} \tag{13.9}$$

Depending on the initial guess, the method is able to converge to any eigenpair $(\lambda_i, \mathbf{u}_i)$. An algorithm that was based on the Rayleigh quotient iteration and was developed especially for the eigenvalue analysis of power systems is the dominant pole algorithm [107, 108].

For the sake of simplicity, numerical methods were presented in (13.6), (13.8), (13.9) for the solution of the LEP, however, they can be modified to handle also the GEP. For example, considering the GEP with matrix pencil $s\mathbf{E} - \mathbf{A}$, $\mathbf{A} \in \mathbb{R}^{n \times n}$, $\mathbf{E} \in \mathbb{R}^{n \times n}$, the inverse power method for finding an eigenvector of the pencil $s\mathbf{E} - \mathbf{A}$ can be described as:

$$\mathbf{b}^{(k)} = \frac{(\mathbf{A} - \sigma\mathbf{E})^{-1}\mathbf{E}\mathbf{b}^{(k-1)}}{||(\mathbf{A} - \sigma\mathbf{E})^{-1}\mathbf{E}\mathbf{b}^{(k-1)}||}, \quad k = 1, 2, 3, \ldots \tag{13.10}$$

The methods described above calculate a single eigenvector in one iteration, and thus, they are referred in the literature as single vector iteration methods. Calculating multiple eigenpairs with a single vector iteration method is possible by applying a deflation technique and repeating the iteration. Regarding the dominant pole algorithm, a deflated version was proposed in [59]. Still, convergence of deflated single vector iterations can be very slow, and thus in practice, if not completely avoided, single vector iteration based algorithms should be used only for the solution of simple problems.

Simultaneous Vector Iteration

The simultaneous vector iteration or subspace method [11] is a vector iteration method that calculates multiple eigenvectors simultaneously in one iteration.

Suppose that we want to compute the p largest-magnitude eigenvalues of the LEP with pencil $s\mathbf{I}_n - \mathbf{A}$, $\mathbf{A} \in \mathbb{R}^{n\times n}$. Extending the idea of the power method, consider the subsequence:

$$\mathbf{A}\mathbf{b},\ \mathbf{A}^2\mathbf{b},\ \mathbf{A}^3\mathbf{b}, \ldots \tag{13.11}$$

where $\mathbf{b} \in \mathbb{C}^{n\times p}$. As it is, the subsequence (13.11), does not converge, as one would desire, to a matrix with columns the p eigenvectors of $s\mathbf{I}_n - \mathbf{A}$. However, convergence to the correct eigenvectors can be achieved by ensuring that the columns of the k-th element $\mathbf{A}^k\mathbf{b}$ of (13.11) are orthonormal vectors. Then, we say that the columns of $\mathbf{A}^k\mathbf{b}$ span the subspace:

$$\mathbf{v}_p = \mathrm{span}(\mathbf{A}^k\mathbf{b}) \ . \tag{13.12}$$

Algorithm 13.1 (Subspace iteration). The steps of the subspace iteration algorithm are the following:

Step 1 The iteration starts with an initial matrix $\mathbf{b}^{(0)}$, $\mathbf{b}^{(0)} \in \mathbb{C}^{n\times p}$, with orthonormal columns, i.e. $(\mathbf{b}^{(0)})^{\mathrm{H}}\mathbf{b}^{(0)} = \mathbf{I}_p$.

Step 2 At the k-th step of the iteration, $k = 1, 2, 3, \ldots$:

(a) Matrix $\mathbf{C}$, $\mathbf{C} \in \mathbb{C}^{n\times p}$, is formed as:

$$\mathbf{C}^{(k)} = \mathbf{A}\mathbf{b}^{(k-1)} \ . \tag{13.13}$$

(b) Matrix $\mathbf{b}^{(k)}$ is updated by maintaining column-orthonormality, through QR decomposition of matrix $\mathbf{C}$:

$$\begin{aligned} \mathbf{C}^{(k)} &= \mathbf{Q}\mathbf{R} \ , \\ \mathbf{b}^{(k)} &= \mathbf{Q} \ , \end{aligned} \tag{13.14}$$

where $\mathbf{Q}$ is unitary and $\mathbf{R}$ is upper triangular.

The columns of $\mathbf{b}^{(k)}$ eventually converge to p right eigenvectors that correspond to the p largest-magnitude eigenvalues.

Note that the largest-magnitude eigenvalues may not be the important ones for the needs of small-signal stability analysis. Similarly to the discussion of the inverse power method, finding the eigenvalues of smallest magnitude is possible by forming an inverse subspace iteration.

Convergence of the subspace iteration is in general slow and thus, its use is avoided for complex problems. Still, acceleration of the speed of convergence is possible by combining the method with the Rayleigh-Ritz procedure. Since the utility of the Rayleigh-Ritz procedure goes far beyond the subspace iteration, we describe it in a separate section.

13.2.2 Rayleigh-Ritz Procedure

The Rayleigh-Ritz procedure is a numerical technique used by many solvers to extract eigenvalue approximations from an associated to these eigenvalues subspace. The eigenvalues are extracted by projecting the matrix pencil onto the constructed subspace. In particular, given a subspace associated to p eigenvalues of $s\mathbf{I}_n - \mathbf{A}$, $\mathbf{A} \in \mathbb{R}^{n \times n}$, the Rayleigh-Ritz procedure for the LEP consists of the following steps:

1. Construct an orthonormal basis $\mathbf{Q}_p$, $\mathbf{Q}_p \in \mathbb{C}^{n \times p}$, which represents the subspace $\mathcal{Q}_p$ associated to p eigenvalues.
2. Compute $\tilde{\mathbf{A}} = \mathbf{Q}_p^{\mathrm{H}} \mathbf{A} \mathbf{Q}_p$. Matrix $\tilde{\mathbf{A}}$ is the projection of $\mathbf{A}$ onto the subspace $\mathcal{Q}_p$.
3. Compute the eigenvalues $\lambda_1, \lambda_2, \dots \lambda_p$ and eigenvectors $\mathbf{v}_1, \mathbf{v}_2, \dots \mathbf{v}_p$, of $\tilde{\mathbf{A}}$. The eigenvalues are called Ritz values and are also eigenvalues of $\mathbf{A}$.
4. Form the vector $\mathbf{u}_i = \mathbf{Q}_p \mathbf{v}_i$, for $i = 1, 2, \dots, p$. Each vector $\mathbf{u}_i$ is called Ritz vector and corresponds to the eigenvalue λ_i.

In case that a GEP is to be solved, then the Rayleigh-Ritz procedure consists in finding the p eigenvalues of $s\mathbf{E} - \mathbf{A}$, $\mathbf{A} \in \mathbb{R}^{n \times n}$, $\mathbf{E} \in \mathbb{R}^{n \times n}$. Then, steps 2 and 3 have to be modified as follows:

2. Compute $\tilde{\mathbf{A}} = \mathbf{Q}_p^{\mathrm{H}} \mathbf{A} \mathbf{Q}_p$ and $\tilde{\mathbf{E}} = \mathbf{Q}_p^{\mathrm{H}} \mathbf{E} \mathbf{Q}_p$.
3. Compute the eigenvalues $\lambda_1, \lambda_2, \dots \lambda_p$ and eigenvectors $\mathbf{v}_1, \mathbf{v}_2, \dots \mathbf{v}_p$, of $s\tilde{\mathbf{E}} - \tilde{\mathbf{A}}$. The p eigenvalues found are called Ritz values and are also eigenvalues of $s\mathbf{E} - \mathbf{A}$.

13.2.3 Schur Decomposition Methods

Schur decomposition-based methods take advantage of the fact that, for any $n \times n$ matrix $\mathbf{A}$, there always exists a unitary matrix $\mathbf{Q} \in \mathbb{C}^{n \times n}$, such that:

$$\mathbf{T} = \mathbf{Q}^{\mathrm{H}} \mathbf{A} \mathbf{Q} \,, \tag{13.15}$$

where $\mathbf{T}$, $\mathbf{T} \in \mathbb{C}^{n \times n}$, is an upper triangular matrix with diagonal elements all the eigenvalues of $\mathbf{A}$. Decomposition (13.15) is known as the Schur decomposition of $\mathbf{A}$, and matrix $\mathbf{T}$ as the Schur form of $\mathbf{A}$.

It appears that the most efficient algorithm to numerically compute the Schur decomposition of a given matrix is the QR algorithm [48]. In an analogy to previously described methods, the QR algorithm can be understood as a nested subspace iteration, or as a nested sequence of n power iterations [137].

Algorithm 13.2 (QR algorithm). The steps of the QR algorithm are as follows:

1. The algorithm starts with the matrix $\mathbf{A}^{(0)} = \mathbf{A}$.
2. At the k-th iteration, $k = 1, 2, 3, \ldots$:
   (a) The QR decomposition of $\mathbf{A}^{(k-1)}$ is calculated:

$$\mathbf{A}^{(k-1)} = \mathbf{Q}^{(k)}\mathbf{R}^{(k)} , \tag{13.16}$$

   where $\mathbf{Q}^{(k)}$ is orthogonal and $\mathbf{R}^{(k)}$ is upper triangular. The QR decomposition (13.16) is numerically computed by applying Householder transformations [72].

   (b) The new matrix $\mathbf{A}^{(k)}$ is calculated as follows:

$$\mathbf{A}^{(k)} = \mathbf{R}^{(k)}\mathbf{Q}^{(k)} . \tag{13.17}$$

If $\mathbf{A}^{(k)}$ converges, say after ν steps, then $\mathbf{A}^{(\nu)} = \mathbf{T}$. The diagonal elements of $\mathbf{A}^{(\nu)}$ are the eigenvalues of $\mathbf{A}$.

The corresponding right eigenvectors are found as the columns of the matrix:

$$\mathbf{Q} = \mathbf{Q}^{(1)}\mathbf{Q}^{(2)} \ldots \mathbf{Q}^{(\nu)} , \tag{13.18}$$

where $\mathbf{Q} \in \mathbb{C}^{n \times n}$.

Throughout the algorithm, matrices $\mathbf{A}^{(k)}$, $\mathbf{A}^{(k-1)}$ are similar and thus, they have the same eigenvalues with the same multiplicities. To clearly see the similarity, rewrite (13.17) as:

$$\begin{aligned} \mathbf{A}^{(k)} = \mathbf{R}^{(k)}\mathbf{Q}^{(k)} &= \left(\mathbf{Q}^{(k)}\right)^{-1}\mathbf{Q}^{(k)}\mathbf{R}^{(k)}\mathbf{Q}^{(k)} \\ &= \left(\mathbf{Q}^{(k)}\right)^{-1}\mathbf{A}^{(k-1)}\mathbf{Q}^{(k)} . \end{aligned} \tag{13.19}$$

Overall, the QR algorithm uses a complete basis of vectors and computes all eigenvalues of the matrix pencil $s\mathbf{I}_n - \mathbf{A}$. In practice, the QR algorithm is not used as it is, but multiple shifts are applied to accelerate the convergence rate, which leads to the implicitly shifted QR algorithm [58].

For the sake of completeness, we outline here how Householder transformations are applied to an arbitrary matrix $\mathbf{\Phi}_o$, $\mathbf{\Phi}_o \in \mathbb{R}^{n \times n}$, so that $\mathbf{\Phi}_o = \mathbf{QR}$. The procedure is completed in n steps. In the following, $\boldsymbol{\phi}_j$ denotes the first column of the sub-matrix $(\mathbf{\Phi}_j)_{(j+1):n,(j+1):n}$, $j = 1, 2, 3, \ldots, n$, and has elements $\phi_{j,i}$; $\mathbf{e}_{n,1}$ denotes the first column of the identity matrix $\mathbf{I}_n$.

At the j-th step, $j = 1, 2, 3, \ldots, n$, the Householder matrix is defined as:

$$\mathbf{H}_j = \mathbf{I}_{n-j+1} - 2\frac{\boldsymbol{u}_j\boldsymbol{u}_j^{\mathrm{T}}}{\boldsymbol{u}_j^{\mathrm{T}}\boldsymbol{u}_j} , \tag{13.20}$$

where

$$\boldsymbol{u}_j = \boldsymbol{\phi}_{j-1} - \mathrm{sign}(\phi_{j-1,1})\mathbf{e}_{n-j+1}||\boldsymbol{\phi}_{j-1}||_2 . \tag{13.21}$$

The new matrix $\mathbf{\Phi}_j$ is then calculated as follows:

$$\mathbf{\Phi}_j = \tilde{\mathbf{H}}_j \mathbf{\Phi}_{j-1} \ , \tag{13.22}$$

where $\tilde{\mathbf{H}}_j = \mathbf{I}_{j-1} \oplus \mathbf{H}_j$. Finally, $\mathbf{Q}$ and $\mathbf{R}$ are derived as:

$$\begin{aligned} \mathbf{Q} &= \tilde{\mathbf{H}}_1 \tilde{\mathbf{H}}_2 \dots \tilde{\mathbf{H}}_n \ , \\ \mathbf{R} &= \mathbf{B}_n \ . \end{aligned} \tag{13.23}$$

The analog of the QR algorithm for the GEP is the QZ algorithm [126]. The QZ algorithm takes advantage of the fact that for any $n \times n$ matrices $\mathbf{A}$, $\mathbf{E}$, there always exist unitary matrices $\mathbf{Q} \in \mathbb{C}^{n \times n}$, $\mathbf{Z} \in \mathbb{C}^{n \times n}$, such that:

$$\begin{aligned} \mathbf{T}_1 &= \mathbf{Q}^{\mathrm{H}} \mathbf{A} \mathbf{Z} \ , \\ \mathbf{T}_2 &= \mathbf{Q}^{\mathrm{H}} \mathbf{E} \mathbf{Z} \ , \end{aligned} \tag{13.24}$$

are upper triangular, and the eigenvalues of the pencil $s\mathbf{E} - \mathbf{A}$ are the ratios $w_{1,ii}/w_{2,ii}$ of the diagonal elements of $\mathbf{T}_1$, $\mathbf{T}_2$, respectively. Decomposition (13.24) is known as generalized Schur decomposition.

From the implementation viewpoint, the algorithm computes all eigenvalues of the matrix pencil $s\mathbf{E} - \mathbf{A}$ by first reducing $\mathbf{A}$, $\mathbf{E}$, to upper Hessenberg and upper triangular form, respectively, through Householder transformations. The method is equivalent to applying the implicitly shifted QR algorithm to $\mathbf{A}\mathbf{E}^{-1}$, without, nonetheless, explicitly computing $\mathbf{E}^{-1}$ [58].

A disadvantage of Schur decomposition-based methods is that they are computationally expensive. In addition, they are dense matrix methods, i.e. they generate complete fill-in in general sparse matrices and therefore, can not be applied to large sparse matrices simply because of massive memory requirements. Even so, for small to medium size problems, the QR algorithm is the standard solution for finding the full spectrum for the conventional LEP.

13.2.4 Krylov Subspace Methods

The power method is based on calculating the products $\mathbf{Ab}$, $\mathbf{A}^2\mathbf{b}$, $\mathbf{A}^3\mathbf{b}$ and so on, until convergence. The method utilizes the converged vector, let's say $\mathbf{A}^{\nu}\mathbf{b}$, which corresponds to the largest-magnitude eigenvalue. However, the power iteration discards the information of all vectors calculated in the meantime. The main idea of Krylov subspace eigenvalue methods is exactly to utilize this intermediate information to calculate the p largest-magnitude eigenvalues of $s\mathbf{I}_n - \mathbf{A}$.

The Krylov subspace that corresponds to $\mathbf{A}$ and vector $\mathbf{b}$ is defined as:

$$\mathcal{K}_p(\mathbf{A}, \mathbf{b}) = \mathrm{span}\{\mathbf{b}, \mathbf{Ab}, \mathbf{A}^2\mathbf{b}, \dots, \mathbf{A}^{p-1}\mathbf{b}\} \ . \tag{13.25}$$

Subspace (13.25) is spanned by the columns of the Krylov matrix, which is defined as:

$$\mathbf{K}_p = [\mathbf{b} \ \mathbf{Ab} \ \mathbf{A}^2\mathbf{b} \ \dots \ \mathbf{A}^{p-1}\mathbf{b}] \ . \tag{13.26}$$

The matrix-vector multiplications typically render the columns of $\mathbf{K}_p$ linearly dependent. Thus, these columns require orthonormalization. The orthonormalized vectors can be then employed to provide approximations of the eigenvectors that correspond to the p eigenvalues of largest magnitude. The eigenvalues are extracted by projecting $\mathbf{A}$ onto the Krylov subspace, typically through the Rayleigh-Ritz procedure.

Arnoldi iteration

The Arnoldi iteration finds the p largest-magnitude eigenvalues of $s\mathbf{I}_n - \mathbf{A}$, $\mathbf{A} \in \mathbb{R}^{n\times n}$. In particular, it employs the Gram-Schmidt process to find an orthonormal basis of the Krylov subspace. Then, the eigenvalues are extracted by applying the Rayleigh-Ritz procedure.

Algorithm 13.3 (Arnoldi iteration)**.** The steps of the Arnoldi iteration algorithm are the following.

1. The algorithm starts with a random vector $\mathbf{b}$. The vector is normalized as $\boldsymbol{q}_1 = \dfrac{\mathbf{b}}{||\mathbf{b}||}$.
2. An orthonormal basis is produced by the Gram-Schmidt process. At the k-step, $k = 1, 2, \ldots$, the vector $\boldsymbol{q}_{k+1}$, is calculated and normalized as follows:

$$\begin{aligned} \hat{\boldsymbol{q}}_k &= \mathbf{A}\boldsymbol{q}_k - \sum_{i=1}^{k} h_{i,k}\boldsymbol{q}_i \,, \\ \boldsymbol{q}_{k+1} &= \frac{\hat{\boldsymbol{q}}_k}{h_{k+1,k}} \,, \end{aligned} \tag{13.27}$$

   where $h_{i,k} = \boldsymbol{q}_i^{\mathrm{H}}\mathbf{A}\boldsymbol{q}_k$, $h_{k+1,k} = ||\hat{\boldsymbol{q}}_k||$.

After k steps, the Arnoldi relation is formulated as follows:

$$\mathbf{A}\mathbf{Q}_k = \mathbf{Q}_k\mathbf{H}_k + \hat{\boldsymbol{q}}_k\mathbf{e}_k^{\mathrm{H}} \,, \tag{13.28}$$

where $\mathbf{Q}_k$, $\mathbf{Q}_k \in \mathbb{C}^{n\times k}$, is the matrix with columns $\boldsymbol{q}_k$; $\mathbf{H}_k$, $\mathbf{H}_k \in \mathbb{R}^{k\times k}$, is in Hessenberg form and has elements $h_{i,j}$; $\mathbf{e}_k$ denotes the k-th column of the identity matrix $\mathbf{I}_k$. The algorithm stops when $h_{k+1,k} = 0$, suppose when $k = p$. Then, relation (13.28) becomes:

$$\mathbf{A}\mathbf{Q}_p = \mathbf{Q}_p\mathbf{H}_p \,. \tag{13.29}$$

In (13.29), $\mathbf{Q}_p = [\boldsymbol{q}_1, \boldsymbol{q}_2, \ldots, \boldsymbol{q}_p]$ is an orthonormal basis of the Krylov subspace. Following the Rayleigh-Ritz procedure, the eigenvalue λ_i, $i = 1, 2, \ldots, p$ of $\mathbf{H}_p$ is also eigenvalue of $s\mathbf{I}_n - \mathbf{A}$. Moreover, if $\mathbf{v}_i$ is eigenvector of $\mathbf{H}_p$, the Ritz vector $\mathbf{u}_i = \mathbf{Q}_p\mathbf{v}_i$ is eigenvector that corresponds to λ_i.

The Arnoldi iteration converges fast if the initial vector has larger entries at the direction of desired eigenvalues. This is usually not the case and thus, it is common that many iterations are required. On the other hand, the ability to compute the columns of $\mathbf{Q}_p$ is constrained because of its high memory requirements. For this reason, in practice, the Arnoldi iteration is typically restarted after a number of steps, using a new, improved initial vector. The restarts can be explicit or implicit. The idea of explicit restarts is to utilize the information of the most recent factorization to produce a better initial vector. This is usually combined with a locking technique, i.e. a technique to ensure that converged vectors do not change in successive runs of the algorithm. On the other hand, the idea of the Implicitly Restarted (IR) Arnoldi iteration, proposed in [92], is to combine the Arnoldi iteration with the implicitly shifted QR algorithm.

Krylov-Schur Method

The Krylov-Schur method [158] is a generalization of the Lanczos thick restart [179] for non-Hermitian matrices. It is based on the idea to achieve the effect of the implicit restart technique in a simpler way.

The method considers the Arnoldi relation (13.28) and applies the QR algorithm to bring matrix $\mathbf{H}_k$ to its Schur form, i.e. $\mathbf{T}_k = \mathbf{W}_k^{\mathrm{H}}\mathbf{H}_k\mathbf{W}_k$, where $\mathbf{T}_k$ is upper triangular with diagonal elements the eigenvalues of $\mathbf{H}_k$ (Ritz values). Then, (13.28) becomes:

$$\mathbf{A}\mathbf{Q}_k\mathbf{W}_k = \mathbf{Q}_k\mathbf{W}_k\mathbf{T}_k + \hat{q}_k\mathbf{e}_k^{\mathrm{H}}\mathbf{W}_k \ , \tag{13.30}$$

or equivalently,

$$\mathbf{A}\mathbf{G}_k = \mathbf{G}_k\mathbf{T}_k + \hat{q}_k\boldsymbol{w}_k^{\mathrm{H}} \ , \tag{13.31}$$

where $\mathbf{G}_k = \mathbf{Q}_k\mathbf{W}_k$, $\boldsymbol{w}_k^{\mathrm{H}} = \mathbf{e}_k^{\mathrm{H}}\mathbf{W}_k$. Decomposition (13.31) is reordered to separate undesired Ritz values from the desired ones. The part that corresponds to the desired Ritz values is kept and the rest is discarded:

$$\mathbf{A}\mathbf{G}_{k,d} = \mathbf{G}_{k,d}\mathbf{T}_{k,d} + \hat{q}_k\boldsymbol{w}_{k,d}^{\mathrm{H}} \ , \tag{13.32}$$

where $\mathbf{G}_{k,d}$, $\mathbf{T}_{k,d}$ and $\boldsymbol{w}_{k,d}^{\mathrm{H}}$ represent the parts of $\mathbf{G}_k$, $\mathbf{T}_k$ and $\boldsymbol{w}_k^{\mathrm{H}}$ that correspond to the desired Ritz values after the reordering, respectively. Then, the reduced decomposition (13.32) is expanded to order k. The above process is repeated until convergence to an invariant subspace is achieved.

Generalized Eigenvalue Problem

Krylov subspace methods above were described for the solution of the LEP. Using Krylov subspace methods for the solution of the GEP is usually done by employing a spectral transform. For example, by applying the shift & invert transform to the pencil $s\mathbf{E} - \mathbf{A}$, the Arnoldi relation becomes:

$$(\mathbf{A} - \sigma\mathbf{E})^{-1}\mathbf{E}\mathbf{Q}_k = \mathbf{Q}_k\mathbf{H}_k + \hat{q}_k\mathbf{e}_k^{\mathrm{H}} \ , \tag{13.33}$$

where in this case it is $\mathbf{Q}_k^{\mathrm{H}}\mathbf{E}\mathbf{Q}_k = \mathbf{I}_k$. Each eigenvalue μ obtained after the convergence of (13.33) to an invariant subspace, is then transformed to find the corresponding eigenvalue λ of the original problem as $\lambda = 1/\mu + \sigma$.

13.2.5 Contour Integration Methods

Contour integration methods find the eigenvalues of the pencil $s\mathbf{E} - \mathbf{A}$ that are inside a given, user-defined domain of the complex plane. The method proposed by Sakurai and Sugiura [148] and its variants [9, 149] are the most characteristic examples of this class.

Hankel matrix-based Contour Integration

Suppose that p distinct eigenvalues $\lambda_1, \lambda_2, \ldots, \lambda_p$ are located inside positively oriented simple closed curve Γ in $\mathbb{C}$.

The contour integration method with Hankel matrices (CI-Hankel), proposed in [148], applies a moment-based approach to reduce the problem of finding the p eigenvalues of $s\mathbf{E} - \mathbf{A}$ to finding the eigenvalues of a $p \times p$ matrix pencil. For a non-zero vector $\mathbf{b} \in \mathbb{R}^{n\times 1}$, consider the moments:

$$\mu_k = \frac{1}{2\pi i}\int_\Gamma (s-\gamma)^k (\mathbf{E}\mathbf{b})^{\mathrm{T}}(s\mathbf{E} - \mathbf{A})^{-1}\mathbf{E}\mathbf{b}ds \ , \ k = 0, 1, \ldots, p-1 \ , \tag{13.34}$$

where γ is a user-defined point that lies in Γ. A standard case is that Γ is defined as a circle with center γ and radius ρ.

The moments μ_k are used to construct the $p \times p$ Hankel matrices $\mathbf{H}_p := [\mu_{i+j-2}]_{i,j=1}^{p}$ and $\mathbf{H}_p^{<} := [\mu_{i+j-1}]_{i,j=1}^{p}$. Then, the eigenvalues of the matrix pencil $s\mathbf{H}_p^{<} - \mathbf{H}_p$ are given by $\lambda_1 - \gamma$, $\lambda_2 - \gamma$, ..., $\lambda_p - \gamma$.

The corresponding eigenvectors are found through the contour integral:

$$\boldsymbol{\sigma}_k = \frac{1}{2\pi i}\int_\Gamma (s-\gamma)^k (s\mathbf{E} - \mathbf{A})^{-1}\mathbf{E}\mathbf{b}ds \ , \ k = 0, 1, \ldots, p-1 \ . \tag{13.35}$$

Given the eigenvectors associated to $\lambda_1 - \gamma$, $\lambda_2 - \gamma$, ..., $\lambda_p - \gamma$, the eigenvectors of $s\mathbf{E} - \mathbf{A}$ are retrieved through a simple vector manipulation.

In practice, the integrals (13.34), (13.35) are approximated through a numerical integration technique. For example, using the N-point trapezoidal rule when Γ is defined as the circle with center γ and radius ρ, the following approximations of (13.34), (13.35) are obtained:

$$\mu_k \approx \frac{1}{N}\sum_{j=0}^{N-1} (w_j - \gamma)^{k+1}(\mathbf{E}\mathbf{b})^{\mathrm{T}}(w_j\mathbf{E} - \mathbf{A})^{-1}\mathbf{E}\mathbf{b} \ , \tag{13.36}$$

$$\boldsymbol{\sigma}_k \approx \frac{1}{N}\sum_{j=0}^{N-1} (w_j - \gamma)^{k+1}(w_j\mathbf{E} - \mathbf{A})^{-1}\mathbf{E}\mathbf{b} \ , \ k = 0, 1, \ldots, p-1 \ , \tag{13.37}$$

where $w_j = \gamma + \rho e^{(2\pi i/N)(j+1/2)}$, $j = 0, 1, \ldots, N-1$. In order to compute (13.36), (13.37), the following linear systems need to be solved:

$$(w_j\mathbf{E} - \mathbf{A})\boldsymbol{y}_j = \mathbf{Eb}\ , \ j = 0, 1, \ldots, N-1\ . \tag{13.38}$$

Systems (13.38) are independent for each j, and hence, they can be solved in parallel. A parallel implementation can significantly improve computational efficiency, especially if the dimensions of $\mathbf{A}$, $\mathbf{E}$ are large.

Finally, in practice, Hankel matrices with dimensions larger than $p \times p$ may be used to reduce the influence of the quadrature error and improve accuracy.

Contour Integration with Rayleigh-Ritz Procedure

In case that some eigenvalues in the defined region are very close, the accuracy of the CI-Hankel method decreases, as the associated Hankel matrices become ill-conditioned. An alternative approach is to use contour integration to construct a subspace associated to the eigenvalues in Γ, and then extract the eigenpairs from such subspace using the Rayleigh-Ritz procedure. The resulting method is the CI-RR (Contour Integration with Rayleigh-Ritz), and was proposed in [149].

Starting from (13.35), it can be proved that, if the column vectors $\{\boldsymbol{q}_1, \boldsymbol{q}_2, \ldots, \boldsymbol{q}_p\}$ form an orthonormal basis of span$\{\boldsymbol{\sigma}_o, \boldsymbol{\sigma}_1, \ldots, \boldsymbol{\sigma}_{p-1}\}$, then the eigenvalues λ_i and the corresponding eigenvectors $\mathbf{u}_i$, $i = 1, 2, \ldots, p$, can be extracted by using the basis $\mathbf{Q}_p = [\boldsymbol{q}_1, \boldsymbol{q}_2, \ldots, \boldsymbol{q}_p]$ and by applying the Rayleigh-Ritz procedure for the pencil $s\mathbf{E} - \mathbf{A}$. To construct the orthonormal basis $\mathbf{Q}_p$, the contour integral (13.35) is first approximated through the N-point trapezoidal rule.

In practice, an orthonormal basis $\mathbf{Q}_p$ with size larger than p may be used, to improve accuracy. Compared to the Hankel method, the CI-RR method is typically more accurate and thus is most of the times preferred.

FEAST Algorithm

FEAST is a contour integration algorithm first proposed in [142] for the solution of symmetric eigenvalue problems, but was later extended to include non-symmetric eigenvalue problems. The algorithm can be understood as an accelerated subspace iteration method combined with the Rayleigh-Ritz procedure [160], and from this point of view, it has some similarities with the CI-RR method. The unique characteristic of the FEAST algorithm is that it implements an acceleration through a rational matrix function that approximates the spectral projector onto the subspace.

14

Open Source Libraries

14.1 Overview

This chapter provides an overview of open-source numerical eigensolvers that implement state-of-art numerical algorithms for non-symmetric eigenvalue problems.

Six open-source libraries are considered, namely LAPACK and its GPU-based version MAGMA, ARPACK, SLEPc, FEAST and z-PARES. These are not, by any means, the only open-source libraries available on the Internet but are those that we managed to compile and install on Linux and Mac OS X operating systems and that worked for one or more eigenvalue problems considered in this book. The libraries considered here are only those that were successful to solve relatively "large" eigenvalue problems. Libraries that do not scale well or, simply, that did not work for us, are not presented below.

14.1.1 LAPACK

- *Summary:* LAPACK (Linear Algebra PACKage) [3] is a standard library aimed at solving problems of numerical linear algebra, such as systems of linear equations and eigenvalue problems.
- *Eigenvalue Methods:* QR and QZ algorithms.
- *Programming Language:* FORTRAN 90.
- *Arithmetic:* Can handle both real and complex types.
- *Matrix Formats:* Cannot handle general sparse matrices, but is functional with dense matrices. In fact, LAPACK is the standard dense matrix data interface used by all other eigenvalue libraries.
- *Returned Eigenvectors:* LAPACK algorithms are two-sided, i.e. return both right and left eigenvectors.
- *Dependencies:* A large part of the computations required by the routines of LAPACK are performed by calling the BLAS (Basic Linear Algebra Subprograms) [90]. In general, BLAS functionality is classified in three levels.

Level 1 defines routines that carry out simple vector operations; Level 2 defines routines that carry out matrix-vector operations; and Level 3 defines routines that carry out general matrix-matrix operations. Modern optimized BLAS libraries, such as ATLAS (Automatically Tuned Linear Algebra Software) [33] and Intel MKL (Math Kernel Library), typically support all three levels for both real and complex data types.

- *GPU-based version:* MAGMA (Matrix Algebra for GPU and Multicore Architectures) [163] provides hybrid CPU/GPU implementations of LAPACK routines. It depends on NVidia CUDA. For general non-symmetric matrices, MAGMA includes the QR algorithm for the solution of the LEP but does not support the solution of the GEP.

- *Parallel version:* ScaLAPACK (Scalable LAPACK) [17] provides implementations of LAPACK routines for parallel distributed memory computers. Similarly to the dependence of LAPACK on BLAS, ScaLAPACK depends on PBLAS (Parallel BLAS), which in turn depends on BLAS for local computations and BLACS (Basic Linear Algebra Communication Subprograms) for communication between nodes.

- *Development:* As an eigensolver, LAPACK is the successor of EISPACK [54]. The first version of LAPACK was released in 1992. Compared to EISPACK, LAPACK was restructured to include the block forms of QR and QZ algorithms, which allows exploiting Level 3 BLAS and leads to improved efficiency [3]. The latest version of LAPACK is 3.7 and was released in 2016.

14.1.2 ARPACK

- *Summary:* ARPACK (ARnoldi PACKage) [93] is a library developed for solving large-scale eigenvalue problems using the IR-Arnoldi iteration.

- *Eigenvalue Methods:* IR-Arnoldi iteration.

- *Programming Language:* FORTRAN 77.

- *Arithmetic:* Both real and complex types.

- *Matrix Formats:* ARPACK supports the Reverse Communication Interface (RCI), which provides to the user the freedom to customize the matrix data format as desired. In particular, with RCI, whenever a matrix operation has to take place, control is returned to the calling program with an indication of the task required and the user can, in principle, choose the solver for the specific task independently from the library.

- *Returned Eigenvectors:* Only right eigenvectors are calculated.

- *Dependencies:* ARPACK depends on a number of subroutines from LAPACK/BLAS. Moreover, ARPACK requires to be linked to a library that factorizes matrices. This can be either dense or sparse. In the simulations described in this book, we linked ARPACK to the efficient library KLU, which is part of SuiteSparse [44], and that is particularly suited for sparse matrices whose structure originates from an electrical circuit.

- *GPU-based version:* To the best of the authors' knowledge, a library that provides a functional GPU-based implementation of ARPACK is not available to date.

- *Parallel version:* PARPACK (Parallel ARPACK) is an implementation of ARPACK for parallel computers. The message parsing layers supported by PARPACK are MPI (Message Passing Interface) [156] and BLACS.

- *Development:* The first version of ARPACK became available on Netlib in 1995. The last few years, ARPACK has stopped getting updated by Rice University. The library has been forked into ARPACK-NG, a collaborative effort among software developers, including Debian, Octave and Scilab, to put together their own improvements and fixes of ARPACK. The latest version of ARPACK-NG is 3.7.0 and was released in 2019.

14.1.3 SLEPc

- *Summary:* SLEPc (Scalable Library for Eigenvalue Problem computations) [69] is a library that focuses on the solution of large sparse eigenproblems.

- *Eigenvalue Methods:* SLEPc includes a variety of methods, both for symmetric and non-symmetric problems. For non-symmetric problems, it provides the following methods: power/inverse power/Rayleigh quotient iteration with deflation, in a single implementation; Subspace iteration with Rayleigh-Ritz projection and locking; Explicitly Restarted and Deflated (ERD) Arnoldi; Krylov-Schur; Generalized Davidson (GD); Jacobi-Davidson (JD); CI-Hankel and CI-RR methods.

- *Programming Language:* C.

- *Arithmetic:* Both real and complex types.

- *Matrix Formats:* SLEPc depends on LAPACK as an interface for dense matrix and on MUMPS as an interface for sparse Compressed Sparse Row (CSR) matrix formats. In addition, it supports custom data formats, enabled by RCI.

- *Returned Eigenvectors:* Only the power method and Krylov-Schur method implementations are two-sided. All other algorithms return only right eigenvectors.

- *Dependencies:* SLEPc depends on PETSc (Portable, Extensible Toolkit for Scientific Computation) [7]. By default the matrix factorization routines provided by PETSc are utilized by SLEPc but, at the compilation stage, SLEPc can be linked to other more efficient solvers, e.g. MUMPS, which is recommended by SLEPc developers and exploits parallelism.
- *GPU-based version:* SLEPc supports GPU computing, which depends on NVidia CUDA.
- *Parallel version:* SLEPc includes a parallel version which depends on MPI. The parallel version employs MUMPS as its linear sparse solver.
- *Development:* The first version of SLEPc (2.1.1) was released in 2002. The latest version is SLEPc 3.13 and was released in 2020.

14.1.4 FEAST

- *Summary:* FEAST [143] is the eigensolver that implements the FEAST algorithm, first proposed in [142]. Among other characteristics, the package includes the option to switch to IFEAST, which uses an inexact iterative solver to avoid direct matrix factorizations. This feature is particularly useful if the sparse matrices are very large and carrying out direct factorization is very expensive.
- *Eigenvalue Methods:* FEAST.
- *Programming Language:* FORTRAN 90, C.
- *Arithmetic:* Both real and complex types.
- *Matrix Formats:* FEAST depends on LAPACK as an interface for dense matrix, on SPIKE [144] as an interface for banded matrix and on MKL-PARDISO [153] for sparse CSR matrix formats. In addition, FEAST includes RCI and thus, data formats can be customized by the user. Using the sparse interface requires linking FEAST with Intel MKL. Linking the library with BLAS/LAPACK and not with MKL is possible, but seriously impacts the performance and the behavior of the library.
- *Returned Eigenvectors:* The FEAST algorithm implementation is two-sided.
- *Dependencies:* FEAST requires LAPACK/BLAS.
- *GPU-based version:* Currently not supported.
- *Parallel version:* PFEAST and PIFEAST are the parallel implementations of FEAST and IFEAST, respectively. Both support three-level MPI message parsing layer, with available options for the MPI library being Intel MPI, OpenMPI and MPICH. PFEAST and PIFEAST employ MKL-Cluster-PARDISO and PBiCGStab, respectively, as their parallel linear sparse solvers.

- *Development:* The first version of FEAST (v1.0) was released in 2009 and supported only symmetric matrices. Through the years FEAST added many functionalities, such as support for non-symmetric matrices, parallel computing and support for polynomial eigenvalue problems. Since 2013, Intel MKL has adopted FEAST v2.1 as its extended eigensolver. The latest version is FEAST v4.0, which was released in 2020.

14.1.5 z–PARES

- *Summary:* z-PARES [50] is a complex moment-based contour integration eigensolver for GEPs that finds the eigenvalues (and corresponding eigenvectors) that lie into a contour path defined by the user.
- *Eigenvalue Methods:* CI-Hankel, CI-RR.
- *Programming Language:* FORTRAN 90/95.
- *Arithmetic:* Both real and complex types.
- *Matrix Formats:* z-PARES depends on LAPACK for dense matrices and on MUMPS (MUltifrontal Massively Parallel Solver) [2] for sparse CSR matrices. In addition, it supports custom data formats, enabled by RCI.
- *Returned Eigenvectors:* Only right eigenvectors are calculated.
- *Dependencies:* z-PARES requires BLAS/LAPACK to be installed.
- *GPU-based version:* Currently not supported.
- *Parallel version:* z-PARES includes a parallel version, which exploits two-level MPI layer and employs MUMPS as its sparse solver.
- *Development:* The latest version of z-PARES is v0.9.6a and was released in 2014.

14.1.6 Summary of Library Features

Tables 14.1, 14.2, and 14.3 provide a synoptic summary of the methods, relevant features and versions and dependencies of the open-source libraries considered in this chapter.

14.2 Left Eigenvectors

As discussed in Chapter 8 both right and left eigenvectors, $\mathbf{u}$ and $\mathbf{w}$, respectively, are required to calculate participation factors. However, in general,

TABLE 14.1: Methods of open-source libraries for non-symmetric eigenvalue problems.

Library	Method
LAPACK	QR, QZ
ARPACK	IR-Arnoldi
SLEPc	Power/Inverse Power/Rayleigh Quotient Iteration, Subspace, ERD-Arnoldi, Krylov-Schur, GD, JD, CI-Hankel, CI-RR
FEAST	FEAST
z-PARES	CI-Hankel, CI-RR

TABLE 14.2: Relevant features of open-source libraries for non-symmetric eigenvalue problems.

Library	Data formats				Computing		Two-sided	Real/	Releases	
	dense	CSR	band	RCI	GPU	paral.		compl.	first	latest
LAPACK	✓	✗	✗	✗	✓[a]	✓[b]	✓	✓	1992	2016
ARPACK	✗	✗	✗	✓	✗	✓	✗	✓	1995	2019[c]
SLEPc	✓	✓	✗	✓	✓	✓	✓[d]	✓	2002	2020
FEAST	✓	✓	✓	✓	✗	✓	✓	✓	2009	2020
z-PARES	✓	✓	✗	✓	✗	✓	✗	✓	2014	2014

[a]With MAGMA.
[b]With ScaLAPACK.
[c]Now as ARPACK-NG.
[d]In SLEPc, only the power and the Krylov-Schur method implementations are two-sided.

numerical libraries return by default only the right eigenvectors. A handful of libraries allow calculating both left and right eigenvectors at once. Among these we cite LAPACK, MAGMA and FEAST. But these are the exception rather than the rule. A mechanism to calculate left eigenvectors, thus, has to be put in place.

The very definition of left and right eigenvectors gives a hint. From (1.167) we observe that, if λ is an eigenvalue of the pencil $s\mathbf{I}_n - \mathbf{A}$ and $\mathbf{u}$, $\mathbf{w}$, are its associated right and left eigenvectors, respectively, then $\mathbf{w}^{\mathrm{T}}$ is also a right eigenvector associated to λ for the pencil $s\mathbf{I}_n - \mathbf{A}^{\mathrm{T}}$, i.e.:

$$\mathbf{A}^{\mathrm{T}}\,\mathbf{w}^{\mathrm{T}} = \lambda\,\mathbf{w}^{\mathrm{T}}\,. \tag{14.1}$$

Thus, to calculate both right and left eigenvectors, one just needs to solve

TABLE 14.3: Versions and dependencies of open-source libraries for non-symmetric eigenvalue problems.

Library (Version)	Dependencies (Version)
LAPACK (3.8.0)	ATLAS (3.10.3)
MAGMA (2.2.0)	NVidia CUDA (10.1)
ARPACK-NG (3.5.0)	SuiteSparse KLU (1.3.9)
FEAST (v4.0)	PARDISO (5.0.0)
z-PARES (0.9.6a)	OpenMPI (3.0.0), MUMPS (5.1.2)
SLEPc (3.8.2)	PETSc (3.8.4), MUMPS (5.1.2)

twice the eigenvalue problem, one for $\mathbf{A}$ and one for $\mathbf{A}^{\mathrm{T}}$. Of course, this approach doubles the computational burden and might not be viable for large eigenvalue problems.

If $\mathbf{Q}$ and $\mathbf{P}$ are the matrices of right and left eigenvectors of a regular matrix $\mathbf{A}$, respectively, and if $\mathbf{J}_n$ is a diagonal matrix with the eigenvalues of $\mathbf{A}$ as diagonal elements, then the following expressions hold:

$$\begin{aligned}\mathbf{AQ} &= \mathbf{QJ}_n\,,\\ \mathbf{PA} &= \mathbf{J}_n\mathbf{P}\,,\end{aligned}$$

or, equivalently:

$$\begin{aligned}\mathbf{PAQ} &= \mathbf{PQJ}_n\,,\\ \mathbf{PAQ} &= \mathbf{J}_n\mathbf{PQ}\,.\end{aligned}$$

Equaling the last two expressions gives:

$$\mathbf{PQJ}_n = \mathbf{J}_n\mathbf{PQ}\,,$$

which implies that the matrix product $\mathbf{PQ}$ must be a diagonal matrix. A byproduct of this result is that, if $\mathbf{A}$ is symmetric, then $\mathbf{P} = \mathbf{Q}^{\mathrm{T}}$, and if $\mathbf{A}$ is Hermitian, $\mathbf{P} = \mathbf{Q}^{\mathrm{H}}$. For a non-symmetric matrix $\mathbf{A}$ with non-degenerate eigenvalues, then, it is always possible to normalize $\mathbf{P}$ and $\mathbf{Q}$ such that $\mathbf{PQ}^{-1} = \mathbf{I}_n$. This property can, in principle be utilized to calculate the left eigenvectors from the right ones. However, the calculation of the inverse of a (dense) matrix quickly become an intractable problem to solve as n increases. Moreover, $\mathbf{Q}$ has to be square for the inverse to be calculated, which implies that $\mathbf{A}$ must be square and that *all* eigenvalues have to be calculated. Since, for the small-signal stability analysis of physical systems, one is generally interested only in the critical eigenvalues, the calculation of $\mathbf{P}$ through the inverse of $\mathbf{Q}$ appears quite impractical if not even unattainable for very large problems.

In the classical paper [154], Semlyen and Wang suggest a formula to calculate directly left eigenvectors from right ones. The formula is based on the Matrix Modification Lemma [68] and reads:

$$\mathbf{w}^{\mathrm{T}} = \mathbf{H} - \frac{\mathbf{H}\,|\mathbf{u}^{\mathrm{T}}\,\mathbf{H}^{\mathrm{T}}|}{1 + \mathbf{u}^{\mathrm{T}}\,\mathbf{H}^{\mathrm{T}}}\,, \tag{14.2}$$

where

$$\mathbf{H} = \mathbf{e}^{\mathrm{T}}\,\mathbf{K}^{-1}\,, \tag{14.3}$$

where $\mathbf{e}$ is a unit vector with unity at the location, say j, corresponding to the largest element of $\mathbf{u}$,[1] and $\mathbf{K}$ is equal to $\mathbf{A} - \lambda\,\mathbf{I}_n$ except for the j-th column which is substituted for $\mathbf{e}$. An expression similar to (14.2) can be obtained for pencils in the form $s\mathbf{E} - \mathbf{A}$.

From a computational point of view, (14.2) is less demanding than the calculation of the inverse of $\mathbf{Q}$ as it requires the factorization of $\mathbf{K}$, not the calculation of its inverse, although it still requires that $\mathbf{K}$ is invertible. Moreover, (14.2) does not require the knowledge of all eigenvalues and of the full matrix $\mathbf{Q}$. The method given in [154], however, is not necessarily cheaper than the re-calculation of the eigenvalues for $\mathbf{A}^{\mathrm{T}}$, especially if the numerical library does not search all eigenvalues and allows providing an initial guess of the eigenvalues to be found.

Ultimately, which is the most efficient solution for the calculation of the left eigenvalues depends in turn on the library and the eigenvalue problem. The best solution has to be determined case by case.

14.3 Spectral Transforms

We have discussed the application of linear spectral (Möbius) transforms to matrix pencils of linear systems in Chapter 7. These transforms are very commonly utilized in eigenvalue numerical methods usually for one of the following reasons:

- *To find the eigenvalues of interest.* Some numerical methods, for example vector iteration-based methods, find the Largest Magnitude (LM) eigenvalues, whereas the eigenvalues of interest in small-signal stability analysis are typically the ones with Smallest Magnitude (SM) or Largest Real Part (LRP). Thus, it is necessary to apply a transform, e.g. the invert or shift & invert.

- *Address a singularity issue.* A GEP with singular left hand side coefficient matrix $\mathbf{E}$ can create problems to many numerical methods. Applying a spectral transform can help address singularity issues.

[1] In [154], $\mathbf{u}$ is normalized in such a way that its largest element is equal to 1.

- *Accelerate convergence.* Eigenvalues that are not very close to each other can lead to large errors and slow convergence of eigensolvers. Spectral transforms can help magnify the eigenvalues of interest and speedup convergence.

The choice of the best transform for a specific system and eigenvalue problem is a challenging task to solve. In our experience, the selection of shift values is always heuristic.

Example 14.1 (Test of numerical methods for eigenvalue problems). We illustrate how different numerical methods work for the three-bus system presented in Appendix A and linearized around the base-case operating point. The LEP and GEP, in all previous examples of the book, were solved using exclusively the Schur decomposition method implementations provided by LAPACK. In this example, we discuss a variety of methods: LAPACK's QR and QZ algorithms; SLEPc's Subspace, ERD-Arnoldi and Krylov-Schur methods; ARPACK's IR-Arnoldi; and z-PARES' CI-RR method.

The eigenvalues of the three-bus system for the LEP and GEP obtained by LAPACK are shown in Table 14.4. Both QR and QZ algorithms find all 19 finite eigenvalues of the system. For the GEP, the QZ algorithm also finds the additional infinite eigenvalue with its algebraic multiplicity. Since LAPACK is the most mature software tool among those considered in this chapter, the accuracy of the eigenvalues obtained with all other numerical libraries is evaluated by comparing them with the reference solution shown in Table 14.4.

The eigenvalues obtained by the Subspace iteration method are shown in Table 14.5. The implementation of SLEPc only finds the desired number of LM eigenvalues. However, in the s-domain, the critical for the stability eigenvalues are typically not the LM ones, but the ones with LRP or SM. Especially for the GEP, the LM eigenvalue is infinite and, hence, does not provide any meaningful information on the system dynamics.

The first two columns of Table 14.5 indicate that, for the needs of power system small-signal stability analysis, the methods that look for LM eigenvalues must always be combined with a spectral transform. For example, considering the GEP, we can apply the invert transform and pass to SLEPc the pencil of the dual system, i.e. $z\mathbf{A} - \mathbf{E}$. The obtained eigenvalues are transformed back to the original problem to obtain the SM eigenvalues of $s\mathbf{E} - \mathbf{A}$. The results are shown in the third column of Table 14.5.

Table 14.6 shows the eigenvalues found by solving the LEP with Krylov subspace methods and in particular, with ARPACK's IR-Arnoldi iteration, SLEPc's ERD-Arnoldi method and SLEPc's Krylov-Schur method. ARPACK is set to find the 10 LRP eigenvalues. On the other hand, Krylov subspace methods in SLEPc can find all eigenvalues for small problems. ARPACK returns a spurious zero eigenvalue but, overall, the three methods are able to find the desired eigenvalues with good accuracy.

The eigenvalues found by solving the GEP with Krylov subspace methods are presented in Table 14.7. In this case, we were unable to find any

TABLE 14.4: Solution of the LEP and GEP for the three-bus system: QR and QZ algorithms.

Library	LAPACK	
Problem	LEP	GEP
Method	QR	QZ
Spectrum	All	All
Eigs. found	0.0000 (2)	0.0000 (2)
	$-0.2516 \pm \jmath 4.4309$	$-0.2516 \pm \jmath 4.4309$
	$-0.8211 \pm \jmath 3.4132$	$-0.8211 \pm \jmath 3.4132$
	-1.0096	-1.0096
	-1.0147	-1.0147
	$-1.1003 \pm \jmath 2.5390$	$-1.1003 \pm \jmath 2.5390$
	-3.6091	-3.6091
	-13.6561	-13.6561
	$-19.6722 \pm \jmath 4.1469$	$-19.6722 \pm \jmath 4.1469$
	$-20.5133 \pm \jmath 5.9284$	$-20.5133 \pm \jmath 5.9284$
	-23.7196	-23.7196
	-33.7327	-33.7327
	-35.2064	-35.2064
		∞ (37)

TABLE 14.5: Solution of the LEP, GEP, and dual GEP for the three-bus system: subspace method.

Library	SLEPc		
Method	Subspace		
Spectrum	10 LM		
Problem	LEP	GEP	GEP
Transform	–	–	Invert
Eigs. found	$-0.2516 \pm \jmath 4.4309$	∞ (10)	0.0000 (2)
	-13.6561		$-0.2516 \pm \jmath 4.4309$
	$-19.6722 \pm \jmath 4.1469$		$-0.8211 \pm \jmath 3.4131$
	$-20.5133 \pm \jmath 5.9284$		-1.0096
	-23.7196		-1.0147
	-33.7327		$-1.1003 \pm \jmath 2.5390$
	-35.2064		-3.6091

correct eigenvalues with ARPACK. For this size, SLEPc's ERD-Arnoldi and Krylov-Schur methods allow obtaining all eigenvalues. In order to obtain the eigenvalues with good accuracy, we use the option "Target Real Part" (TRP) which allows targeting eigenvalues with specified real part. We set TRP to -0.01 and apply the shift & invert transform with $\sigma = -0.01$.

TABLE 14.6: Solution of the LEP for the three-bus system: Krylov subspace methods.

Library	ARPACK	SLEPc	SLEPc
Method	IR-Arnoldi	ERD-Arnoldi	Krylov-Schur
Spectrum	10 LRP	All	All
Eigs.	0.0000 (3)	0.0000 (2)	0.0000 (2)
found	$-0.2516 \pm \jmath 4.4309$	$-0.2516 \pm \jmath 4.4309$	$-0.2516 \pm \jmath 4.4309$
	$-0.8211 \pm \jmath 3.4132$	$-0.8211 \pm \jmath 3.4132$	$-0.8211 \pm \jmath 3.4132$
	-1.0096	-1.0096	-1.0096
	-1.0147	-1.0147	-1.0147
	$-1.1003 \pm \jmath 2.5390$	$-1.1003 \pm \jmath 2.5390$	$-1.1003 \pm \jmath 2.5390$
	-3.6091	-3.6091	-3.6091
		-13.6561	-13.6561
		$-19.6722 \pm \jmath 4.1469$	$-19.6722 \pm \jmath 4.1469$
		$-20.5133 \pm \jmath 5.9284$	$-20.5133 \pm \jmath 5.9284$
		-23.7196	-23.7196
		-33.7327	-33.7327
		-35.2064	-35.2064

For the sake of completeness, we mention that the eigenvalues obtained with SLEPc, when compared to the ones found by LAPACK, appeared to be shifted by a constant offset $-\sigma$, i.e. 0.0100 was returned instead of 0.0000, and so on. The results shown in Table 14.7 take into account such a shift by adding σ to all output values returned by SLEPc.

Finally, we discuss the results obtained using the contour integration based methods by FEAST and z-PARES. For both FEAST and z-PARES' CI-RR, we define the search contour to be a circle with center the point $c = (-0.01, 4)$ and radius $\rho = 8$. Note that, for the GEP, FEAST could not handle the singularity of $\mathbf{E}$ in the pencil $s\mathbf{E} - \mathbf{A}$ and returned an error. To overcome this issue, we applied the Cayley transform with $\sigma = -1$ to the original pencil and passed to SLEPc the transformed pencil $z(\mathbf{A} + \mathbf{E}) - (-\mathbf{A} + \mathbf{E})$ to FEAST. This is another utility of spectral transforms, i.e. they can be properly utilized to address singularity issues.

For the setup described above, both methods obtain the same eigenvalues for both LEP and GEP. The search contour and the location of the characteristic roots are depicted in Figure 14.1.

Finally, the obtained eigenvalues are summarized in Table 14.8. Compared to the results by LAPACK, both methods accurately find the eigenvalues located in the defined region. □

TABLE 14.7: Solution of the GEP for the three-bus system: Krylov subspace methods.

Library	SLEPc	SLEPc
Method	ERD-Arnoldi	Krylov-Schur
Spectrum	All	All
Transform	Shift & invert	Shift & invert
	$\sigma = -0.01$	$\sigma = -0.01$
Eigs.	0.0000 (2)	0.0000 (2)
found	$-0.2516 \pm \jmath 4.4310$	$-0.2516 \pm \jmath 4.4310$
	$-0.8211 \pm \jmath 3.4132$	$-0.8211 \pm \jmath 3.4132$
	-1.0096	-1.0096
	-1.0147	-1.0147
	$-1.1003 \pm \jmath 2.5390$	$-1.1003 \pm \jmath 2.5390$
	-3.6091	-3.6091
	-13.6561	-13.6561
	$-19.6722 \pm \jmath 4.1469$	$-19.6722 \pm \jmath 4.1469$
	$-20.5133 \pm \jmath 5.9284$	$-20.5133 \pm \jmath 5.9284$
	-23.7196	-23.7196
	-33.7327	-33.7327
	-35.2064	-35.2064

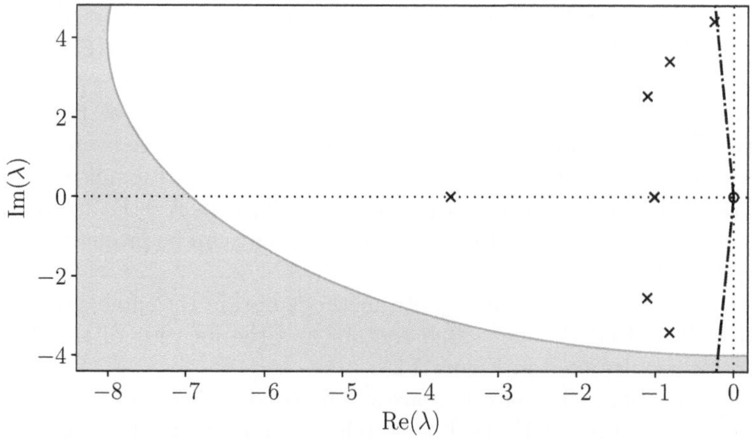

FIGURE 14.1: Eigenvalues obtained with z-PARES and FEAST for the three-bus system.

TABLE 14.8: Solution of the LEP and GEP for the three-bus system: contour integration methods.

Library	FEAST	z-PARES
Method	FEAST	CI-RR
Spectrum	$c = (-0.01, 4), \rho = 8$	$c = (-0.01, 4), \rho = 8$
Transform (GEP)	Cayley, $\sigma = -1$	–
Eigs.	0.0000 (2)	0.0000 (2)
found	$-0.2516 + \jmath 4.4309$	$-0.2516 + \jmath 4.4309$
	$-0.8211 \pm \jmath 3.4132$	$-0.8211 \pm \jmath 3.4132$
	-1.0096	-1.0096
	-1.0147	-1.0147
	$-1.1003 \pm \jmath 2.5390$	$-1.1003 \pm \jmath 2.5390$
	-3.6091	-3.6091

15

Large Eigenvalue Problems

The largest ever computer-based matrix eigenvalue calculations with dense eigensolvers to date were carried out in Japan in 2014. In particular, the analysis was carried out with the K computer in Riken, to evaluate the Japanese eigensolver EigenExa. EigenExa was compared with other parallel dense eigensolvers (such as ScaLAPACK). EigenExa managed to find the eigenvalues of a $10^6 \times 10^6$ problem in about 1 hour. This result has a high cost. The K computer includes 88,000 processors which draw a peak power of 12.6 MW (see Figure 15.1).[1] Its operation costs annually US$10 million.

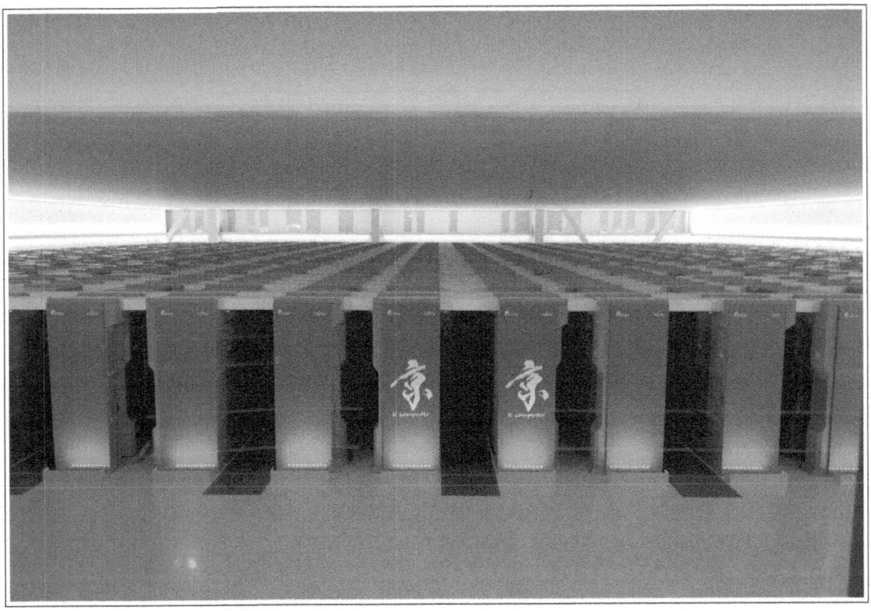

FIGURE 15.1: The K computer built by Fujitsu and installed at the Riken Advanced Institute for Computational Science. The letter "K" sounds as the Chinese word 京, which means 10^{16}.

[1]Photo by Toshihiro Matsui, Tsukuba and Yokohama, Japan, licensed under the Creative Commons Attribution 2.0 Generic license. The text of the license is available at: https://creativecommons.org/licenses/by/2.0/deed.en

Even using sparse matrices and limiting the search to a subset of the spectrum, the solution of large eigenvalue problems is challenging. Simply put, when it comes to eigenvalue analysis, the computing capability does matter.

This chapter discusses the scalability of the numerical solution of eigenvalue problems. The examples presented below illustrate the scalability and performance of the algorithms and libraries described in previous chapters when applied to large eigenvalue problems based on real-world electrical power networks.

Example 15.1 (All-island Irish transmission system). This example considers the detailed dynamic model of the all-island Irish transmission system (AIITS) considered in Example 8.4. The system consists of 1,479 buses, 1,851 transmission lines and transformers, 245 loads, 22 synchronous generators with AVRs and TGs, 6 PSSs; and 176 wind generators. In total, the dynamic model has $n = 1,443$ state variables and $l = 7,197$ algebraic variables. The pencils and sizes of the LEP and GEP problems for the AIITS are shown in Table 15.1. In addition, the dynamic model has been validated using frequency data from a severe event that occurred in the real system in 2018, see [131].

TABLE 15.1: Dimensions of the LEP and GEP for the AIITS.

Problem	Pencil	Size
LEP	$s\,\mathbf{I}_n - \mathbf{A}_{\rm S}$	$1,443 \times 1,443$
GEP	$s\,\mathbf{E} - \mathbf{A}$	$8,640 \times 8,640$

Results of the eigenvalue analysis of the AIITS are discussed for different numerical method implementations, namely, QR and QZ algorithms by LAPACK, QR algorithm by MAGMA, Subspace iteration, ERD-Arnoldi and Krylov-Schur methods by SLEPc, IR-Arnoldi by ARPACK; and contour integration methods by z-PARES and FEAST. In particular, we provide for each method the rightmost eigenvalues, as well as the time required to complete the solution of the eigenvalue problem.

Two remarks are relevant regarding the computing times reported in the remainder of this example. First, all simulations were executed on a server mounting two quad-core Intel Xeon 3.50 GHz CPUs, 1 GB NVidia Quadro 2,000 GPU, 12 GB of RAM, and running a 64-bit Linux OS. Second, since, not all method implementations include two-sided versions and in order to provide as a fair comparison as possible, all eigensolvers are called so as to return only the calculated eigenvalues and not eigenvectors.

The results obtained with Schur decomposition methods are presented in Table 15.2. The obtained rightmost eigenvalues are the same for both LEP and GEP. The system root loci plot is shown in Figure 15.2. It is worth mentioning that, as opposed to all other numerical methods, obtaining these eigenvalues requires no parameter tuning. In the remainder of this example, the eigenvalues of Table 15.2 are considered the reference when evaluating the accuracy of the rest of the methods.

Regarding the computational time, we see that, for the LEP, both LAPACK and the GPU-based MAGMA are very efficient at this scale, with MAGMA providing only a marginal speedup. On the other hand, when it comes to solving the GEP with LAPACK's QZ method, scalability becomes a serious issue, since the problem is solved in 3,669.77 s, which is computationally cumbersome.

TABLE 15.2: Schur decomposition methods, LEP and GEP for the AIITS.

Library	LAPACK	MAGMA	LAPACK
Problem	LEP	LEP	GEP
Method	QR	QR	QZ
Spectrum	All	All	All
Time [s]	3.94	3.54	3,669.77
Found	1,443 eigs.	1,443 eigs.	8,640 eigs.
LRP eigs.	0.0000	0.0000	0.0000
	-0.0869	-0.0869	-0.0869
	$-0.1276 \pm \jmath 0.1706$	$-0.1276 \pm \jmath 0.1706$	$-0.1276 \pm \jmath 0.1706$
	$-0.1322 \pm \jmath 0.4353$	$-0.1322 \pm \jmath 0.4353$	$-0.1322 \pm \jmath 0.4353$
	-0.1376	-0.1376	-0.1376
	-0.1382	-0.1382	-0.1382
	-0.1386	-0.1386	-0.1386
	-0.1390	-0.1390	-0.1390
	-0.1391	-0.1391	-0.1391
	-0.1393	-0.1393	-0.1393
	-0.1394	-0.1394	-0.1394

The eigenvalues found by the subspace iteration are shown in Table 15.3. We were not able to obtain any eigenvalues for the LEP. Regarding the GEP, we have set the method to look for the 50 LM eigenvalues of the dual system, which correspond to the 50 SM eigenvalues of the prime system. With this setup, the pair $-0.1322 \pm \jmath 0.4353$ is not captured, since its magnitude is larger than the magnitudes of the 50 SM eigenvalues. To obtain also this pair, we could customize the spectral transform or simply increase the number of the eigenvalues to be returned. However, the best setup is not known a priori and thus, some heuristic parameter tuning is required. Finally, the method does not scale well, since solution of the GEP is completed in 6,807.24 s.

The rightmost eigenvalues found with Krylov subspace methods for the LEP and GEP are shown in Table 15.4 and Table 15.5, respectively. For the LEP, ARPACK is set up to find the 50 LRP eigenvalues. Although all obtained eigenvalues are actual eigenvalues of the system, some of the LRP ones are missed. No correct eigenvalues were found for the GEP, which we attribute to

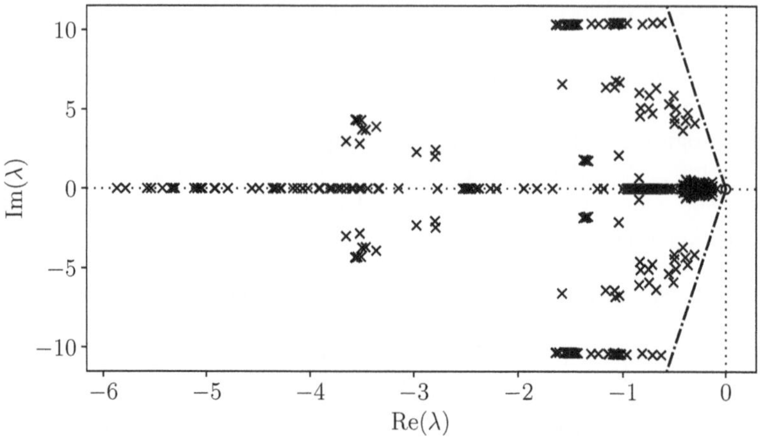

FIGURE 15.2: Root loci computed with LAPACK for the AIITS.

TABLE 15.3: Subspace iteration, LEP and GEP for the AIITS.

Library	SLEPc
Method	Subspace
Spectrum	50 LM
Transform	Invert
Time [s]	6, 807.24
Found	50
LRP eigs.	−0.0000
	−0.0869
	$-0.1276 \pm \jmath 0.1706$
	−0.1376
	−0.1382
	−0.1386
	−0.1390
	−0.1391
	−0.1393
	−0.1394
	−0.1397

the fact that a non-symmetric $\mathbf{E}$ is not supported. In SLEPc methods, both for LEPs and GEPs we have set the TRP parameter to -0.01, also combining with a shift & invert transform with $\sigma = -0.01$. The two methods are able to capture all rightmost eigenvalues. However, to obtain the correct results, the output values of SLEPc had to be shifted by σ.

Finally, the Krylov subspace methods by SLEPc appear to be more efficient than ARPACK's IR-Arnoldi. Compared to Schur decomposition methods, at this scale, Krylov methods, although they require some tuning, appear to be by far more efficient for the GEP, but less efficient for the LEP.

TABLE 15.4: Krylov subspace methods and LEP for the AIITS.

Library Method Spectrum	ARPACK IR-Arnoldi 50 LRP	SLEPc ERD-Arnoldi 50 TRP	SLEPc Krylov-Schur 50 TRP
Transform	-	Shift & invert $\sigma = -0.01$	Shift & invert $\sigma = -0.01$
Time [s] Found	76.96 26 eigs.	17.84 54 eigs.	16.58 55 eigs.
LRP eigs.	-0.0000	0.0000	0.0000
	-0.0869	-0.0869	-0.0869
	$-0.1276 \pm \jmath 0.1706$	$-0.1276 \pm \jmath 0.1706$	$-0.1276 \pm \jmath 0.1706$
	$-0.1322 \pm \jmath 0.4353$	$-0.1322 \pm \jmath 0.4353$	$-0.1322 \pm \jmath 0.4353$
	$-0.1615 \pm \jmath 0.2689$	-0.1376	-0.1376
	$-0.1809 \pm \jmath 0.2859$	-0.1382	-0.1382
	$-0.2042 \pm \jmath 0.3935$	-0.1386	-0.1386
	$-0.2172 \pm \jmath 0.2646$	-0.1390	-0.1390
	$-0.2335 \pm \jmath 0.3546$	-0.1391	-0.1391
	$-0.2344 \pm \jmath 0.3644$	-0.1393	-0.1393
	$-0.2503 \pm \jmath 0.4363$	-0.1394	-0.1394

The results produced by z-PARES' CI-RR for the LEP and GEP, are presented in Table 15.6. The method is set to look for solutions in the circle with center $c = (-0.01, 4)$ and radius $\rho = 8$. In both cases, the eigenvalues found by z-PARES are actual eigenvalues of the system, although the eigenvalues found for the GEP include noticeable errors, when compared to LAPACK.

The most relevant issue is that the eigenvalues obtained are not the most important ones for the stability of the system, which means that critical eigenvalues are missed. This issue occurs despite the defined search contour being reasonable. Of course, there may be some region for which the critical eigenvalues are captured but, this can not be known *a priori*. Regarding the simulation time, the method for the AIITS is faster than SLEPc's Krylov subspace methods for the LEP, but slower for the GEP.

The search contour and the location of the characteristic roots found by z-PARES for the LEP are depicted in Figure 15.3.

Finally, we were not able to obtain any eigenvalues for the AIITS using the installed version of FEAST. Results obtained with an older version of the library are reported in [120]. In particular for the AIITS, FEAST solved the GEP in 232.06 s, providing approximations of 40 eigenvalues.

TABLE 15.5: Krylov subspace methods and GEP for the AIITS.

Library	SLEPc	SLEPc
Method	ERD-Arnoldi	Krylov-Schur
Spectrum	50 TRP	50 TRP
Transform	Shift & invert	Shift & invert
	$\sigma = -0.01$	$\sigma = -0.01$
Time [s]	8.93	7.64
Found	51 eigs.	53 eigs.
LRP eigs.	0.0000	0.0000
	-0.0869	-0.0869
	$-0.1276 \pm \jmath 0.1706$	$-0.1276 \pm \jmath 0.1706$
	$-0.1322 \pm \jmath 0.4353$	$-0.1322 \pm \jmath 0.4353$
	-0.1376	-0.1376
	-0.1382	-0.1382
	-0.1386	-0.1386
	-0.1390	-0.1390
	-0.1391	-0.1391
	-0.1393	-0.1393
	-0.1394	-0.1394

TABLE 15.6: Contour integration method for, LEP and GEP for the AIITS.

Library	z-PARES	
Method	CI-RR	
Spectrum	$c = (-0.01, 4), \rho = 8$	
Problem	LEP	GEP
Time [s]	10.81	17.10
Found	49 eigs.	52 eigs.
LRP eigs.	$-0.3041 + \jmath 4.1425$	$-0.3040 + \jmath 4.1429$
	$-0.3720 + \jmath 4.7773$	$-0.3715 + \jmath 4.7774$
	$-0.3945 + \jmath 4.3121$	$-0.3947 + \jmath 4.3122$
	$-0.4184 \pm \jmath 3.6794$	$-0.4187 \pm \jmath 3.6794$
	$-0.4866 + \jmath 5.0405$	$-0.4865 + \jmath 5.0405$
	$-0.5011 + \jmath 4.1276$	$-0.5007 + \jmath 4.1274$
	$-0.5022 + \jmath 4.4417$	$-0.5018 + \jmath 4.4417$
	$-0.5077 + \jmath 5.8727$	$-0.5097 + \jmath 5.8747$
	$-0.5555 + \jmath 5.3444$	$-0.5542 + \jmath 5.3436$
	$-0.6765 + \jmath 6.3426$	$-0.6761 + \jmath 6.3412$

Assume that time-delays are introduced into 5 PSSs ranging from 0.09 s to 0.15 s. The rightmost eigenvalues of the AIITS can be obtained through three

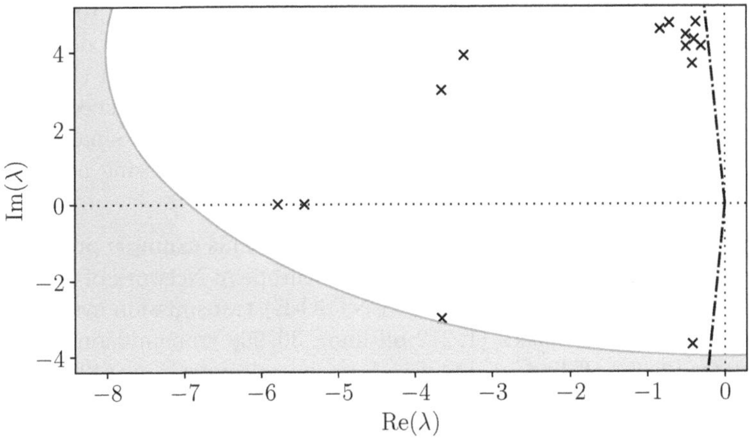

FIGURE 15.3: Root loci obtained with z-PARES and LEP for the AIITS.

different discretization schemes, namely the Chebyshev polynomial interpolation, Krylov method and the sparse delay matrices discussed in Chapter 10. Table 15.7 shows the results obtained for these three techniques, which are obtained by employing SLEPc's ERD-Arnoldi method combined with the shift & invert transform and $\sigma = -0.01$.

TABLE 15.7: Spectrum discretization and time delays for the AIITS.

Method	Chebyshev interpolation	Krylov method	Sparse Delay matrices
Order	7,290	8,748	1,478
Problem	LEP	GEP	GEP
Time [s]	57.72	88.23	19.88
Found	33	35	27
LRP eigs.	0.0000	0.0000	0.0000
	-0.0669	-0.0681	-0.0680
	$-0.1076 + \jmath 0.1706$	$-0.0866 + \jmath 2.1927$	$-0.0866 + \jmath 2.1927$
	$-0.1123 + \jmath 0.4349$	$-0.1124 + \jmath 0.4374$	$-0.1124 + \jmath 0.4377$
	-0.1176	$-0.1281 + \jmath 0.1754$	$-0.1280 + \jmath 0.1723$
	-0.1182	-0.1286	$-0.1316 + \jmath 0.4363$
	-0.1186	-0.1321	-0.1376
	-0.1190	-0.1354	-0.1383
	-0.1191	-0.1372	-0.1386

According to Table 15.7, the rightmost eigenvalues (except for the redundant null ones) obtained by the three methods are similar, namely -0.0669,

−0.0681 and −0.0680. This indicates that all methods are able to solve the small-signal stability analysis of a large system with inclusions of delays. The Krylov method shows a much larger computational burden than the Chebyshev polynomial interpolation, which, fortunately, can be reduced through the sparse discretization. The sparse discretization achieves the smallest order of the resulting eigenvalue problem, and therefore is a promising method for application to large systems. □

Example 15.2 (ENTSO-E transmission system). This example presents the simulation results for a dynamic model of the European Network of Transmission System Operators for Electricity (ENTSO-E) transmission system. The system includes 21,177 buses (1,212 off-line); 30,968 transmission lines and transformers (2,352 off-line); 1,144 zero-impedance connections (420 off-line); 4,828 power plants represented by sixth-order and second-order synchronous machine models; and 15,756 loads (364 off-line), modeled as constant active and reactive power consumption. Synchronous machines represented by sixth-order models are also equipped with dynamic AVR and TG models. The system also includes 364 PSSs.

As summarized in Table 15.8, the system has in total $n = 49,396$ state variables and $l = 96,770$ algebraic variables. The pencil $s\mathbf{E}-\mathbf{A}$ has dimensions $146,166 \times 146,166$ and the matrix $\mathbf{A}$ has 654,950 non-zero elements, which represent the 0.003% of the total number of elements of the matrix.

TABLE 15.8: ENTSO-E system: system dimensions and statistics.

n	49,396
l	96,770
Dimensions of $\mathbf{A}$	$146,166 \times 146,166$
Sparsity degree of $\mathbf{A}$ [%]	99.997

Neither the LEP or GEP could be solved using Schur decomposition methods. At this scale, the dense matrix representation required by LAPACK and MAGMA libraries leads to massive memory requirements, and a segmentation fault error is returned by the CPU.

Among the algorithms that support sparse matrices, we test here only the contour-integration based methods which, in fact, were the ones that were able to tackle this large eigenvalue problem on the available hardware.

The effect of changing the search region of the z-PARES CI-RR method on the eigenvalue analysis of the ENTSO-E system is shown in Table 15.9. Interestingly, simulations showed that shrinking the defined contour may lead to a marginal increase of the computation time. Although not intuitive, this result indicates that the mass of the computational burden is mainly determined by the large size of the ENTSO-E system, and that, at this scale, smaller subspaces are not necessarily constructed faster by the CI-RR algorithm. Regarding the number of eigenvalues obtained, using a region that is too small leads, as expected, to missing an important number of critical eigenvalues.

TABLE 15.9: Impact of the search region of the CI-RR method on the eigenvalue analysis of the ENTSO-E system.

Library	z-PARES		
Problem	GEP		
Method	CI-RR		
c	$(-0.01, 4)$	$(-0.01, 3)$	$(-0.01, 3)$
ρ	8	4	2
Time [s]	364.85	375.67	378.71
Found	349 eigs.	350 eigs.	110 eigs.

We test the impact of applying spectral transforms to the matrix pencil $s\mathbf{E} - \mathbf{A}$, on the eigenvalue analysis of the ENTSO-E system with z-PARES' CI-RR. In particular, we test the invert transform that yields the dual pencil $z\mathbf{A} - \mathbf{E}$; and the inverted Cayley transform, i.e. $s = (z+1)/(\sigma z - \sigma)$, which yields the pencil $z(\mathbf{E} - \sigma\mathbf{A}) - (-\sigma\mathbf{A} - \mathbf{E})$.

The results are shown in Table 15.10. Passing the transformed matrices to z-PARES provides a marginal speedup to the eigenvalue computation. In addition, considering either the prime system or the inverted Cayley transform with $\sigma = -1$, results in finding the same number of eigenvalues, whereas when the dual system is considered a number of eigenvalues is missed.

Finally, results of the eigenvalue analysis for the ENTSO-E system with FEAST using an older version of the library were reported in [120]. FEAST solved the prime and dual GEPs of the ENTSO-E system in $11,155.69$ s and 853.44 s, respectively, finding 476 eigenvalues in the circle with center $c = (-2.5, 2.5)$ and radius $\rho = 3.54$. □

TABLE 15.10: Impact of spectral transforms of the CI-RR method on the eigenvalue analysis of the ENTSO-E system.

Library	z-PARES		
Problem	GEP		
Method	CI-RR		
Spectrum	$c = (-0.01, 4), \rho = 8$		
Transform	-	Invert	Inverted Cayley
Time [s]	364.85	350.82	337.43
Found	349 eigs.	297 eigs.	349 eigs.

Part V

Appendices

A

Three-Bus System

The three-bus system that is utilized in several examples of the book is a simplification of the well-known two-area system described in [87]. The single-line diagram of the system is shown in Figure A.1. We did not use the original two-area system as its number of variables, while not huge, is still too large to properly visualize the admittance, Jacobian and state matrices of the system in the pages of a book. On the other hand, the two-area system is an excellent example of system that shows inter-area electromechanical oscillations and is thus particularly relevant for small-signal stability analysis. This is the feature that has been preserved in the three-bus system described in this appendix.

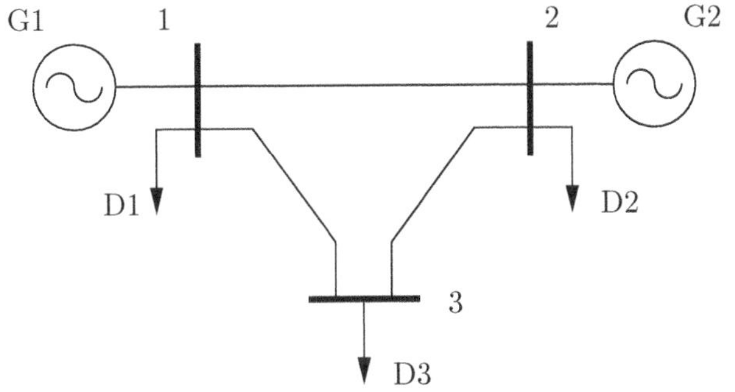

FIGURE A.1: Three-bus system: single-line diagram.

Then, we have modified the dynamic data to obtain also a Hopf bifurcation and, as a byproduct, poorly damped as well as unstable eigenvalues due to the interaction of the AVRs with the swing equation of the synchronous machines. With this aim, we have utilized a fifth-order model for one of the machines, in the same vein as the model of machine 1 of the well-known IEEE 14-bus system (see, for example, [113]) that, as it is well-known, also shows a Hopf bifurcation for given topology and loading conditions that can be eliminated by means of a properly tuned PSS.

A.1 Network Data

The transmission system operates at nominal voltage 230 kV and nominal frequency 60 Hz. Table A.1 shows the transmission line parameters. Per unit data are calculated assuming $S_b = 100$ MVA power base.

TABLE A.1: Three-bus system: branch data.

Branch #	From (h)	To (k)	R_{hk} [pu(Ω)]	X_{hk} [pu(Ω)]	B_{hk} [pu(Ω^{-1})]
1	1	2	0.022	0.220	0.385
2	1	3	0.010	0.110	0.385
3	2	3	0.011	0.110	0.385

A.2 Static Data

The power flow data and solution for the base-case operating point are shown in Tables A.2 and A.3, respectively. The nominal voltage of all buses of the network is 230 kV. The total generation, demand and losses at the base-case operating point are shown in Table A.4. Per unit data are calculated assuming $S_b = 100$ MVA as power base.

TABLE A.2: Three-bus system: base-case power flow solution.

Bus #	v [pu(kV)]	θ [rad]	P_G [pu(MW)]	Q_G [pu(MVar)]	P_D [pu(MW)]	Q_D [pu(MVar)]
1	1.0100	−0.2959	6.6700	0.4838	3.6700	0.7000
2	1.0200	−0.4577	7.7397	0.6916	7.6700	0.7000
3	0.9922	−0.5432	0.0000	0.0000	3.0000	0.2000

TABLE A.3: Three-bus system: base-case power flows and losses in the branches.

Branch #	From (h)	To (k)	P_{hk} [pu(MW)]	Q_{hk} [pu(MVar)]	P_{loss} [pu(MW)]	Q_{loss} [pu(MVar)]
1	1	2	0.7486	−0.2559	0.0122	−0.2750
2	1	3	2.2514	0.0397	0.0502	0.1667
3	2	3	0.8062	0.0106	0.0073	−0.3164

TABLE A.4: Three-bus system: base-case total generation, demand and losses.

Type	Active power [pu(MW)]	Reactive power [pu(MVar)]
Generation	14.4100	1.1753
Demand	14.3400	1.6000
Losses	0.0697	−0.4247

A.3 Dynamic Data

The three-bus system includes two synchronous machines. The parameters of the synchronous machine models are shown in Table A.5. Note that the q-axis transient time constant for G1 is $T'_{\text{qo},1} = 0$ s, i.e. the dynamics of the d-axis transient voltage $e'_{\text{r,d},1}$ of G1 are neglected. The dynamic data of the synchronous machine AVRs are shown in Table A.6.

A.4 Reduction to OMIB

For the purpose of some examples of this book, the equivalent OMIB of the three-bus system is derived here. In particular, the OMIB system includes G1 which is connected to a node of constant voltage and frequency. With this aim, the EMF $e'_{\text{r,d},2}$ of G2 is seen as an ideal voltage source. Each synchronous generator is assumed to be represented by a classical second-order electromechanical model. The parameters of the classical model for the two machines are shown in Table A.7.

TABLE A.5: Three-bus system: data of the synchronous machines.

Parameter	Units	Machine 1	Machine 2
$S_{\rm n}$	MVA	1800	1800
$V_{\rm n}$	kV	230	230
Model		M1	M2
M	MWs/MVA	13.00	12.35
$T'_{\rm do}$	s	8.00	8.00
$T''_{\rm do}$	s	0.03	0.03
$T'_{\rm qo}$	s	0	0.40
$T''_{\rm qo}$	s	0.03	0.05
$X_{\rm d}$	pu(Ω)	1.80	1.80
$X'_{\rm d}$	pu(Ω)	0.30	0.30
$X''_{\rm d}$	pu(Ω)	0.25	0.25
$X_{\rm q}$	pu(Ω)	1.70	1.70
$X'_{\rm q}$	pu(Ω)	0.55	0.55
$X''_{\rm q}$	pu(Ω)	0.25	0.25
X_ℓ	pu(Ω)	0.20	0.20
$R_{\rm a}$	pu(Ω)	0.0025	0.0025

Recalling equation (1.77), the relation between current injections $\bar{\boldsymbol{\imath}}_{\rm dq}$ and voltages $\bar{\boldsymbol{v}}_{\rm dq}$ at the network buses is given by:

$$\bar{\boldsymbol{\imath}}_{\rm dq} = \bar{\mathbf{Y}}_{\rm bus}\,\bar{\boldsymbol{v}}_{\rm dq}\,, \tag{A.1}$$

where $\bar{\mathbf{Y}}_{\rm bus}$ is the network admittance matrix as defined in Section 1.3.2.3. Particularizing (1.77) for the three-bus system yields:

$$\begin{bmatrix} \bar{\imath}_{1,\rm dq} \\ \bar{\imath}_{2,\rm dq} \\ \bar{\imath}_{3,\rm dq} \end{bmatrix} = \begin{bmatrix} \bar{Y}_{11} & \bar{Y}_{12} & \bar{Y}_{13} \\ \bar{Y}_{21} & \bar{Y}_{22} & \bar{Y}_{23} \\ \bar{Y}_{31} & \bar{Y}_{32} & \bar{Y}_{33} \end{bmatrix} \begin{bmatrix} \bar{v}_{1,\rm dq} \\ \bar{v}_{2,\rm dq} \\ \bar{v}_{3,\rm dq} \end{bmatrix} ,$$

where the numerical values of the elements of $\bar{\mathbf{Y}}_{\rm bus}$ are given in Table 1.1.

To obtain an equivalent OMIB, we first convert the loads to constant admittances using base-case operating conditions, as follows:

$$\bar{Y}_{\rm D,1} = \frac{\bar{S}^*_{\rm D,1}}{v_1^2} = \frac{3.67 - \jmath 0.70}{1.01^2} = 3.598 - \jmath 0.686 \text{ pu}\,,$$

$$\bar{Y}_{\rm D,2} = \frac{\bar{S}^*_{\rm D,2}}{v_2^2} = \frac{7.67 - \jmath 0.70}{1.02^2} = 7.372 - \jmath 0.673 \text{ pu}\,,$$

$$\bar{Y}_{\rm D,3} = \frac{\bar{S}^*_{\rm D,3}}{v_3^2} = \frac{3.0 - \jmath 0.2}{0.992^2} = 3.047 - \jmath 0.203 \text{ pu}\,,$$

TABLE A.6: Three-bus system: data of the AVRs.

Parameter	Units	AVR 1	AVR 2
$K_{\rm a}$		40.0	40.0
$T_{\rm a}$	s	0.055	0.055
$K_{\rm f}$		0.001	0.001
$T_{\rm f}$	s	1.0	1.0
$K_{\rm ef}$		1.00	1.00
$T_{\rm ef}$	s	0.36	0.30
$T_{\rm R}$	s	0.05	0.05
$A_{\rm ef}$		0.0056	0.0056
$B_{\rm ef}$		1.075	1.075
$v_{\rm a}^{\rm max}$	pu(kV)	5.0	5.0
$v_{\rm a}^{\rm min}$	pu(kV)	−5.0	−5.0

TABLE A.7: Three-bus system: data of the synchronous machines with classical model.

Parameter	Units	Machine 1	Machine 2
$S_{\rm n}$	MVA	1800	1800
$V_{\rm n}$	kV	230	230
M	MWs/MVA	13.00	12.35
$X'_{\rm d}$	pu(Ω)	0.30	0.30

Next, we include load admittances in the $\bar{\mathbf{Y}}_{\rm bus}$:

$$\begin{bmatrix} \bar{\imath}_{\rm s,dq,1} \\ \bar{\imath}_{\rm s,dq,2} \\ 0 \end{bmatrix} = \begin{bmatrix} \bar{Y}_{11} + \bar{Y}_{\rm D,1} & \bar{Y}_{12} & \bar{Y}_{13} \\ \bar{Y}_{21} & \bar{Y}_{22} + \bar{Y}_{\rm D,2} & \bar{Y}_{23} \\ \bar{Y}_{31} & \bar{Y}_{32} & \bar{Y}_{33} + \bar{Y}_{\rm D,3} \end{bmatrix} \begin{bmatrix} \bar{v}_{1,\rm dq} \\ \bar{v}_{2,\rm dq} \\ \bar{v}_{3,\rm dq} \end{bmatrix} , \qquad (A.2)$$

where $\bar{\imath}_{\rm s,dq,1}$ and $\bar{\imath}_{\rm s,dq,2}$ are the stator currents of the machines 1 and 2, respectively.

Eliminating $\bar{v}_{3,\rm dq}$ from (A.2) leads to:

$$\begin{bmatrix} \bar{\imath}_{\rm s,dq,1} \\ \bar{\imath}_{\rm s,dq,2} \end{bmatrix} = \begin{bmatrix} \bar{Y}_1 - \bar{Y}_{13}^2/\bar{Y}_3 & \bar{Y}_{12} - \bar{Y}_{13}\bar{Y}_{23}/\bar{Y}_3 \\ \bar{Y}_{12} - \bar{Y}_{13}\bar{Y}_{23}/\bar{Y}_3 & \bar{Y}_2 - \bar{Y}_{23}^2/\bar{Y}_3 \end{bmatrix} \begin{bmatrix} \bar{v}_{1,\rm dq} \\ \bar{v}_{2,\rm dq} \end{bmatrix} = \bar{\mathbf{Y}}_{\rm bus}^{\rm R} \begin{bmatrix} \bar{v}_{1,\rm dq} \\ \bar{v}_{2,\rm dq} \end{bmatrix} ,$$

where:

$$\bar{Y}_1 = \bar{Y}_{11} + \bar{Y}_{\rm D,1} ,$$

$$\bar{Y}_2 = \bar{Y}_{22} + \bar{Y}_{\mathrm{D},2} \,,$$
$$\bar{Y}_3 = \bar{Y}_{33} + \bar{Y}_{\mathrm{D},3} \,,$$

and

$$\bar{\mathbf{Y}}^{\mathrm{R}}_{\mathrm{bus}} = \begin{bmatrix} \bar{Y}^{\mathrm{R}}_{11} & \bar{Y}^{\mathrm{R}}_{12} \\ \bar{Y}^{\mathrm{R}}_{21} & \bar{Y}^{\mathrm{R}}_{22} \end{bmatrix} = \begin{bmatrix} 5.222 - \jmath 9.392 & -0.136 + \jmath 8.925 \\ -0.136 + \jmath 8.925 & 8.996 - \jmath 9.365 \end{bmatrix} .$$

The elements of $\bar{\mathbf{Y}}^{\mathrm{R}}_{\mathrm{bus}}$ can be assumed to be obtained from an equivalent transmission line that connects buses 1 and 2, as shown in Figure A.2.a. The values of such an equivalent line satisfy the following conditions:

$$\bar{Y}^{\mathrm{R}}_{\mathrm{L},1} + \bar{Y}^{\mathrm{R}}_{\mathrm{L},12} = \bar{Y}^{\mathrm{R}}_{11} \,,$$
$$\bar{Y}^{\mathrm{R}}_{\mathrm{L},2} + \bar{Y}^{\mathrm{R}}_{\mathrm{L},12} = \bar{Y}^{\mathrm{R}}_{22} \,,$$
$$-\bar{Y}^{\mathrm{R}}_{\mathrm{L},12} = \bar{Y}^{\mathrm{R}}_{12} \,,$$

or, equivalently,

$$\bar{Y}^{\mathrm{R}}_{\mathrm{L},1} = 5.086 - \jmath 0.467 \ \mathrm{pu}(\Omega^{-1}) \,,$$
$$\bar{Y}^{\mathrm{R}}_{\mathrm{L},2} = 8.860 - \jmath 0.440 \ \mathrm{pu}(\Omega^{-1}) \,,$$
$$\bar{Y}^{\mathrm{R}}_{\mathrm{L},12} = 0.136 - \jmath 8.925 \ \mathrm{pu}(\Omega^{-1}) \,.$$

The corresponding impedances of the equivalent line are:

$$\bar{Z}^{\mathrm{R}}_{\mathrm{L},1} = \frac{1}{\bar{Y}^{\mathrm{R}}_{\mathrm{L},1}} = 0.1950 + \jmath 0.0179 \ \mathrm{pu}(\Omega) \,,$$
$$\bar{Z}^{\mathrm{R}}_{\mathrm{L},2} = \frac{1}{\bar{Y}^{\mathrm{R}}_{\mathrm{L},2}} = 0.1126 + \jmath 0.0056 \ \mathrm{pu}(\Omega) \,,$$
$$\bar{Z}^{\mathrm{R}}_{\mathrm{L},12} = \frac{1}{\bar{Y}^{\mathrm{R}}_{\mathrm{L},12}} = 0.0017 + \jmath 0.1120 \ \mathrm{pu}(\Omega) \,.$$

The delta-star transformation of the impedances of the equivalent line yields (see Figure A.2.b):

$$\bar{Z}_0 = \frac{\bar{Z}^{\mathrm{R}}_{\mathrm{L},1}\bar{Z}^{\mathrm{R}}_{\mathrm{L},1}}{\bar{Z}^{\mathrm{R}}_{\mathrm{L},1} + \bar{Z}^{\mathrm{R}}_{\mathrm{L},2} + \bar{Z}^{\mathrm{R}}_{\mathrm{L},12}} = 0.0630 - \jmath 0.0175 \ \mathrm{pu}(\Omega) \,,$$
$$\bar{Z}_1 = \frac{\bar{Z}^{\mathrm{R}}_{\mathrm{L},1}\bar{Z}^{\mathrm{R}}_{\mathrm{L},12}}{\bar{Z}^{\mathrm{R}}_{\mathrm{L},1} + \bar{Z}^{\mathrm{R}}_{\mathrm{L},2} + \bar{Z}^{\mathrm{R}}_{\mathrm{L},12}} = 0.0215 + \jmath 0.0613 \ \mathrm{pu}(\Omega) \,,$$
$$\bar{Z}_2 = \frac{\bar{Z}^{\mathrm{R}}_{\mathrm{L},2}\bar{Z}^{\mathrm{R}}_{\mathrm{L},12}}{\bar{Z}^{\mathrm{R}}_{\mathrm{L},1} + \bar{Z}^{\mathrm{R}}_{\mathrm{L},2} + \bar{Z}^{\mathrm{R}}_{\mathrm{L},12}} = 0.0138 + \jmath 0.0348 \ \mathrm{pu}(\Omega) \,.$$

In Figure A.2.b, the impedance $\bar{Z}'_2$ represents the series of $\bar{Z}_2$ and $\jmath X'_{\mathrm{d},2}$:

$$\bar{Z}'_2 = \bar{Z}_2 + \jmath \frac{S_{\mathrm{b}}}{S_{\mathrm{n},2}} X'_{\mathrm{d},2}$$

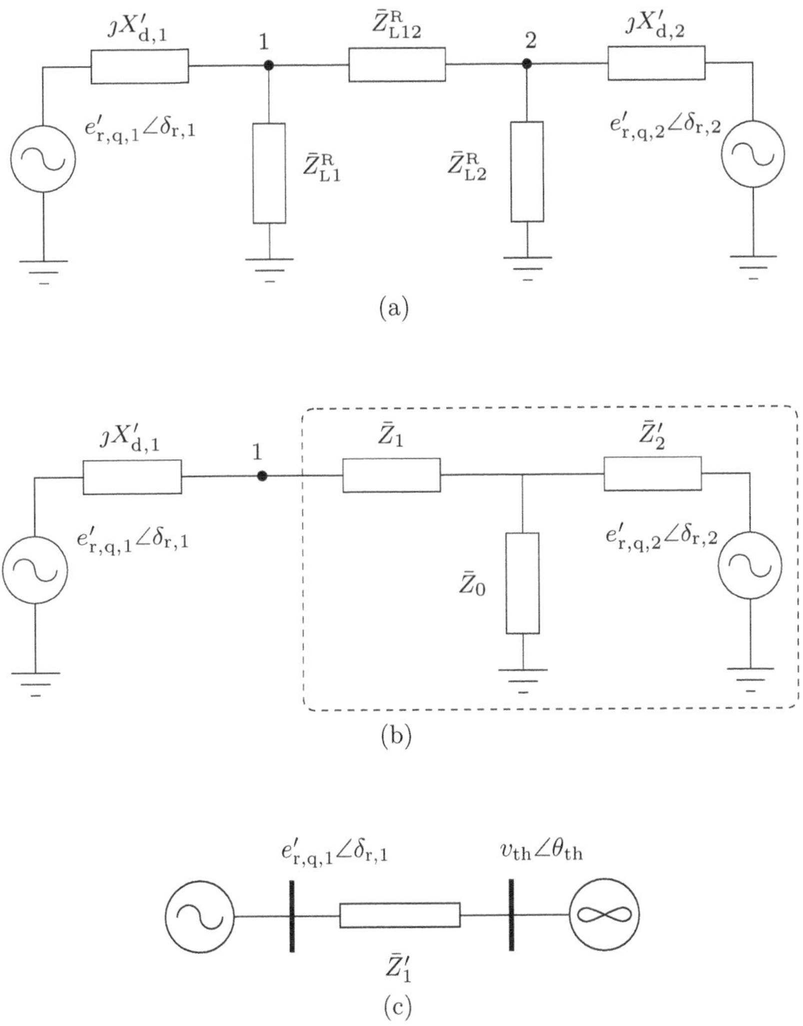

FIGURE A.2: Three-bus system reduction to OMIB.

$$
\begin{aligned}
&= 0.0134 + \jmath 0.0348 + \jmath \frac{0.3}{18} \\
&= 0.0134 + \jmath 0.0514 \ \text{pu}(\Omega) \ ,
\end{aligned}
$$

where the machine d-axis transient reactance is referred to the system base.

The EMFs of the synchronous machines 1 and 2 obtained with the classical model described in Section 1.3.4.7 and for the base-case loading conditions are:

$$
\begin{aligned}
e'_{\rm r,q,1}\angle\delta_{\rm r,1} &= 1.025 \ \text{pu(kV)} \angle -0.1883 \ \text{rad} \, , \\
e'_{\rm r,q,2}\angle\delta_{\rm r,2} &= 1.040 \ \text{pu(kV)} \angle -0.3359 \ \text{rad} \, ,
\end{aligned}
$$

The Thevenin equivalent of the network seen from bus 1 has voltage:

$$\begin{aligned}\bar{v}_{\rm th} &= \frac{\bar{Z}_0}{\bar{Z}_0 + \bar{Z}_2'}\, e'_{\rm r,q,2} \angle \delta_{\rm r,2} \\ &= 0.715 \ {\rm pu(kV)} \angle -1.086 \ {\rm rad} \ ,\end{aligned}$$

and impedance:

$$\begin{aligned}\bar{Z}_{\rm th} &= \bar{Z}_0 || \bar{Z}_2' + \bar{Z}_1 \\ &= 0.055 + \jmath 0.0854 \ {\rm pu}(\Omega) \, .\end{aligned}$$

Finally, the equivalent total impedance $\bar{Z}_1'$ referred to system bases that accounts for the series of the internal reactance $X'_{\rm d,1}$ of machine 1 and the Thevenin impedance $\bar{Z}_{\rm th}$ of the rest of the system is:

$$\begin{aligned}\bar{Z}_1' &= \jmath \frac{S_{\rm b}}{S_{\rm n,1}} X'_{\rm d,1} + \bar{Z}_{\rm th} \\ &= 0.0552 + \jmath 0.1021 \ {\rm pu}(\Omega) \ ,\end{aligned}$$

or, equivalently, referring the network impedance $\bar{Z}_{\rm th}$ to the synchronous machine nominal capacity:

$$\begin{aligned}\bar{Z}_1' &= \jmath X'_{\rm d,1} + \frac{S_{\rm n,1}}{S_{\rm b}} \bar{Z}_{\rm th} \\ &= 0.9933 + \jmath 1.8380 \ {\rm pu}(\Omega) \ .\end{aligned}$$

The resulting OMIB system is shown in Figure A.2.c.

B

LEP Matrices

The state matrix $\mathbf{A}_S$ as well as the right and left eigenvector matrices, $\mathbf{Q}$ and $\mathbf{P}$, respectively, for the LEP – explicit DAE formulation – for the three-bus system are reported below. The matrices are calculated at the base-case operating point of the system. All matrices are 19×19.

$$\mathbf{A}_S[:,1:6] = \begin{bmatrix} 0 & 376.99 & 0 & 0 & 0 & 0 \\ -0.0314 & 0 & -0.0134 & -0.0134 & -0.0205 & 0.0314 \\ -0.0272 & 0 & -1.0692 & 0.9308 & 0.0073 & 0.0272 \\ -0.967 & 0 & 33.095 & -33.572 & 0.2605 & 0.967 \\ -3.6217 & 0 & 0.1095 & 0.1095 & -14.412 & 3.6217 \\ 0 & 0 & 0 & 0 & 0 & 0 \\ 0.032 & 0 & 0.014 & 0.014 & 0.0206 & -0.032 \\ -0.9214 & 0 & -0.1332 & -0.1332 & -0.838 & 0.9214 \\ 0.0326 & 0 & 0.0496 & 0.0496 & -0.0114 & -0.0326 \\ 0.3177 & 0 & 0.4836 & 0.4836 & -0.1108 & -0.3177 \\ -2.4827 & 0 & -0.359 & -0.359 & -2.258 & 2.4827 \\ -0.4774 & 0 & 8.2775 & 8.2775 & -8.0582 & 0.4774 \\ -0.19 & 0 & 0.9808 & 0.9808 & -1.0938 & 0.19 \\ 0 & 0 & 0 & 0 & 0 & 0 \\ 0 & 0 & 0 & 0 & 0 & 0 \\ 0 & 0 & 0 & 0 & 0 & 0 \\ 0 & 0 & 0 & 0 & 0 & 0 \\ 0 & 0 & 0 & 0 & 0 & 0 \\ 0 & 0 & 0 & 0 & 0 & 0 \end{bmatrix}$$

$$\mathbf{A}_{\mathrm{S}}[:,7{:}12] = \begin{bmatrix} 0 & 0 & 0 & 0 & 0 & 0 \\ 0 & 0 & 0 & 0.0185 & -0.0245 & 0 \\ 0 & 0 & 0 & 0.0504 & -0.0033 & 0 \\ 0 & 0 & 0 & 1.7936 & -0.1156 & 0 \\ 0 & 0 & 0 & 0.0647 & -3.9117 & 0 \\ 376.99 & 0 & 0 & 0 & 0 & 0 \\ 0 & 0 & 0 & -0.0188 & 0.0251 & 0 \\ 0 & -2.50 & 0 & -0.2253 & -1.9007 & 0 \\ 0 & 0 & -0.125 & 0.009 & -0.0272 & 0 \\ 0 & 0 & 33.333 & -33.246 & -0.2654 & 0 \\ 0 & 20.00 & 0 & -0.6072 & -25.122 & 0 \\ 0 & 0 & 0 & 2.3165 & 0.6912 & -20.00 \\ 0 & 0 & 0 & 16.666 & 8.5076 & 0 \\ 0 & 0 & 0 & 0 & 0 & 0 \\ 0 & 0 & 0 & 0 & 0 & 0 \\ 0 & 0 & 0 & 0 & 0 & -727.27 \\ 0 & 0 & 0 & 0 & 0 & 0 \\ 0 & 0 & 0 & 0 & 0 & 0 \\ 0 & 0 & 0 & 0 & 0 & 0 \end{bmatrix}$$

$$\mathbf{A}_{\mathrm{S}}[:,\ 13{:}19] = \begin{bmatrix} 0 & 0 & 0 & 0 & 0 & 0 & 0 \\ 0 & 0 & 0 & 0 & 0 & 0 & 0 \\ 0 & 0.125 & 0 & 0 & 0 & 0 & 0 \\ 0 & 0 & 0 & 0 & 0 & 0 & 0 \\ 0 & 0 & 0 & 0 & 0 & 0 & 0 \\ 0 & 0 & 0 & 0 & 0 & 0 & 0 \\ 0 & 0 & 0 & 0 & 0 & 0 & 0 \\ 0 & 0 & 0 & 0 & 0 & 0 & 0 \\ 0 & 0 & 0.125 & 0 & 0 & 0 & 0 \\ 0 & 0 & 0 & 0 & 0 & 0 & 0 \\ 0 & 0 & 0 & 0 & 0 & 0 & 0 \\ 0 & 0 & 0 & 0 & 0 & 0 & 0 \\ -20.00 & 0 & 0 & 0 & 0 & 0 & 0 \\ 0 & -2.917 & 0 & 2.7778 & 0 & 0 & 0 \\ 0 & 0 & -3.5219 & 0 & 3.3333 & 0 & 0 \\ 0 & 0 & 0 & -18.182 & 0 & -727.27 & 0 \\ -800.00 & 0 & 0 & 0 & -20.00 & 0 & -800.0 \\ 0 & -0.0029 & 0 & 0.0028 & 0 & -1 & 0 \\ 0 & 0 & -0.0035 & 0 & 0.0033 & 0 & -1 \end{bmatrix}$$

$$\mathbf{Q}[:,1{:}6] = \begin{bmatrix} 0 & 0 & 0 & 0 & 0 & 0 \\ 0 & 0 & 0 & 0 & 0 & 0 \\ 0 & 0 & 0 & 0 & 0 & 0 \\ 0.04 & 0.02 & 0 & 0 & 0 & 0 \\ 0 & 0 & 0 & 0.01 & 0 & 0 \\ 0 & 0 & 0 & 0 & 0 & 0 \\ 0 & 0 & 0 & 0 & 0 & 0 \\ 0 & 0 & 0 & 0 & 0 & 0 \\ 0 & 0 & 0 & 0 & 0 & 0 \\ -0.01 & 0.01 & 0 & 0 & 0 & 0 \\ 0 & 0 & -0.01 & 0 & 0 & 0 \\ -0.02 & -0.01 & 0 & -0.01 & 0 & 0 \\ 0.01 & -0.01 & 0 & 0 & -\jmath 0.01 & \jmath 0.01 \\ 0.07 & 0.05 & 0.07 & -0.25 & -0.04 + \jmath 0.01 & -0.04 - \jmath 0.01 \\ -0.06 & 0.09 & -0.14 & -0.03 & -0.17 - \jmath 0.06 & -0.17 + \jmath 0.06 \\ -0.83 & -0.52 & -0.53 & 0.96 & 0.23 - \jmath 0.15 & 0.23 + \jmath 0.15 \\ 0.55 & -0.85 & 0.83 & 0.10 & 0.94 & 0.94 \\ 0 & 0 & 0 & 0 & 0 & 0 \\ 0 & 0 & 0 & 0 & 0 & 0 \end{bmatrix}$$

$$\mathbf{Q}[:,7{:}11] = \begin{bmatrix} 0 & 0 & 0.02 & -0.06 - \jmath 0.06 & -0.06 + \jmath 0.06 \\ 0 & 0 & 0 & 0 & 0 \\ 0 & 0 & 0 & -\jmath 0.01 & \jmath 0.01 \\ 0 & 0 & 0.01 & 0 & 0 \\ 0 & 0 & 0.01 & 0.03 + 0.02\jmath & 0.03 - 0.02\jmath \\ 0 & 0 & -0.02 & 0.06 + \jmath 0.07 & 0.06 - \jmath 0.07 \\ 0 & 0 & 0 & 0 & 0 \\ 0 & 0 & -0.07 & 0.02 & 0.02 \\ 0 & 0 & 0.03 & 0.01 + \jmath 0.01 & 0.01 - \jmath 0.01 \\ 0 & 0 & 0.04 & 0.01 + \jmath 0.01 & 0.01 - \jmath 0.01 \\ 0 & 0 & -0.07 & 0.03 + \jmath 0.01 & \\ \jmath 0.01 & -\jmath 0.01 & 0 & -0.01 - \jmath 0.01 & -0.01 + \jmath 0.01 \\ 0 & 0 & 0 & 0.02 & 0.02 \\ 0.15 + \jmath 0.04 & 0.15 - \jmath 0.04 & -0.12 & 0.2 - \jmath 0.15 & 0.2 + \jmath 0.15 \\ -0.03 & -0.03 & -0.98 & -0.26 + 0.35\jmath & -0.26 - 0.35\jmath \\ -0.98 & -0.98 & 0.03 & 0.43 + \jmath 0.18 & 0.43 - \jmath 0.18 \\ 0.14 - \jmath 0.01 & 0.14 + \jmath 0.01 & 0.03 & -0.72 & -0.72 \\ 0 & 0 & 0 & 0 & 0 \\ 0 & 0 & 0 & 0 & 0 \end{bmatrix}$$

$$\mathbf{Q}[:,12:15] = \begin{bmatrix} -0.07 & 0 & 0 & -0.03 - \jmath 0.01 \\ 0 & 0 & 0 & 0 \\ 0.07 & -0.01 & -0.01 & 0.02 + \jmath 0.01 \\ 0.07 & -0.01 & -0.01 & 0.02 \\ 0.03 & 0 & 0 & 0.01 \\ 0.08 & 0 & 0 & 0.03 + \jmath 0.01 \\ 0 & 0 & 0 & 0 \\ 0.03 & 0 & 0 & 0.01 \\ -0.09 & -0.02 - \jmath 0.01 & -0.02 + \jmath 0.01 & -0.02 \\ -0.09 & -0.02 - \jmath 0.01 & -0.02 + \jmath 0.01 & -0.02 \\ 0.04 & 0 & 0 & 0.01 \\ 0.04 & -0.01 & -0.01 & 0.01 \\ -0.05 & -0.02 & -0.02 & -0.01 \\ -0.50 & 0.13 - 0.32\jmath & 0.13 + 0.32\jmath & -0.29 + 0.4\jmath \\ 0.61 & 0.3 - 0.38\jmath & 0.3 + 0.38\jmath & 0.13 - 0.44\jmath \\ -0.34 & 0.49 - \jmath 0.08 & 0.49 + \jmath 0.08 & -0.56 \\ 0.46 & 0.63 & 0.63 & 0.43 - 0.22\jmath \\ 0 & 0 & 0 & 0 \\ 0.04 & 0 & 0 & 0 \end{bmatrix}$$

$$\mathbf{Q}[:,16:19] = \begin{bmatrix} -0.03 + \jmath 0.01 & 0 & 0.71 & 0.71 \\ 0 & 0 & 0 & 0 \\ 0.02 - \jmath 0.01 & -0.07 & 0 & 0 \\ 0.02 & -0.08 & 0 & 0 \\ 0.01 & -0.01 & 0 & 0 \\ 0.03 - \jmath 0.01 & -0.02 & 0.71 & 0.71 \\ 0 & 0 & 0 & 0 \\ 0.01 & 0.01 & 0 & 0 \\ -0.02 & -0.07 & 0 & 0 \\ -0.02 & -0.07 & 0 & 0 \\ 0.01 & 0.01 & 0 & 0 \\ 0.01 & -0.07 & 0 & 0 \\ -0.01 & -0.07 & 0 & 0 \\ -0.29 - 0.4\jmath & 0.57 & 0 & 0 \\ 0.13 + 0.44\jmath & 0.55 & 0 & 0 \\ -0.56 & 0.39 & 0 & 0 \\ 0.43 + 0.22\jmath & 0.42 & 0 & 0 \\ 0 & 0 & 0 & 0 \\ 0 & 0.06 & 0 & 0 \end{bmatrix}$$

$$
\mathbf{P}^{\mathrm{T}}[:,1:6] = \begin{bmatrix}
0.01 & 0 & -0.04 & 0.03 & -0.03 - \jmath 0.01 & -0.03 + \jmath 0.01 \\
-0.13 & -0.03 & 0.58 & -0.7 & 0.47 & 0.47 \\
-0.4 & -0.22 & -0.01 & 0 & 0.09 - \jmath 0.01 & 0.09 + \jmath 0.01 \\
0.42 & 0.21 & 0 & 0 & -0.03 - \jmath 0.01 & -0.03 + \jmath 0.01 \\
-0.01 & 0 & -0.07 & 0.12 & -0.09 + \jmath 0.03 & -0.09 - \jmath 0.03 \\
-0.01 & 0 & 0.04 & -0.03 & 0.03 + \jmath 0.01 & 0.03 - \jmath 0.01 \\
0.13 & 0.03 & -0.58 & 0.7 & -0.47 & -0.47 \\
0.02 & -0.03 & 0.39 & 0.1 & 0.24 + \jmath 0.18 & 0.24 - \jmath 0.18 \\
0.55 & -0.67 & -0.01 & 0 & 0.46 - \jmath 0.01 & 0.46 + \jmath 0.01 \\
-0.58 & 0.67 & 0 & 0 & -0.28 - \jmath 0.08 & -0.28 + \jmath 0.08 \\
-0.03 & 0.05 & -0.41 & -0.05 & -0.16 - \jmath 0.23 & -0.16 + \jmath 0.23 \\
-0.01 & -0.01 & 0 & 0 & -0.03 & -0.03 \\
0.02 & -0.04 & -0.01 & 0 & -0.24 + \jmath 0.04 & -0.24 - \jmath 0.04 \\
0 & 0 & 0 & 0 & 0 & 0 \\
0 & 0 & 0 & 0 & 0 & 0 \\
0 & 0 & 0 & 0 & 0 & 0 \\
0 & 0 & 0 & 0 & 0 & 0 \\
-0.01 & 0 & 0 & 0 & -\jmath 0.01 & \jmath 0.01 \\
0.01 & -0.02 & 0 & 0 & -0.04 - \jmath 0.06 & -0.04 + \jmath 0.06
\end{bmatrix}
$$

$$
\mathbf{P}^{\mathrm{T}}[:,7:11] = \begin{bmatrix}
0.03 + \jmath 0.01 & 0.03 - \jmath 0.01 & -0.01 & -\jmath 0.01 & \jmath 0.01 \\
-0.63 & -0.63 & 0.7 & 0.71 & 0.71 \\
-0.05 - \jmath 0.22 & -0.05 + \jmath 0.22 & 0 & -\jmath 0.01 & \jmath 0.01 \\
-0.01 + \jmath 0.09 & -0.01 - \jmath 0.09 & 0 & 0 & 0 \\
0.15 + \jmath 0.1 & 0.15 - \jmath 0.1 & 0 & 0 & 0 \\
-0.03 - \jmath 0.01 & -0.03 + \jmath 0.01 & 0.01 & \jmath 0.01 & -\jmath 0.01 \\
0.63 & 0.63 & -0.7 & -0.71 & -0.71 \\
-0.02 + \jmath 0.2 & -0.02 - \jmath 0.2 & -0.11 & 0 & 0 \\
-0.03 + \jmath 0.05 & -0.03 - \jmath 0.05 & 0 & \jmath 0.01 & -\jmath 0.01 \\
0.02 - \jmath 0.02 & 0.02 + \jmath 0.02 & 0 & 0 & 0 \\
0.06 - \jmath 0.17 & 0.06 + \jmath 0.17 & 0.01 & 0 & 0 \\
0.03 + \jmath 0.18 & 0.03 - \jmath 0.18 & 0 & 0 & 0 \\
0.01 - \jmath 0.06 & 0.01 + \jmath 0.06 & 0 & 0 & 0 \\
0 & 0 & 0 & 0 & 0 \\
0 & 0 & 0 & 0 & 0 \\
0 & 0 & 0 & 0 & 0 \\
0 & 0 & 0 & 0 & 0 \\
-0.04 + \jmath 0.01 & -0.04 - \jmath 0.01 & -0.01 & -\jmath 0.01 & \jmath 0.01
\end{bmatrix}
$$

$$\mathbf{P}^{\mathrm{T}}[:,12:15] = \begin{bmatrix} 0 & -\jmath 0.01 & \jmath 0.01 & 0 \\ 0.05 & 0.67 & 0.67 & 0.7 \\ -0.01 & 0.03-\jmath 0.13 & 0.03+\jmath 0.13 & -0.02+\jmath 0.04 \\ 0 & -0.01 & -0.01 & 0 \\ 0 & 0.02+\jmath 0.01 & 0.02-\jmath 0.01 & -0.01 \\ 0 & \jmath 0.01 & -\jmath 0.01 & 0 \\ -0.05 & -0.67 & -0.67 & -0.7 \\ 0 & -0.05-\jmath 0.04 & -0.05+\jmath 0.04 & -0.01-\jmath 0.01 \\ 0.01 & -\jmath 0.15 & \jmath 0.15 & 0.02-\jmath 0.03 \\ 0 & -0.01 & -0.01 & 0 \\ 0 & -0.01 & -0.01 & 0 \\ 0 & -0.02 & -0.02 & 0.01 \\ 0 & -0.03+\jmath 0.01 & -0.03-\jmath 0.01 & -0.01 \\ 0 & 0 & 0 & 0 \\ 0 & 0 & 0 & 0 \\ 0 & 0 & 0 & 0 \\ 0 & 0 & 0 & 0 \\ -0.7 & -0.02-\jmath 0.14 & -0.02+\jmath 0.14 & 0.01+\jmath 0.08 \\ 0.72 & -0.08-\jmath 0.14 & -0.08+\jmath 0.14 & 0.01-\jmath 0.07 \end{bmatrix}$$

$$\mathbf{P}^{\mathrm{T}}[:,16:19] = \begin{bmatrix} 0 & 0 & 0 & 0 \\ 0.7 & 0 & 0.71 & 0.71 \\ -0.02-\jmath 0.04 & 0.01 & 0 & 0 \\ 0 & 0 & 0 & 0 \\ -0.01 & 0 & 0 & 0 \\ 0 & 0 & 0 & 0 \\ -0.7 & 0 & 0.7 & 0.7 \\ -0.01+\jmath 0.01 & 0 & 0 & 0 \\ 0.02+\jmath 0.03 & 0.01 & 0 & 0 \\ 0 & 0 & 0 & 0 \\ 0 & 0 & 0 & 0 \\ 0.01 & 0 & 0 & 0 \\ -0.01 & 0 & 0 & 0 \\ 0 & 0 & 0 & 0 \\ 0 & 0 & 0 & 0 \\ 0 & 0 & 0 & 0 \\ 0 & 0 & 0 & 0 \\ 0.01-\jmath 0.08 & 0.78 & 0 & 0 \\ 0.01+\jmath 0.07 & 0.63 & 0 & 0 \end{bmatrix}$$

C

GEP Matrices

The matrices **A** and **E** for the GEP – implicit DAE formulation – for the three-bus system are reported below. The matrices are calculated at the base-case operating point of the system. Both matrices are 56×56. The columns of matrix **E** that are not shown are null.

0	1	0	0	0	0	0	0	0	0	0	0	0	0	0	0	0	0	0	0
0	0	0	0	0	0	−5.443	3.7703	0	0	0	0	0	0	0	0	0	0	0	0
0	0	−1	0	0	0	0	0	0	0	0	0	0	0	0	0	0	0	0	0
0	0	0	−1	0	0	0	0	0	0	0	0	0	0	0	0	0	0	0	0
0	0	0	1	−1	0	0	0	0	0	0	0	0	0	0	0	0	0	0	0
0	0	−1	0	0	−1	0	0	0	0	0	0	0	0	0	0	0	0	0	0
0	0	0	0	0	0	0	1	0	0	0	0	0	0	0	0	0	0	0	0
0	0	0	0	0	0	−1	0	0	0	0	0	0	0	0	0	0	0	0	0
0	0	0	0	0	0	0	0	0	1	0	0	0	0	0	0	0	0	0	0
0	0	0	0	0	0	0	0	0	0	0	0	0	0	−5.9583	4.7471	0	0	0	0
0	0	0	0	0	0	0	0	0	0	−1	0	0	0	0	0	0	0	0	0
0	0	0	0	0	0	0	0	0	0	0	−1	0	0	0	0	0	0	0	0
0	0	0	0	0	0	0	0	0	0	0	1	−1	0	0	0	0	0	0	0
0	0	0	0	0	0	0	0	0	0	1	0	0	−1	0	0	0	0	0	0
0	0	0	0	0	0	0	0	0	0	0	0	0	0	0	1	0	0	0	0
0	0	0	0	0	0	0	0	0	0	0	0	0	0	−1	0	0	0	0	0
0	0	0	0	0	0	0	0	0	0	0	0	0	0	0	0	−1	0	0	0
0	0	0	0	0	0	0	0	0	0	0	0	0	0	0	0	0	−1	0	0
0	0	0	0	0	0	0	0	0	0	0	0	0	0	0	0	0	0	−1.0501	0
0	0	0	0	0	0	0	0	0	0	0	0	0	0	0	0	0	0	0	−1.0566
0	0	0	0	0	0	0	0	0	0	0	0	0	0	0	0	−40.0	0	0	0
0	0	0	0	0	0	0	0	0	0	0	0	0	0	0	0	0	−40.0	0	0
0	0	0	0	0	0	0	0	0	0	0	0	0	0	0	0	0	0	0	0
0	0	0	0	0	0	0	0	0	0	0	0	0	0	0	0	0	0	0	0
−5.6545	0	0	0	0	0	0	0	0	0	0	0	0	0	0	0	0	0	0	0
0	0	0	0	0	0	0	0	−7.1064	0	0	0	0	0	0	0	0	0	0	0
0	0	0	0	0	0	0	0	0	0	0	0	0	0	0	0	0	0	0	0
−3.445	0	0	0	0	0	0	0	0	0	0	0	0	0	0	0	0	0	0	0
0	0	0	0	0	0	0	0	−2.7451	0	0	0	0	0	0	0	0	0	0	0
0	0	0	0	0	0	0	0	0	0	0	0	0	0	0	0	0	0	0	0
0	0	0	0	0	0	0	0	0	0	0	0	0	0	0	0	0	0	0	0
0	0	0	0	0	0	0	0	0	0	0	0	0	0	0	0	0	0	0	0
0	0	0	0	0	0	0	0	0	0	0	0	0	0	0	0	0	0	0	0
0	0	0	0	0	0	0	0	0	0	0	0	0	0	0	0	0	0	0	0
0	0	0	0	0	0	0	0	0	0	0	0	0	0	0	0	0	0	0	0
0	0	0	0	0	0	0	0	0	0	0	0	0	0	0	0	0	0	0	0
0	0	0	0	0	0	0	0	0	0	0	0	0	0	0	0	0	0	0	0
0	0	0	0	0	0	0	0	0	0	0	0	0	0	0	0	0	0	1	0
0.8697	0	0	0	0	0	0	0	0	0	0	0	0	0	0	0	0	0	0	0
−0.5135	0	0	0	0	0	0	0	0	0	0	0	0	0	0	0	0	0	0	0
0	0	0	−0.5	−0.5	0	1	0	0	0	0	0	0	0	0	0	0	0	0	0
0	0	0.1429	0	0	−0.8571	0	1	0	0	0	0	0	0	0	0	0	0	0	0
0	0	0	0	0	0	0	0	0	0	0	0	0	0	0	0	0	0	0	0
0	0	0	0	0	0	0	0	0	0	0	0	0	0	0	0	0	0	0	0
0	0	0	0	0	0	0	0	0	0	0	0	0	0	0	0	0	0	0	0
0	0	0	0	0	0	0	0	0	0	0	0	0	0	0	0	0	0	0	1
0	0	0	0	0	0	0	0	0.8512	0	0	0	0	0	0	0	0	0	0	0
0	0	0	0	0	0	0	0	−0.5621	0	0	0	0	0	0	0	0	0	0	0
0	0	0	0	0	0	0	0	0	0	0	0	−1	0	1	0	0	0	0	0
0	0	0	0	0	0	0	0	0	0	0	0	0	1	0	1	0	0	0	0
0	0	0	0	0	0	0	0	0	0	0	0	0	0	0	0	0	0	0	0
0	0	0	0	0	0	0	0	0	0	0	0	0	0	0	0	0	0	0	0
0	0	0	0	0	0	0	0	0	0	0	0	0	0	0	0	0	0	0	0
0	0	0	0	0	0	0	0	0	0	0	0	0	0	0	0	0	0	0	0
0	0	0	0	0	0	0	0	0	0	0	0	0	0	0	0	0	0	0	0
0	0	0	0	0	0	0	0	0	0	0	0	0	0	0	0	0	0	0	0

Columns 1 to 20 of matrix **A** for the implicit DAE formulation for the three-bus system.

$$
\begin{bmatrix}
0 & 0 & 0 & 0 & 0 & 0 & 0 & 0 & 0 & 0 & 0 & 0 & 0 & 0 & 0 & 0 & 0 & 0 \\
0 & 0 & 0 & 0 & 0 & 0 & 0 & 0 & 0 & 0 & 0 & 0 & 0 & 0 & 0 & 0 & 1 & 0 \\
0 & 0 & 0 & 0 & 0 & 0 & 0 & 0 & 0 & 0 & 0 & 0 & 0 & 0 & 0 & 0 & 0 & 0 \\
0 & 0 & 0 & 0 & 0 & 0 & 0 & 0 & 0 & 0 & 0 & 0 & 0 & 0 & 0 & 0 & 0 & 1 \\
0 & 0 & 0 & 0 & 0 & 0 & 0 & 0 & 0 & 0 & 0 & 0 & 0 & 0 & 0 & 0 & 0 & 0 \\
0 & 0 & 0 & 0 & 0 & 0 & 0 & 0 & 0 & 0 & 0 & 0 & 0 & 0 & 0 & 0 & 0 & 0 \\
0 & 0 & 0 & 0 & 0 & 0 & 0 & 0 & 0 & 0 & 0 & 0 & 0 & 0 & 0 & 0 & 0 & 0 \\
0 & 0 & 0 & 0 & 0 & 0 & 0 & 0 & 0 & 0 & 0 & 0 & 0 & 0 & 0 & 0 & 0 & 0 \\
0 & 0 & 0 & 0 & 0 & 0 & 0 & 0 & 0 & 0 & 0 & 0 & 0 & 0 & 0 & 0 & 0 & 0 \\
0 & 0 & 0 & 0 & 0 & 0 & 0 & 0 & 0 & 0 & 0 & 0 & 0 & 0 & 0 & 0 & 0 & 0 \\
0 & 0 & 0 & 0 & 0 & 0 & 0 & 0 & 0 & 0 & 0 & 0 & 0 & 0 & 0 & 0 & 0 & 0 \\
0 & 0 & 0 & 0 & 0 & 0 & 0 & 0 & 0 & 0 & 0 & 0 & 0 & 0 & 0 & 0 & 0 & 0 \\
0 & 0 & 0 & 0 & 0 & 0 & 0 & 0 & 0 & 0 & 0 & 0 & 0 & 0 & 0 & 0 & 0 & 0 \\
0 & 0 & 0 & 0 & 0 & 0 & 0 & 0 & 0 & 0 & 0 & 0 & 0 & 0 & 0 & 0 & 0 & 0 \\
0 & 0 & 0 & 0 & 0 & 0 & 0 & 0 & 0 & 0 & 0 & 0 & 0 & 0 & 0 & 0 & 0 & 0 \\
0 & 0 & 0 & 0 & 0 & 0 & 0 & 0 & 0 & 0 & 0 & 0 & 0 & 0 & 0 & 0 & 0 & 0 \\
0 & 0 & 0 & 0 & 0 & 0 & 0 & 0 & 0 & 0 & 0 & 0 & 0 & 0 & 0 & 0 & 0 & 0 \\
0 & 0 & 0 & 0 & 0 & 0 & 0 & 0 & 0 & 0 & 0 & 0 & 0 & 0 & 0 & 0 & 0 & 0 \\
1 & 0 & 0 & 0 & 0 & 0 & 0 & 0 & 0 & 0 & 0 & 0 & 0 & 0 & 0 & 0 & 0 & 0 \\
0 & 1 & 0 & 0 & 0 & 0 & 0 & 0 & 0 & 0 & 0 & 0 & 0 & 0 & 0 & 0 & 0 & 0 \\
-1 & 0 & -40.0 & 0 & 0 & 0 & 0 & 0 & 0 & 0 & 0 & 0 & 0 & 0 & 0 & 0 & 0 & 0 \\
0 & -1 & 0 & -40.0 & 0 & 0 & 0 & 0 & 0 & 0 & 0 & 0 & 0 & 0 & 0 & 0 & 0 & 0 \\
0 & 0 & -1 & 0 & 0 & 0 & 0 & 0 & 0 & 0 & 0 & 0 & 0 & 0 & 0 & 0 & 0 & 0 \\
0 & 0 & 0 & -1 & 0 & 0 & 0 & 0 & 0 & 0 & 0 & 0 & 0 & 0 & 0 & 0 & 0 & 0 \\
0 & 0 & 0 & 0 & -1.2697 & 0.45 & 0.8197 & -13.132 & 4.5005 & 9.0164 & 1 & 0 & 0 & 0 & 0 & 0 & 0 & 0 \\
0 & 0 & 0 & 0 & 0.45 & -1.3501 & 0.9001 & 4.5005 & -13.116 & 9.0009 & 0 & 1 & 0 & 0 & 0 & 0 & 0 & 0 \\
0 & 0 & 0 & 0 & 0.8197 & 0.9001 & -1.7198 & 9.0164 & 9.0009 & -17.632 & 0 & 0 & 1 & 0 & 0 & 0 & 0 & 0 \\
0 & 0 & 0 & 0 & 13.132 & -4.5005 & -9.0164 & -1.2697 & 0.45 & 0.8197 & 0 & 0 & 0 & 1 & 0 & 0 & 0 & 0 \\
0 & 0 & 0 & 0 & -4.5005 & 13.116 & -9.0009 & 0.45 & -1.3501 & 0.9001 & 0 & 0 & 0 & 0 & 1 & 0 & 0 & 0 \\
0 & 0 & 0 & 0 & -9.0164 & -9.0009 & 17.632 & 0.8197 & 0.9001 & -1.7198 & 0 & 0 & 0 & 0 & 0 & 1 & 0 & 0 \\
0 & 0 & 0 & 0 & 1.5459 & 0 & 0 & -3.3607 & 0 & 0 & -0.5865 & 0 & 0 & 0.8223 & 0 & 0 & 0 & 0 \\
0 & 0 & 0 & 0 & 0 & 2.7075 & 0 & 0 & -7.0487 & 0 & 0 & -0.4508 & 0 & 0 & 0.915 & 0 & 0 & 0 \\
0 & 0 & 0 & 0 & 0 & 0 & 0.9121 & 0 & 0 & -2.8898 & 0 & 0 & -0.3609 & 0 & 0 & 0.9242 & 0 & 0 \\
0 & 0 & 0 & 0 & 3.3607 & 0 & 0 & 1.5459 & 0 & 0 & 0.8223 & 0 & 0 & 0.5865 & 0 & 0 & 0 & 0 \\
0 & 0 & 0 & 0 & 0 & 7.0487 & 0 & 0 & 2.7075 & 0 & 0 & 0.915 & 0 & 0 & 0.4508 & 0 & 0 & 0 \\
0 & 0 & 0 & 0 & 0 & 0 & 2.8898 & 0 & 0 & 0.9121 & 0 & 0 & 0.9242 & 0 & 0 & 0.3609 & 0 & 0 \\
0 & 0 & 0 & 0 & 0 & 0 & 0 & 0 & 0 & 0 & 0 & 0 & 0 & 0 & 0 & 0 & -1 & 0 \\
0 & 0 & 0 & 0 & 0 & 0 & 0 & 0 & 0 & 0 & 0 & 0 & 0 & 0 & 0 & 0 & 0 & -1 \\
0 & 0 & 0 & 0 & 0.4058 & 0 & 0 & 0.914 & 0 & 0 & 0 & 0 & 0 & 0 & 0 & 0 & 0 & 0 \\
0 & 0 & 0 & 0 & -0.914 & 0 & 0 & 0.4058 & 0 & 0 & 0 & 0 & 0 & 0 & 0 & 0 & 0 & 0 \\
0 & 0 & 0 & 0 & 0 & 0 & 0 & 0 & 0 & 0 & 0 & 0 & 0 & 0 & 0 & 0 & 0 & 0 \\
0 & 0 & 0 & 0 & 0 & 0 & 0 & 0 & 0 & 0 & 0 & 0 & 0 & 0 & 0 & 0 & 0 & 0 \\
0 & 0 & 0 & 0 & 0 & 0 & 0 & 0 & 0 & 0 & 0 & 0 & 0 & 0 & 0 & 0 & 0 & 0 \\
0 & 0 & 0 & 0 & 0 & 0 & 0 & 0 & 0 & 0 & 0 & 0 & 0 & 0 & 0 & 0 & 0 & 0 \\
0 & 0 & 0 & 0 & 0 & 0 & 0 & 0 & 0 & 0 & 0 & 0 & 0 & 0 & 0 & 0 & 0 & 0 \\
0 & 0 & 0 & 0 & 0 & 0 & 0 & 0 & 0 & 0 & 0 & 0 & 0 & 0 & 0 & 0 & 0 & 0 \\
0 & 0 & 0 & 0 & 0 & 0.505 & 0 & 0 & 0.8631 & 0 & 0 & 0 & 0 & 0 & 0 & 0 & 0 & 0 \\
0 & 0 & 0 & 0 & 0 & -0.8631 & 0 & 0 & 0.505 & 0 & 0 & 0 & 0 & 0 & 0 & 0 & 0 & 0 \\
0 & 0 & 0 & 0 & 0 & 0 & 0 & 0 & 0 & 0 & 0 & 0 & 0 & 0 & 0 & 0 & 0 & 0 \\
0 & 0 & 0 & 0 & 0 & 0 & 0 & 0 & 0 & 0 & 0 & 0 & 0 & 0 & 0 & 0 & 0 & 0 \\
0 & 0 & 0 & 0 & 0 & 0 & 0 & 0 & 0 & 0 & 0 & 0 & 0 & 0 & 0 & 0 & 0 & 0 \\
0 & 0 & 0 & 0 & 0 & 0 & 0 & 0 & 0 & 0 & 0 & 0 & 0 & 0 & 0 & 0 & 0 & 0 \\
0 & 0 & 0 & 0 & -1.173 & 0 & 0 & 1.6445 & 0 & 0 & 0 & 0 & 0 & 0 & 0 & 0 & 0 & 0 \\
0 & 0 & 0 & 0 & 0 & -0.9015 & 0 & 0 & 1.83 & 0 & 0 & 0 & 0 & 0 & 0 & 0 & 0 & 0 \\
0 & 0 & 0 & 0 & 0 & 0 & 0 & 0 & 0 & 0 & 0 & 0 & 0 & 0 & 0 & 0 & 0 & 0 \\
0 & 0 & 0 & 0 & 0 & 0 & 0 & 0 & 0 & 0 & 0 & 0 & 0 & 0 & 0 & 0 & 0 & 0
\end{bmatrix}
$$

Columns 21 to 38 of matrix **A** for the implicit DAE formulation for the three-bus system.

0	0	0	0	0	0	0	0	0	0	0	0	0	0	0	0	0	0
0	0	−0.5141	−0.8705	0	0	0	0	0	0	0	0	0	0	0	0	0	0
0	0	0	0.0639	0	0	0	0	0	0	0	0	0	0	0	0	0	0
0	0	−0.0833	0	0	0	0	0	0	0	0	0	0	0	0	0	0	0
0	0	−0.0056	0	0	0	0	0	0	0	0	0	0	0	0	0	0	0
0	0	0	−0.0194	0	0	0	0	0	0	0	0	0	0	0	0	0	0
1	0	0.0001	0	0	0	0	0	0	0	0	0	0	0	0	0	0	0
0	1	0	0.0001	0	0	0	0	0	0	0	0	0	0	0	0	0	0
0	0	0	0	0	0	0	0	0	0	0	0	0	0	0	0	0	0
0	0	0	0	0	0	1	0	0	0	−0.5627	−0.852	0	0	0	0	0	0
0	0	0	0	0	0	0	0	0	0	0	0.0603	0	0	0	0	0	0
0	0	0	0	0	0	0	1	0	0	−0.0831	0	0	0	0	0	0	0
0	0	0	0	0	0	0	0	0	0	−0.003	0	0	0	0	0	0	0
0	0	0	0	0	0	0	0	0	0	0	0.0203	0	0	0	0	0	0
0	0	0	0	0	0	0	0	1	0	0.0001	0	0	0	0	0	0	0
0	0	0	0	0	0	0	0	0	1	0	0.0001	0	0	0	0	0	0
0	0	0	0	0	0	0	0	0	0	0	0	0	0	1	0	0	0
0	0	0	0	0	0	0	0	0	0	0	0	0	0	0	1	0	0
0	0	0	0	0	0	0	0	0	0	0	0	0	0	0	0	0	0
0	0	0	0	0	0	0	0	0	0	0	0	0	0	0	0	0	0
0	0	0	0	0	0	0	0	0	0	0	0	0	0	0	0	40.0	0
0	0	0	0	0	0	0	0	0	0	0	0	0	0	0	0	0	40.0
0	0	0	0	0	0	0	0	0	0	0	0	0	0	0	0	0	0
0	0	0	0	0	0	0	0	0	0	0	0	0	0	0	0	0	0
0	0	0.4058	−0.914	0	0	0	0	0	0	0	0	0	0	0	0	0	0
0	0	0	0	0	0	0	0	0	0	0.505	−0.8631	0	0	0	0	0	0
0	0	0	0	0	0	0	0	0	0	0	0	0	0	0	0	0	0
0	0	0.914	0.4058	0	0	0	0	0	0	0	0	0	0	0	0	0	0
0	0	0	0	0	0	0	0	0	0	0.8631	0.505	0	0	0	0	0	0
0	0	0	0	0	0	0	0	0	0	0	0	0	0	0	0	0	0
0	0	0	0	0	0	0	0	0	0	0	0	0	0	0	0	0	0
0	0	0	0	0	0	0	0	0	0	0	0	0	0	0	0	0	0
0	0	0	0	0	0	0	0	0	0	0	0	0	0	0	0	0	0
0	0	0	0	0	0	0	0	0	0	0	0	0	0	0	0	0	0
0	0	0	0	0	0	0	0	0	0	0	0	0	0	0	0	0	0
0	0	0	0	0	0	0	0	0	0	0	0	0	0	0	0	0	0
0	0	0	0	0	0	0	0	0	0	0	0	0	0	0	0	0	0
0	0	0	0	0	0	0	0	0	0	0	0	0	0	0	0	0	0
−1	0	0	0	0	0	0	0	0	0	0	0	0	0	0	0	0	0
0	−1	0	0	0	0	0	0	0	0	0	0	0	0	0	0	0	0
0	0	0.0139	0	0	0	0	0	0	0	0	0	0	0	0	0	0	0
0	0	0	0.0139	0	0	0	0	0	0	0	0	0	0	0	0	0	0
3.7703	5.443	0.5135	0.8697	−1	0	0	0	0	0	0	0	0	0	0	0	0	0
−5.443	3.7703	0.8697	−0.5135	0	−1	0	0	0	0	0	0	0	0	0	0	0	0
0	0	0	0	0	0	−1	0	0	0	0	0	0	0	0	0	0	0
0	0	0	0	0	0	0	−1	0	0	0	0	0	0	0	0	0	0
0	0	0	0	0	0	0	0	−1	0	0	0	0	0	0	0	0	0
0	0	0	0	0	0	0	0	0	−1	0	0	0	0	0	0	0	0
0	0	0	0	0	0	0	0	0	0	0.0139	0	0	0	0	0	0	0
0	0	0	0	0	0	0	0	0	0	0	0.0139	0	0	0	0	0	0
0	0	0	0	0	0	0	0	4.7471	5.9583	0.5621	0.8512	−1	0	0	0	0	0
0	0	0	0	0	0	0	0	−5.9583	4.7471	0.8512	−0.5621	0	−1	0	0	0	0
0	0	0	0	0	0	0	0	0	0	0	0	0	0	−2.02	0	0	0
0	0	0	0	0	0	0	0	0	0	0	0	0	0	0	−2.04	0	0
0	0	0	0	0	0	0	0	0	0	0	0	0	0	0	0	−1	0
0	0	0	0	0	0	0	0	0	0	0	0	0	0	0	0	0	−1

Columns 39 to 56 of matrix **A** for the implicit DAE formulation for the three-bus system.

$$
\begin{bmatrix}
0.0027 & 0 & 0 & 0 & 0 & 0 & 0 & 0 & 0 & 0 & 0 & 0 & 0 & 0 & 0 & 0 & 0 & 0 & 0 & 0 & 0 & 0 & 0 & 0 \\
0 & 234.0 & 0 & 0 & 0 & 0 & 0 & 0 & 0 & 0 & 0 & 0 & 0 & 0 & 0 & 0 & 0 & 0 & 0 & 0 & 0 & 0 & 0 & 0 \\
0 & 0 & 0 & 0 & 0 & -0.0929 & 0 & 0 & 0 & 0 & 0 & 0 & 0 & 0 & 0 & 0 & 0 & 0 & 0 & 0 & 0 & 0 & 0 & 0 \\
0 & 0 & 0 & 8.0 & 0.225 & 0 & 0 & 0 & 0 & 0 & 0 & 0 & 0 & 0 & 0 & 0 & 0 & 0 & 0 & 0 & 0 & 0 & 0 & 0 \\
0 & 0 & 0 & 0 & 0.03 & 0 & 0 & 0 & 0 & 0 & 0 & 0 & 0 & 0 & 0 & 0 & 0 & 0 & 0 & 0 & 0 & 0 & 0 & 0 \\
0 & 0 & 0 & 0 & 0 & 0.033 & 0 & 0 & 0 & 0 & 0 & 0 & 0 & 0 & 0 & 0 & 0 & 0 & 0 & 0 & 0 & 0 & 0 & 0 \\
0 & 0 & 0 & 0 & 0 & 0 & 0 & 0 & 0 & 0 & 0 & 0 & 0 & 0 & 0 & 0 & 0 & 0 & 0 & 0 & 0 & 0 & 0 & 0 \\
0 & 0 & 0 & 0 & 0 & 0 & 0 & 0 & 0 & 0 & 0 & 0 & 0 & 0 & 0 & 0 & 0 & 0 & 0 & 0 & 0 & 0 & 0 & 0 \\
0 & 0 & 0 & 0 & 0 & 0 & 0 & 0 & 0.0027 & 0 & 0 & 0 & 0 & 0 & 0 & 0 & 0 & 0 & 0 & 0 & 0 & 0 & 0 & 0 \\
0 & 0 & 0 & 0 & 0 & 0 & 0 & 0 & 0 & 222.3 & 0 & 0 & 0 & 0 & 0 & 0 & 0 & 0 & 0 & 0 & 0 & 0 & 0 & 0 \\
0 & 0 & 0 & 0 & 0 & 0 & 0 & 0 & 0 & 0 & 0.4 & 0 & 0 & 0 & 0 & 0 & 0 & 0 & 0 & 0 & 0 & 0 & 0 & 0 \\
0 & 0 & 0 & 0 & 0 & 0 & 0 & 0 & 0 & 0 & 0 & 8.0 & 0 & 0 & 0 & 0 & 0 & 0 & 0 & 0 & 0 & 0 & 0 & 0 \\
0 & 0 & 0 & 0 & 0 & 0 & 0 & 0 & 0 & 0 & 0 & 0 & 0.03 & 0 & 0 & 0 & 0 & 0 & 0 & 0 & 0 & 0 & 0 & 0 \\
0 & 0 & 0 & 0 & 0 & 0 & 0 & 0 & 0 & 0 & 0 & 0 & 0 & 0.05 & 0 & 0 & 0 & 0 & 0 & 0 & 0 & 0 & 0 & 0 \\
0 & 0 & 0 & 0 & 0 & 0 & 0 & 0 & 0 & 0 & 0 & 0 & 0 & 0 & 0 & 0 & 0 & 0 & 0 & 0 & 0 & 0 & 0 & 0 \\
0 & 0 & 0 & 0 & 0 & 0 & 0 & 0 & 0 & 0 & 0 & 0 & 0 & 0 & 0 & 0 & 0 & 0 & 0 & 0 & 0 & 0 & 0 & 0 \\
0 & 0 & 0 & 0 & 0 & 0 & 0 & 0 & 0 & 0 & 0 & 0 & 0 & 0 & 0 & 0 & 0.05 & 0 & 0 & 0 & 0 & 0 & 0 & 0 \\
0 & 0 & 0 & 0 & 0 & 0 & 0 & 0 & 0 & 0 & 0 & 0 & 0 & 0 & 0 & 0 & 0 & 0.05 & 0 & 0 & 0 & 0 & 0 & 0 \\
0 & 0 & 0 & 0 & 0 & 0 & 0 & 0 & 0 & 0 & 0 & 0 & 0 & 0 & 0 & 0 & 0 & 0 & 0.36 & 0 & 0 & 0 & 0 & 0 \\
0 & 0 & 0 & 0 & 0 & 0 & 0 & 0 & 0 & 0 & 0 & 0 & 0 & 0 & 0 & 0 & 0 & 0 & 0 & 0.3 & 0 & 0 & 0 & 0 \\
0 & 0 & 0 & 0 & 0 & 0 & 0 & 0 & 0 & 0 & 0 & 0 & 0 & 0 & 0 & 0 & 0 & 0 & 0 & 0 & 0.055 & 0 & 0 & 0 \\
0 & 0 & 0 & 0 & 0 & 0 & 0 & 0 & 0 & 0 & 0 & 0 & 0 & 0 & 0 & 0 & 0 & 0 & 0 & 0 & 0 & 0.05 & 0 & 0 \\
0 & 0 & 0 & 0 & 0 & 0 & 0 & 0 & 0 & 0 & 0 & 0 & 0 & 0 & 0 & 0 & 0 & 0 & -0.001 & 0 & 0 & 0 & 1 & 0 \\
0 & 0 & 0 & 0 & 0 & 0 & 0 & 0 & 0 & 0 & 0 & 0 & 0 & 0 & 0 & 0 & 0 & 0 & 0 & -0.001 & 0 & 0 & 0 & 1 \\
0 & 0 & 0 & 0 & 0 & 0 & 0 & 0 & 0 & 0 & 0 & 0 & 0 & 0 & 0 & 0 & 0 & 0 & 0 & 0 & 0 & 0 & 0 & 0 \\
0 & 0 & 0 & 0 & 0 & 0 & 0 & 0 & 0 & 0 & 0 & 0 & 0 & 0 & 0 & 0 & 0 & 0 & 0 & 0 & 0 & 0 & 0 & 0 \\
0 & 0 & 0 & 0 & 0 & 0 & 0 & 0 & 0 & 0 & 0 & 0 & 0 & 0 & 0 & 0 & 0 & 0 & 0 & 0 & 0 & 0 & 0 & 0 \\
0 & 0 & 0 & 0 & 0 & 0 & 0 & 0 & 0 & 0 & 0 & 0 & 0 & 0 & 0 & 0 & 0 & 0 & 0 & 0 & 0 & 0 & 0 & 0 \\
0 & 0 & 0 & 0 & 0 & 0 & 0 & 0 & 0 & 0 & 0 & 0 & 0 & 0 & 0 & 0 & 0 & 0 & 0 & 0 & 0 & 0 & 0 & 0 \\
0 & 0 & 0 & 0 & 0 & 0 & 0 & 0 & 0 & 0 & 0 & 0 & 0 & 0 & 0 & 0 & 0 & 0 & 0 & 0 & 0 & 0 & 0 & 0 \\
0 & 0 & 0 & 0 & 0 & 0 & 0 & 0 & 0 & 0 & 0 & 0 & 0 & 0 & 0 & 0 & 0 & 0 & 0 & 0 & 0 & 0 & 0 & 0 \\
0 & 0 & 0 & 0 & 0 & 0 & 0 & 0 & 0 & 0 & 0 & 0 & 0 & 0 & 0 & 0 & 0 & 0 & 0 & 0 & 0 & 0 & 0 & 0 \\
0 & 0 & 0 & 0 & 0 & 0 & 0 & 0 & 0 & 0 & 0 & 0 & 0 & 0 & 0 & 0 & 0 & 0 & 0 & 0 & 0 & 0 & 0 & 0 \\
0 & 0 & 0 & 0 & 0 & 0 & 0 & 0 & 0 & 0 & 0 & 0 & 0 & 0 & 0 & 0 & 0 & 0 & 0 & 0 & 0 & 0 & 0 & 0 \\
0 & 0 & 0 & 0 & 0 & 0 & 0 & 0 & 0 & 0 & 0 & 0 & 0 & 0 & 0 & 0 & 0 & 0 & 0 & 0 & 0 & 0 & 0 & 0 \\
0 & 0 & 0 & 0 & 0 & 0 & 0 & 0 & 0 & 0 & 0 & 0 & 0 & 0 & 0 & 0 & 0 & 0 & 0 & 0 & 0 & 0 & 0 & 0 \\
0 & 0 & 0 & 0 & 0 & 0 & 0 & 0 & 0 & 0 & 0 & 0 & 0 & 0 & 0 & 0 & 0 & 0 & 0 & 0 & 0 & 0 & 0 & 0 \\
0 & 0 & 0 & 0 & 0 & 0 & 0 & 0 & 0 & 0 & 0 & 0 & 0 & 0 & 0 & 0 & 0 & 0 & 0 & 0 & 0 & 0 & 0 & 0 \\
0 & 0 & 0 & 0 & 0 & 0 & 0 & 0 & 0 & 0 & 0 & 0 & 0 & 0 & 0 & 0 & 0 & 0 & 0 & 0 & 0 & 0 & 0 & 0 \\
0 & 0 & 0 & 0 & 0 & 0 & 0 & 0 & 0 & 0 & 0 & 0 & 0 & 0 & 0 & 0 & 0 & 0 & 0 & 0 & 0 & 0 & 0 & 0 \\
0 & 0 & 0 & 0 & 0 & 0 & 0 & 0 & 0 & 0 & 0 & 0 & 0 & 0 & 0 & 0 & 0 & 0 & 0 & 0 & 0 & 0 & 0 & 0 \\
0 & 0 & 0 & 0 & 0 & 0 & 0 & 0 & 0 & 0 & 0 & 0 & 0 & 0 & 0 & 0 & 0 & 0 & 0 & 0 & 0 & 0 & 0 & 0 \\
0 & 0 & 0 & 0 & 0 & 0 & 0 & 0 & 0 & 0 & 0 & 0 & 0 & 0 & 0 & 0 & 0 & 0 & 0 & 0 & 0 & 0 & 0 & 0 \\
0 & 0 & 0 & 0 & 0 & 0 & 0 & 0 & 0 & 0 & 0 & 0 & 0 & 0 & 0 & 0 & 0 & 0 & 0 & 0 & 0 & 0 & 0 & 0 \\
0 & 0 & 0 & 0 & 0 & 0 & 0 & 0 & 0 & 0 & 0 & 0 & 0 & 0 & 0 & 0 & 0 & 0 & 0 & 0 & 0 & 0 & 0 & 0 \\
0 & 0 & 0 & 0 & 0 & 0 & 0 & 0 & 0 & 0 & 0 & 0 & 0 & 0 & 0 & 0 & 0 & 0 & 0 & 0 & 0 & 0 & 0 & 0 \\
0 & 0 & 0 & 0 & 0 & 0 & 0 & 0 & 0 & 0 & 0 & 0 & 0 & 0 & 0 & 0 & 0 & 0 & 0 & 0 & 0 & 0 & 0 & 0 \\
0 & 0 & 0 & 0 & 0 & 0 & 0 & 0 & 0 & 0 & 0 & 0 & 0 & 0 & 0 & 0 & 0 & 0 & 0 & 0 & 0 & 0 & 0 & 0 \\
0 & 0 & 0 & 0 & 0 & 0 & 0 & 0 & 0 & 0 & 0 & 0 & 0 & 0 & 0 & 0 & 0 & 0 & 0 & 0 & 0 & 0 & 0 & 0 \\
0 & 0 & 0 & 0 & 0 & 0 & 0 & 0 & 0 & 0 & 0 & 0 & 0 & 0 & 0 & 0 & 0 & 0 & 0 & 0 & 0 & 0 & 0 & 0 \\
0 & 0 & 0 & 0 & 0 & 0 & 0 & 0 & 0 & 0 & 0 & 0 & 0 & 0 & 0 & 0 & 0 & 0 & 0 & 0 & 0 & 0 & 0 & 0 \\
0 & 0 & 0 & 0 & 0 & 0 & 0 & 0 & 0 & 0 & 0 & 0 & 0 & 0 & 0 & 0 & 0 & 0 & 0 & 0 & 0 & 0 & 0 & 0 \\
0 & 0 & 0 & 0 & 0 & 0 & 0 & 0 & 0 & 0 & 0 & 0 & 0 & 0 & 0 & 0 & 0 & 0 & 0 & 0 & 0 & 0 & 0 & 0 \\
0 & 0 & 0 & 0 & 0 & 0 & 0 & 0 & 0 & 0 & 0 & 0 & 0 & 0 & 0 & 0 & 0 & 0 & 0 & 0 & 0 & 0 & 0 & 0 \\
0 & 0 & 0 & 0 & 0 & 0 & 0 & 0 & 0 & 0 & 0 & 0 & 0 & 0 & 0 & 0 & 0 & 0 & 0 & 0 & 0 & 0 & 0 & 0 \\
0 & 0 & 0 & 0 & 0 & 0 & 0 & 0 & 0 & 0 & 0 & 0 & 0 & 0 & 0 & 0 & 0 & 0 & 0 & 0 & 0 & 0 & 0 & 0
\end{bmatrix}
$$

Columns 1 to 24 of matrix **E** for the implicit DAE formulation for the three-bus system.

Bibliography

[1] M. E. Aboul-Ela, A. A. Sallam, J. D. McCalley, and A. A. Fouad, 'Damping controller design for power system oscillations using global signals,' *IEEE Transactions on Power Systems*, **11** (*2*), 767–773, 1996.

[2] P. Amestoy, I. S. Duff, J. Koster, and J.-Y. L'Excellent, 'A fully asynchronous multifrontal solver using distributed dynamic scheduling,' *SIAM Journal on Matrix Analysis and Applications*, **23** (*1*), 15–41, 2001.

[3] E. Anderson, Z. Bai, C. Bischof, L. S. Blackford, J. Demmel, J. J. Dongarra, J. Du Croz, S. Hammarling, A. Greenbaum, A. McKenney, and D. Sorensen, *LAPACK Users' Guide*, 3rd ed. Philadelphia, PA: SIAM, 1999.

[4] P. M. Anderson and A. A. Fouad, *Power system control and stability*, 2nd ed. IEEE Press: Wiley-Interscience, 2003, previous ed. Ames, Iowa: Iowa State University Press, 1977.

[5] B. Appasani and D. K. Mohanta, 'A review on synchrophasor communication system: communication technologies, standards and applications,' *Protection and Control of Modern Power Systems*, **3**, 1–17, 2018.

[6] T. Aprille and T. Trick, 'A computer algorithm to determine the steady-state response of nonlinear oscillators,' *IEEE Transactions on Circuit Theory*, **19** (*4*), 354–360, Jul. 1972.

[7] Argonne National Laboratory, 'PETSc users manual,' 2020. [Online]. Available: https://www.mcs.anl.gov/petsc.

[8] P. Aristidou, D. Fabozzi, and T. Van Cutsem, 'Dynamic simulation of large-scale power systems using a parallel Schur-complement-based decomposition method,' *IEEE Transactions on Parallel and Distributed Systems*, **25** (*10*), 2561–2570, Oct. 2014.

[9] J. Asakura, T. Sakurai, H. Tadano, T. Ikegami, and K. Kimura, 'A numerical method for nonlinear eigenvalue problems using contour integrals,' *JSIAM Letters*, **1**, 52–55, 2009.

[10] G. A. Baker Jr. and P. Graves-Morris, *Padé Approximants - Part I: Basic Theory*. Boston, MA: Addison-Wesley, 1981.

[11] K. J. Bathe and E. L. Wilson, 'Solution methods for large generalized eigenvalue problems in structural engineering,' *International Journal for Numerical Methods in Engineering*, **6**, 213–226, 1973.

[12] J. Bélair and M. C. Mackey, 'Consumer memory and price fluctuations in commodity markets: An integrodifferential model,' *Journal of Dynamics and Differential Equations*, **1** (*3*), 299–325, Jul. 1989.

[13] A. Bellen, N. Guglielmi, and A. Ruehli, 'Methods for linear systems of circuit delay differential equations of neutral type,' *IEEE Transactions on Circuits and Systems - I: Fundamental Theory and Applications*, **1**, 212–216, 1999.

[14] A. Bellen and S. Maset, 'Numerical solution of constant coefficient linear delay differential equations as abstract cauchy problems,' *Numerische Mathematik*, **84** (*3*), 351–374, Jan 2000.

[15] G. L. Berg, 'Power system load representation,' *Proceedings of the IEEE*, **120** (*3*), 344–348, 1973.

[16] F. Bizzarri, A. Brambilla, and G. Storti Gajani, 'Steady state computation and noise analysis of analog mixed signal circuits,' *IEEE Transactions on Circuits and Systems - I: Regular Papers*, **59** (*3*), 541–554, Mar. 2012.

[17] L. S. Blackford, J. Choi, A. Cleary, E. D'Azevedo, J. Demmel, I. Dhillon, J. Dongarra, S. Hammarling, G. Henry, A. Petitet, K. Stanley, D. Walker, and R. C. Whaley, *ScaLAPACK Users' Guide*. Philadelphia, PA: SIAM, 1997.

[18] V. Bokharaie, R. Sipahi, and F. Milano, 'Small-signal stability analysis of delayed power system stabilizers,' in *Proceedings of the Power Systems Computation Conference*, Wroclaw, Poland, Aug. 2014.

[19] Y. Boyarintsev, *Methods of Solving Singular Systems of Ordinary Differential Equations*. John Wiley & Sons, 1992.

[20] D. Breda, 'Solution operator approximations for characteristic roots of delay differential equations,' *Applied Numerical Mathematics*, **56** (*3*), 305–317, 2006, selected Papers, The 3rd International Conference on the Numerical Solutions of Volterra and Delay Equations.

[21] D. Breda, S. Maset, and R. Vermiglio, 'Pseudospectral approximation of eigenvalues of derivative operators with non-local boundary conditions,' *Applied Numerical Mathematics*, **56** (*3*), 318–331, 2006, selected Papers, The 3rd International Conference on the Numerical Solutions of Volterra and Delay Equations.

[22] K. E. Brenan, S. L. Campbell, and L. R. Petzold, *Numerical Solution of Initial-Value Problems in Differential-Algebraic Equations.* Philadelphia, PA: SIAM, 1996.

[23] R. T. Byerly, R. J. Bennon, and D. E. Sherman, 'Eigenvalue analysis of synchronizing power flow oscillations in large electric power systems,' *IEEE Transactions on Power Apparatus and Systems*, **PAS-101** (*1*), 235–243, 1982.

[24] S. L. Campbell, *Singular Systems of Differential Equations.* Pitman Advanced Publishing Program, 1980.

[25] C. Campos and J. E. Roman, 'Parallel Krylov solvers for the polynomial eigenvalue problem in SLEPc,' *SIAM J. Sci. Comput.*, **38** (*5*), S385–S411, 2016.

[26] C. Campos and J. E. Roman, 'Restarted Q-Arnoldi-type methods exploiting symmetry in quadratic eigenvalue problems,' *BIT Numer. Math.*, **56** (*4*), 1213–1236, 2016.

[27] C. Campos and J. E. Roman, 'Inertia-based spectrum slicing for symmetric quadratic eigenvalue problems,' *Numer. Linear Algebra Appl.*, **(in press)** (*x*), 1–17, 2020.

[28] C. Campos and J. E. Roman, 'A polynomial Jacobi-Davidson solver with support for non-monomial bases and deflation,' *BIT Numer. Math.*, **60**, 295–318, 2020.

[29] M. Caputo and M. Fabrizio, 'A new definition of fractional derivative without singular kernel,' *Progress in Fractional Differentiation and Applications*, **1** (*2*), 73–85, 2015.

[30] F. E. Cellier and E. Kofman, *Continuous System Simulation.* Springer, 2006.

[31] N. R. Chaudhuri, B. Chaudhuri, R. Majumder, and A. Yazdani, *Multi-terminal Direct-current Grids: Modeling, Analysis, and Control.* New Jersey, US: John Wiley & Sons, 2014.

[32] J. H. Chow, *Power system coherency and model reduction*, 1st ed., ser. Power Electronics and Power Systems 94. Berlin, Germany: Springer-Verlag, 2013.

[33] R. Clint Whaley, A. Petitet, and J. J. Dongarra, 'Automated empirical optimizations of software and the ATLAS project,' *Parallel Computing*, **27** (*1*), 3–35, 2001.

[34] L. Dai, *Singular control systems*, ser. Lecture Notes in Control and Information Sciences. Berlin, Germany: Springer, 1988.

[35] I. Dassios, 'On non-homogeneous linear generalized linear discrete time systems,' *Circuits, Systems, and Signal Processing*, **31** (*5*), 699–1712, 2012.

[36] I. Dassios and D. Baleanu, 'Duality of singular linear systems of fractional nabla difference equations,' *Applied Mathematical Modelling*, **39** (*14*), 4180–4195, 2015.

[37] I. Dassios and D. Baleanu, 'Caputo and related fractional derivatives in singular systems,' *Applied Mathematics and Computation*, **337**, 591–606, 2018.

[38] I. Dassios, D. Baleanu, and G. Kalogeropoulos, 'On non-homogeneous singular systems of fractional nabla difference equations,' *Applied Mathematics and Computation*, **227**, 112–131, 2014.

[39] I. Dassios and G. Kalogeropoulos, 'On a non-homogeneous singular linear discrete time system with a singular matrix pencil,' *Circuits, Systems and Signal Processing*, **32** (*4*), 1615–1635, 2013.

[40] I. Dassios, G. Tzounas, and F. Milano, 'The Möbius transform effect in singular systems of differential equations,' *Applied Mathematics and Computation*, **361**, 338–353, 2019.

[41] I. Dassios, G. Tzounas, and F. Milano, 'Generalized fractional controller for singular systems of differential equations,' *Journal of Computational and Applied Mathematics*, 2020, in press.

[42] I. Dassios, G. Tzounas, and F. Milano, 'Participation factors for singular systems of differential equations,' *Circuits, Systems and Signal Processing*, **39** (*1*), 83–110, 2020.

[43] E. R. Davidson, 'The iterative calculation of a few of the lowest eigenvalues and corresponding eigenvectors of large real-symmetric matrices,' *Journal of Computational Physics*, **17** (*1*), 87–94, 1975.

[44] T. A. Davis, *Direct Methods for Sparse Linear Systems*. SIAM, 2006.

[45] M. Di Bernardo, C. Budd, A. Champneys, and P. Kowalczyk, *Piecewise-smooth Dynamical Systems, Theory and Applications*. Springer-Verlag, 2008.

[46] D. Fabozzi and T. Van Cutsem, 'On angle references in long-term time-domain simulations,' *IEEE Transactions on Power Systems*, **26** (*1*), 483–484, Feb. 2011.

[47] M. Farkas, *Periodic Motions*. Springer-Verlag, 1994.

[48] J. G. F. Francis, 'The QR transformation a unitary analogue to the LR transformation – Part 1,' *Comput. J.*, **4** (*3*), 265–271, 1961.

[49] E. Fridman, 'New Lyapunov-Krasovskii functionals for stability of linear retarded and neutral type systems,' *Journal of systems and control Letters*, **43**, 309–319, Feb. 2001.

[50] Y. Futamura and T. Sakurai, *z-Pares Users' Guide Release 0.9.5.* University of Tsukuba, 2014.

[51] R. Gantmacher, *The Theory of Matrices I, II.* Chelsea, 1959.

[52] B. Gao, G. K. Morison, and P. Kundur, 'Voltage stability evaluation using modal analysis,' *IEEE Transactions on Power Systems*, **7** (*4*), 1529–1542, 1992.

[53] B. Gao, G. K. Morison, and P. Kundur, 'Towards the development of a systematic approach for voltage stability assessment of large-scale power systems,' *IEEE Transactions on Power Systems*, **11** (*3*), 1314–1324, 1996.

[54] B. S. Garbow, 'EISPACK – a package of matrix eigensystem routines,' *Computer Physics Communications*, **7** (*4*), 179–184, Nov. 1974.

[55] S. Geng and I. A. Hiskens, 'Second-order trajectory sensitivity analysis of hybrid systems,' *IEEE Transactions on Circuits and Systems - I: Regular Papers*, **66** (*5*), 1922–1934, 2019.

[56] M. Gibbard, P. Pourbeik, and D. Vowles, *Small-Signal Stability, Control and Dynamic Performance of Power Systems.* Adelaide, South Australia: University of Adelaide Press, 2015.

[57] I. Gohberg, P. Lancaster, and L. Rodman, *Matrix Polynomials.* Cambridge, MA, US: Academic Press, 1983.

[58] G. H. Golub and C. F. Van Loan, *Matrix Computations,* 4th ed. Baltimore, US: The Johns Hopkins University Press, 2013.

[59] S. Gomes, Jr., N. Martins, and C. Portela, 'Sequential computation of transfer function dominant poles of s-domain system models,' *IEEE Transactions on Power Systems*, **24** (*2*), 776–784, 2009.

[60] I. Grattan-Guiness, *Companion Encyclopedia of the History and Philosophy of the Mathematical Sciences: Volume One.* Routledge, 2004.

[61] C. M. Grinstead and J. L. Snell, *Introduction to Probability.* AMS, 2003.

[62] E. Hairer and G. Wanner, *Solving Ordinary Differential Equations II – Stiff and Differential-Algebraic Problems*, 2nd ed. Springer, 1996.

[63] J. K. Hale, *Theory of Differential Equations.* Springer, 1977.

[64] A. M. A. Hamdan, 'Coupling measures between modes and state variables in power-system dynamics,' *International Journal of Control*, **43** (*3*), 1029–1041, 1986.

[65] H. M. A. Hamdan and A. M. A. Hamdan, 'On the coupling measures between modes and state variables and subsynchronous resonance,' *Electric Power Systems Research*, **13** (*3*), 165–171, 1987.

[66] N. Hatziargyriou, Ed., *Microgrids: Architectures and Control.* Wiley IEEE Press, 2014.

[67] B. Hayes and F. Milano, 'Viable computation of the largest Lyapunov characteristic exponent for power systems,' in *IEEE PES Innovative Smart Grid Technologies Conference Europe (ISGT-Europe)*, 2018, 1–6.

[68] H. V. Henderson and S. R. Searle, 'On deriving the inverse of a sum of matrices,' *SIAM Review*, **23**, 53–60, Jan. 1981.

[69] V. Hernandez, J. E. Roman, and V. Vidal, 'SLEPc: A scalable and flexible toolkit for the solution of eigenvalue problems,' *ACM Transactions on Mathematical Software*, **31** (*3*), 351–362, 2005.

[70] I. A. Hiskens, 'Power system modeling for inverse problems,' *IEEE Transactions on Circuits and Systems - I: Regular Papers*, **51** (*3*), 539–551, Mar. 2004.

[71] I. A. Hiskens and M. A. Pai, 'Trajectory sensitivity analysis of hybrid systems,' *IEEE Transactions on Circuits and Systems I: Fundamental Theory and Applications*, **47** (*2*), 204–220, 2000.

[72] A. S. Householder, 'Unitary triangularization of a nonsymmetric matrix,' *Journal of the ACM*, **5** (*4*), 339–342, 1958.

[73] S. Hu, M. Dunlavey, S. Guzy, and N. Teuscher, 'A distributed delay approach for modeling delayed outcomes in pharmacokinetics and pharmacodynamics studies,' *Journal of Pharmacokinetics and Pharmacodynamics*, **45** (*2*), 285–308, Apr 2018.

[74] IEEE Power System Dynamic Performance Committee, 'Dynamic Models Turbine-Governors in Power System Studies,' IEEE, Tech. Rep., 2013, PES-TR1.

[75] IEEE Working Group on Computer Modelling of Excitation Systems, 'Excitation system models for power system stability studies,' *IEEE Transactions on Power Apparatus and Systems*, **100** (*2*), 494–509, Feb. 1981.

[76] IEEE/CIGRE Joint Task Force on stability Terms and Definitions, 'Definition and classification of power system stability,' *IEEE Transactions on Power Systems*, **19** (*2*), 1387–1401, May 2004.

[77] M. Ilić and J. Zaborszky, *Dynamic and Control of Large Electric Power Systems*. John Wiley & Sons, 2000.

[78] D. Ivanescu, S.-I. Niculescu, L. Dugar, J. M. Dion, and E. I. Verriest, 'On delay-dependent stability for linear neutral systems,' *Automatica*, **39**, 255–261, Feb. 2003.

[79] E. Jarlebring, K. Meerbergen, and W. Michiels, 'A Krylov method for the delay eigenvalue problem,' *SIAM J. Scientific Computing*, **32**, 3278–3300, 2010.

[80] H. R. Karimi, 'Robust delay-dependent H_∞ control of uncertain time-delay systems with mixed neutral discrete and distributed time-delays and markovian switching parameters,' *IEEE Transactions on Circuits and Systems - I: Regular Papers*, **58**, 1910–1923, 2011.

[81] A. V. Knyazev, 'Toward the optimal preconditioned eigensolver: Locally optimal block preconditioned conjugate gradient method,' *SIAM Journal on Scientific Computing*, **23** (*2*), p. 517–541, 2001.

[82] D. N. Kosterev, C. W. Taylor, and W. A. Mittelstadt, 'Model validation for the August 10, 1996 WSCC system outage,' *IEEE Transactions on Power Systems*, **14** (*3*), 967–979, 1999.

[83] P. Krause, O. Wasynczuk, S. Sudhoff, and S. Pekarek, *Analysis of Electric Machinery and Drive Systems*, 3rd ed., ser. IEEE Press Series on Power Engineering. John Wiley & Sons, 2013.

[84] D. Kressner, *Numerical Methods for General and Structured Eigenvalue Problems*, 4th ed. Springer, 2015.

[85] E. V. Krishnamurthy, 'A stability test for sampled data systems using matrix power method,' *Proceedings of the IEEE*, **53** (*1*), 91–92, 1965.

[86] G. Kron, 'A new theory of hunting [includes discussion],' *Transactions of the American Institute of Electrical Engineers. Part III: Power Apparatus and Systems*, **71** (*4*), 859–866, 1952.

[87] P. Kundur, *Power System Stability and Control*. Mc-Grall Hill, 1994.

[88] A. C. Laboratory, 'QPmR - Quasi-Polynomial Mapping Based Rootfinder,' 2014. [Online]. Available: http://www.cak.fs.cvut.cz/algorithms/qpmr

[89] R. Lamour, 'Floquét-theory for differential-algebraic equations (DAE),' *ZAMM - Journal of Applied Mathematics and Mechanics / Zeitschrift für Angewandte Mathematik und Mechanik*, **78** (*S3*), 989–990, 1998.

[90] C. L. Lawson, R. J. Hanson, D. R. Kincaid, and F. T. Krogh, 'Basic linear algebra subprograms for FORTRAN usage,' ACM Transactions on Mathematical Software, 5 (3), 308–323, 1979.

[91] B. Lee and V. Ajjarapu, 'Period-doubling route to chaos in an electrical power system,' *IEE Proceedings C - Generation, Transmission and Distribution*, **140** (*6*), 490–496, 1993.

[92] R. B. Lehoucq and D. C. Sorensen, 'Deflation techniques for an implicitly restarted Arnoldi iteration,' *SIAM Journal on Matrix Analysis and Applications*, **17** (*4*), 1996.

[93] R. B. Lehoucq, D. C. Sorensen, and C. Yang, 'ARPACK users guide: Solution of large scale eigenvalue problems by implicitly restarted Arnoldi methods.' Software, Environments, and Tools, SIAM, Philadelphia, US, 1998.

[94] C. Li, Y. Chen, T. Ding, Z. Du, and F. Li, 'A sparse and low-order implementation for discretization-based eigen-analysis of power systems with time-delays,' *IEEE Transactions on Power Systems*, **34** (*6*), 5091–5094, Nov 2019.

[95] J. Liu, W. Yao, J. Wen, J. Fang, L. Jiang, H. He, and S. Cheng, 'Impact of power grid strength and PLL parameters on stability of grid-connected DFIG wind farm,' *IEEE Transactions on Sustainable Energy*, **11** (*1*), 545–557, Jan 2020.

[96] M. Liu, F. Bizzarri, A. M. Brambilla, and F. Milano, 'On the impact of the dead-band of power system stabilizers and frequency regulation on power system stability,' *IEEE Transactions on Power Systems*, **34** (*5*), 3977–3979, 2019.

[97] M. Liu, I. Dassios, G. Tzounas, and F. Milano, 'Stability analysis of power systems with inclusion of realistic-modeling WAMS delays,' *IEEE Transactions on Power Systems*, **34** (*1*), 627–636, Jan 2019.

[98] M. Liu and F. Milano, 'Small-signal stability analysis of power systems with inclusion of periodic time-varying delays,' in *Proceedings of the Power Systems Computation Conference*, June 2018, 1–7.

[99] J. Louisell, 'New examples of quenching in delay differential-delay equations having time-varying delay,' in *Proceedings of the 4th European Control Conference*, Karlsruhe, Germany, Sep. 1999.

[100] C. Lu, X. Zhang, X. Wang, and Y. Han, 'Mathematical expectation modeling of wide-area controlled power systems with stochastic time delay,' *IEEE Transactions on Smart Grid*, **6** (*3*), 1511–1519, May 2015.

[101] J. Machowski, J. W. Bialek, and J. R. Bumby, *Power System Dynamics and Stability*, 2nd ed. John Wiley & Sons, 2008.

[102] M. S. Mahmoud and A. Ismail, 'Delay-dependent robust stability and stabilization of neutral systems,' *Apply Mathematics Science*, **1** (*10*), 471–490, 2007.

[103] T. Maly and L. R. Petzold, 'Numerical methods and software for sensitivity analysis of differential-algebraic systems,' *Applied Numerical Mathematics*, **20** (*1*), 57–79, 1996, Method of Lines for Time-Dependent Problems.

[104] T. J. Manetsch, 'Time-varying distributed delays and their use in aggregative models of large systems,' *IEEE Transactions on Systems, Man, and Cybernetics*, **SMC-6** (*8*), 547–553, Aug 1976.

[105] T. J. Manetsch, 'On the role of systems analysis in aiding countries facing acute food shortages,' *IEEE Transactions on Systems, Man, and Cybernetics*, **7** (*4*), 264–273, Apr. 1977.

[106] R. Marconato, *Electric Power Systems.* CEI, Italian Electrotechnical Committee, 2002, **2**.

[107] N. Martins, 'The dominant pole spectrum eigensolver [for power system stability analysis],' *IEEE Transactions on Power Systems*, **12** (*1*), 245–254, 1997.

[108] N. Martins, L. T. G. Lima, and H. J. C. P. Pinto, 'Computing dominant poles of power system transfer functions,' *IEEE Transactions on Power Systems*, **11** (*1*), 162–170, 1996.

[109] W. Michiels, V. V. Assche, and S. Niculescu, 'Stabilization of time-delay systems with a controlled time-varying delay and application,' *IEEE Transactions on Automatic Control*, **50** (*4*), 493–504, Apr. 2005.

[110] W. Michiels and S. I. Niculescu, *Stability and Stabilization of Time-delay Systems: An Eigenvalue-based Approach.* Philadelphia, PA: SIAM, 2007.

[111] W. Michiels and T. Vyhlídal, 'An eigenvalue based approach for the stabilization of linear time-delay systems of neutral type,' *Automatica*, **41** (*6*), 991–998, 2005.

[112] F. Milano, 'Continuous Newton's method for power flow analysis,' *IEEE Transactions on Power Systems*, **24** (*1*), 50–57, 2009.

[113] F. Milano, *Power System Modelling and Scripting.* Springer, 2010.

[114] F. Milano, 'A Python-based software tool for power system analysis,' in *Proceedings of the IEEE PES General Meeting*, Vancouver, BC, July 2013.

[115] F. Milano, 'Semi-implicit formulation of differential-algebraic equations for transient stability analysis,' *IEEE Transactions on Power Systems*, **31** (*6*), 4534–4543, Nov. 2016.

[116] F. Milano, 'Small-signal stability analysis of large power systems with inclusion of multiple delays,' *IEEE Transactions on Power Systems*, **31** (*4*), 3257–3266, Jul 2016.

[117] F. Milano, 'Implicit continuous Newton method for power flow analysis,' *IEEE Transactions on Power Systems*, **34** (*4*), 3309–3311, 2019.

[118] F. Milano and M. Anghel, 'Impact of time delays on power system stability,' *IEEE Transactions on Circuits and Systems - I: Regular Papers*, **59** (*4*), 889–900, Apr 2012.

[119] F. Milano and I. Dassios, 'Small-signal stability analysis for non-index 1 Hessenberg form systems of delay differential-algebraic equations,' *IEEE Transactions on Circuits and Systems - I: Regular Papers*, **63** (*9*), 1521–1530, Sep. 2016.

[120] F. Milano and I. Dassios, 'Primal and dual generalized eigenvalue problems for power systems small-signal stability analysis,' *IEEE Transactions on Power Systems*, **32** (*6*), 4626–4635, 2017.

[121] F. Milano, F. Dörfler, G. Hug, D. Hill, and G. Verbič, 'Foundations and challenges of low-inertia systems,' in *Proceedings of the Power Systems Computation Conference*, Dublin, Ireland, June 2018, 1–22.

[122] F. Milano and Á. Ortega, *Converter-Inerfaced Energy Storage Systems.* Cambridge University Press, 2019.

[123] F. Milano and Á. Ortega, *Frequency Variations in Power Systems – Modeling, State Estimation and Control.* Wiley IEEE Press, 2020.

[124] F. Milano, Ed., *Advances in Power System Modelling, Control and Stability Analysis*, ser. Energy Engineering. Institution of Engineering and Technology, 2016.

[125] L. Mirkin, 'On the approximation of distributed-delay control laws,' *Systems and Control Letters*, **51** (*5*), 331–342, 2004.

[126] C. B. Moler and G. W. Stewart, 'An algorithm for generalized matrix eigenvalue problems,' *SIAM Journal on Numerical Analysis*, **10** (*2*), 241–256, 1973.

[127] D. K. Molzahn, B. C. Lesieutre, and H. Chen, 'Counterexample to a continuation-based algorithm for finding all power flow solutions,' *IEEE Transactions on Power Systems*, **28** (*1*), 564–565, 2013.

[128] J. A. Momoh, *Smart Grid: Fundamentals of Design and Analysis.* Wiley IEEE Press, 2012.

[129] C. A. Monje, Y. Chen, B. M. Vinagre, D. Xue, and V. Feliu, *Fractional-order Systems and Controls, Fundamentals and Applications.* Springer, 2010.

[130] L. Moreau and D. Aeyels, 'Trajectory-based global and semi-global stability results,' *Modern Applied Mathematics Techniques in Curcuits, Systems and Control*, **1**, 71–76, 1999.

[131] M. A. A. Murad, G. Tzounas, M. Liu, and F. Milano, 'Frequency control through voltage regulation of power system using SVC devices,' in *Proceedings of the IEEE PES General Meeting*, 2019.

[132] Y. Obata, S. Takeda, and H. Suzuki, 'An efficient eigenvalue estimation technique for multimachine power system dynamic stability analysis,' *IEEE Transactions on Power Apparatus and Systems*, **PAS-100** (*1*), 259–263, 1981.

[133] S. Okubo, H. Suzuki, and K. Uemura, 'Modal analysis for power system dynamic stability,' *IEEE Transactions on Power Apparatus and Systems*, **PAS-97** (*4*), 1313–1318, 1978.

[134] F. L. Pagola, I. J. Perez-Arriaga, and G. C. Verghese, 'On sensitivities, residues and participations: applications to oscillatory stability analysis and control,' *IEEE Transactions on Power Systems*, **4** (*1*), 278–285, Feb 1989.

[135] A. Pai, *Energy Function Analysis for Power System Stability.* Kluwer Academic Publisher, 1989.

[136] A. Papachristodoulou, M. M. Peet, and S. Niculescu, 'Stability analysis of linear system with time-varying delays: Delay uncertainty and quenching,' in *Proceedings of the 46th IEEE Conference on Decision and Control*, New Orleans, LA. USA, Dec. 2007.

[137] B. N. Parlett and W. G. Poole, 'A geometric theory for QR, LU and power iteration,' *SIAM Journal on Numerical Analysis*, **10** (*2*), 1973.

[138] M. Pavella, D. Ernst, and D. Ruiz Vega, *Transient Stability of Power Systems – A Unified Approach to Assessment and Control.* Boston: Kluwer Academic Publishers, 2000.

[139] T. P. Peixoto, 'The graph-tool Python library,' *figshare*, 2014, available at graph-tool.skewed.de.

[140] I. J. Perez-Arriaga, G. C. Verghese, and F. C. Schweppe, 'Selective modal analysis with applications to electric power systems, Part I: Heuristic introduction,' *IEEE Transactions on Power Apparatus and Systems*, **PAS-101** (*9*), 3117–3125, Sep. 1982.

[141] I. Podlubny, 'Fractional-order systems and $PI^{\lambda}D^{\mu}$-controllers,' *IEEE Transactions on Automatic Control*, **44** (*1*), 208–214, Jan. 1999.

[142] E. Polizzi, 'Density-matrix-based algorithm for solving eigenvalue problems,' *Physical Review B, American Physical Society*, **79** (*11*), 2009.

[143] E. Polizzi, 'FEAST eigenvalue solver v4.0 user guide,' 2020.

[144] E. Polizzi and A. H. Sameh, 'A parallel hybrid banded system solver: the SPIKE algorithm,' *Parallel Computing*, **32** (*2*), 177–194, 2006.

[145] R. Preece, J. V. Milanović, A. M. Almutairi, and O. Marjanovic, 'Damping of inter-area oscillations in mixed AC/DC networks using WAMS based supplementary controller,' *IEEE Transactions on Power Systems*, **28** (*2*), 1160–1169, May 2013.

[146] W. J. Rugh, *Linear system theory*. Prentice Hall International, 1996.

[147] Y. Saad, *Numerical Methods for Large Eigenvalue Problems – Revised Edition*. SIAM, 2011.

[148] T. Sakurai and H. Sugiura, 'A projection method for generalized eigenvalue problems using numerical integration,' *Journal of Computational and Applied Mathematics*, **159** (*1*), 119–128, 2003.

[149] T. Sakurai and H. Tadano, 'CIRR: a rayleigh-ritz type method with contour integral for generalized eigenvalue problems,' *Hokkaido Mathematical Journal*, **36** (*4*), 745–757, 2007.

[150] G. Sánchez Ayala, V. Centeno, and J. Thorp, 'Gain scheduling with classification trees for robust centralized control of PSSs,' *IEEE Transactions on Power Systems*, **31** (*3*), 1933–1942, May 2016.

[151] P. W. Sauer and M. A. Pai, *Power System Dynamics and Stability*. Prentice Hall, 1998.

[152] P. W. Sauer, C. Rajagopalan, and M. A. Pai, 'An explanation and generalization of the AESOPS and PEALS algorithms (power system models),' *IEEE Transactions on Power Systems*, **6** (*1*), 293–299, 1991.

[153] O. Schenk and K. Gartner, 'Solving unsymmetric sparse systems of linear equations with PARDISO,' *Journal of Future Generation Computer Systems*, **20** (*3*), 2004.

[154] A. Semlyen and L. Wang, 'Sequential computation of the complete eigensystem for the study zone in small signal stability analysis of large power systems,' *IEEE Transactions on Power Systems*, **3** (*2*), 715–725, 1988.

[155] R. Seydel, *Practical Bifurcation and Stability Analysis: From Equilibrium to Chaos*, 2nd ed. Springer-Verlag, 1994.

[156] M. Snir, S. Otto, S. Huss-Lederman, D. Walker, and J. Dongarra, *MPI-The Complete Reference, Volume 1: The MPI Core*. MIT Press, 1998.

[157] D. C. Sorensen, 'Implicit application of polynomial filters in a k-step Arnoldi method,' *SIAM Journal on Matrix Analysis and Applications*, **13** (*1*), 357–385, 1992.

[158] G. W. Stewart, 'A Krylov–Schur algorithm for large eigenproblems,' *SIAM Journal on Matrix Analysis and Applications*, **23** (*3*), 601–614, 2002.

[159] E. Süli and D. F. Mayers, *An Introduction to Numerical Analysis.* Cambridge University Press, 2003.

[160] P. Tak, P. Tang, and E. Polizzi, 'FEAST as a subspace iteration eigensolver accelerated by approximate spectral projection,' *SIAM Journal on Matrix Analysis and Applications*, **35** (*2*), 354–390, 2014.

[161] C. J. Tavora and O. J. M. Smith, 'Characterization of equilibrium and stability in power systems,' *IEEE Transactions on Power Apparatus and Systems*, **PAS-91** (*3*), 1127–1130, May 1972.

[162] Y.-C. Tian and D. Levy, 'Compensation for control packet dropout in networked control systems,' *Information Sciences*, **178** (*5*), 1263–1278, 2008.

[163] S. Tomov, J. Dongarra, and M. Baboulin, 'Towards dense linear algebra for hybrid GPU accelerated manycore systems,' *Parallel Computing*, **36** (*5-6*), 232–240, Jun. 2010.

[164] Y. Tsividis and C. McAndrew, *Operation and Modeling of the MOS Transistor*, 3rd ed. Oxford University Press, 2011.

[165] G. Tzounas, I. Dassios, and F. Milano, 'Modal participation factors of algebraic variables,' *IEEE Transactions on Power Systems*, **35** (*1*), 742–750, 2020.

[166] G. Tzounas, I. Dassios, M. A. A. Murad, and F. Milano, 'Theory and implementation of fractional order controllers for power system applications.' *IEEE Transactions on Power Systems*, 2020, in press.

[167] J. M. Uudrill, 'Dynamic stability calculations for an arbitrary number of interconnected synchronous machines,' *IEEE Transactions on Power Apparatus and Systems*, **PAS-87** (*3*), 835–844, 1968.

[168] L. S. Vargas and C. A. Cañizares, 'Time dependence of controls to avoid voltage collapse,' *IEEE Transactions on Power Systems*, **15** (*4*), 1367–1375, 2000.

[169] R. von Mises and H. Pollaczek-Geiringer, 'Praktische verfahren der gleichungsauflösung,' *Zeitschrift für Angewandte Mathematik und Mechanik*, **9** (*2*), 152–164, 1929.

[170] T. L. Vu and K. Turitsyn, 'Lyapunov functions family approach to transient stability assessment,' *IEEE Transactions on Power Systems*, **31** (*2*), 1269–1277, 2016.

[171] T. Vyhlídal, P. Zítek, QPmR - Quasi-Polynomial root-finder: algorithm update and examples. In: T. Vyhlídal , J.F. Lafay, R. Sipahi (eds) Delay Systems. Advances in Delays and Dynamics, vol 1. Springer, Berlin, Germany, 2014.

[172] T. Vyhlídal and P. Zítek, 'Mapping based algorithm for large-scale computation of quasi-polynomial zeros,' *IEEE Transactions on Automatic Control*, **54** (*1*), 171–177, Jan 2009.

[173] T. Vyhlídal and P. Zítek, 'Quasipolynomial mapping based rootfinder for analysis of time delay systems,' *IFAC Proceedings Volumes*, **36** (*19*), 227–232, 2003, 4th IFAC Workshop on Time Delay Systems (TDS 2003), Rocquencourt, France, 8-10 September 2003.

[174] C. F. Wagner, 'Effect of armature resistance upon hunting of synchronous machines,' *Transactions of the American Institute of Electrical Engineers*, **49** (*3*), 1011–1024, 1930.

[175] C. F. Wagner, 'Damper windings for water-wheel generators,' *Transactions of the American Institute of Electrical Engineers*, **50** (*1*), 140–151, 1931.

[176] S. Wang, X. Meng, and T. Chen, 'Wide-area control of power systems through delayed network communication,' *IEEE Transactions on Control Systems Technology*, **20** (*2*), 495–503, Mar. 2012.

[177] J. Winkelman, J. Chow, B. Bowler, B. Avramovic, and P. Kokotovic, 'An analysis of interarea dynamics of multi-machine systems,' *IEEE Transactions on Power Apparatus and Systems*, **PAS-100** (*2*), 754–763, Feb. 1981.

[178] D. Y. Wong, G. J. Rogers, B. Porretta, and P. Kundur, 'Eigenvalue analysis of very large power systems,' *IEEE Transactions on Power Systems*, **3** (*2*), 472–480, 1988.

[179] K. Wu and H. Simon, 'Thick-restart Lanczos method for large symmetric eigenvalue problems,' *SIAM Journal on Matrix Analysis and Applications*, **22** (*2*), 602–616, 2000.

[180] X. Xu, R. M. Mathur, J. Jiang, G. J. Rogers, and P. Kundur, 'Modeling of generators and their controls in power system simulations using singular perturbations,' *IEEE Transactions on Power Systems*, **13** (*1*), 109–114, Feb. 1998.

[181] A. Yazdani and R. Iravani, *Voltage-Sourced Converters in Power Systems – Modeling, Control and Applications.* Wiley IEEE Press, 2010.

[182] H. Ye, Y. Liu, and P. Zhang, 'Efficient eigen-analysis for large delayed cyber-physical power system using explicit infinitesimal generator discretization,' *IEEE Transactions on Power Systems*, **31** (*3*), 2361–2370, May 2016.

[183] D. Yue and Q. L. Han, 'A delay-dependent stability criterion of neutral systems and its application to a partial element equivalent circuit model,' *IEEE Transactions on Circuits and Systems - II: Express Briefs*, **51**, 685–689, Sep. 2004.

[184] Y. Zheng, M. J. Brudnak, P. Jayakumar, J. L. Stein, and T. Ersal, 'A predictor-based framework for delay compensation in networked closed-loop systems,' *IEEE/ASME Transactions on Mechatronics*, **23** (*5*), 2482–2493, Oct 2018.

Index